Materials Forming, Machining and Tribology

Series editor

J. Paulo Davim, Aveiro, Portugal

More information about this series at http://www.springer.com/series/11181

Golam Kibria · B. Bhattacharyya
J. Paulo Davim
Editors

Non-traditional Micromachining Processes

Fundamentals and Applications

Springer

Editors
Golam Kibria
Department of Mechanical Engineering
Aliah University
Kolkata
India

J. Paulo Davim
Department of Mechanical Engineering
University of Aveiro
Aveiro
Portugal

B. Bhattacharyya
Department of Production Engineering
Jadavpur University
Kolkata
India

ISSN 2195-0911 ISSN 2195-092X (electronic)
Materials Forming, Machining and Tribology
ISBN 978-3-319-52008-7 ISBN 978-3-319-52009-4 (eBook)
DOI 10.1007/978-3-319-52009-4

Library of Congress Control Number: 2016963424

Printed on acid-free paper

This Springer imprint is published by Springer Nature
The registered company is Springer International Publishing AG
The registered company address is: Gewerbestrasse 11, 6330 Cham, Switzerland

Preface

Since the last decade, the major research challenges and developments around the globe are to manufacture microcomponents with high accurate of microfeatures which have geometrical dimensions even less than tens of microns. It is obvious that the demand of microcomponents is diversely increasing day by day in the field of aviation, automotive, biomedical, optical, electronics, and energy sectors. This wide range of applications of microcomponents brings the necessity to develop promising micromachining technologies which can yield high degree of precision, improved machining rate with accurate geometrical dimensions, reliability, flexibility, and after all environment friendly. Moreover, the developments of material science provide materials which have enormous material properties especially in their machinability and hardness. In this situation, it is the task of manufacturer engineers to develop micromachining processes, both traditional and non-traditional, which can fulfill the demands of microproduct having several micro- and nanofeatures.

In the area of non-traditional micromachining processes, few books are available. These books mostly describe numerical and experimental research details at different stages. However, it is the most important task at this stage to gain the fundamental knowledge about different non-traditional micromachining processes. Not only that, but it is also important to be acquainted with the capabilities and applications of these micromachining processes in various product manufacturing. Moreover, several hybrid micromachining processes have also been developed to improve the surface integrity, geometrical accuracy, and machining efficiency. The present book entitled "Non-traditional Micromachining Processes—Fundamentals and Applications" will definitely fulfill the gap between the demand of microfeatured components and successful micromachining of these products.

This book starts with introduction of various micromachining processes, especially, non-traditional, with recent research and developments in each processes. Further, mechanical type micromachining process such as micro-USM is presented with mechanism of material removal, types and technologies of micro-USM and recent research advancements. Then, electro-thermal processes such as micro-EDM, wire-EDM, micromachining using laser beam (micro-turning, micro-cutting, etc) were discussed. Further, electrochemical micromachining process was discussed for micromachining and micro fabrication of various complex and high accuracy profiles. Then, hybrid type micromachining such as micro electrochemical discharge machining processes were presented. Chapter 1 gives an overview of various micromachining processes including various hybrid and nanofinishing processes with their applications, working principles, challenges in scaling down a macro machining process to a micromachining process, and some recent developments in each category of process. Chapter 2 describes the overview of microultrasonic machining process with brief description of process parameters, system details, and recent development and research challenges for micromachining of intricate parts with high degree of accuracy and surface profile. Chapter 3 describes the overview of EDM and micro-EDM process with details of significant process parameters and experimental investigation with analysis of microhole machining process utilizing some novel machining strategies such as microtool rotation, polarity changing, utilization of deionized water as dielectric, implementation of boron carbide additives into dielectrics. In Chap. 4, a brief introduction of microwire electrical discharge machining (MWEDM) is presented with details of micromachining system, wire failure prevention, detailed analysis of charging and discharging circuits, and details of wire lag and vibration phenomena during micromachining of complex geometrical components. Chapter 5 deals with the experimental investigation and analysis on pulsed Nd:YAG laser microturning process of aluminum oxide ceramics. This process is new in the category of laser micromachining process, and a lot of research and developments are still needed for exploring the capabilities of pulsed Nd:YAG laser for obtaining high precision microparts with required surface finish. In Chap. 6, the overall concept of fiber laser micromachining and influence of machining process parameters on geometrical accuracy of microparts is presented. The chapter also describes various micromachining processes such as microcutting, microdrilling, engraving, and marking using fiber laser systems on wide range of materials from polymer to ceramics. In Chap. 7, in-depth experimental study and analysis are carried out during laser beam cutting of Inconel 625 superalloy at dry condition to explore the influence of various process parameters on kerf width utilizing pulsed Nd:YAG laser micromachining system. In Chap. 8, fundamental of anodic dissolution of microelectrochemical machining is presented with brief overview of significant process parameters and important techniques for improving the machining accuracy and reliability by implementing novel geometry of microtool and its insulation during microchannel formation using EMM process. Chapter 9 presents a detailed overview of electrochemical micromachining of titanium and its alloy and the effect of process parameters on feature profile during maskless and through mask EMM. Chapter 10 provides the fundamental aspect of microelectrochemical discharge machining process and description of machining system and process parameters. The chapter also describes the research and development carried out around the globe in microdrilling and microcutting using micro-ECDM. Chapter 11 presents the overview of travel wire electrochemical spark machining (TWECSM), the development of micromachining setup, brief discussion of process parameters, and parametric effect on machining rate and spark gap width during machining of glass fiber epoxy composite.

This book offers a comprehensive treatment of various non-traditional micromachining techniques and future directions representing a valuable reference to engineers and R&D researchers involved in micromachining. The present book can also be used as a research book for final undergraduate engineering course or as a topic on micromachining processes at the postgraduate level. Also, this book can serve as a useful reference for academics, researchers, mechanical, manufacturers, industrial and materials engineers, professionals in micromachining processes and related industries. The scientific interest in this book is evident from many important centers of the research, laboratories, universities as well as industries. Therefore, it is hoped that this book will inspire and enthuse others to undertake research in non-traditional micromachining processes.

The editors acknowledge Springer for this opportunity and for their enthusiastic and professional support. Finally, we would like to thank all the chapter authors for their availability for this work.

Kolkata, India — Golam Kibria
Kolkata, India — B. Bhattacharyya
Aveiro, Portugal — J. Paulo Davim

Acknowledgements

This book has become a reality due to the constant inspirations and encouragement received from the senior professors and colleagues like Dr. B. Doloi, Dr. D. Banerjee, and Dr. B.R. Sarkar of Production Engineering Department, Jadavpur University, Kolkata; Dr. B.B. Pradhan and Mr. I. Shivakoti of Sikkim Manipal Institute of Technology, Sikkim, India. The editors would like to convey warm regards to Dr. A. Manna, Dr. S. Dhobe, Dr. V.U. Rathod, Dr. Sandip S. Anasane, Dr. Josephine Peter, Dr. Mukandar Sekh, Dr. Shamim Haider, and Mr. Omar Faruk Biswas for constant support and active participation in preparing the manuscripts of this book.

Financial support from the University Grants Commission (UGC), All India Council for Technical Education (AICTE), Department of Science and Technology (DST), and Council of Scientific and Industrial Research (CSIR) for carrying out research in this area has proved to be useful for utilizing research outcomes to enrich this book.

The editors acknowledge Springer for this opportunity and for their enthusiastic and professional support. The team members of Springer like Guido Zosimo Landolfo, Editorial Director; Dr. Dieter Merkle, Springer International Publishing (Physical Sciences and Engineering); and Dr. Mayra Castro, Applied Sciences Editorial and Senior Publishing Editor of Springer have put their constant effort in transforming this book into its final shape. Finally, I would like to thank all the chapter authors for their availability for this work.

Kolkata, India — Golam Kibria
Kolkata, India — B. Bhattacharyya
Aveiro, Portugal — J. Paulo Davim
January 2017

Contents

Editors and Contributors

About the Editors

Dr. Golam Kibria is an assistant professor in the Department of Mechanical Engineering in Aliah University, Kolkata, India. He completed his M.Tech. in Production Engineering from Jadavpur University, Kolkata in 2008 and Ph.D. from Jadavpur University, Kolkata, in 2014. He has worked as a senior research fellow (SRF) in Council of Scientific and Industrial Research (CSIR)-sponsored project from 2008 to 2011. His research interests include non-conventional machining processes, micromachining, and advanced manufacturing and forming technology. He is a life member of The Institution of Engineers (IEI), India. He is the author of several chapters on internationally recognized book publishers such as Elsevier, Springer, and Nova Publishers. He has also published 26 International and national research papers in various reputed journals and 37 research papers in reputed national and international conference proceedings. He is an editorial board member as well as a reviewer of a number of reputed international journals, namely Optics and Laser Technology, International Journal of Advanced Manufacturing Technology, Manufacturing Review, and International Journal of Physical Sciences. In his credit, he has received I.S.T.E. National Award for Best M.Tech. Thesis in Mechanical Engineering in 2008. He also received the Institution Prize (Gold Medal) of 2008–2009 from The Institution of Engineers (India) for the best paper.

B. Bhattacharyya is a professor and ex-head of the Production Engineering Department and coordinator of Center of Advance Study Programme under University Grants Commission and Quality Improvement Programme under AICTE of Jadavpur University. His research areas include non-traditional machining, micromachining, and advanced manufacturing systems. He has published 108 research papers in National and International Journals and 271 research papers in National and International Conferences. He has published several chapters and conference proceedings. Recently, he has published one book titled "Electrochemical Micromachining for Nanofabrication, MEMS and Nanotechnology," William Andrew Applied Science Publishers, Micro and Nano Technologies Series, Elsevier Inc, USA. Several Ph.D. theses have been completed under his guidance. He has completed several research projects. He is recipient of various awards, e.g., Gold Medal and Certificate of Achievements for research papers and thesis as well as the Career Award of the UGC, New Delhi.

J. Paulo Davim received his Ph.D. in Mechanical Engineering in 1997, M.Sc. degree in Mechanical Engineering (materials and manufacturing processes) in 1991, Licentiate degree (5 years) in Mechanical Engineering in 1986, from the University of Porto (FEUP), the Aggregate title from the University of Coimbra in 2005, and a D.Sc. from London Metropolitan University in 2013. He is Eur Ing and Senior Chartered Engineer by the Portuguese Institution of Engineers with a MBA and Specialist title in Engineering and Industrial Management. Currently, he is a professor at the Department of Mechanical Engineering of the University of Aveiro. He has more than 30 years of teaching and research experience in Manufacturing, Materials, and Mechanical Engineering with special emphasis in Machining and Tribology. Recently, he has also interest in Management/Industrial Engineering and Higher Education for Sustainability/Engineering Education. He has received several scientific awards. He has worked as a evaluator of projects for international research agencies as well as an examiner of Ph.D. thesis for many universities. He is the editor in chief of several international journals, guest editor of journals, books editor, book series editor, and scientific advisory for many international journals and conferences. Presently, he is an editorial board member of 30 international journals and acts as a reviewer for more than 80 prestigious Web of Science journals. In addition, he has also published as editor (and co-editor) more than 80 books and as author (and co-author) more than 10 books, 60 chapters, and 350 articles in journals and conferences (more than 200 articles in journals indexed in Web of Science/h-index 35+ and SCOPUS/h-index 44+).

Contributors

Sandip S. Anasane Department of Production Engineering and Industrial Management, College of Engineering Pune, Pune, India

B. Bhattacharyya Production Engineering Department, Jadavpur University, Kolkata, India

S. Das Production Engineering Department, Jadavpur University, Kolkata, India

S. Debnath Production Engineering Department, Jadavpur University, Kolkata, India

B. Doloi Production Engineering Department, Jadavpur University, Kolkata, India

Golam Kibria Mechanical Engineering Department, Aliah University, Kolkata, India

A.S. Kuar Production Engineering Department, Jadavpur University, Kolkata, India

S. Kunar Production Engineering Department, Jadavpur University, Kolkata, India

Anup Malik Department of Mechanical Engineering, PEC University of Technology, Chandigarh, India

Alakesh Manna Department of Mechanical Engineering, PEC University of Technology, Chandigarh, India

S. Mitra Production Engineering Department, Jadavpur University, Kolkata, India

B.B. Pradhan Mechanical Engineering Department, Sikkim Manipal Institute of Technology, Sikkim, India

Asit Baran Puri Mechanical Engineering Department, National Institute of Technology Durgapur, Durgapur, West Bengal, India

V. Rathod Department of Mechanical Engineering, Government Polytechnic Ratnagiri, Maharashtra, India

N. Roy Production Engineering Department, Jadavpur University, Kolkata, India

B.R. Sarkar Production Engineering Department, Jadavpur University, Kolkata, India

A. Sen Production Engineering Department, Jadavpur University, Kolkata, India

I. Shivakoti Mechanical Engineering Department, Sikkim Manipal Institute of Technology, Sikkim, India

Chapter 1
Non-traditional Micromachining Processes: Opportunities and Challenges

S. Debnath, S. Kunar, S.S. Anasane and B. Bhattacharyya

Abstract The high demand and stringent design requirements in developing fields of microengineering as well as various needs of society and nation require the utilization of suitable techniques of non-traditional machining processes on different existing and newly developed metals, non metals, alloys, polymers, ceramics, rubber and composites, etc. Presently, nontraditional machining techniques have expanded their applicability in the field of micromachining and offer better opportunities with several inherent advantages that make these processes superior as well as more efficient than conventional one. Non-traditional mechanical micromachining processes include abrasive jet machining (AJM), water jet machining (WJM), ultrasonic machining (USM), ion beam machining (IBM), etc. Non-traditional thermal micromachining processes include micromachining by electro discharge machining (EDM), laser beam machining (LBM), electron beam machining (EBM), etc. Non-traditional chemical and electrochemical micromachining processes have been used successfully to generate micro features of high quality. Hybrid micromachining can also be utilized effectively for generating more intricate shapes and complex parts. Advanced finishing processes using non-traditional machining like abrasive flow finishing (AFF), magnetic abrasive finishing (MAF), etc. are also gaining popularity to cope up the steep demand in finishing intricate, complex, durable and sophisticated shapes that are highly economical and posses better surface quality and property. The opportunities and challenges of each non-traditional micromachining and finishing processes are to be

S. Debnath · S. Kunar · B. Bhattacharyya (✉)
Production Engineering Department, Jadavpur University, Kolkata 700032, India
e-mail: bb13@rediffmail.com

S. Debnath
e-mail: subhrajit.me32@gmail.com

S. Kunar
e-mail: sandip.sandip.kunar@gmail.com

S.S. Anasane
Department of Production Engineering and Industrial Management,
College of Engineering Pune, Pune 411005, India
e-mail: anasane@gmail.com

G. Kibria et al. (eds.), *Non-traditional Micromachining Processes*,
Materials Forming, Machining and Tribology,
DOI 10.1007/978-3-319-52009-4_1

investigated considering various practical applications in different micro engineering fields. These non-traditional micromachining techniques as well as advanced finishing processes can be improved further and utilized more successfully in the near future for numerous microengineering applications.

1.1 Introduction

Recent advancement in technology has created a distinctive opportunity in the fabrication of desired products with specific criteria. This available as well as developing technology is being used to produce various parts, structures, components, shapes etc. on different materials with required geometrical and dimensional characteristics. With the increasing use of different new materials and alloys having superior mechanical properties like higher strength, increased hardness, better corrosion resistance as well as temperature resistance, the need for components of these high performing materials and alloys are continuously increasing in various sectors like thermal, energy, medical, space, avionics, automobile, textile, food, material processing, etc. [1].

To meet this high demand and stringent requirements in such developing fields of modern fabrication industries as well as to satisfy various needs of society and nation, various traditional manufacturing techniques have been utilized such as forming, casting, machining, deposition, joining as well as different non-traditional processes on different existing and newly developed metals, non metals, alloys, polymers, ceramics, rubber and composites etc. Among the mentioned methods, non-traditional machining techniques have expanded their applicability and offer better opportunities with several inherent advantages that make these processes superior as well as more efficient than conventional one. Both traditional and non-traditional machining processes can further be divided into two domains: macro machining and micro machining. Macro machining deals with the machining of structures or parts which can easily be viewed by naked eye and can be measured by simple techniques as well as machines that are usual and readily available. On the other hand, micro machining involves those machining techniques and material removal mechanisms where the produced parts cannot be observed clearly by naked eye and have dimensions within micron (1–999 μm) ranges. Such micro features, after fabrication, need sophisticated techniques and instruments for their measurements.

1.1.1 Overview of Non-traditional Machining Processes

Non-traditional machining (NTM) processes can be defined as a group of processes that remove excess material from workpiece by various techniques constituting mechanical, thermal, electrical, chemical energies or combinations of these energies without the involvement of sharp cutting tools which are used for traditional manufacturing techniques [2]. The traditional machining methods such as milling, drilling, shaping etc. are not suitable for the machining of tremendously hard and brittle materials. But, the non-conventional machining methods are utilized for the machining of complex and intricate 3D shapes on harder materials as well as thinner materials, where usual machining processes are neither feasible nor satisfactory due to various unavoidable reasons like difficulties in clamping very delicate materials and machining of very flexible workpiece with complex shapes.

Various types of non-traditional machining processes have been developed for specific requirements to fabricate desired complex features. The material removal mechanisms of non-traditional machining processes are basically different than those in the traditional processes that involves different form of energies, other than the excessive forces exercised by a cutting tool, (which is in physical contact with the work piece) that are applied to remove the excess material from the work surface to generate desired shape. Non-traditional machining processes which are extensively exercised now-a-days can be characterized by material removal mechanism, type of energy involved in removal of material, forms of material removal, type and material of tooling, etc. Moreover, different aspects which are responsible for the correct selection of non-traditional machining methods include physical parameters of the processes, capability of the processes in machining desired material and product with required geometry, dimension and size of product to be machined, process applicability, feasibility, operational characteristics as well as economics of the process involved. Physical parameters of the processes involve machining voltage, current, power, machining gap, work medium and other various specific parameters for specific techniques. For example, Electrochemical Machining (ECM) requires power as high as 100 kW, Electron Beam Machining (EBM) and Laser Beam Machining (LBM) require high voltages in the order of 150 and 5 kV respectively and require careful handling, Electro Discharge Machining (EDM) and Ultrasonic Machining (USM) require medium power whereas, EBM and Ion Beam Machining (IBM) require vacuum, etc. NTM processes involve generation of various shapes on work material. For example, EBM and LBM are used for micro drilling and cutting of parts having slenderness ratio, $l/D < 20$. USM and EDM are useful for sinking cavities and standard hole drilling. ECM is useful for fine hole drilling and contour machining. Both EDM and ECM processes have good capability to make shallow and deep pockets as well as contoured surfaces whereas, AJM is useful for shallow pocketing. As discussed earlier, applicability and usability of any machining process are other aspects which need to be taken care of. For example, to machine electrically non-conducting materials, both ECM and EDM are unsuitable, whereas other various mechanical methods can achieve the desired results. LBM is not suitable for machining of reflective surfaces whereas, USM is suitable for machining of refractory ceramic material and AJM can be used for super alloys also.

The machining characteristics of different non-conventional processes can be analyzed with respect to material removal rate, tolerance achieved, surface finish obtained, depth of machined surface, etc. Economics of the processes is another important factor for their end use in fabrication industries. This includes capital cost, efficiency of material removal, cost for tooling, cost for power, etc. ECM has no tool wear but the capital cost of ECM is very high when compared with traditional mechanical contour grinding and other non-conventional machining processes whereas, capital cost for AJM is comparatively low. EDM has higher tooling cost than other machining processes.

In contrast to conventional processes, NTM processes have lower feed rates and require systematic approach for implementation. However, some processes permit batch processing which increases the overall output. NTM processes are used to machine hard and tough materials and these processes have capability to machine the materials with good surface finish and machining accuracy where, some processes have the ability to machine larger feature sizes at lower capital cost compared to conventional approaches. For example, Abrasive Jet Machining (AJM) uses high velocity stream of gases with added abrasives to remove material by brittle fracture from the workpiece. Water Jet Cutting (WJC) removes the material using high-velocity fluid jet impinging on the workpiece during slitting operation. In Ultrasonic Machining (USM), vibrating tool is used to propel the abrasives in slurry at high velocity against workpiece. In ECM, the material is removed by anodic dissolution where tool and workpiece are dipped in electrolyte solution. EDM process removes material by electric discharge from a pulsating DC power supply across narrow machining gap between the tool and the workpiece separated by a dielectric fluid. Electron Beam Machining (EBM) is a thermal process that uses a beam of high-energy electrons focused on the workpiece to melt and vaporize the metal. Laser Beam Machining (LBM) uses laser that is intensely focused and coherent stream of light for vaporizing and thermally ablate the materials. Lasers are also used for joining and heat treating of materials. Chemical Machining (CHM) process is a controlled chemical dissolution (CD) of a workpiece material with the application of strong reagent (etchant). Special coatings called maskants protect the areas in which the machining is not desired.

1.1.2 Need of Micromachining

The demand of miniaturized parts is increasing day by day in modern world. Micromachining utilizes miniature milling, drilling, cutting, turning, controlled dissolution, polishing and other various operations to produce three dimensional features and formed surfaces in micron and submicron ranges. This scaling down of processes from macro to micro regime also requires lowering down of magnitudes of involving operating parameters. Also, functional requirements of many devices demand very tight tolerances and the use of a wide variety of engineering materials. Although geometric and material capabilities of micromachining have been demonstrated [3], industrial application of micromachining has been hindered by the lack of experience and knowledge on the micro-machinability of materials [4]. But, emerging miniaturization technologies have been perceived by many potential key technologies of the future that will reveal completely different ways how people and machines interact with the physical world. Various existing miniaturized parts that can be produced by micromachining processes include micro-electromechanical systems (MEMS), micro-sensors, micro fuel cells, micro channels, micro holes, micro grooves, micro pump, micro valves, medical implants and devices, micro-actuators and various other different micron sized features. Another application of micromachining process involves fabrication of integrated circuits (IC) for electronic devices. These ICs have also large scale use in CD players, computer components, etc. Relays and switches are required to be assembled to produce functional micro sized mechanical parts. For making of lighter aircrafts and satellites, miniaturized components are necessary. These type of aircrafts poses increasing demand in this rapidly changing world.

In this new era, the miniaturization in medical fields is another effective approach for health inspection and microsurgery without pain. Micromachining is the key technology which facilitates the above all and different other requirements of microproducts to keep pace with the rapidly growing needs. Various advantages of microproducts offer less consumption of energy for their actuation; they need lesser places, lesser requirement of materials etc. To cope up with the huge demand in miniaturized parts for modern fabrication industries, scientists and engineers are concentrated towards development of suitable micromachining processes for fabrication of required microproducts. Hence, as per requirement of product miniaturization and manufacturing of the miniaturized products, emphasis should be placed on advanced micromachining processes in addition to the accompanying research trends in miniaturization of equipments and metrology systems.

1.1.2.1 Condition of Micromachining

Micromachining generally adopts a top down approach, which may involve various conventional as well as non-conventional machining techniques [5]. Micromachining can easily generate various microfeatures such as microholes, microslots, microchannels, etc. as well as microproducts like micro sensors, micro thermal devices and computer chips even down to nano regime. Approaches towards micromachining need to be specific and systematic for various methods. However, these approaches can be correlated and directed towards micromachining by following two guidelines: (i) concept of unit removal (UR) which is defined as the material removal per pulse or per unit time for generating the smallest adjustable dimensions of the micro-product and (ii) requirement of precise machining equipments to obtain the desired microproduct with higher accuracy by reducing the dimensional error in proportion to the size of the micro-products.

1.1.2.2 Application Opportunities

In recent years, manufacturing industries have observed the rapid increase in demand for micro-products and micro-components in various emerging and developing fields including electronics, optics, biomedical, medical, opto-electronics, thermal, power, semiconductor, biotechnology, automotive, aerospace, communication, etc. For example, micro-EDM is used for the production of micro gears, nozzles of diesel fuel injectors, etc. Micro-EDM milling can be used for production of miniature three dimensional gas turbine impeller on Si3N4-TiN ceramic composite which serves as a key component for micro power generation system. Moreover, electroformed nozzles are presently used in number of commercially produced ink-jet printers. The electroformed nozzles are formed by plating nickel on a mandrel (mould), which defines the pattern of the nozzle, and after that the finished product is removed. Also, the pulsating current/voltage gives better control over micro-ECM for fabrication of microgrooves for self-acting fluid film bearings and complex shapes in thin films and foils. Another potential application is fabrication of cooling holes in turbine blades. The main objective of cooling blade in turbine is usually to achieve maximum heat transfer coefficients while minimizing the coolant flow rate. Usually the turbine blades are made of a super alloy with a very high melting temperature. By using the lost wax method, these cavities are cast. With this technique, it is possible to produce cavities with a serpentine shape. But, other micromachining processes can be easily employed for drilling of those cooling holes in the blades that are positioned along the span of the turbine blades.

1.1.3 Importance of Non-traditional Machining in Micromachining Domain

The development of miniaturized products in large number has inevitably intended a revolution in the field of micro-manufacturing technologies. This applies both to the development and application of new non-conventional technologies i.e. EDM, laser, EBM, PAM, ECM and so on, as well as the adaptation and optimization of conventional machining technologies of milling, turning, etc., in order to comply with the dimensional exigencies of precise micro parts with good surface quality. In this aspect, non-conventional machining methods offer greater flexibility, precision, rapidity and quality in producing micro-three-dimensional complex features on a wide range of materials, in contrast to the limitations offered by more conventional techniques.

In the last few years, the interest in micromachining technology has increased in the various industrial sectors. The potential of miniaturized products persists to grow numerous technical challenges gradually. With continuous miniaturization, companies are developing new and appropriate machining technologies to meet the unique challenges and to keep pace with the rising demand. The fabrication of miniature parts is not new. Many companies have used various machining technologies such as EDM and laser to produce micro shapes for many years. Today the difference lies in the sheer volume of products that require micro machining. Many manufacturers are developing micro machining technologies and techniques to minimize this gap. Companies are looking for parts having feature sizes of less than 100 microns, or somewhat larger than a human hair. At this dimensional regime, the slightest variation in the manufacturing process caused by material or cutting tool, temperature variations in the machine, vibration and any other phenomenon will have direct impact on the ability to produce features in micron and submicron ranges on a large scale. With this continued miniaturization, companies are developing new products and technologies where the limit of tolerances are mind boggling and would have been unthinkable just a few years ago. For example, using end mills of 50 μm diameter and EDM wire of diameter 20 μm or electrodes smaller than a few tenths micron are becoming more common. The applications related to micro-meso machining technologies are being employed in the manufacture of wide variety of products and devices. The trend is growing continuously in ultra precision and micromachining domain that still requires a fresh look at new machining technologies and process mechanics.

1.2 Classification of Non-traditional Micromachining Processes

From the ancient times people are using tools driven by necessary energy sources to meet the daily needs and to make the life more comfortable and enjoyable. In the prehistoric ages tools were made of stones for the item being made. Later, the invention of iron tools and its implementation lead to the fabrication of more sophisticated articles of required metals. But with further progress in the field of sophistication, recent twentieth century products need to be made of most durable and unmachinable materials. To overcome the difficulties created by these harder materials, tools are now made of alloy steel, carbide, diamond and ceramics, etc. where different, newer and more advanced methods are utilized for powering up these tools.

Each time tools with new material and power sources are used, the effectiveness and capability of fabricators greatly increase. Since 1940's, a revolution in fabrication industries has been going on allowing fabricators to cope up high demand for complicated designs of products made up of tougher and difficult-to-cut materials. This revolution is centered primarily on implementation of numerous new tools and types of energy which lead to the introduction of various non-traditional manufacturing processes for joining, forming as well as removal of materials. Apart from this, the conventional techniques of machining depends on electric motors and harder tool materials for carrying out operations like drilling, sawing, broaching,

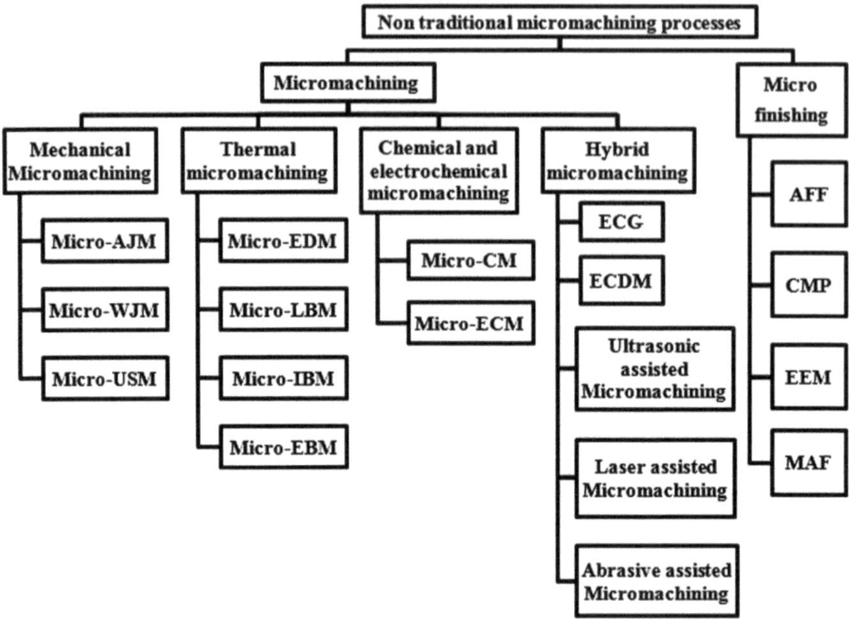

Fig. 1.1 Classification of non-traditional micromachining processes

etc. Conventional process like forming depends on energy from electrical motor as well as hydraulics and gravitational forces. Moreover, removal of material can now be done with electrochemical reaction, high temperature plasmas and high-velocity jets of liquids and abrasives. Different harder materials can be easily formed with magnetic fields, explosives and the shock waves from powerful electric sparks.

The non-traditional micromachining processes are not affected by mechanical properties of material and can generate complex shapes by controlling different physical parameters of the processes. Based on the type of energy required, medium of energy transfer, inherent mechanism and basic requirements, non-traditional micromachining processes can be classified as mechanical micromachining, thermal micromachining, chemical and electrochemical micromachining, hybrid micromachining. Various types of non-traditional micromachining processes have been shown in Fig. 1.1.

1.2.1 Mechanical Micromachining

Mechanical Micromachining process includes abrasive jet machining (AJM), water jet machining (WJM), ultrasonic machining (USM), abrasive flow finishing (AFF), magnetic abrasive finishing (MAF), etc. The process mechanics and mechanism of material removal of these micromachining techniques have been stated briefly here under.

AJM uses kinetic energy of loose abrasives to remove material from workpiece. Before operation, the workpiece surface is covered with wear resistant mask leaving the portions on to which machining needs to be carried out. Hence, mask also determines the geometric and dimensional accuracy of the micro feature to be produced. During operation the entire workpiece surface remains exposed to particle beam and gas or air jet containing microscopic loose abrasive particles of around 50 μm grit size impinges the work material in the unmasked area with high velocity (typically between 150 and 400 m/s) leading to removal of material by brittle fracture and formation of desired patterns. Material removal rate during machining depends on other parameters like the size of abrasive particles, type of abrasives, abrasive flow rate, pressure of gas, type of work material, etc. which can also be controlled more effectively than standard sand blasting. Popular abrasives for AJM include aluminum oxide, silicon carbide, etc. and the erosion action of AJM can especially be applied for cleaning, cutting, etching, deburring or polishing of various brittle materials such as glass and ceramics where very fine intricate features can also be generated [6]. Machining of ceramics by other traditional process like wet chemical etching is relatively expensive for which AJM is being used. Furthermore, researchers have fabricated channels and holes on polymer like polymethyl methacrylate (PMMA) [7, 8] by AJM. Other existing applications of AJM include machining of devices used in micro-electro- mechanical-systems (MEMS) [9] and opto-electronics, fabrication of complex 3D and free standing micro features on glass [10], marking on rubber stencils and electronic equipments, frosting glass surfaces, engraving codes, numbers on car windows, etc.

Unlike AJM, WJM uses kinetic energy of water to remove material from the target workpiece. Here, water jet of high velocity (up to 900 m/s) strikes the workpiece surface which, upon impact, converts to pressure energy. The magnitude of the pressure generated reaches up to three times than that of normal stagnant pressure. When the generated pressure is higher enough than the binding force of the material making up the workpiece, erosion occurs and the material removal takes place. A typical WJM setup consists of pumping unit for generating high pressure of water, high pressure hydraulic accumulator to store pressure energy during idle periods and to provide smoother flow during operations, various types of nozzles, tube fittings, valves, etc. WJM is not thermally activated and is highly economical and eco friendly process for cutting porous materials like corrugated boards, woods, and soft materials like rubber, lead, asbestos, etc. with high surface finish and straightness. Also, the process is applicable for rock excavation [11], cutting features on sandstones, cutting fibers and polyethylene for automotive industries, hydraulic mining and drilling of coals, cleaning and descaling in oil refineries, chemical plants, etc. Moreover, micro features of higher thickness can be produced by multi pass cutting whereas, various products of metals like aluminum, copper and alloys like brass, steel, etc. can be machined, although for such case very high jet pressure is required which limits its uses for commercial intensifiers.

USM is another process especially for fabricating micro structures of both electrically conductive and non conductive materials, semi conductive, hard and brittle materials like ceramics, glass, etc. Here, co-axial microtool is primarily produced of ductile and tougher materials and then implemented for machining. The tool vibrates at ultrasonic frequency and the tiny cutting gap between microtool and workpiece is filled up with constantly flowing abrasive slurry. The repetitive linear mechanical motion of microtool from very high frequency drives the abrasive particles towards the workpiece. With each downward motion, thousands of abrasive particles accelerate and strike simultaneously resulting in removal of millions of chips in the form of craters within that time. USM for micromachining is a combined machining technique for fabricating intricate 3D micro features on almost any type of material. Hence, the process is capable to fabricate micro devices used in MEMS [12], inclined and square holes, trenches [13], micro structures on quartz and synthetic ruby and other features, to be used in the fields of medical, aerospace, electronics, etc.

FIB uses beam of ions with specific diameter and intensity for removing work material atom by atom. Whole process is carried out in a vacuum chamber to avoid interaction of ions with surrounding gas molecules. Of the several ion sources, liquid metal consisting of gallium is used widely. Machining occurs when gallium ion strikes the work surface with high potential. The sputtered and accelerated ions dislocate the atoms, positive ions and secondary electrons forming the workpiece surface. The dislodged atom on the workpiece transfers its momentum to surrounding atoms and material removal continues. The material removal rate of the process is highly dependent on sputtering yield which is influenced by several other parameters like types of substrate atom and ions, energy of accelerated ions, their direction of impingement and density [14]. FIB has negligible thermal effects but the process is not at all economical. However, the process is applicable for producing parts in micro and even in nano ranges like tools and dies for micro EDM, micro forming, nano machining, fabrication of end mill cutters for micromachining, diamond micro indenters, surgical micro tool, tissue stabilizer [15] and other various micro and nano components used in bio-technology.

AFF involves final and fine machining of metallic as well as non metallic parts or micro parts produced leading to higher surface quality and lower surface roughness achievable in the range of 0.05–1.0 μm [16] for improved performance, closer tolerance, greater life span and durability of the product. In AFF, materials in terms of very small quantities are removed by forceful moving of semisolid abrasive media over the workpiece through a passage by extruding them from a cylinder. The applied force is driven hydraulically or mechanically and the flow of media is directed to desired work surfaces by proper tooling. The velocity by which the media can be moved over the work surface depends upon the complexity, type, geometry and cross sectional area of passage where higher amount of force is required if restrictions are too many. The large number of abrasive particles in the media acts as a multi point cutting tool leading to slower removal of material during finishing process. Also, other factors like type and geometry of work material, type of media, abrasive grit size, etc. greatly influence the material removal rate during operations. AFF is more economical, offers better efficiency, flexibility as well as accuracy and is capable enough for machining complex internal surfaces which are inaccessible by other conventional finishing techniques.

MAF constitutes finishing of ferromagnetic as well as non ferromagnetic flat and cylindrical work surfaces by the help of magnetic force. Here, magnetic abrasive particles act as a cutting tool which constitutes ferromagnetic components such as iron particles and abrasive grains such as silicon carbide, alumina, diamond, etc. The magnetic abrasive particles remain stick together in the line of magnetic force due to the application of electromagnetic field forming flexible magnetic abrasive brushes. The necessary finishing force is generated and maintained by electromagnetic field and finishing occurs when magnetic abrasive brushes are rotated and are made to contact with vibratory or moving work surface. The finishing efficiency and the quality of surface finish obtained are greatly influenced by the type, size and shape of work material, strength of the magnetic field, diameter of the magnetic abrasive particle, size and type of abrasive grains, mixing proportion of ferromagnetic components with abrasive grains, relative motion and speed of magnetic abrasive particles with work surface, etc. Finishing with MAF can achieve surface roughness in the range of 0.04–1.0 μm [16], where the process is also capable of polishing internal cylindrical surfaces, deburring and chamfering of edges [17].

1.2.2 Thermal Micromachining

Various micromachining processes involve the application of intense localized heating for machining of workpiece and producing intricate micro features. These processes remove material by melting and vaporization. Different operating principles of these processes include micromachining by electro discharge machining (EDM), laser beam machining (LBM), electron beam machining (EBM), etc. which have been discussed briefly as follows.

EDM removes material by electrical discharge or sparks. During the operation, cathode and anode are separated by a tiny gap of few microns and are immersed in dielectric fluid like kerosene, deionized water, air, etc. When DC pulse power is applied between cathode and anode, plasma channel forms due to very high electric field gradient resulting in formation of electrical discharge or microspark. This microspark melts and vaporizes the workpiece as well as tool material from where some of the molten material is flushed away and removed by flowing dielectric and the remaining part resolidifies and forms recast layer. The tool is gradually fed forward into workpiece by maintaining a suitable discharge gap; as a result thermal erosive action continues and machining occurs. In contrast to macro counterpart, micro EDM has become a popular process for precise machining of intricate micro features of various geometries on various difficult-to-cut but electrically conductive materials with better accuracy and surface finish by decreasing the size of the spark, increasing the sparking frequency and maintaining as well as controlling other operating parameters. The process is quiet capable of fabricating micro nozzles, drilling high aspect ratio holes [18], making high precision metal masking and other numerous micro components required in automobile and electronics industries. However, the process suffers from thermal effects which may deteriorate the surface integrity, quality and property of the produced micro feature. Also, after each discharge, EDM removes a small crater on the tool electrode which may degrade the final machined shape and decrease the accuracy.

LBM is another process for machining complex patterns and intricate micro features where pulsating laser is irradiated and the target work material residing in the focal region absorbs laser energy and heats up to its melting temperature. The net result is the melting and vaporizing of work material at that focal point and material removal takes place via ablation process. The melting and vaporization of work material greatly depends on the intensity of laser as well as properties of workpiece and duration of pulses. LBM is a thermally induced process which generates heat affected zone (HAZ) in the peripheral area of the machining zone resulting in generation of thermal stress and surface as well as subsurface cracking leading to minimizing the life span of the micro product. Though commercial uses comprise of long pulse laser, but ultra fast or short pulse laser is very much effective to machine features within micro ranges. Other applications of this process include machining of printer and hard disk drive heads, different types of micro holes and channels, grating on optical fiber [19], different micro components for microfluidics [20], engraving, marking, etc.

EBM uses kinetic energy of electrons for removing material from the work surface. Here, very high potential gradient is applied to the cathode which emits the electron. The ejected electrons are further accelerated by the potential of anode. The beam of electrons strikes the workpiece surface converting its kinetic energy into very high heat energy upon impact. This localized heating causes melting and vaporization at the beam area causing removal of material. This melting and vaporization as well as material removal rate depends upon beam energy, beam width, thermal property as well as thickness of workpiece material, etc. Alike FIB, EBM is carried out in vacuum chamber to avoid collision of electron beam with air and other airborne particles. Though the process is thermal induced, but still the smaller beam area restricts the span of heat affected zone resulting in minimized detrimental effects on the final product. EBM is a faster material removal technique for producing micro slots, micro holes of different shapes and contoured internal profiles and various other complex micro features applicable in thermal and nuclear sectors, aerospace and automotive industries, etc.

1.2.3 Chemical and Electrochemical Micromachining

Chemical micro machining (CMM) includes chemical reactions required to oxidize and remove work piece material where the oxidization products are carried away from the surface by suitable medium. Here no external power is used and anode and cathode, separated by proper medium, are placed in such a way that the rate of removal of material by oxidation equals to the rate of reduction of the etchant species. Material removal by CMM involves sequential steps where very accurate control of process parameters like composition, temperature and type of solution are required to enhance the mass transfer.

CMM is highly applicable for fabricating metallic parts such as mask, recording heads, instrument panels, enclosure screens, etc., printed circuit boards (PCB), semiconductor device, integrated circuit (IC) and other microcircuit devices to be used in electronics industry in addition with photosensitive resist. But, further in-depth research work is needed to be carried out to minimize the limitations of CMM caused from lower material removal rate, lower accuracy of parts being produced, inability in machining chemically resistive materials, use of hazardous materials like highly corrosive electrolyte that may cause a risk to environmental safety, etc.

Electrochemical micromachining (EMM) is an electrochemical material removal process that implements the method of anodic dissolution and can easily fabricate metallic features in micron and submicron ranges by offering enhanced process control. On the other hand, most of the non-conventional machining processes are thermally activated; examples include Electron Beam Machining (EBM), Laser Beam Machining (LBM), etc. Hence, thermal distortion, deterioration of surface and material property, formation of heat affected zone are of prime concern that need to be taken care of [21]. CMM and EMM are thermal free processes. But, alike EMM, CMM cannot be controlled adequately in micromachining domain. However, accuracy of EMM can be highly improved by controlling required process parameter settings and maintaining very narrow gap in the order of less than fifty microns between microtool and workpiece resulting in generation of ultra precision micro features with desired dimensions. EMM is a very promising micro manufacturing process that offers several advantages like higher material removal rate, greater surface finish, better precision and control, flexibility, reliability, environment acceptability and can even machine chemically resistive materials and alloys like titanium and titanium alloys, stainless steel, copper alloys and super alloys, etc. [22] which have wide use in biomedical, optics, electronics, and MEMS. EMM can be also used as a cost effective manufacturing process for fabricating micro parts that were previously produced by CMM.

1.2.4 Hybrid Micromachining

Hybrid micromachining denotes combination and application of two micro machining techniques simultaneously, thereby utilizing the merits of both processes for generating more intricate shapes and complex parts with better material removal rate, surface quality and integrity, dimensional accuracy and with shorter production time. As an example, conventional grinding offers better surface quality, better dimensional accuracy with lower tolerance values but the process suffers from formation of burrs, residual stresses and various other thermal effects like generation of heat affected zone after machining that may lead to degradation of product durability. On the other hand, electrochemical material removal technique does not have the detrimental effects of high heat generation. Hence, a combined electrochemical grinding (ECG) process has been introduced for utilizing the advantages of both constituent processes as well as avoiding their inherent limitations for generation of micro product. Although primary research in hybrid micro machining techniques was focused on the implementation of electrical and electrochemical phenomena and processes like electro chemical discharge machining (ECDM) is used to fabricate micro features of non conductive and ceramic materials like quartz, glass, etc.; but research is still needed for generating more complex and intricate 3D micro shapes. Some of the other hybrid processes which are applied for machining of desired micro features include ultrasonic, laser and abrasive assisted micromachining. ECDM milling is used in Pyrex glass, glass etc. to fabricate fine 3D microstructures of non conductive materials. Simultaneous ED-EC milling is suitable to acquire better surface finish of about 22 nm on SS304 material. ECM grinding is also applicable for SS321 material to achieve surface roughness of about 0.21 μm. EDM-electro rheological fluid assisted hybrid micromachining is used on aluminum alloy, SUS 304 etc. to obtain fine surface finish in the range of 0.06–0.08 μm. Also, laser drilling Jet-ECM is able to reduce formation of recast layer more than 90% in SS321.

1.3 Introduction to Different Non-traditional Micromachining Processes

Now-a-days non-traditional micromachining techniques are being used for the fabrication of ultra precise parts i.e. micro slots, grooves, intricate 3D shapes, etc. for large scale production. As, conventional machining techniques to be used in micro domain encounter various problems like requirement of high cutting force, tool wear, heat generation between tool and workpiece, less machining accuracy etc.; various non-traditional micromachining techniques are gaining popularity due to particular advantages and have been used extensively by modern fabricators [23]. Numerous types of micromachining techniques are available today which have their own merits and limitations and can be implemented based on specific requirements of the part to be produced. Various types of micromachining processes are used for milling microgrooves, drilling microholes, etc. [24] and to obtain a more clear idea of the processes involved, some of them have been discussed briefly in the following sub sections.

1.3.1 Micro Ultrasonic Machining (USM)

Micro ultrasonic machining (USM) is one of the important micromachining techniques for fabricating high aspect ratio 3D micro features on harder and stronger materials like titanium and its alloys, nickel alloys, etc. and brittle materials like ceramics, glass, quartz, composites, precious stones, etc. [5]. Micro USM can be utilized even on silicon for fabricating micro holes of diameter as small as 5 μm [25] to be used in circuit board with integrated circuits. Moreover, fabricating micro holes not only facilitate the electronic industries, but these types of holes are also well suited to be used in biomedical filters, grids, fluidic filters, fuel injection nozzles, ink-jet printer nozzles, optical apertures, micropipettes, high pressure orifices, manipulators, pneumatic sensors, etc.

1.3.1.1 Mechanism of Material Removal

The process of micro USM combines consecutive application of processes for carrying out machining operations right from production of micro tool to fabrication of final micro product. As discussed earlier, the main difference between micro USM process from usual USM is in the application of micro tool, means for attaching it with tool holder or horn and fabrication of precise micro features in micron or submicron ranges that are not possible in conventional USM. In macro or conventional USM, the tool is fabricated on another system and is then attached by brazing, soldering or screwing but these processes cannot guarantee the concentricity or the accuracy with which tool approaches if the same processes are used for mounting micro tool in micro USM. Hence, for micro USM, the coaxial microtool of various shapes like polygonal, spiral or cylindrical is first prepared by wire electro discharge grinding(WEDG)/wire electro discharge machining (WEDM) on-machine and is then used directly for carrying out required machining operations. Besides, the combination of WEDG with EDM can fabricate 3D micro tools whereas, micro tool with diameter as low as 3 μm can also be fabricated [26]. However, micro tools can also be fabricated like electrochemical micromachining or chemical micromachining.

Micro USM process starts with converting low frequency signal to high frequency. This high frequency signal is then converted to high frequency linear motion or vibration in ultrasonic range (equal or beyond 20 kHz) via piezoelectric or magnetostrictive transducer. The transducer then feeds this ultrasonic vibration to tool holder and tool or sonotrode which is the mirror image of the micro feature to be generated and has the required design criteria for generating resonance to achieve greater machining efficiency and constant wave with high amplitude. Before starting the operation, a narrow machining gap is set between the micro tool and work surface. Moreover, during machining this narrow gap is flooded with abrasive slurry consisting of loose abrasive particles like silicon carbide, boron carbide, alumina, diamond dust, etc. suspended in oils but mainly in water due to its

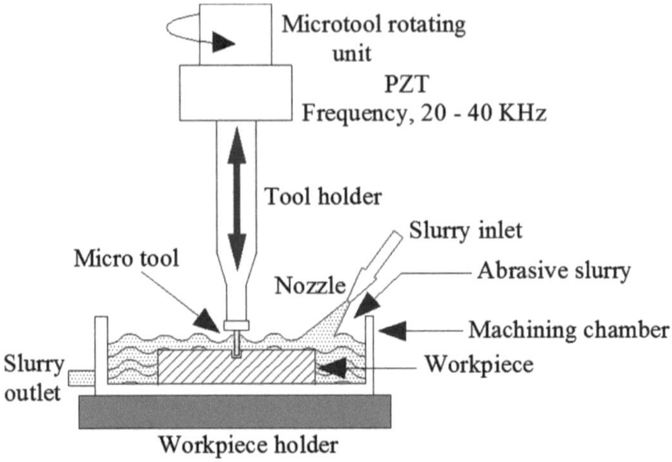

Fig. 1.2 Schematic of micro USM system

good property of transferring ultrasonic wave. The cyclic and ultrasonic linear vibration of microtool gives downward stroke to the abrasive particles. The tiny abrasive particles accelerate and simultaneously strike the work surface resulting in removal of tiny chips from the work material in the form of craters. The tool is then moved into the workpiece with constant force for continuation of machining operations and generation of required micro feature. A schematic diagram of micro USM setup has been shown in Fig. 1.2.

However, micro USM can be done in two ways; one is by vibrating the micro tool and another is by vibrating the workpiece itself which has been introduced recently. In workpiece vibration method the tool is not vibrated, rather piezoelectric vibrations are provided on the workpiece. Here, no horn is needed to amplify vibrations during machining which makes the setup simple and compact. Moreover, the process eradicates the problem like tool wear which is frequent in first method. The principle operating parameters that need to be maintained for achieving better material removal rate, dimensional accuracy and surface finish during micro USM include frequency and amplitude of vibration, load, rotary motion if it has been implemented on tool head or workpiece, type of abrasive particles and liquid component, size of abrasive particles, concentration and flow rate of slurry, type, geometry and thickness of work material, etc.

1.3.1.2 Opportunities

Micro USM can fabricate micro features both on electrically conductive and non conductive materials and posses numerous applications in the fields of medical, electronics, aerospace, etc. The geometric capabilities of micro USM for producing micro slots, holes and 3D cavities can be enhanced by fabricating in situ micro tool followed by machining by it in the same setup. With this process, researchers have developed micro holes as small as 20 μm diameter and 50 μm depth on silicon and quartz [27]. Moreover, rotary micro tool can be introduced which is one of the cost effective process for fabricating micro holes. Here the tool is given ultrasonic vibrations as well as rotation. The combination of this grinding mechanism and rotation leads to sliding and rolling of abrasive grains. As a result, impact and indentation increase that give better material removal rate [28]. In addition, USM can be assisted with turning for strain hardening, reducing residual stress of work material, minimizing machining time, improving surface quality of workpiece, increment of tool life, etc. [29]. Moreover, researchers have fabricated multi micro tool and used this tool for drilling micro holes having 20 μm diameter on soda lime glass by micro USM. For this, a tungsten micro tool was fabricated by wire electro discharge grinding. With this tungsten micro tool array of microholes was fabricated on copper foil by EDM which was later used for fabricating multi micro tool for USM by reverse EDM [30]. In addition, the capability of micro USM can be extended more considering its implementation in the emerging area like microfluidics. Microfluidics extensively uses micro channels and micro cavities for delivering and storing liquid or gas. Also micro channels are used for micro heat exchanger applications and for micro sensors. These micro channels and micro cavities can effectively be produced by micro USM. Micro USM has enough potential for producing 3D micro channels or 3D micro features on silicon and on other various newer materials like different types of polymers, ceramics and glass that can readily be applied for micro-electromechanical system (MEMS).

1.3.1.3 Challenges

Micro USM is well established and promising micromachining technique for generating intricate microfeatures in non conductive brittle and harder materials. However, various challenges still exhibit in feasibility and geometrical ability of machining. Micro USM does not only refer to generation of micro features, but the size of the tool should be made small as well. Microtools suffer from complex fabrication, improper handling and fixing to the tool holder, etc. which may deteriorate the accuracy and performance of the process. Moreover, the alignment of the tool should be accurate enough for confirming the dimensional accuracy of features. Making a micro tool with traditional process such as micro milling and then fixing it to the tool holder with silver brazing or soft soldering does not confirm its alignment because the tool is tiny enough for producing and assembling with traditional methods. Hence, in situ micro tool is first prepared by WEDG/WEDM and then micro USM is carried by the developed micro tool in the same setup.

Micro USM process also suffers from tool wear which leads to gradual shortening of tool tip resulting in inconsistent vibration amplitude, change in tool position and deterioration of geometric accuracy of the produced micro feature. For this, implementation of ultrasonic vibration on workpiece rather than tool has been found to be advantageous. Further, workpiece vibration helps in better removal of machining products by enhancing flushing of slurry and produces better stirring leading to increase in machining efficiency. But this process also restricts the size of workpiece which may influence the vibration amplitude and accuracy.

In addition, various other research issues concerning better online monitoring and controlling, proper tooling technology, measurement accuracy, surface integrity, extended process capability, etc. are there, which need to be strengthened further. Another concerned area of micro USM is interaction between abrasive particles with workpiece for which a well formulated study for understanding and analyzing the material removal mechanism needs to be carried out and phenomenon like cavitation, chemical reaction, micro chipping, etc. during micro machining need to be investigated. In the situations like micro cutting, localized temperature may play a significant role. All the issues need to be resolved in the context of high frequency low amplitude vibration as well as noise reduction that call for further research.

1.3.1.4 Micro Engineering Applications

Micro USM has extensively been used for different applications like making small holes in helicopter power transmission shafts and gears, fabrication of 3D micro-air turbines, machining of watch bearings and jewels, generation of high aspect ratio micro holes on silicon and glass wafers for pressure and flow sensors [31] and numerous other micro features which are required in MEMS. Moreover, micro-features can be generated on the micro tool tip to be used for micro USM. Researchers have fabricated 3D micro gear which acts as a micro tool during USM [32]. The patterned surface on the tool replicates itself when fed towards the workpiece leading to parallel production of identical micro features with lower machining time. Also, researchers have fabricated high aspect ratio micro pillars on Lead Zirconate Titanate PZT [33], spiral groove, contoured 3D micro cavity with tool path designed using CAD/CAM [34], micro feature in alumina to transfer this pattern for generating intricate shaped micro tool [35], wafers in borosilicate glass, etc. Micro USM can effectively machine and polish steel tool surface [36], micro channels for bio-MEMS, whereas semiconductor industries greatly rely on Micro USM as the industry needs like slicing of semiconductor components and processing of various characteristically brittle materials. Also, Micro USM can be utilized on relatively ductile metal-based materials with appropriate alteration, which is a sheer need in microfluidics and heat transfer industries.

1.3.2 Micro Electro-Discharge Machining (EDM)

Micro electro-discharge machining (EDM) is one of the promising and cost-effective methods of machining high strength conductive and brittle materials. In comparison to other methods this micromachining process is relatively new.

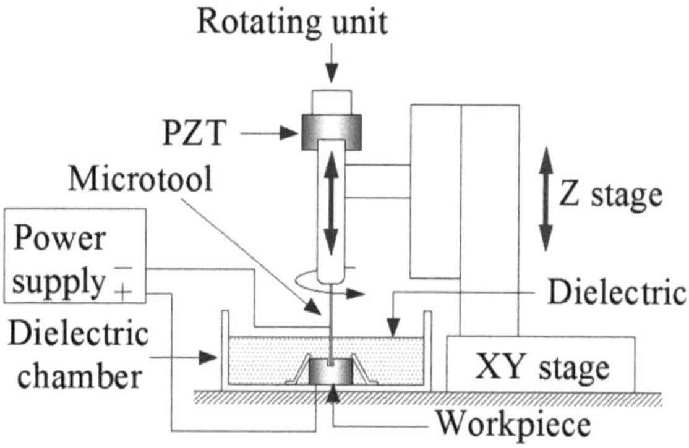

Fig. 1.3 Basic arrangements and setup of micro electro-discharge machining process

Here, a conductive tool and workpiece are submerged in dielectric medium and are separated by a small gap. The tool is made cathode and workpiece is made anode that are connected to a resistance capacitance (RC) circuit. When the tool moves toward the workpiece, discharge occurs resulting in the formation of plasma column. The flux of electron generates ample amount of heat which melts the workpiece material. With the feed of the tool, discharge continues making a replica of the electrode on the work surface. Micro EDM has similar characteristics as EDM except that the energy range is lower and size of the tool, discharge energy and resolution of axis movement are in micron ranges. Micro electro-discharge machining is an efficient process for producing micron sized shapes even on extremely hard tungsten carbide and stainless steel as no cutting force is involved during operation. But as the process is thermally activated, hence it suffers from limitations like formation of heat affected zones, change in material property and formation of micro cracks on the workpiece surface. Moreover, during each discharge the tool material also melts which limits its use for further applications. Figure 1.3 shows basic arrangement and setup of the Micro electro-discharge machining process.

1.3.2.1 Mechanism of Material Removal

As discussed earlier, micro electro-discharge machining is a thermoelectric process that melts and removes the workpiece material from various extremely hard and conductive materials by applying and maintaining controlled discharges or sparks between cathode tool and anode workpiece that are separated by tiny spark gap and are dipped in suitable dielectric medium [37] like kerosene, deionized water, air, etc. When machining starts, due to the application of voltage from suitable DC pulse power source, high electric field generates at the points where the tool and workpiece are closest. As a result, electrons start moving from cathode to anode and collide with the neutral molecules of dielectric during this movement. This collision results in formation of more free electrons that get loose from neutral molecules. The ionization gradually increases to such extent that a narrow channel of constant conductivity or plasma channel creates leading to continuous flow of electrons towards anode and ions towards cathode. Upon striking the anode and cathode surfaces their kinetic energy changes to intense heat energy resulting in generation of momentarily current impulse forming discharge or spark. These discharges or sparks generate intense heat leading to melting and evaporation of workpiece as well as tool material.

Actually, during pulse on time the pressure in the plasma channel gets very high that restricts the evacuation of molten metal. During pulse off time this pressure drops instantly leading to evaporation of molten metal from the work surface. The rate of metal removal from both workpiece and tool is highly dependent on the kinetic energy of electrons and ions respectively. But, as electrons are lighter in weight than that of ions; it has been seen that anode workpiece corrodes at faster rate than cathode if both are made up of same material. Hence, workpiece is always chosen as anode. As the material is removed from that machining zone; the spark gap between tool as well as workpiece increases resulting in formation of next spark at next closest points between tool and workpiece. Thus, the discharges travel all over the tool surface leading to uniform material removal and generation of approximate mirror image on the work surface. This phenomenon continues when the tool is fed into the workpiece leading to formation of required feature on work surface.

Both straight and reverse polarities can be used during micro EDM, where, in straight polarity workpiece is made positive and tool is negative which is simply opposite in reverse polarity. During machining the sparking frequency is very high whereas, spark radius is small and temperature is very high. As a result, small amount of material melts and evaporates where a portion of this is removed by dielectric medium and the remaining portion gets solidified on the workpiece surface forming recast layer. This thermally activated micro electrical discharge machining process also forms and leaves small amount of crater on both workpiece and tool surface which hinders further use of same tool and destroys surface integrity of the produced micro shapes [38].

The operating principle of micro EDM is almost similar to conventional EDM process. In micro EDM, for maintaining the tiny spark gap of as small as 1–5 μm [39], servo controlled feed mechanism of enhanced positional accuracy and superior sensitivity is used that can be used both for conventional precision manufacturing as well as fabrication of various micro features like micro inserts, micro moulds, etc. in micron and submicron ranges [40]. But the main difference between macro and micro EDM lies in the process of eroding the material. In macro EDM the motto is to maximize the erosion and minimize the machining time by setting the pulse and current parameters accordingly where in micro EDM efforts are made to increase the erosion efficiency required for micromachining by minimizing the energy of electro-spark and erosion through single capacitor discharge that also make the power circuitry of this process different.

1.3.2.2 Opportunities

Micro EDM is a popular and efficient process for fabricating various complex micro features. The setup is simple and the process has added flexibility which can be modified further with ease in the context of specific requirements and applications. For fabricating and sharpening of tiny tool electrode in single discharge, researchers have developed micro EDM setup with two drives of different response frequencies, one is lower response stepper motor controlled for coarser movement of workpiece and positioning while another has piezoelectric transducer having higher response for finer movement of electrode [41]. Moreover, researchers have developed micro EDM setup with two types of local actuator module. The first module consists of two electromagnetic actuators for holding the electrode, controlling the thrust and better positioning, whereas the second module consists of eight electromagnetic actuators for motion in radial direction for smoother, accurate and stable machining and minimization of tool wear [42]. In addition, micro EDM has been combined with other processes like micro USM and micro punching for fabricating micro holes for MEMS in borosilicate glass and on stainless steel sheet [43, 44].

Newer methods of fabrication by micro EDM involve use of variable capacitance for minimizing electrode wear, implementation of sequential machining like wire EDG for in situ fabrication of electrode and generate microfeature with same electrode in micro EDM followed by electrochemical polishing to enhance geometric accuracy and surface quality, etc. [45]. Furthermore, the efficiency, process stability, material removal rate, flushing and surface quality of micro EDM can be enhanced by rotating the electrode, rotating the workpiece, etc. Moreover, micro EDM, as a potential process for fabricating micro shapes of stronger and wear resistant tungsten carbide, will aid to modern fabricators in their respective areas and will be of immense help in understanding the basics and fundamental science of the process, thereby enlarging its implementation field. But further research work is still needed to cope up with the challenges that are encountered frequently. For example, development and implementation of new pulse generator is needed that can produce lower discharge energy suitable for micro scale productions. Moreover, micro EDM milling can be automated to cope up the steep needs in producing micro-molds and dies. But superior control of operating parameters with involvement of proper optimization technique is required to achieve quality machining with required dimensions, accuracy and surface finish. Also, multi-purpose and miniature sized machine tool can be fabricated by micro EDM which will disclose more advanced research areas in the concerned field.

1.3.2.3 Challenges

More recent advancements in the field of sophistication and miniaturization lead to increased demand where micro EDM has been playing an important role in fabricating micron sized products. Though some research challenges in the area of micro EDM are still prevalent and need more concentration such as lower material removal rate found in machining of tungsten carbide micro-structures which make it time-consuming and difficult to be applied for mass scale production. The difficulties are also there in the circulation of dielectric medium and proper removal of machining products, particularly when the spark gap is in the order of few microns during machining a complex shaped micro hole or micro cavity with high aspect ratio. Such improper flushing will eventually degrade the micro machining efficiency and will restrict the formation of high aspect ratio micro holes.

Wear of micro tool is another problem in micro EDM that can stop the machining process at the middle, if not taken care of at the first time. For machining square, rectangular, triangular and specific shaped micro features as well as complex shaped 3D micro mould, rotating electrode cannot be used in die sinking micro EDM. As a result, the machining products deposit and stick between tool and workpiece that effect badly on the circulation of dielectric and deteriorate quality of the machined shapes. Other disadvantage of micro EDM includes formation of relatively poorer and imperfect surface after machining. The surface roughness increases with increase in machining current, pulse on time and voltage [46]. The thermal actions during machining also give rise to micro cracks, residual stress; it changes the microstructure as well as microhardness of surface and subsurface layers that leads to hydrogen and carbon diffusion. Moreover, each machining leaves a crater on microtool as well as on work surface which is influenced by pulse energy. Furthermore, the generated recast layer on the workpiece after machining needs additional operations like precision grinding for removal. Due to such research challenges, micro EDM parameters require to be carefully selected [47] and further attention and investigations are needed for successful utilization of micro EDM for fabrication of micro features.

1.3.2.4 Micro Engineering Applications

Micro EDM is extensively applied for the fabrication of inject nozzle for bubble jet color printer, micro-slots, high aspect ratio micro-holes, dies for precision blanking, micro-gear wheels with complex shape, miniature neurosurgical instruments like micro-forceps made up of nickel-titanium wire, etc. and several other micro components for aerospace industries. Moreover, researchers have fabricated micro-electrode by block-micro EDM process [48], high aspect ratio WC–Co micro-structures with gear pattern by combining LIGA with micro EDM [49], micro compressor and micro turbine impeller by micro EDM milling [50], high aspect ratio micro array with reverse micro EDM [51], micro ball joint of bio compatible magnesium alloy for use in medical devices [52], micro electrode by single spark during micro EDM [53], etc. Presently, micro-EDM is a readily used technique in industries for high-precision machining and generating micro shapes with higher accuracy on various types of conductive metals and alloys, graphite and different harder ceramic materials.

1.3.3 Micro Laser Beam Machining (LBM)

In modern days, micro machining using laser radiation has become very popular and is acting as a versatile fabrication process by means of its different superb features like contact and wear less machining, better flexibility and possibility for high degree of automation, etc. As a result, micro laser beam machining is being widely used for fabricating micro components on various metals, ceramics and polymer to be applied in medical industries, automobile sectors, etc. Micro laser beam machining process is also being used for producing semiconductors and solar cell. Laser for micro machining attributes wide range of pulse duration from femto-second to microsecond, variety of wavelengths and different pulse repetition rates from single to megahertz allowing the process to fabricate complex machined micro parts with high aspect ratio. Various micromanufacturing domains like cutting, drilling, welding, surface texturing also use micro laser beam machining process for removing the material by rapid heating, melting and evaporation. Also, the detrimental effects of high heat generation like burr formation, melting, etc. during micromachining can be minimized to some extent by using lasers of extremely short pulse duration. Figure 1.4 shows the schematic of laser micromachining system.

1.3.3.1 Mechanism of Material Removal

Micro LBM uses thermal ablation technique to remove material from the workpiece [54]. Typical laser micromachining setup consists of several features like a CNC machine tool. It has a central computer for controlling and moving the stages and positioning the workpiece at exact focal point. Moreover, the controller can halt the laser pulses as per requirement, adjust the pulse and change the pulse repetition rate

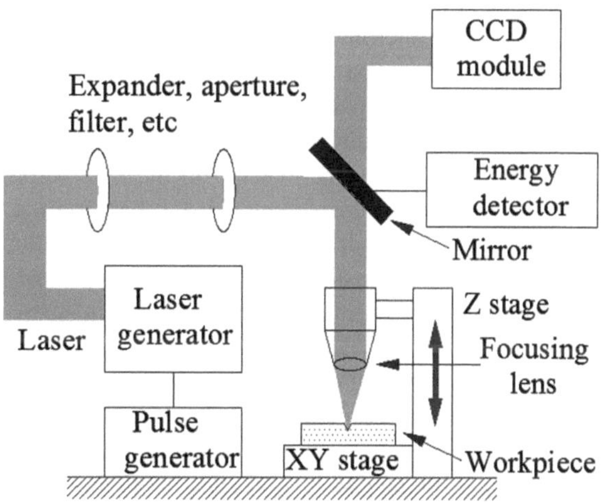

Fig. 1.4 Schematic of laser micromachining system

during machining. Video monitoring system is implemented during operations whereas power monitor adjusts optical attenuation for minimizing or maximizing the power of the focused beam. When laser beam is focused on the work surface, the energy is absorbed and generates heat. The penetration depth of laser increases with increase in absorptivity whereas the absorptivity of any material depends upon its optical property. Upon interaction with metal, light couples with the free electrons of the workpiece which depends on the transfer of photonic energy to electron vibrational energy. The electrons get excited and start colliding with other atoms in their vicinity, thus knocking them out. The knocked off atoms repeat the same procedure resulting in extreme raise in temperature which conducts and produces a temperature distribution. If the material is strong absorber, high temperature will cause in melting and vaporization leading to thermal ablation. If the absorptivity of material is low, the coupling will be very low and temperature will not rise to the extent of removal of material. The power density during machining greatly affects the vaporization rate. If the vaporization rate is high enough, it causes a shockwave. The pressure of this shockwave ejects the molten metal at high speed after the laser pulse.

The mechanism of material removal during laser micromachining can be achieved by three ways: direct writing, mask projection technique, interference technique. In direct writing manner desired micro features are generated by moving the workpiece with respect to fixed laser beam or by scanning the laser beam by keeping the workpiece fixed. In mask projection technique a mask is patterned with the desired shape of the workpiece and is illuminated by laser which is shaped and homogenized for uniform intensity distribution across the uncovered region of mask. The pattern is then projected on the work surface using projection lens. In interference technique laser beam is split and super-positioned with beam splitter for producing interference patterns. These interference patterns have unique intensity variation that enables periodic micromachining over work surface. Material removal by thermal ablation is very useful method for machining metal as well as ceramics where smaller thermal diffusivity and higher value of absorption coefficient is required to maintain for achieving better ablation efficiency.

Pulsed laser micromachining can be carried out in two time regions, one is long pulse and another one is ultra fast or ultra short pulse. In long pulse laser, time required for heat diffusion is shorter than pulse duration. This results in easy heat diffusion in the surrounding leading to decrease in overall machining efficiency. Also, the boiling of work material in the machining zone drives the molten metal away in the form of droplets which fall back on the work surface and form debris that are very difficult to remove and need additional machining operations. Moreover, heat diffusion results in formation of heat affected zone near the peripheral area of the machining zone and subsequent heating and cooling waves propagate through it leading to formation of thermal stress and micro cracks. On the other hand, machining efficiency in short pulse laser like nano second, pico second and femto second is higher and better ablation occurs as the time interval between two subsequent pulses is very short which entraps the laser energy. The heat generation instantaneously goes past to melting and evaporation point leading to machining of various refractory materials like rhenium, molybdenum, etc. Moreover, there is no debris or droplets on the work surface, negligible heat affected zone and no surface or subsurface damage after machining. Other parameters like ablation threshold and reflectivity of workpiece material, wavelength, pulse duration as well as pulse repetition rate greatly influence the process of ablation during machining.

1.3.3.2 Opportunities

Micro LBM is a versatile process for fabricating various simple as well as intricate micro features in metal, glass, plastic, as well as machining very thin and delicate foils with desired precision and dimensional accuracy. The process includes machining of required micro shapes by drilling, cutting, marking, and even involves operations like threading, turning, grooving by utilizing multiple laser beams. The process can be utilized for controlled fracture of delicate items. Laser drilling of high aspect ratio micro holes can be enhanced further by implementing newer techniques like helical drilling that removes work material in helical pass splitting the total machining depth into multiple single steps; trepanning that uses moving laser spot in circular orientation and polarization control that can be achieved by changing linear laser polarization into circular one [55, 56]. Additionally, helical drilling can be combined with polarization control for fabricating high quality and high aspect ratio precise micro holes. Laser cutting can be done by trepanning. Trepanning is done by optomechanically moving the laser on the work surface in a predefined path which helps in machining larger sized holes even on ceramics. CNC cutting is used for machining nonlinear micro holes with programmable shape whereas laser marking is used to machine very low depth grooves to form desired patterns.

Micro LBM is well known and flexible process to fabricate microfeatures in submicron or even in nano ranges where possibilities are still there to carry out further research works by using fiber laser and also by using very smaller pulse duration in nano, pico and femto second ranges on different materials. Nano second laser can further be used for excimer laser, CO_2 laser, copper vapor laser, diode pumped solid state laser, etc. [57]. Other research scopes include study of change in surface morphology due to heating, successful implementation of lasers for obtaining high resolution, intricate 3D and high aspect ratio micro features on numerous materials, generation of micro textured surfaces to be used in micro fluidics and micro mechanics. Micro LBM is also utilized in areas like biomedical and processing materials for semiconductor, photovoltaic cell, etc.

1.3.3.3 Challenges

Micro LBM is a thermally activated process. Hence, during machining it generates machining debris which require involvement of further machining techniques for removal. Moreover, some amount of molten material resolidifies and forms recast layer around the machined feature. Formation of heat effected zone and modification of grain structure is prominent in laser machining. Also, the work surface after machining undergoes thermal and mechanical stress as well as surface and subsurface cracking. All the above mentioned problems can be encountered by using ultra short pulse laser. However, formation of recast layer does not disappear completely even for femto second lasers. But these techniques are not at all economical and give rise to optical problems. On the other hand, ultra high laser intensity generates which leads to strong deformation of laser light in the focal plane that may affect the machining accuracy [58].

Furthermore, newer techniques like underwater laser machining have been introduced which need extensive research for confirming its feasibility in encountering several thermal related issues. Another key problem in laser machining is fabrication of micro features on reflective surfaces that restricts its application in fabricating features of copper, aluminum, etc. for which further research works are needed. Other research issues of micro LBM include engraving lines for larger areas, minimizing machine-to-machine variation and generation of fumes and odors, decreasing longer machining time needed by pulsed laser systems, characterizing the ablation behavior and heat conductibility of various fiber reinforced composites, etc.

1.3.3.4 Micro Engineering Applications

Micro LBM processes are being used for micro-texturing as well as machining of silicon wafers, thin glass sheet [59, 60], micromachining of optical material like lithium niobate, Gallium Arsenide and lithium tantalite, accurate machining of fluoropolymers to be used for miniature sized lab-on-chip technologies, drilling tiny holes in chip packages and polymers used for microcircuit boards, micro-texturing of magnetic hard disk drives, video heads, scribing, patterning, dicing of developed optical wafer as well as substrates and other numerous micro as well as nano range applications like smoothing of synthetic CVD diamond for optical applications like thermal as well as infrared sensors and detectors, etc. [61]. Micro LBM is also used in various biomedical applications like fabrication of stents, micro holes in patient monitoring probes, various micro components in MEMS, rapid prototyping of micro fluidic channels with optical waveguides, micro-textured surfaces for cell adhesion [62], etc. Further applications involve texturing of magnetic disk and diamond films, fabrication of various polymeric micro structures for transferring these patterns on metallic substrate, patterning and repairing of electronic circuits and devices, fabricating micro patterns of inorganic, organic and biological materials, fabricating bio sensors, thin film transistors, OLED display [63] as well as drilling of meshes in metallic foils, trimming of machined devices, percussion drilling and underwater ablation of silicon [64, 65], fabrication of nickel intravascular rotor micro turbine [66], mold inserts for photo molding, etc. [67].

1.3.4 Micro Ion Beam Machining (IBM)

Micro Ion Beam Machining is another atom by atom material removal technique for fabricating precise parts in micro as well as nano ranges. In this machining process, ions of inert gas such as argon are bombarded with very high kinetic energy of about 10 keV on the workpiece surface resulting in ejection of atoms by elastic collision. A schematic diagram denoting the micro IBM setup has been shown in Fig. 1.5. Due to rising demand in fabricating micro and nano components, micro IBM is gaining popularity in manufacturing semiconductor devices, maskless implantation as well as failure analysis and other respective fields [68]. Apart from traditional techniques like X-ray or UV lithography that are used as intermediate process for primarily producing micro and nano scale features, micro ion beam can efficiently generate high resolution structures on any conductive component by direct writing manner. This micromachining process is also helpful to fabricate microtools, etching of IC-patterns, generation of micro and nano features on silicon substrates, sharpening of diamond cutting tools, etc.

1.3.4.1 Mechanism of Material Removal

In micro IBM, ion with specific intensity is moved towards work surface for fabricating required micro feature. The operation of IBM is carried out in vacuum by using vacuum pump for avoiding collision with surrounding air molecules and airborne particles. Other accessories of IBM setup includes stage, detectors, ion column, liquid metal ion source for providing ion, inlet and containment for gas

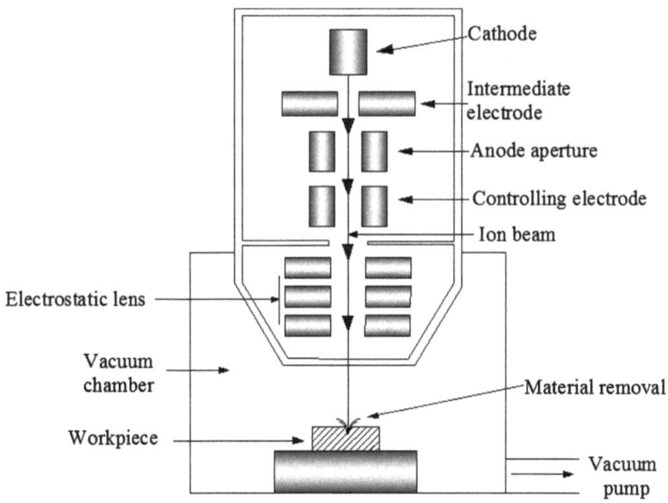

Fig. 1.5 Schematic of micro IBM setup

[69] as well as controlling and monitoring unit. Among several species, Ga is primarily used for liquid metal ion source due to its several advantages like low volatility, low melting point, low surface energy, etc. [70]. Liquid metal ion source again consists of extraction electrode, capillary tube with tungsten needle, shielding, etc. The electrical coil melts the gallium which wets the tungsten tip. When very high electric field with voltage of about 7 kV is applied in extraction electrode, molten Ga deforms and forms a cone like shape which is known as Taylor cone. Moreover, the strong electric field pulls out the positively charged ions and accelerates them through the ion column with voltages ranging from 5 to 50 keV, whereas continue supply of ions are there as molten Ga constantly flows to the Taylor cone. The ions form parallel ion beams with the help of probe forming lens and is made to pass through mass separator which maintains mass-charge ratio. A drift tube is positioned following the mass separator that removes the ions which are not directed vertically. The lens focuses this beam on the work surface which is also accompanied by an electrostatic beam deflector for controlling the striking direction and position of beam on the work surface. The generated ion beam collides as well as impinges with high kinetic energy and controlled manner that dislocates the atoms, positive ion and secondary electrons in work surface and subsurface. The energy of this ion sputtering ranges from 10 to 100 keV and if this energy is higher than the binding energy of workpiece atom, dislodging of substrate atom occurs. The procedure repeats itself and the dislodged atoms contribute to further removal resulting in eroding of work material. The number of atoms ejected per ion is known as sputtering yield that depends upon types of ion and atoms in work surface, ion energy and density, incident angle, etc.

Various types of interaction occurs when an accelerated ion strikes the workpiece. These are swelling, backscattering, implantation, redeposition, etc. With the increase in ion energy, sputtering yield increases, however, the work surface may swell. Further increase in ion energy leads to undesirable effects like ion implantation. Back scattering is the removal of incident atom either by direct or multiple collisions where the back scattered atom are available for redeposition. The parameters that defines the characteristic of ion beam include ion flux, which is flow of energy or particles passing through a specific area per unit time; fluence, that can be defined as the number of incident particles on a defined cross sectional area for a certain time interval; dose, which denotes the amount of particles absorbed after collision in the work surface, beam current; current density, dwell time, beam spot size, incidence angle, etc.

1.3.4.2 Opportunities

IBM is a popular process for precise and accurate fabrication of micro as well as nano features. In contrast to highly precise and conventional lithography for machining micro/nano features, IBM, as a maskless process [71], offers several advantages like finer focusing of ion beam, availability of various ion masses, better energy density, etc. which make the process appropriate even for accurate nanoscale patterning and fabrication of intricate 3D micro/nano features. IBM constitutes various techniques for material removal like milling, which involves sputtering and etching, implantation, deposition, imaging, etc. Ion beam milling involves sputtering and removal of work material by scanning the work piece leading to maskless patterning and formation of any predefined shaped micro/nano feature. The process of milling, chemical reaction, machining rate and efficiency can be enhanced further by using suitable precursor gas like xenon difluoride which is fed into the work chamber. Moreover, ion beam milling with precursor gas like SiO_2 and nitride for aluminum, phenanthrene for carbon materials can be used in additive manufacturing where the gas contains the material required to be deposited is injected by capillary gas feeding nozzle. The process can be used for micro/nano surface slotting, hole drilling, ring core milling, h bar milling, etc. As a supplementary of ion milling, ion implantation can modify work surface in which specific mass of ions are accelerated with high kinetic energy that bombards the work surface and are implanted [72].

Moreover, the process can be combined with other processes like deep reactive ion etching, wet etching, [73] etc. Similar to chemical vapor deposition, ion beam deposition can contribute to maskless deposition of metals like Pt and W as well as insulators like SiO_2 on various surfaces, where better resolution can be obtained but the deposition rate is slower. Ion beam imaging does not involves deposition or removal of material but alike scanning electron microscope, ion beam can be scanned over a sample surface for producing images and revealing surface topography that further unbound wide applicability and research opportunity for the process. Additionally, IBM is a potential process which can effectively be utilized for nano scale production that discloses its superior usability in micro as well as nano regime. In thin films, single-crystal whiskers and in layered structured thin graphite flake which also poses superior electronic and electrical properties, IBM can easily produce perfect nano stacks along the c-axis [74] by 3D etching. Also, this process has become popular for precision fabrication of ultra fine devices in submicron as well as nano ranges; thus enabling IBM for further investigation and extending research scope and opportunity.

1.3.4.3 Challenges

The major challenge in micro IBM relies in minimizing the damage caused by ion beam on the wokpiece surface. With the increase in ion implantation, individual disordered cascade regions get overlap forming a damaged layer over the surface that generates residual stress in the component. The ion beam can also amorphize the workpiece surface by creating point defects, dislocations, phase transformation and grain size modification; thus requiring further study, extensive research and systematic investigation in this area. Moreover, samples prepared for characterization and produced by IBM contains impurities like oxygen, galium, carbon, etc., where during machining, oxygen can get into vacuum chamber due to its relatively poor build quality. Several other problems concerning IBM include ion-beam induced grain growth in fine-grained Ni as well as in Ni alloys [75], formation of Ga containing surface phases [76], change in chemical properties of work surface in ion beam milling, etc. that requires systematic research.

Though the process is being utilized for fabricating micro components and other micro-nano applications, the cost of micro ion beam machine is very high and requires skilled operators for carrying out machining operations. Apart from the ability in fabrication of precise parts, micro IBM suffers from spreading radiation from propagating ions that requires added protection for fabricators during operation, formation of void, bubble, etc. in the final machined shape. Moreover, getting exposed to radiation in the machining zone for prolonged duration can change the material property of the workpiece that should be taken care of.

1.3.4.4 Micro Engineering Applications

Micro IBM has lots of applications in the fields of micro as well as nano manufacturing and can even develop feature of about 5 nm size. IBM is very popular in semi conductor industries for failure analysis and modification of electronic circuits, debugging ICs, repairing mask, preparing samples for TEM, rapid prototyping of micro/nano features to be used in MEMS, sensors [77], fabrication of nano pillars, AFM and SPM tips [78], micro tools [79], fabrication of nano patterns [80], and several trenches [81], micro gear by FIB-LIGA, etc. Furthermore, researchers have fabricated nano-hand [82], 3D honey comb structure [83], lens moulds [84], stamps [85], different types of manipulator like as two pillar, nano actuator with coil structure, 3D nano rotors [86], nano tools, bio nano tools, various bio metric structures like glass capillary for bio medical treatment, neural interface for connecting nervous system with external devices [87], nano structured thin film magnetic head, nano cup [88] and other various 3D curved nano structure [89], etc. Moreover, the process can be used for machining micro cutting tools made up of diamond, tungsten carbide, high-speed steel and various other harder materials where cutting dimensions within the range of 15–100 μm [90] can easily be achieved and other complex geometric features like tool nose radius, rake and relief angles can also be made within micron range by micro IBM process.

1.3.5 Micro Electron Beam Machining (EBM)

Micro EBM is a thermal activated micro machining process where electrons with high speed strike the workpiece surface, thereby transferring the kinetic energy into heat energy that melts and vaporizes the workpiece material. During micro EBM process, very high velocity (228,478 km/s) of electron beam is achieved by using voltage of about 150 kV and is focused over a point of 10–200 μm diameter that results in very high power density in the order of 7000 billion W/sq.mm. This high power density is enough to vaporize any type of material instantaneously.

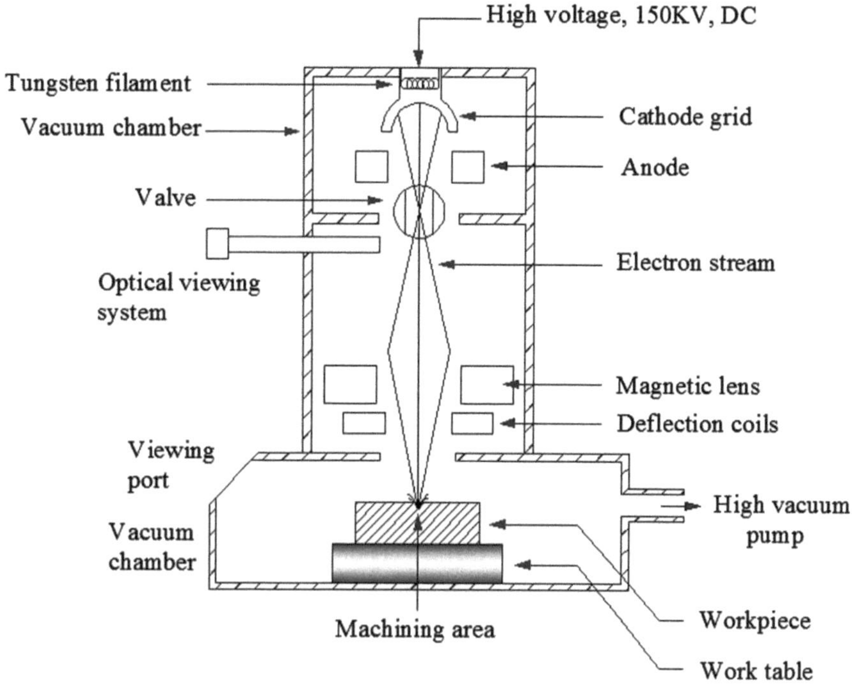

Fig. 1.6 Schematic of micro EBM setup

Whole micro EBM process is conducted in vacuum chamber without any contamination to evade any possible collision of accelerated electrons with air molecules. This restricts the process to be used over large work pieces as the vacuum chamber is very confined. But the process has advantages in having very less thermal effects on the workpiece because the machining zone is very small, typically in the order of only 25–50 μm. Furthermore, the process is able to generate different intricate shapes and contours by directing the electron beam using magnetic deflection coil. A schematic diagram of the micro EBM setup has been shown in Fig. 1.6.

1.3.5.1 Mechanism of Material Removal

In micro EBM process, high velocity of electron beam is generated from an electron beam gun which consists of tungsten or tantalum cathode filament. When large potential difference is applied between cathode and anode, cathode gets heated up to around 2500 °C and thermo ionic emission of free electron occurs that is further facilitated by maintaining vacuum inside the beam gun chamber. The negative biased cathode repels the electrons causing them to move towards annular cathode grid. The cathode grid is also made negative so that the electrons can approach to annular anode. Due to the application of potential field and attraction of annular anode the stream of electrons gets accelerated leaving the anode with very high velocity that is about half the velocity of light. The cathode grid is made concave for easy concentration of beam. The beam is then made to pass through a series of magnetic lenses and apertures that shapes the beam, reduce divergence and improves the quality of that beam. The beam then comes to electromagnetic lens and deflection coil. The electromagnetic lens is used to focus the electron beam over a very tiny spot whereas the deflection coil can maneuver and move this beam in desired direction. A valve is also situated in the electron beam machine for controlling the beam duration.

The entire EBM process needs to be carried out in a vacuum chamber [91] as electrons travelling at very high velocity may collide with air molecules in the way and may lose their energy. Moreover, the beam will get disturbed which will make the control of required operating parameters difficult and will adversely affect the machining quality. During machining, the electron beam is focused on the workpiece. The beam interacts with the work surface and the electrons impinge into the work material. This penetration depth is highly dependent upon work material properties like melting temperature, thermal conductivity, vaporization temperature, etc. and the characteristics of the converging beams [92]. Upon striking the work surface, the kinetic energy of electron coverts to tremendous heat energy that melts and vaporizes the localized area of the workpiece in the focal region. Finally the molten material is wiped off from the machining zone by using high pressure of gas formed by the reaction of beam and backing material that is applied at the underside of workpiece. However, if some molten metal is left, it will form a recast layer but this will not oxidize as the process is carried out in vacuum. The electron beam gun has the facility to even operate in pulse mode during machining where single pulse is used to drill holes on thin sheets and multiple pulses are used to drill hole on thicker plates. Other than the pulse duration, control of various process parameters is necessary for satisfactory machining which include beam current, working distance, beam shape and size, incident angle etc.

1.3.5.2 Opportunities

Micro EBM can drill very high aspect ratio (25:1) micro holes on various low strength materials with very high drilling rates leading to minimization of production cost. Moreover, tapered, reverse tapered and barrel shaped holes can also be fabricated be maneuvering the deflection coil and introducing CNC system [93]. The process is suitable for machining wide range of electrically conductive and non conductive materials like Ni, Ti and other super alloys, stainless steel, plastic, aluminum, leathers, ceramics, etc. Also, the detrimental thermal effect like formation of heat affected zone expands over only few microns due to the application of shorter pulses. Moreover, in general performance of the machining does not depend on physical, mechanical or metallurgical property of the work material and requires very simple work holding due to the absence of any cutting force. As a result, the fixturing and holding cost is lower and the process can be used for machining of very brittle and fragile materials. The process also offers several opportunities that can be extended further by applying it in fabricating complex micro components.

1.3.5.3 Challenges

Alike other discussed micro machining techniques, micro EBM process is not devoid of any difficulties or problems. Though no mechanical distortion is there, but generation of very high thermal energy in a confined zone may introduce residual thermal stresses and the quality of the edges may be influenced by thermal property of the workpiece as well as pulse energy. Moreover, the process is not at all economical, requires high operating cost, initial investment, regular maintenance and involvement of skilled operators. Additionally, for creating vacuum the time required is much more, resulting in increase of non productive idle periods and overall machining time. However, the influence of this incidence can be minimized to some extent by introducing vacuum load locks. Though the formation of heat affected zone is negligible, the formation of recast layer cannot be avoided. Hence, further research is still necessary in order to encounter various challenges and to enhance the process feasibility as well as applicability.

1.3.5.4 Micro Engineering Applications

Micro EBM has diverse application field starting from the fabrication of bi-metal saw blades to transmission assemblies. On flat as well as contoured surfaces, CNC EBM drilling is used for rapid fabrication of large number of holes that are faster than other processes. Moreover, the technique is capable of drilling micro holes by single millisecond pulse of electron beam making the total process much faster [94]. Various types and shapes of precise slots and holes like tapered, bowed wall, straight wall with various contoured internal profiles can be produced by micro EBM. Moreover, holes can be drilled on workpiece of thickness 9.5 mm and on harder metals like titanium and its alloy [95] by precisely controlling the beam energy which can be used extensively in aerospace application. Electron beam milling can be used for cutting of different shaped apertures in thin foils and wafers. Also, the process is used for selective etching of thin films on glass or ceramics. Electron beam drilling can contribute ultrafine filter production for numerous purposes. Moreover, the holes can be fabricated on different ceramic materials. Micro EBM can be used for making holes in wire drawing dies, fuel injector nozzles and fabrication of fine gas orifices in nuclear space reactors, etc. Also, micro EBM can be used in pattern generation for IC fabrication, polishing of metal mold [96], repairing of defects in masks [97] and fabricating various other micro features in chemical, textile, food processing and semiconductor industries [98].

1.3.6 Micro Chemical Machining (CM)

Micro Chemical Machining (CM) or Chemical Micro Machining is a popular process where material from the workpiece is removed by chemical reactions from strong acidic or alkaline chemical reagent. The process involves precise and accurate contouring of any shape and size of metals where material removal occurs by the action of chemical cell. In ancient days, chemical machining was used by artisans to engrave metals. But in recent years, the applicability of CM has largely increased where it can be used for milling of pockets and can generate complex geometric and delicate features as no cutting forces and thermal distortion is involved here. The machining depth in micro CM process ranges from 2.54 to 12.27 mm.

1.3.6.1 Mechanism of Material Removal

Micro chemical machining refers removal of material over selected areas of workpiece by controlled chemical reaction where the workpiece material is dipped inside strong chemical reagent or etchant such as HF, HNO_3, $FeCl_3$, etc. and different other acids and alkaline solutions. Etchant reacts with the work metal and produces metallic salt that dissolves in the solution. The selection of suitable etchant depends upon various factors like type, surface finish of work material and its interaction with etchant, type of maskant to be used, material removal rate during machining and required depth of penetration, etc. Prior to machining the workpiece is covered in chemically resistive maskant excluding the areas on to where the machining is supposed to occur. Maskants restrict the etchants in penetrating and dissolving the undesired area. The etchant removes the material by controlled chemical reaction from all the exposed areas of the workpiece where the material removal rate may vary from 0.0025 to 0.1 mm/min.

Maskants are mainly three types, photoresist, screen print and cut and peel. Photoresist maskant is used for photochemical machining where etching is carried out by photosensitive resist and can produce complicate and accurate shapes. In screen printing technique, a screen usually of stainless steel is used to press the work surface. The screen has the openings on to which the etching is needed to be carried out. Etching occurs in the exposed area when the maskant is rolled over. The process is used for high volume production but the accuracy of the product is low. Cut and peel maskants are primarily applied to the entire work surface by dipping the workpiece, spraying or flow coating and then cut and peeled off from the areas needed to be exposed for etching. The accuracy of the process is low but very deep etching can be obtained. Chemical micro machining is a time consuming process, but it can generate contours, pockets, etc. Various other forms of micro CM include chemical blanking where etching is carried out through thin sheets; chemical or electrochemical polishing, etc. The parameters that affect the performance of chemical micromachining process include type and temperature of etchant, type and

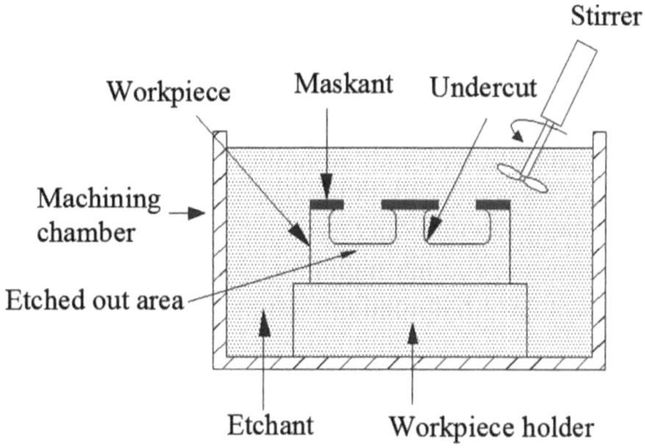

Fig. 1.7 Schematic of micro CM process

method of applying maskant, method for circulation of etchant, etc. A schematic diagram of chemical micro machining process is shown in Fig. 1.7.

1.3.6.2 Opportunities

Uniform material removal can be achieved by chemical micro machining. Moreover, highly skilled operators are not at all required during machining and the process can even fabricate various tapered surfaces with good surface finish within low tolerance limit. As discussed earlier, chemical micro machining process can easily be utilized for blanking, milling, cleaning, etc. Chemical blanking refers to fabrication of features that penetrates the entire work material like slots, holes, etc. Chemical milling produces blind details like channels, pockets and also reduces weight of work material whereas chemical cleaning ensures suitable adhesion of maskant to work material. Additionally, chemical micro machining offers several opportunities that can further be extended which involve removal of recast layer from parts machined with EDM and sharp burrs of machined portions, removal of decarburized layer from low alloy steel forgings, etc. Moreover, after machining by other method, extruded and oversized parts can be machined by chemical micro machining over selected and desired area and multiple designs on the same work piece can be machined simultaneously.

1.3.6.3 Challenges

Alike other process, chemical micro machining is not devoid of limitations. This process is limited to fewer metals as selection and finding of appropriate etchant and maskant for a range of materials is difficult. In addition, the process is highly dependent on the availability of required etchant and maskant. The gases produced during machining may entrap under the masking leading to uneven etching, decrease in etching efficiency and generation of shapes with poor surface quality.

Moreover, when such gases come out, they may spread and attack other equipments. In chemical micro machining, material removal rate is very low which can further deteriorate as it gets contaminated with machining products. Other challenges in micro CM include difficulties in machining sharp corners, thicker as well as chemically resistive material, limitation in scribing accuracy, degradation of dimensional accuracy in the final product, etc. The etchants used in chemical micro machining are toxic and hazardous for environment that require expensive methods for disposal. For machining of alloys, surface quality gets degraded due to differential etching that requires further study and research work.

1.3.6.4 Micro Engineering Applications

Chemical micro machining is used to machine various types of metals like copper, aluminum and its alloys, lead, steel, titanium, nickel as well as various types of non metals like ceramics, glass, plastic, etc. [99]. It is used to produce shallow holes and cavities over large surface area [100], contours, pockets, depression [101], parts with stepped features, machining of edges of sheet metal, engraving on metal workpiece, machining of huge work surfaces like airplane wing to tiny workpiece like IC chips [102], fabrication of stainless steel edge filter by photochemical machining [103], and batch size production of various other components in chemical and missile industries. Other popular applications of chemical micro machining involves minimizing thickness of walls, ribs, webs of components produced by conventional process, narrow cuts in large thin sheets applicable in aerospace industries, etc.

1.3.7 Micro Electrochemical Machining (ECM)

Micro Electrochemical Machining or Electrochemical Micromachining has become more popular with recent advancements to fulfill the needs of the fabrication of macro and micro components. With rapid developments in the fields of automotive, aerospace, electronics, optics, medical devices and much more, this process finds wide applications in the machining of titanium and titanium alloys, super alloys and stainless steel structures which are difficult to machine and pose many challenges for conventional machining processes. This process provides added advantages in the shaping of complex features with no thermal stress, burr formation and tool wear and can be implemented on metals regardless of their hardness which facilitates it to execute precision jobs with accuracy in the order of ± 1 μm on 50 μm [104]. Governed by the laws of electrolysis of Michael Faraday, this process removes material by the mechanism of anodic dissolution during electrolysis where it is possible to achieve unit material removal as the size of ions that makes the process extremely useful for machining delicate and complex micro features. But still, research is needed to explore further opportunities and overcome the

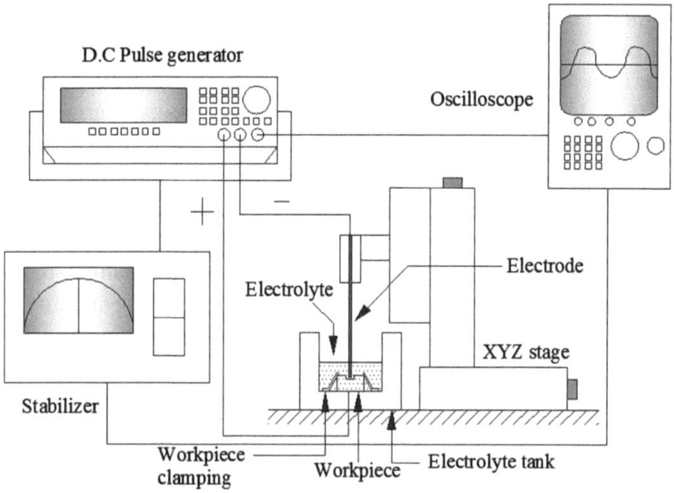

Fig. 1.8 Schematic diagram of EMM

challenges in machining different newer materials with greater accuracy for successful generation of more intricate micro features. A schematic diagram of Electrochemical Micromachining setup is shown in Fig. 1.8.

1.3.7.1 Mechanism of Material Removal

As discussed earlier, Electrochemical Micromachining is an electrochemical material dissolution process where both cathode microtool and anode workpiece remains submerged in proper electrolytic solution separated by narrow inter-electrode gap (IEG). IEG controls the dimensional accuracy as well as resolution of the micro feature to be produced and the lower will be the IEG, the better will be the localization of anodic dissolution. There are two types of electrolyte available for EMM, one is passivating that contains oxidizing anions which give better accuracy such as sodium nitrate [105], sodium chlorate, etc. and other one is non-passivating such as sodium chloride which contains aggressive anions. Also, different acid electrolytes such as sulfuric acid, hydrochloric acid, etc. are preferable which does not produces any insoluble reaction products. The electrolyte needs to be in static condition during operation as introducing high electrolyte flow like conventional ECM may vibrate the micro tool tip and deteriorate the machining accuracy and stability. During machining, for achieving instantaneous high current density in the order of 75–100 A/cm^2 as well as high mass transport and to confine the dissolution process with in micron or sub-micron ranges, for enhancing the machining accuracy and for reducing the IEG in the range of 5–20 μm, low voltage in the range of 1–10 V with ultra short D.C pulse current is applied that also ensures effective removal of machined products like sludge and hydrogen bubbles from the IEG during pulse off time [106]. For machining to occur, the micrometer scale tool electrode is then feed forward or feed toward the workpiece by maintaining a predefined IEG. From the anodic workpiece, surface metal starts dissolving into metallic ions generating an approximate mirror image of the tool by corresponding electrochemical reactions. The corrupting position of the workpiece varies with the location of tool electrode and anodic dissolution continues during whole the time resulting in generation of desired feature on the workpiece.

Throughout the material removal process, for maintaining uniform current flow as well as removal of material and to achieve better machining stability, the feed of tool is kept balance with the rate of dissolution where increase in gap will hinder the machining process by increasing electrolyte resistance and decrease in gap will result in improper flushing and coagulation of reaction products leading to occurrence of frequent short circuits in the machining zone. However, still during machining, machining products like gas bubbles, sludge, etc. resulting from ongoing electrochemical reactions may stick between the microtool and the workpiece which can clog the small IEG that may lead to formation of micro sparks and deterioration of machining quality and accuracy. For such cases, to remove the electrolysis products from the IEG and to supply more fresh electrolytes, use of rotary electrodes and vibrations by piezoelectric transducer (PZT) [107] may be used either in tool or in workpiece which generates pressure waves that can clean the machining gap and maintain machining efficiency. Other parameters which effect the dissolution process involve concentration of electrolyte, duration of each pulse as well as pulse on time [108], applied frequency, temperature of electrolyte, etc.

1.3.7.2 Opportunities

Electrochemical Micromachining is very economical and eco friendly process for machining micro shapes effectively. Moreover, the process offers higher precision, control and burr free machining, flexibility, reliability and machining of different chemically resistive harder materials like titanium and its alloys, stainless steel, various super alloys, etc. Furthermore, as EMM is not thermally activated process hence, the process involves no adverse thermal effects like formation of heat effected zone in the machined surface, wearing out of microtools, generation of thermal stress and distortion, formation of surface as well as subsurface cracks and deterioration of surface property, quality as well as integrity. Additionally, EMM is a versatile process and involves different techniques for generating desired micro features which involve jet EMM, capillary EMM, 3D EMM, through mask EMM, etc. Furthermore, various types of precise and high aspect ratio micro tool having good surface finish and with different tip shapes like conical, cylindrical, spherical, disk shape [109], wedge-shape [110], etc. for EMM can be in situ fabricated as per micro product requirement by reverse EMM process which can later be used for machining of micro features in same setup that leads in reducing the tool handling and fixing issues. Another advantage of EMM is the use of multi micro tool which can machine arrays of micro feature simultaneously and decreases the production time. But, this material removal process offers several opportunities which are still unexplored and require extensive research, further investigation and detailed study [24]. Hence, researchers are concentrating towards appropriate designing and development of various types of microtools, proper selection of suitable electrolyte, improvement of micro tool feeding system, proper control and monitoring of IEG during machining and achieving better material removal rate, elimination of micro spark generated in IEG, achieving better surface finish and dimensional accuracy, etc.

1.3.7.3 Challenges

Electrochemical Micromachining process takes higher processing time for fabrication of micro features. Moreover, the inherent feature of this micro machining process restricts its application for electrically conductive materials only. Except this, there are other research scopes that need particular attention like achieving better machining performance, machining smaller and high aspect ratio 3D micro features as well as nano components, better optimization of operating parameters as well as achieving proper control and stability during micromachining.

1.3.7.4 Micro Engineering Applications

As discussed earlier, electrochemical micromachining is an effective and versatile process for generating microfeatures applicable in electronics, optoelectronics, medical, biomedical, optics, semiconductor, aerospace and different other ultra precision industries [111]. The process is applied for fabricating micro components for MEMS, micro nozzles, plate for ink-jet printer head, finishing of print bands for high-speed impact printers [112], mask less metal cutting by electrochemical saw, deburring of micro parts, fabrication of diaphragm for electro optical systems, various complex, high aspect ratio, intricate and contoured 3D micro components, structure, profiles and shapes, different high aspect ratio micro structures on silicon [113] for MEMS and on different composite materials [114], etc. Moreover, researchers have fabricated single as well as arrays of micro grooves, holes and slots with complex geometry and internal features [115–117], different micro milled [118], micro turned structures [119], micro reactor, micro gear [120], square helix, micro patterns in flat as well as on cylindrical surfaces [121], sub micrometer cavity with controlled roughness at two different scales [122], nano structured electrode surface with the help of STM tip [123], different 3D nano structures in nickel [124] and various other micro/nano features.

1.4 Introduction to Various Hybrid Micromachining Processes

Hybrid micromachining processes are combination of two or more machining processes and these processes are becoming more and more attractive for fabricators as well as researchers. The development of hybrid micromachining processes improves the advantages of characteristic features of constituent processes while reducing their disadvantages [125]. The CNC machining and additive process propose more benefits due to more accurate machining features in contrast to the limitations of every other process. The addition of ultrasonic vibration and drilling decreases the tool wear rate and cutting force [126]. The recast layer and heating zone are removed by the integration of laser drilling and electrochemical machining process. The hybrid micromachining processes have enormous potentials because these processes produce more intricate parts with good dimensional features, higher machining accuracy, lower machining time, etc. These processes have huge research potential for improving machining characteristics, enlarging the application areas, enhancing the process characteristics, etc.

1.4.1 Electrochemical Grinding (ECG)

Electrochemical grinding (ECG) is one of the micromachining techniques, in which electrolyte solution and mechanical grinding process are combined to remove the material from the machining zone with concurrent action of electrochemical reaction and mechanical abrasive action for providing burr free and stress free machined surface with good machining quality [127, 128]. ECG eradicates a separate deburring operation due to electrolyte circulation. In ECG process, the shape of workpiece is formed with analogous shape of grinding wheel. The schematic diagram of electrochemical grinding is shown in Fig. 1.9. The main characteristic of ECG is that the insulating abrasive particles are inserted into metallic grinding wheel. For instance, the diamond abrasive particles are non conductive materials which are embedded into conductive grinding wheel. The embedded carbon brush with electrolyte commutator spindle grips the grinding wheel which is connected with negative terminal of DC power supply and workpiece is connected with positive terminal of DC power supply. During machining the electrolyte is supplied to the grinding wheel near the work piece. The nozzle is used to supply the flow of electrolyte into the machining zone and the grinding wheel carries the electrolyte during machining.

The material is removed by electrochemical dissolution. The oxide layer is formed on the workpiece during machining. The grinding wheel removes the fresh metal with oxide layer. Major material removal takes place due to electrochemical action and rest of materials is removed by the mechanical grinding action. It eliminates the metallurgical damage of the machined surface which is involved in

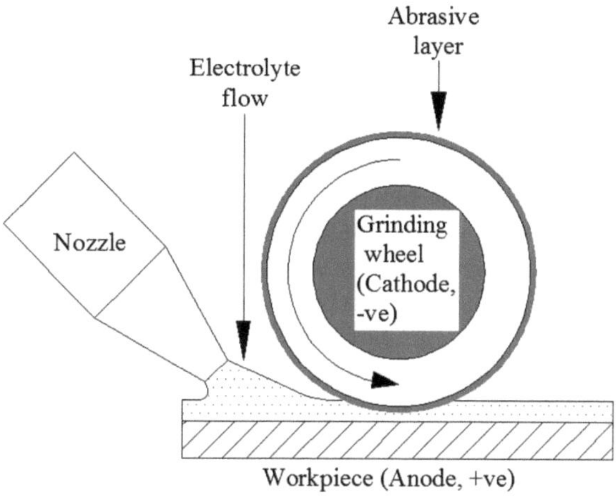

Fig. 1.9 Schematic diagram of electrochemical grinding

conventional grinding process that includes cracking and heat distortion. ECG can produce sharp edges within the range of 0.001–0.005 μm radius by varying the power settings. The process can machine thin and delicate metal sheets with better machining accuracy and feature characteristics. Very less tolerance can be acquired in the order of 0.005 mm by controlling the micro ECG characteristics such as type of material being cut, its size and depth of cut. But, this process is applied only for conductive materials i.e. stainless steel, aluminum, hastelloy, inconel, etc. Presently, researchers are concentrating to increase the machining rate, machining accuracy, wheel life, etc. with the development of proper ECG set up. Lower pulsed voltage and lower duty ratio are more suitable for enhancing the machining accuracy while the grit size has no effect during machining [129].

1.4.2 Electrochemical Discharge Micromachining

Electrochemical discharge micromachining (ECDM) process is a hybrid machining process, in which the mechanisms of electrochemical machining (ECM) and electro-discharge machining (EDM) are combined to obtain good quality features particularly on electrically non conducting materials. In ECDM thermal erosive effect of electro-discharge action (ED) follows the electrochemical (EC) reaction and electrochemical action that help in the generation of positively charged ionic gas bubbles. During machining, DC or pulsed DC voltage is applied between cathode tool and auxiliary electrode which are immersed in suitable electrolyte solution of KOH, NaOH, etc. As a result, hydrogen bubbles form on the tool surface and oxygen bubbles generate on the counter electrode due to electrolysis. With increase in machining voltage, current density increases. The increased current density increases the quantity as well as size of these bubbles leading to generation of a layer of gas bubbles surrounding the tool electrode. These evolved gas bubbles gradually combine with each other and form a gas film which serves as an insulating media and generate enough resistance creating higher potential difference between these two electrodes. When the voltage becomes high enough and exceeds a critical value, electrical discharges occur between the tool electrode and the upper layer of electrolyte. If the workpiece is placed in the discharge zone, machining does take place in the form of melting, vaporization and thermal erosion due to the heat generated by the discharges. The material is eroded by electro chemical dissolution as well as electro-discharging between two electrodes [130]. The combined material removal mechanism facilitates higher material removal rate in ECDM than ECM or EDM process. The schematic diagram of electrochemical discharge machining is shown in Fig. 1.10. It is mostly used for micromachining and scribing out hard and brittle materials, conductive, non conductive materials, etc. This hybrid process is successfully used in non conductive materials such as ceramics and glass for drilling of micro holes and slots by smaller electrodes efficiently and economically. Machining of complex microfeatures is not possible by ECDM process with better accuracy and surface finish on harder materials. So, more systematic investigation is necessary for machining non conductive materials because of their key applications in MEMS field. For example, semiconductor materials can be bonded with quartz or glass because of their transparent property. Also, the process is utilized for machining ceramic materials that are used in high-tech apparatus [131]. ECDM is utilized to machine fine 3D micro-structures on glass and to obtain good surface finish with better machining accuracy in drilling and milling operations. In ECDM, micro structures with higher depth are fabricated using high aspect ratio micro tool in lower electrolyte concentration.

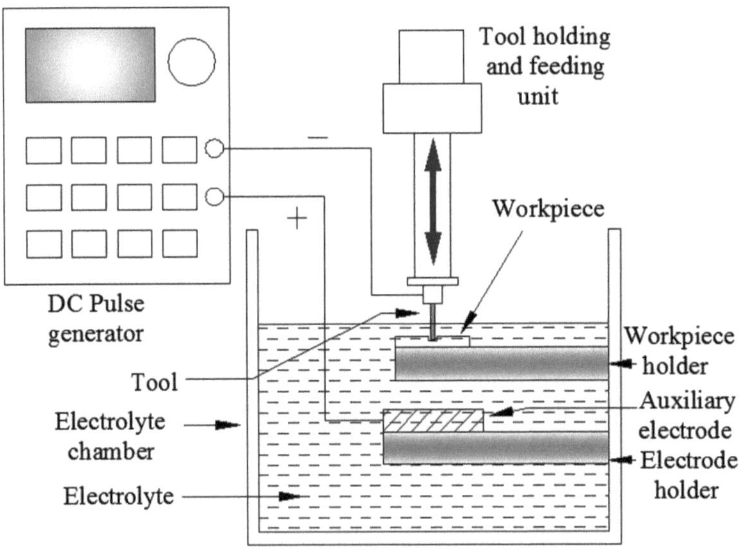

Fig. 1.10 Schematic diagram of electrochemical discharge machining

ECDM process is well familiar in several names such as electrochemical discharge machining [132], spark assisted chemical engraving [133], electrochemical spark machining [134], micro electrochemical discharge machining [135], electrochemical arc machining [136], discharge machining of non conductors [137] and electro erosion dissolution machining [138]. ECDM process has numerous advantages that make it able to drill glass substrates, ceramics, composites and stainless steel materials. Other various applications of ECDM involve fabrication of complex micro profiles [139], micro dies [138], deep drilling [140], machining of cylindrical parts [141], dressing of micro-grinding tools [142], slicing of glass rods [143], etc.

1.4.3 Abrasive Assisted Micromachining

Abrasive assisted micromachining (AM) is an effective hybrid micromachining process in which the material is removed through the injection of jet of small particles for the fabrications of microsized features such as micro-channels, holes, etc. These types of fabricated microfeatures are used in micro-fluidic devices, micro-electro-mechanical-systems (MEMS) and optoelectronic components [144, 145]. A pressurized and high speed air jet containing abrasive particles wears away the materials from the workpiece through the mask and obtains the desired shape of the feature. Abrasive assisted micromachining machines the hard and brittle materials such as glass and ceramics [146, 147] with desired shape and size of microfeatures. The microfeatures of such materials are more important in electronics industry. This hybrid micromachining is also more important in case of polymer materials because these materials offer a wide range of physical and chemical properties. These materials have also other advantages such as low cost, physically robust, good processibility for mass production [148], etc. This process of hybrid micromachining machines easily with desired shape in acrylic polymers such as polymethyl methacrylate (PMMA) [149, 150] which has most valuable application in micro-fluidic devices and micro-electro-mechanical-systems such as those in lab on chip, biomedical devices, heat sinks of electrical devices, capillary electrophoresis chips, etc.

Different types of abrasive assisted micromachining are centrifugal force assisted abrasive flow machining (CFAAFM), abrasive slurry jet micro-machining (ASJM), abrasive enhanced electrochemical slurry jet machining (ESJM), abrasive water jet machining (AWJM), etc. In centrifugal force assisted abrasive flow machining (CFAAFM) process, the machining rate and surface quality increase with lesser number of cycles compared to the capabilities of conventional abrasive flow machining (AFM) [151, 152]. In abrasive slurry jet micro-machining (ASJM), the slurry medium fabricates good microfeatures with relatively high resolution in a variety of brittle and ductile materials and the effects of ASJM operating parameters can be controlled easily on the minimum size of microgrooves [153]. Abrasive enhanced electrochemical slurry jet machining (ESJM) is a new approach in which abrasive slurry-jet machining (ASJM) and electrochemical jet machining (ECJM) are combined for the micromachining of metals with higher material removal rate and good machined surface quality, particularly those that are difficult to machine using ASJM alone, such as titanium, stellite and tungsten carbide [154]. Abrasive water jet machining (AWJM) process offers a high-quality cutting with a good surface finish and the machined components can be used in automotive, aerospace, surgical equipments and defense sector industries owing to its good machining characteristics. AWJM process utilizes the mixture of water and abrasive to erode material from the target surface. The effects of AWJM process parameters, i.e., water pressure, transverse speed, abrasive flow rate, standoff distance, abrasive grit size, etc. can be controlled easily on performance parameters i.e., surface quality, kerf top width, taper angle and material removal rate [155]. In abrasive assisted electro discharge machining (AEDM), thermal interaction is added with abrasion-mechanical action and energy is carried by plasma and also by abrasive grains. Mixing silicon powder into dielectric reduces electrical capacitance across the discharge gap by increasing the gap size. This result in more sparking intensity and improvement in the discharge characteristic, especially in the machining of large workpiece area. Application of powder-mixed working media allows AEDM to obtain crackless affected layer, uniform and mirror finishing of complex shape, etc.

1.4.4 Ultrasonic Assisted Micromachining

Ultrasonic assisted micromachining is another hybrid micromachining technique for machining of brittle materials like, composites, glass, ceramics and graphite with higher machining rate. The workpiece is kept on the fixture and is vibrated in ultrasonic frequency range during micro USM. Abrasive slurry is flowed on the workpiece and the ultrasonic tool strikes the abrasives and removes the material away from the machining zone. The continuous flow of fresh abrasive slurry on vibrated workpiece removes the debris from the tool and the gap between tool and workpiece [156, 157]. During ultrasonic assisted micromachining, the fine abrasive particles used are not safe for human being. Hence, chemically inert common abrasive particles such as boron carbide, silicon carbide, etc. are used in this process. These types of abrasive particles have no toxicity effect on human health during handling. Microtools are necessary to fabricate fine microfeatures on workpiece. These microtools are fabricated by micro electro discharge machining, electrochemical micromachining, etc. But, it is difficult to produce by ultrasonic assisted micromachining as ultrasonic vibration looses the griping of tools in the precise mechanical chuck and the natural frequency of chuck may not be matched with oscillator frequency. The fabricated microtools generate various machined microfeatures i.e. blind or through holes, slots, 3D cavities, etc. using ultrasonic assisted micromachining. This process also fabricates fine micro holes, generates surface textures, micro channels, etc. which are used in ink-jet printer nozzles, biomedical filters, integrated circuit boards, pneumatic sensors, fluidic filters, grids, fuel injection nozzles, high-pressure orifices, optical apertures and manipulators.

Ultrasonic assisted mechanical micromachining is an important hybrid machining in which the material is removed by the simultaneous action of mechanically rotating spindle and higher axial ultrasonic oscillation of workpiece or tool [158]. One of the important ultrasonic assisted mechanical micromachining is ultrasonic assisted grinding where the oscillation of the grinding tool by piezo-electric oscillators is used for achieving better surface integrity of machined products. In ultrasonic assisted EDM, the cavitation and acoustic wave enhance the material removal and flushing condition for microgrooves, slots and micro drilling. The better ejection of molten metal from craters increases due to the slurry circulation. This fact increases the material removal rate as well as generates good surface quality. Ultrasonic assisted abrasive slurry jet micromachining (UAASJM) improves the machining efficiency and quality of the ductile and brittle materials due to the assistance of ultrasonic vibration. Due to ultrasonic vibration the aluminum nitride is polished by the abrasive water jet to get the better polishing efficiency and quality [159].

1.4.5 Laser Assisted Micromachining

Laser can be effectively combined with other micromachining processes to be used as hybrid machining process for obtaining several advantages such as faster machining with improved accuracy and surface quality to generate micro/nano features. Laser assisted micromachining may include laser assisted chemical machining, laser assisted lithography, laser assisted electrochemical micromachining, etc. Laser assisted chemical machining utilizes a maskless photoreactive or thermal reactive process where laser beam enhances the etching of semiconductor, ceramic materials, metals and polymers in an etchant. By directing the path of focused laser beam on the workpiece, the process helps in obtaining maskless patterns in the form of material deposition or dissolution. In Oxide Film Laser Lithography the use of photoresist can be entirely eliminated as because, well focused laser beam (10 μm) forms a pattern of oxide film (200–300 nm) on titanium which acts as a mask and the dissolution is localised only at the irradiated areas. After machining, the freestanding oxide film can easily be removed by ultrasonic cleaning from the edge of machined cavities.

In case of electrochemical micromachining (EMM) or deposition (electroforming)with laser assistance, the localized anodic dissolution or deposition of the process enhances due to laser radiation resulting in better machining accuracy on micromachining zone. The acceleration of electrochemical reactions increases due to laser assistance. The laser radiation boosts this electrochemical process by ten times and localization increases in the micromachining zone. Moreover, in laser assisted electrochemical micromachining, low energy defocused laser beam is applied in the narrow machining zone which does not melt and vaporize the workpiece, but helps in proper circulation of electrolyte in very small IEG by increasing the electrolyte temperature and generating suitable temperature gradient. High temperature increases the electrolyte conductivity. As a result, current density increases that enhances the dissolution rate. Also, by increasing the temperature of the work surface, laser beam helps in reducing the chemical reaction needed for generation of passive metal oxide layer that can minimize the rate of current flow. Hence, due to the application of laser during EMM with non passivating electrolytes, formation of oxide layer gets reduced and material removal rate increases. Furthermore, utilization of laser beam during EMM increases ionic movements and accelerates electrochemical reactions that results in better productivity. Also, laser beam confines the machining in very narrow area and improves the localization that leads to better accuracy and precision.

1.5 Advanced Finishing Processes utilizing Non-traditional Machining

Advanced finishing processes using non-traditional machining is gaining popularity to cope up the steep demand in finishing intricate, complex, durable and sophisticated shapes of various super-alloys, metal-matrix composites, ceramics, polymers, aluminates, etc. that are highly economical and posses better surface quality and property. In advanced finishing process, the type of energies involved and utilization of that energy varies from than that in conventional finishing processes. Where, conventional finishing processes depend on the physical contact between tool and work piece by using the energy from electric motors, hydraulic equipments, etc.; advanced finishing processes implements various non traditional approaches and uses energies from high temperature plasma, electrochemical reactions, jets with high velocity and with abrasives, etc. Hence, advanced finishing processes have wide range of applications particularly in aerospace industries where achieving ultra finished surfaces over various harder and tougher alloys are necessary. Some of the major advanced finishing processes are described in the following sub sections.

1.5.1 Abrasive Flow Finishing (AFF)

Abrasive flow finishing process is one of the super finishing processes in which workpiece and tool moves relatively for accurate finishing of internal shapes of components and complex geometries. The basic arrangements of AFF have been shown in Fig. 1.11. In AFF, the surface is super-finished by the flow of abrasive laden polymer media in single or two-way direction. The abrasive media is passed in one way direction and it comes back from the other end. In two-way process, two

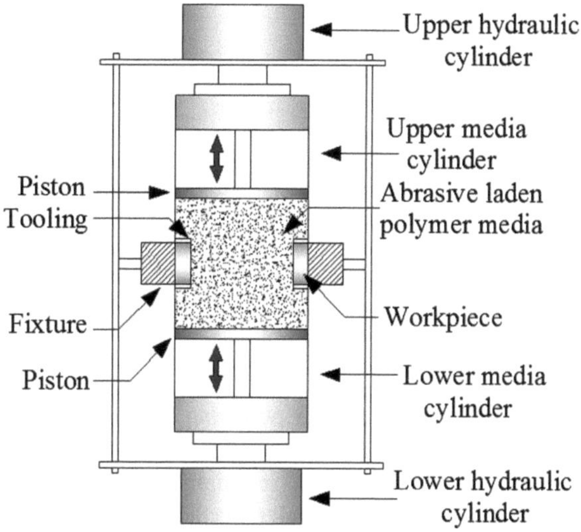

Fig. 1.11 Basic arrangements of abrasive flow finishing

vertically opposite hydraulic cylinders push the abrasive media to and fro. The process is used for complex cavities, edges, etc. At a given time, this process finishes a number of parts or different areas of the same work piece. The atomized AFM systems are capable for handling thousands of parts per day for finishing operation reducing the labor costs and monotonous physical work. Through proper control of the process parameters, this process can be used for variety of super-finishing operations achieving very uniform and precise results.

The significance of all AFF parameters used in process i.e. media, tooling, machine, etc. can be evaluated properly for wide range of applications in finishing process [160]. For instance, the temperature effect on viscosity of AFF medium and rheological properties of grit mixtures are two important factors for finishing of mild steel dies [161]. AFF is also suitable for finishing of inner surface of spheroidal graphite cast iron [162]. The influence of various AFF process parameters i.e., number of AFF cycles, abrasive mesh size, extrusion pressure on the output parameters i.e. surface finish and material removal rate can be controlled properly during finishing of brass and aluminum workpieces [163]. However, the surface finish depends upon the types of materials which should be taken care of before operation [164].

1.5.2 Chemo Mechanical Polishing (CMP)

Chemo mechanical polishing (CMP) is another promising machining technology for the super fishing operation of the integrated circuits (ICs) [165]. The starting wafers for ICs are polished by the CMP process. The schematic diagram of typical CMP setup is shown in Fig. 1.12. In CMP, the abrasive and corrosive chemical

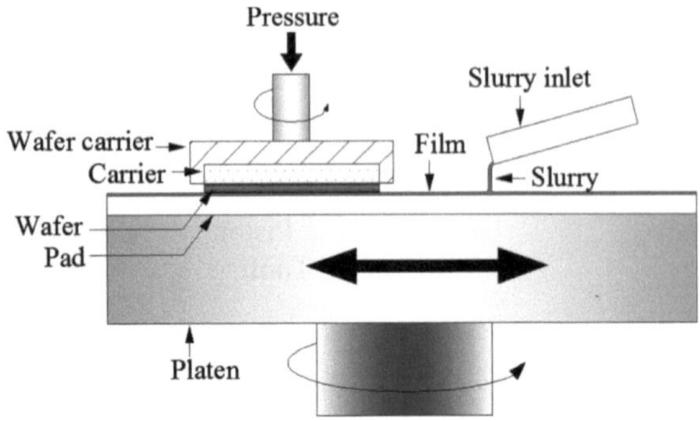

Fig. 1.12 Schematic diagram of chemo mechanical polishing

slurry are combined with polishing pad where the ring diameter is greater than wafer. The dynamic polishing head holds the pad and wafer and these combined parts are kept on a plastic retaining ring. The dynamic polishing head forms the smooth surface where wafer surface is also flat and smooth during operation. Sometimes, this wafer is also used in additional circuit elements. In some cases, CMP mechanisms are used for finishing operation using chemical and mechanical based abrasion. For instance, CMP is able to remove the material from the selective position for any photolithographic system. In some cases, machining by CMP can achieve machining depths as low as few angstrom to 22 nm. CMP is widely used for fine finishing of glass and ceramic surfaces and semiconductor wafers. This process facilitates the control and optimization of the polishing operation for microelectronic devices and can cope up with huge demands of global wise at different levels. Evidently, the process has enough capability in micro and nano ranges and can reduce the difficulties in polishing operation where improvement is still going on in the field of etching and lithography after its introduction in early 80s.

1.5.3 Elastic Emission Machining (EEM)

Elastic emission machining (EEM) is an ultra-precision finishing process for creating smooth flat surfaces as it has the capability to remove material at atomic level by mechanical methods [166]. The finished surface gets completely mirrored and remains crystallographically and physically undisturbed. The system has a rotating tool, slurry and workpiece. The rotating tool and workpiece are immerged in a pool of slurry. Soft material i.e. polyurethane rubber is used as a tool in EEM. The slurry medium contains water and very fine abrasive particles. EEM process uses the hydrodynamic effect to generate the lubrication film between tool and workpiece. The thickness of lubrication film is 1 μm. The size of abrasive particle is much smaller than the film thickness. Material removal occurs at atomic level due to

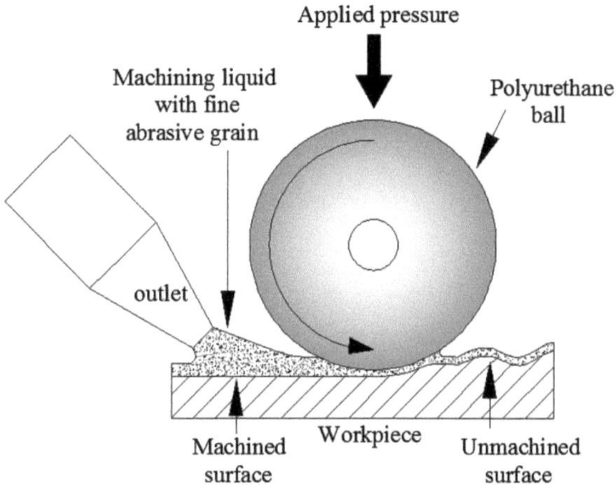

Fig. 1.13 Schematic diagram of elastic emission machining

contact of particles in machining zone. The schematic diagram of elastic emission machining is shown in Fig. 1.13.

Due to the collision between ultra fine particles and workpiece, the workpiece gets finished by atomic level elastic fracture without any plastic deformation [167]. The process finishes the surface with 0.1 nm accuracy without causing any crystallographic damage [168]. This method is successfully used to generate smooth surfaces like mirror finished with surface roughness, Ra of 2 nm for synchrotron radiation optics [169]. The process has capability to perform the surface finishing as well as precision shaping due to its linear relationship between machining volume and machining time. This method has some advantages over traditional precision machining methods such as precision grinding or diamond turning i.e. less requirement of rigidity in machine structure, less sensitivity to temperature variation or system vibration during machining, etc.

1.5.4 Magnetic Abrasive Finishing (MAF)

Magnetic abrasive finishing (MAF) is an efficient method to remove irregular topographies of the surface using both magnetic and abrasive properties. The magnetic particles with abrasive mixture are regarded as magnetic brush. The magnetic field in MAF attracts the magnetic abrasive particles to the inner wall of workpiece. So, concentrated mixture of magnetic abrasive particles increases near the wall of workpiece than the rest of medium. This concentrated mixture increases the radial force acting on magnetic abrasive particles leading to higher depth of cut, better material removal rate and surface finish compared to normal abrasive flow finishing.

During finishing operation this process does not generate vibration or chatter and has no limitations for any type of workpieces. It has also the ability to reduce

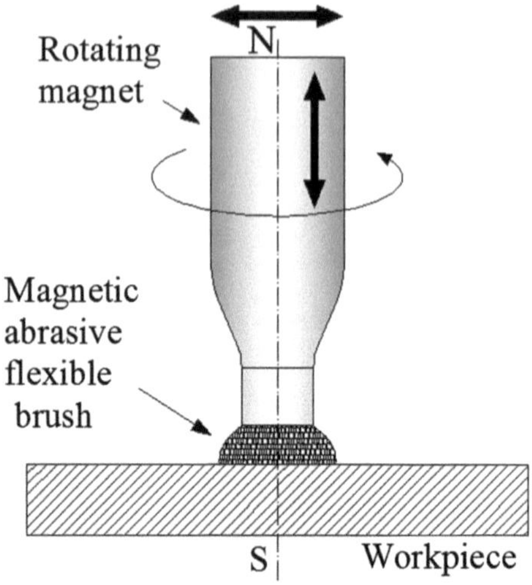

Fig. 1.14 Basic arrangements of magnetic abrasive finishing

surface roughness from few microns to nano levels in just few minutes [170, 171]. Moreover, various complex surfaces can easily be finished by this process [172–175]. The schematic diagram of MAF is shown in Fig. 1.14. The process has capability to produce smoother external surfaces on optics, turbine blades, cutting tools, etc. and internal surfaces on capillary tubes, biopsy needles, curved pipes, etc. Also the process can polish various magnetic materials like iron, cobalt, nickel, etc. as well as non-magnetic materials like glass, ceramic, stainless steel, copper, etc. with economically and effectively. This process is proper combination of homogeneous mixture of abrasive particles and magnetic particles in a magnetic field with machining force for finishing operation. It has numerous advantages such as controllability, self-adaptability, etc. Presently, magnetic field-assisted finishing processes are becoming more attractive for its huge applications in the fields of medical components and optical parts. However, MAF has still some problems in fine finishing in nanometer level of flat and micro-complex surface of harder work materials.

1.6 Conclusions

Latest advancement of micro/nano fabrication techniques is leading to generation of unique features in nm–μm range for their use in many advance fields. The micro and nanometer dimensional spans are used in mechanical, chemical, medical devices, etc. The micromachining technologies have numerous utilities and different applications for human beings. The demands of microproducts are increasing continuously with precise manufactured features ranging between few nanometer to few microns. Micromachining techniques have been developed for the fabrications of miniaturized parts which have numerous applications in our practical lives. Non-traditional machining methods are utilized for the machining of complex and intricate 3D shapes where conventional machining processes are neither feasible nor satisfactory due to various unavoidable reasons.

The above overview of non-traditional micromachining processes will help in conceptual understanding about the physics and chemistry involved. The importance of each technique used and mechanisms of various techniques of non-traditional micromachining, hybrid machining as well as advanced finishing processes have been discussed in the context of feature generation and enhancement of surface quality. The opportunities and challenges of each non-traditional micromachining process have been highlighted considering various practical applications in different micro engineering fields. These non-traditional micromachining techniques as well as advanced finishing processes can be improved further concerning their increasing applications in MEMS where functional requirements of many devices demand very tight tolerances and the use of a wide variety of engineering materials including stainless steel, titanium, brass, aluminum, platinum, iridium, plastics, ceramics and composites, etc.

References

1. Liu X., DeVor R.E., Kapoor S.G., Ehmann K.F., The mechanics of machining at the micro-scale: assessment of the current state of the science, J. Manuf. Sci. Eng. 126 (2004) 666–678.
2. Mecomber J. S., Hurd D., Limbach P.A., Enhanced machining of micron-scale features in microchip molding masters by CNC milling, International Journal of Machine Tools & Manufacture 45(2005) 1542–1550.
3. Chae J., Park S.S., Freiheit T., Investigation of micro-cutting operations, International Journal of Machine Tools & Manufacture 46(2006) 313–332.
4. Filiz S., Conley C.M., Wasserman M.B., Burak Ozdoganlar N., An experimental investigation of micromachinability of copper 101 using tungsten carbide micro-endmills, International Journal of Machine Tools and Manufacture 47(2007)1088–1100.
5. Masuzawa T., State of the art of micromachining, Annals of the CIRP 49 (2000) 473–488.
6. Benedict G. F., Non-traditional Manufacturing Processes, Marcel Dekker Inc., New York, 1987.
7. Getu H., Ghobeity A., Spelt J.K., Papini M., Abrasive jet micromachining of polymethylmethacrylate, Wear 263(2007)1008–1015.
8. Getu H., Ghobeity A., Spelt J. K., Papini M., Abrasive jet micromachining of acrylic and polycarbonate polymers at oblique angles of attack, Wear doi:10.1016/j.wear.2008.01.013.
9. Belloy E., Thurre S., Walckiers E., Sayah A., Gijs, M. A. M., The introduction of powder blasting for sensor and microsystems applications, Sens. Actuators: A: Phys. 84 (2000) 330–337.
10. Pawlowski A., Belloy E., Sayah A., Gijs M. A. M., Powder blasting patterning technology for microfabrication of complex suspended structures in glass, Microelectronic Engineering 67-68 (2003) 557–565.
11. Cooley W.E., Clipp L.L, High pressure water jets for undersea rock excavation, Journal of Engineering for Industry, Trans. ASME, Series B 92(1970)281.

12. Sun X. Q., Masuzawa T., Fujino M., Micro ultrasonic machining and its applications in MEMS, Sensors and Actuators, A 57(1996)159–164.
13. Egashira K., Masuzawa T., Fujino M., Sun, X. Q., Application of USM to micromachining by on-the-machine tool fabrication, International Journal of Electrical Machining, IJEM, 2 (1997)31–36.
14. Tseng A. A., Topical review: Recent developments in micro milling using focused ion beam technology, Journal of Micromechanics and Micro engineering 14(2004)R15-R34.
15. Vasile M. J. et al., Micro fabrication techniques using focused ion beam and emergent applications, Micron 30(1999) 235–244.
16. Jain R.K., Jain V. K., Abrasive Fine Finishing Process-A Review, The Intl. J. Mfg. Sc. Prod. 2(1999)55–68.
17. Shinmura T., Takazawa K., Hatano E. Study on Magnetic Abrasive Finishing, Annals CIRP 39(1990)325–328.
18. Puranik M. S., Joshi S. S., Analysis of accuracy of high-speed aspect ratio holes generated using micro-EDM drilling, Proceedings of Institution of Mechanical Engineering, Part B, Journal of Engineering Manufacture 222 (2008)1453–1464.
19. Alemohammad H. Toyserkani E. Pinkerton A.J., Femtosecond laser micromachining of fiber Bragg gratings for simultaneous measurement of temperature and concentration of liquids, Journal of physics D: Applied Physics 41 (2008) 1–9.
20. Sugioka K. Cheng Y, Midorikawa K., Three-dimensional micromachining of glass using femtosecond laser for lab-on-a-chip device manufacture, Applied Physics A 81(2005) 1–10.
21. Rathod V., Doloi B., Bhattacharyya B., Influence of electrochemical micromachining parameters during generation of microgrooves, Int J Adv Manuf Technol. 76(2015)51–60.
22. Ghoshal B., Bhattacharyya B., Shape control in micro borehole generation by EMM with the assistance of vibration of tool, Precision Engineering 38 (2014) 127–137.
23. McGeough J.A, (Ed.), Micromachining of Engineering Materials, Marcel Dekker Inc, NY (USA) 2002.
24. Bhattacharyya B., Electrochemical Micromachining for Nanofabrication, MEMS and Nanotechnology, William Andrew Applied Science Publishers, Imprint of Elsevier Inc., Massachusetts, pp. 270, 2015.
25. Egashira K., Masuzawa T., Microultrasonic machining by the application of workpiece vibration, CIRP Annals—Manufacturing Technology 48(1999)131–134.
26. Egashira K., Mizutani K., Micro-drilling of Monocrystalline Silicon Using a Cutting Tool, Precision Engineering 26(2002)263–268.
27. Egashira K., Masuzawa T., Fujino M., Sun X. Q., Application of USM to micromachining by on-the-machine tool fabrication, International Journal of Electrical Machining 2(1997) 31–36.
28. Komaraiah M., Reddy P. N., A study on the influence of work piece properties in ultrasonic machining, International Journal of Machine Tools and Manufacture 33(1993) 495–505.
29. Singh R., Khamba J. S., Ultrasonic machining of titanium and its alloys: a review, Journal of Materials Processing Technology 173(2006)125–135.
30. Egashira K., Taniguchi T., Tsuchiya H., Miyazaki M., Microultrasonic machining using multitools, in Proceedings of the 7th International Conference on Progress Machining Technology (ICPMT '04), pp. 297–301, December 2004.
31. Sun X. Q., Masuzawa T., Fujino M., Micro ultrasonic machining and its applications in MEMS, Sensors and Actuators A 57(1996)159–164.
32. Sun X. Q., Masuzawa T., Fujino M., Micro ultrasonic machining and self-aligned multilayer machining/assembly technologies for 3D micromachines, in Proceedings of the IEEE Micro Electro Mechanical Systems (MEMS'96), pp. 312–317, 1996.
33. Boy J. J., Andrey E., Boulouize A., Khan-Malek C., Developments in microultrasonic machining (MUSM) at FEMTO-ST, International Journal of Advanced Manufacturing Technology 47(2010)37–45.
34. Yu Z. Y., Rajurkar K. P., Tandon A., Study of 3D microultrasonic machining, Journal of Manufacturing Science and Engineering, Transactions of the ASME 126(2004)727–732.

35. Zhang C., Rentsch R., Brinksmeier E., Advances in micro ultrasonic assisted lapping of microstructures in hard-brittle materials: a brief review and outlook, International Journal of Machine Tools and Manufacture 45(2005)881–890.
36. Curodeau A., Guay J., Rodrigue D., Brault L., Gagn′e D., Beaudoin L. P., Ultrasonic abrasive μ-machining with thermoplastic tooling, International Journal of Machine Tools and Manufacture 48 (2008) 1553–1561.
37. Jameson E.C., Description and development of electrical discharge machining (EDM), in: Electrical Discharge Machining, Society of Manufacturing Engineers, Dearbern, Michigan, pp. 12, 2001.
38. Allen D.M., Lecheheb A., Micro electro-discharge machining of ink jet nozzles: optimum selection of material and machining parameters, Journal of Materials Processing Technology 58 (1996) 53–66.
39. Newman S. T., Ho K. H., The state of art – Electrical discharge machining, International Journal of machine Tools and Manufacture 43 (2003)1287–1300.
40. Koch O., Ehrfeld W., Michel F., Gruber H. P., Recent progress in micro-electro discharge machining technology – Part 1, Proceedings of the 13th International Symposium for Electromachining ISEM XIII, Bilbao, Spain, 2001.
41. Takezawa H., Hamamatsu H., Mohri N., Saito N., Development of micro-EDM with rapidly sharpened electrode, Journal of Materials Processing Technology 149(2004)112–116.
42. Imai Y., Nakagawa T., Miyake H., Hidai H., Tokura H., Local actuator module for highly accurate micro-EDM, Journal of Materials Processing Technology 149(2004) 328–333
43. Yan B. H., Wang A. C., Huang C.Y., Huang F.Y., Study of precision micro-holes in borosilicate glass using micro-EDM combined with micro ultrasonic vibration machining, International Journal of machine Tools and Manufacture 42(2002)1105–1112
44. Chern G., Wu Y. E., Liu S., Development of micro-punching machine and study on the influence of vibration machining in micro-EDM, Journal of Materials Processing Technology, 180(2006)102–109.
45. Hung J. C., Ling J. K., Yan B. H., Liu H. S., Ho P. H., Using a helical micro-tool in micro-EDM combined with ultrasonic vibration for micro-hole machining, Journal of Micromechanics and Microengineering 16(2006) 2705–2713.
46. Aspiwall D. K., Soo S. L., Berrisford A. E., Walder G., Workpiece Surface Roughness and Integrity after WEDM of Ti-6Al-4 V and Inconel 718 Using Minimum Damage Generator Technology, CIRP Annals - Manufacturing Technology 57(2008) 187–190.
47. Patel K.M., Pandey P.M., Rao P. V., Surface integrity and material removal mechanisms associated with the EDM of Al_2O_3 ceramic composite, International Journal of Refractory Metals and Hard Materials 27(2009) 892–899.
48. Jahan M.P., Rahman M., Wong Y.S., Fuhua L., On-machine fabrication of high- aspect-ratio microelectrodes and application in vibration-assisted micro- EDM drilling of tungsten carbide, Proceedings of the Institution of Mechanical Engineers, Part B: Journal of Engineering Manufacture 224 (2010) 795–814.
49. Takahata K., Shibaike N., Guckel H., High-aspect-ratio WC–Co microstructure produced by the combination of LIGA and micro-EDM, Microsystem Technologies 6 (2000) 175–178.
50. Liu K., Lauwers B., Reynaerts D., Process capabilities of Micro-EDM and its applications, The International Journal of Advanced Manufacturing Technology 47(2010) 11–19.
51. Liao Y.S. et al., Fabrication of high aspect ratio microstructure arrays by micro reverse wire-EDM, Journal of Micromechanics and Microengineering 15(2005) 1547.
52. Lin C.S. et al., Fabrication of micro ball joint by using micro- EDM and electroforming, Microelectronic Engineering 87(2010) 1475–1478.
53. Gao G., Zhao W., Wang Z., Gou Y., Instantaneous fabrication of tungsten microelectrode based on single electrical discharge, Journal of Materials Processing Technology 168(2005) 83–88.
54. Sen A., Doloi B., Bhattacharyya B., Experimental Studies on Fibre Laser Micromachining of Ti-6Al-4V. 5th International & 26th All India Manufacturing Technology, Design and Research Conference (AIMTDR 2014) December 14th, 2014.

55. Föehl C., Dausinger F., High precision deep drilling with ultra short pulses, Proceedings of SPIE2003; 5063: 346–51.
56. Dausinger F., Hugel H., Konov VI., Micromachining with ultra short laser pulses: from basic understanding to technical applications, Proceedings of SPIE 2003; 5147: 106–15.
57. Meijer J., Laser beam machining (LBM) state of the art and new opportunities, J Mater Process Technol 149 (2004)2–17.
58. Breitling D., Ruf A., Dausinger F., Fundamental aspects in machining of metals with short and ultra short laser pulses, Proceedings of SPIE 2004; 5339: 49–63.
59. Rizvi N.H., Femtosecond laser micromachining: current status and applications, Riken Review 50(2003) 107–112.
60. Loeschner U., Mauersberger S., Ebert R. et al., Micromachining of glass with short ns-pulses and highly repetitive fs-laser pulses, in Proceedings of the 27th International Congress on Applications of Lasers and Electro-Optics (ICALEO '08), pp. 193–201, October 2008.
61. Rizvi N., Femtosecond Laser Micromachining: Current Status and Applications. Riken Rev 50(2003) 77–82.
62. Gower M. C., Industrial application of laser micromachining, Opt Express 7(2000) 56–67.
63. Mishra S., Yadava V., Laser Beam Micro Machining (LBMM) – A review, Optics and Lasers in Engineering 73(2015) 89–122
64. Ren J., Kelly M., Hesselink L., Laser ablation of silicon in water with nanosecond and femtosecond pulses, Optics Letters 30(2005) 1740–1742.
65. Kruusing A., Underwater and water-assisted laser processing: part 2—etching, cutting and rarely used methods, Optics and Lasers in Engineering 41(2004) 329–352.
66. Tao Y. L., Chen H., Zhang W., Time scale effects in laser material removal: a review, Int J Adv Manuf Technol 26(2005) 598–608.
67. Hanemann T., Pfleging W., Haubelt J., Gahr K. H. Z., Laser micromachining and light induced reaction injection molding as suitable process sequence for the rapid fabrication of micro component, Microsyst Technol 7(2002) 209–14.
68. Bhavsar S.N., Aravindan S., Rao P.V., A Critical Review on Microtools Fabrication by Focused Ion Beam (FIB) Technology, Proc. of the World Congress on Eng. (2009).
69. Raffa V., Castrataro P., Menciassi A., Dario P., Focused Ion Beam as a Scanning Probe: Methods and Applications, Applied NanoScience and Technology (2006) 361–412.
70. Lucille A., Stevie G., Fred A., Introduction to focused ion beams, instrumentation, theory, techniques and practice, Springer Inc, Boston, USA. 2005.
71. Sadoh T., Eguchi H., Kenjo A., Miyao M., Etching characteristics of SiO_2 irradiated with focused ion beam, Nucl Instrum Meth B 206 (2003) 478.
72. Kang H.J., Ahn S.H., Lee J.S., Lee J.H., Surface modification of aluminum by nitrogen-ion implantation, Int J Precis Eng Man 7(2005) 1.
73. Villanueva G., Plaza J.A., Sanchez-Amores A., Bausells J., Martinez E., Samitier J., et al. Deep reactive ion etching and focused ion beam combination for nanotip fabrication, Mater Sci Eng C-Mater 26(2006) 164.
74. Venugopal G., Fabrication and Characteristics of Submicron Stacked-Junctions on Thin Graphite Flakes, J. Nanosci. Nanotechnol 11(2011) 1405–1408.
75. Park C.M., Bain J.A., Focused ion beam induced grain growth in magnetic materials for recording heads, J. Appl. Phys. 91(2002) 6380–6832.
76. Nastasi M., Mayer J.W., Hirvonen J.K., Ion- Solid Interactions: Fundamentals and Applications (Cambridge University Press, Cambridge, UK, 1996).
77. Nellen P.M., Callegari V., Bronnimann R., FIB-milling photonic structures and sputtering simulation, Microelectron Eng 83(2006)1805.
78. Miller M.K., Russell K.F., Atom probe specimen preparation with a dual beam SEM/FIB miller, Ultramicroscopy 107(2007)761.
79. Ali M.Y., Ong A.S., Fabricating micromilling tool using wire electrodischarge grinding and focused ion beam sputtering, Int J Precis Eng Man 31(2006)501.

80. Hopman W.C.L., Ay F., Hu W., Gadgil V.J, Kuipers L., Pollnau M., et al. Focused ion beam scan routine, dwell time and dose optimizations for submicrometer period planar photonic crystal components and stamps in silicon, Nanotechnology 18(2007)195305.
81. Li W., Minev R., Dimov S., Lalev G., Patterning of amorphous and polycrystalline Ni78B14Si8 with a focused-ion-beam, Appl Surf Sci 253(2007)5404.
82. Malek C.K., Hartley F.T., Neogi J., Fast prototyping of high-aspect ratio, high resolution X-ray masks by gas-assisted focused ion beam, Microsyst Technol 9(2003)409.
83. Hosokawa H., Shimojima K., Chino Y., Yamada Y., Wen C.E., Mabuchi M., Fabrication of nanoscale Ti honeycombs by focused ion beam, Mater Sci Eng A-Struct 344(2003)365.
84. Youn S.W., Okuyama C., Takahashi M., Maeda R., A study on fabrication of silicon mold for polymer hot-embossing using focused ion beam milling, J Mater Process Tech 201(2008) 548.
85. Li H.W., Kang D.J., Blamire M.G., Huck W.T.S., Focused ion beam fabrication of silicon print masters, Nanotechnology 14(2003)220.
86. Igaki J., Kometani R., Nakamatsu K., Kanda K., Haruyama Y., Ochiai Y., et al., Three dimensional rotor fabrication by focused-ion-beam chemical-vapor deposition, Microelectron Eng 83(2006)1221.
87. Kometani R., Funabiki R., Hoshino T., Kanda K., Haruyama Y., Kaito T., et al., Cell wall cutting tool and nano-net fabrication by FIB-CVD for sub-cellular operations and analysis, Microelectron Eng 83(2006)1642.
88. Reyntjens S., Puers R., A review of focused ion beam applications in microsystem technology, Journal of micromechanics and microengineering 11 (2001) 287–300
89. Fujii T., Iwasaski M., Munekane M., Takeuchi T., Hasuda M., Asahata T., et al., A nanofactory by focused ion beam, J Micromech Microeng 11(2005)S286.
90. Ding X., Lim G.C., Cheng C.K., Butler D.L., Shaw K.C., Liu K., Fong W.S., Fabrication of a micro-size diamond tool using a focused ion beam, J. Micromech. Microeng. 18 (2008).
91. Selada A., Manaia A., Vieira M.T., Pouzada A.S., Effect of LBM and large-area EBM finishing on micro-injection moulding surfaces, Int J Adv Manuf Technol 52 (2011) 171–182.
92. Desilets B. H., Mechanism of cavity formation in unfired ceramic by electron beam machining, Journal of vacuum science technology 15(1978)3.
93. Moarrefzadeh A., Finite-Element Simulation of Electron Beam Machining (EBM) Process, International Journal of Multidisciplinary Sciences And Engineering 2(2011).
94. Chryssolouris G., Anifantis N., Karagiannis S., Electron Assisted Machining: an Overview, ASME Journal of Manufacturing Science and Engineering 119 (1997) 766–769.
95. Biamino S., Penna A., Ackelid U., Sabbadini S., Tassa O., Fino P., Pavese M., Gennaro P., Badini C., Electron beam melting of Ti-48Al-2Cr-2Nb alloy: microstructure and mechanical properties investigation, Intermetallics 19 (2010) 776–781.
96. Uno Y., Okada A., Uemura K., Raharjo P., Sano S., Yu Z., Mishima S, A new polishing method of metal mold with large-area electron beam irradiation, J Mat Process Technol 187–188(2007) 77–80.
97. Utke, Hoffmann, and Melngailis, Gas-assisted focused electron beam and ion beam processing, J. Vac. Sci. Technol. 26(2008).
98. Jain V. K., Advanced Machining Processes, Allied Publishers, Delhi, 2002.
99. Blak J.T., et. al., DeGarmo's Materials and Processes in Manufacturing, John Wiley & Sons, Inc, Tenth ed. 2007.
100. Drozda T.J., Tool and Manufacturing Engineers, Handbook (Chapter 14: Nontraditional Machining), SME Publishing, 1989.
101. Fadaei T. A., A New Etchant For The Chemical Machining of St304, Journal of Materials Processing Technology 149(2004) 404–408.
102. Al-Ethari A. H., Alsultani K. F., Dakhil N., Variables Affecting the Chemical Machining of Stainless Steel 420, International Journal of Engineering and Innovative Technology (IJEIT) 3(2013).

103. Allen D.M., Talib T.N., Manufacture of stainless steel edge filters: an application of electrolytic photopolishing, precision engineering (1983) 57–59.
104. Bhattacharyya B., Munda J., Malapati M., Advancement in electrochemical micro-machining, International Journal of Machine Tools & Manufacture 44 (2004) 1577–1589.
105. Lohrengel M. M., Kluppel I., Rosenkranz C., Bettermann H., Schultze J. W., Microscopic investigations of electrochemical machining of Fe in $NaNO_3$, Electrochimia Acta 48 (2003) 3203–3211
106. Kozak J., Rajurkar K. P., Wei B., Modeling and analysis of pulse electrochemical machining, Transactions of ASME, 116 (1994) 316–323.
107. Munda J., Malapati M., Bhattacharyya B., Control of micro-spark and stray –current effect during EMM process, Journal of Materials Processing Technology 194(2007) 151–158.
108. Kock M., Kirchner V., Schuster R., Electrochemical micro machining with ultra short voltage pulses-a versatile method with lithographic precision, Electrochimia Acta 48(2003) 3213–3219.
109. Rathod V., Doloi B., Bhattacharyya B., Parametric Investigation into the Fabrication of Disk Microelectrodes by Electrochemical Micromachining, Journal of Micro- and Nano-Manufacturing, Vol. 1/ 041005-1-041005-11, December 2013
110. Wang J.J.J., et al., Fabrication of wedge-shape tool via electrochemical micromachining with diamond-like carbon coating, Journal of Materials Processing Technology 187–188[2007] 264–269.
111. Landolt D., Chauvy P. F., Zinger O., Electrochemical micro machining, polishing and surface structuring of metals: fundamental aspects and new developments, Electrochimia Acta 48(2003) 3185–3201.
112. Datta M., Landolt D., Fundamental aspects and applications of electrochemical micro fabrication, Electrochimia Acta 45(2000) 2535–2558
113. Bassu M., Strambini L.M., Barillaro G., Advances in Electrochemical Micromachining of Silicon: Towards MEMS Fabrication, Procedia Engineering 25(2011) 1653–1656.
114. Munda J., Malapati M., Bhattacharyya B., Control of microspark and stray-current effect during EMM process, Journal of Materials Processing Technology 194(2007) 151–158.
115. Wang M., et al., Electrochemical machining of the spiral internal turbulator, International Journal of Advanced Manufacturing Technology 49(2010) 969–973.
116. Jo C.H., Kim B.H., Chu C.N., Micro electrochemical machining for complex internal micro features, CIRP Annals -Manufacturing Technology 58(2009) 181–184.
117. Bo Hyun K., Byung Jin P., Chong Nam C., Fabrication of multiple electrodes by reverse EDM and their application in micro ECM, Journal of Micromechanics and Microengineering 16(2006) 843.
118. Nguyen M.D., Rahman M., Wong Y.S., Enhanced surface integrity and dimensional accuracy by simultaneous micro- ED/EC milling, CIRP Annals - Manufacturing Technology 61(2012) 191–194.
119. Kunieda M., et al., Electrochemical micromachining using flat electrolyte jet, CIRP Annals - Manufacturing Technology 60(2011) 251–254.
120. Hong Shik S., Hyun K. B., Chong Nam C., Analysis of the side gap resulting from micro electrochemical machining with a tungsten wire and ultrashort voltage pulses, Journal of Micromechanics and Microengineering 18(2008)075009.
121. Kunieda M., Mizugai K., Watanabe S., Shibuya N., Iwamoto N., Electrochemical Micromachining using Flat Electrolyte Jet, CIRP Annals, Manufacturing Technology 60 (2011) 251–254.
122. Zinger O., Chauvy P.F., Landolt D., Scale-Resolved Electrochemical Surface Structuring of Titanium for Biological Applications, J. Electrochem. Soc., 150(2003) B495.
123. Kock M., Kirchner V., Schuster R., Electrochemical Micromachining with Ultrashort Voltage Pulses - a Versatile Method with Lithographical Precision, Electrochemical Acta 48 (2003) 3213–3219.

124. Trimmer A. L., Hudson J. L., Kock M., Schuster R., Single-step Electrochemical Machining of Complex Nanostructures with Ultrashort Voltage Pulses, Applied Physics Letters 82 (2003) 3327–3329.
125. Karunakaran K., Pushpa V., Akula S., Suryakumar S., Techno-economic analysis of hybrid layered manufacturing, International Journal of Intelligent Systems Technologies and Applications 4(2008) 161–176.
126. Heisel U., Wallaschek J., Eisseler R., Potthast C., Ultrasonic deep hole drilling in electrolytic copper ECu 57, CIRP Annals-Manufacturing Technology 57(2008) 53–56.
127. Lauwers B., Klocke F., Klink A., Hybrid processes in manufacturing, CIRP Ann-Manuf Techn 63(2014) 561–583.
128. Curtis D.T., Soo S.L., Aspinwall D.K., Electrochemical superabrasive machining of a nickel-based aeroengine alloy using mounted grinding points, CIRP Ann-Manuf Techn 58 (2009) 173–176.
129. Tehrani A.F., Atkinson J., Overcut in pulsed electrochemical grinding, P I Mech Eng B-J Eng 214(2000) 259–269.
130. Bhattacharyya B., Doloi B., Sorkhel S.K, Experimental investigations into electrochemical discharge machining (ECDM) of non-conductive ceramic material, Journal of Materials Processing Technology 95(1999) 145–154.
131. Jain V.K., Adhikary S., On the mechanism of material removal in electrochemical spark machining of quartz under different polarity conditions, Journal of Materials Processing Technology 200(2008) 460–470.
132. Allesu K., Ghosh A., Muju M.K., Preliminary qualitative approach of a proposed mechanism of material removal in electrical machining of glass, Eur. J. Mech. Eng. 36 (1992) 202–207.
133. Fascio V., Wüthrich R., Bleuler H., Spark assisted chemical engraving in the light of electrochemistry, Electrochim. Acta 49 (2004) 3997–4003.
134. Tandon S., Jain V.K., Kumar P., Rajurkar K.P., Investigations into machining of composites, Precis. Eng. 12 (1990) 227–238.
135. Langen H., Breguet J.M., Bleuler H., Renaud Ph, Masuzawa T., Micro electrochemical discharge machining of glass, Int. J. Electr. Mach. 3 (1998) 65–69.
136. Crichton I.M., Mcgeough J.A., Studies of the discharge mechanisms in electrochemical arc machining, J. Appl. Electrochem. 15 (1985) 113–119.
137. Cook N.H., Foote G.B., Jordan P., Kalyani B.N., Experimental studies in electro machining, J. Eng. Ind. 95 (1973) 945–950.
138. Khairy A.B.E., Mcgeough J.A., Die-sinking by electro erosion-dissolution machining, CIRP Ann. Manuf. Technol. 39 (1990) 191–195.
139. Didar T.F., Dolatabadi A., Wüthrich R., Characterization and modeling of 2D glass micro-machining by spark-assisted chemical engraving (SACE) with constant velocity, J. Micromech. Micro Eng. 18 (2008) 9.
140. Jain V.K., Chak S.K., Electrochemical spark trepanning of alumina and quartz, Mach. Sci. Technol. 4 (2000) 277–290.
141. Furutani K., Maeda H., Machining a glass rod with a lathe-type electro-chemical discharge machine, J. Micromech. Microeng. 18 (2008) 8.
142. Schöpf M., Beltram I., Boccadoro M., Kramer D., ECDM (Electrochemical discharge machining) a new method for trueing and dressing of metal bonded diamond grinding tools, CIRP Ann. Manuf. Technol. 50 (2001) 125–128.
143. Peng W.Y., Liao Y.S., Study of electrochemical discharge machining technology for slicing non-conductive brittle materials, J. Mater. Process. Technol. 149 (2004) 363–369.
144. Solignac D., Sayah A., Constantin S., Freitag Rand M., Giijs M. A., Powder blasting for realization of microchips for bio-analytic applications, Sensors Actuators A 92(2001) 388–93.

145. Liu C., Chen J., Engel J., Zou J., Wang X., Fan Z., Ryu K., Shaikh K., Bullen D., Polymer micromachining and applications in sensors, microfluidics, and nanotechnology, 226th National Meeting of the American Chemical Society (ACS) 2003 (NewYork, September 11–17).
146. Mills C. A., Martinez E., Bessueille F., Villanueva G., Bausells J., Samitier J., Errachid A., Production of structures for micro fluidic using polymer imprint techniques, Microelectron. Eng. 78–9(2005) 695–700.
147. Holger B., Laurie E. L., Polymer micro fluidic devices, Talanta 56(2002) 267–87.
148. Lai S., Lee L. J., Yu L., Koelling K. W., Madou M. J. Micro-and nano-fabrication of polymer based micro fluidic platforms for bioMEMS applications Proc. Materials Research Society Symp. 729(2002) (U1.7.1-U1.7.11)
149. Getu H., Ghobeity A., Spelt J. K., Papini M. Abrasive jet micromachining of polymethylmethacrylate, Wear 263(2007) 1008–15.
150. Getu H., Ghobeity A., Spelt J. K., Papini M., Abrasive jet micromachining of acrylic and polycarbonate polymers at oblique angles of attack, Wear doi:10.1016/j.wear.2008.01.013.
151. Singh R., Walia R. S., Study the effects of centrifugal force on abrasive flow machining process. International Journal of Research in Mechanical Engineering and Technology 2 (2012) 34–39.
152. Singh R., Walia R. S., Suri N. M., Study of parametric effect on surface roughness improvement for hybrid centrifugal force assisted abrasive flow machining process, International Journal of Latest Research in Science and Technology 1(2012) 198–201.
153. Nouraei H., Kowsari K., Papini M., Spelt J.K., Operating parameters to minimize feature size in abrasive slurry jet micro-machining, Precision Engineering 44 (2016) 109–123.
154. Liu Z., Nouraei H., Papini M., Spelt J.K., Abrasive enhanced electrochemical slurry jet micro-machining: Comparative experiments and synergistic effects, Journal of Materials Processing Technology 214 (2014) 1886–1894.
155. Shukla R., Singh D., Experimentation investigation of abrasive water jet machining parameters using Taguchi and Evolutionary optimization techniques, Swarm and Evolutionary Computation, doi:10.1016/j.swevo.2016.07.002.
156. Thoe T.B., Aspinwall D.K., Wise M. L.H., Review on Ultrasonic Machining, International Journal of Machine Tools & Manufacture. 38(1998) 239–355.
157. Yu Z., Rajurkar K.P., Tandon A., Study of 3D Micro-Ultrasonic Machining, Journal of Manufacturing Science and Engineering Vol. 126; 2004.
158. Markov A.I., Neppiras E., Ultrasonic machining of intractable materials, Iliffe, London, 1966.
159. Lv Z., Huang C.Z., Zhu H.T, Wang J., Wang Y., Yao P., A research on ultrasonic assisted abrasive water jet polishing of hard-brittle materials, Int. J. Adv. Manuf. Technol. 78 (2015) 1361–1369.
160. Rhoades L., Abrasive flow machining: a case study, Journal of Material Processing Technology 28 (1991) 107–116.
161. Davies P.J., Fletcher A.J., The assessment of the rheological characteristics of various polyborosilixane/grit mixtures as utilized in the abrasive flow machining, Proceedings of Institute of Mechanical Engineers 209 (1995) 409–418.
162. Raju H.P., Narayanasamy K., Srinivasa Y.G., Krishnamurthy R., Characteristics of extrude honed SG iron internal primitives, Journal of Materials Processing Technology 166 (2005) 455–464.
163. Rajesha S., G Venkatesh., Sharma A.K., Kumar P., Performance study of a natural polymer based media for abrasive flow machining, Indian Journal of Engineering & Materials Sciences 17 (2010) 407–413.
164. Kumar R., Murtaza Q., Walia R.S., Three Start Helical Abrasive Flow Machining for Ductile Materials, 3rd International Conference on Materials Processing and Characterization 6 (2014) 1884–1890.
165. Steigerwald J. M., Murarka S. P., Gutmann R. J., Chemical Mechanical Planarization of Microelectronic Materials Wiley, New York, 1997, pp. 1–12.

166. Mori Y., Yamauchi K., Endo K., Mechanism of atomic removal in elastic emission machining, Precision Engineering 10(1)(1988) 24–28
167. Hiraoka D., Hamazaki K., Matsumura K., Research on developing a prototype NC EEM machine for machining high-power laser mirrors, Koyo Engineering Journal, 155E(1999) 42–47.
168. Mori Y., Ikawa N., Okuda T., Yamagata K., Numerically Controlled Elastic Emission Machining; Technology reports of the Osaka University 26(1976) 283–294.
169. Endo K., Namba H., New machining method for making precise and very smooth mirror surfaces made from Cu and Al alloys for synchrotron radiation optics, Rev Sci ln- strum 60 (1989)2.120-2.123.
170. Chang G.W., Yan B.H., Hsu R.T., Study on cylindrical magnetic abrasive finishing using unbonded magnetic abrasives, Int. J. of Mach. Tool & Manuf 2002; 42: 575–583.
171. Yan B.H., Chang G.W., Cheng J.T., Hsu R.T., Electrolytic magnetic abrasive finishing, Int J of Mach Tool & Manuf 43(2003) 1355–1365.
172. Yin S., Shinmura T., Vertical vibration-assisted magnetic abrasive finishing and deburring for magnesium alloy, Int J of Mach Tool & Manuf 44(2004) 1297–1303.
173. Yamaguchi H., Shinmura T., Sekine M., Uniform Internal Finishing of SUS304 Stainless Steel Bent Tube Using a Magnetic Abrasive Finishing Process, J of Manuf Sci & Eng. 127 (2005) 605–611.
174. Shinmura T., Takazawa K., Hatano E., Study on magnetic abrasive finishing – effects of various types of magnetic abrasives on finishing characteristics, Bull Japan Soc of Prec Eng 21 (1987) 139–141.
175. Shinmura T., Takazawa K., Hatano E., Matsunaga M., Study on magnetic abrasive finishing, Annals of the CIRP 39(1990) 325–328.

Chapter 2
Recent Advancement on Ultrasonic Micro Machining (USMM) Process

S. Das, B. Doloi and B. Bhattacharyya

Abstract There are a lot of developments in the micro manufacturing methods for the production of the three-dimensional miniaturized products made up of different advanced materials. Ultrasonic micro machining is an essential technique for the fabrication of micro parts on the hard, brittle and non-conductive materials like glass, ceramics and silicon with high aspect ratio. Ultrasonic micro machining is the mechanical type non conventional micro machining process. Material removal mechanism of USMM is similar to macro ultrasonic machining process. Adequate surface finish with stiff tolerances and dimensions can be achieved by ultrasonic micro machining (USMM) on hard and brittle materials. During the last decades, a number of researchers have explored experimentally and theoretically this ultrasonic micro machining (USMM) process technique with different materials. Recent development on ultrasonic micro machining (USMM) process has been highlighted and discussed in details with different types of ultrasonic micro machining (USMM) set up and material removal mechanism. Design and developments of micro-tools for USMM process have also been discussed. Influences of different process parameters on various responses of USMM have been discussed here.

S. Das (✉) · B. Doloi · B. Bhattacharyya
Production Engineering Department, Jadavpur University, Kolkata 700032, India
e-mail: somnath96@gmail.com

B. Doloi
e-mail: bdoloionline@rediffmail.com

B. Bhattacharyya
e-mail: bb13@rediffmail.com

G. Kibria et al. (eds.), *Non-traditional Micromachining Processes*,
Materials Forming, Machining and Tribology,
DOI 10.1007/978-3-319-52009-4_2

2.1 Introduction

Fabrication of the miniaturized products with multi functions are the urgent need of manufacturing industry. Micro-parts are required to meet the increasing demands in various areas like electronics, aerospace, automotive, medical devices, etc. The main advantage of those miniaturized products are small space requirement, less energy and material consumption, easiness in delivery and less price of the product.

Now the products with the extraordinary characteristic of materials such as high hardness, strength, corrosion resistant and heat resistant etc. are machined at the micron range of dimensions. Production of complex 3D miniature components such as fabrication of micro holes in thousand numbers on filters for processing food and textile industries, which are not possible by any conventional machining methods. Such features on those components can get only through the non-conventional micro machining process.

The generation of micro machining features on ceramics, glass, carbides and metallic alloys, etc. by traditional machining method is tremendously difficult and it also takes more time. Adequate surface finish with stiff tolerances and dimensions can be achieved by ultrasonic micro machining (USMM) on hard and brittle materials. Other Non Traditional machining process like μEDM, μECM, and LBM etc. can't machine those materials in proper shape and size [1]. Using other non-traditional machining process it is difficult to machine square, irregular and complex shaped holes and surface impressions. Ultrasonic micro machining (USMM) can be used for those purposes. Any hard and fragile material can be machined using USMM. The main advantage of ultrasonic micro machining process is that it can machine any electrical conductive and non-conductive materials [2]. Micro tool development and fixing on the USMM system is the very much problematic area. This problem can be eliminated by micro tool development with the help of wire electro-discharged grinding (WEDG) method and the same tool can be utilized to make micro-holes 5–20 μm in diameter on a silicon plate [3, 4].

A number of research works have been reported in micro-USM processes. Recent advancement on ultrasonic micro machining (USMM) process has been highlighted and discussed in details with different types of ultrasonic micro machining (USMM) set up and material removal mechanism. Design and developments of micro-tools for USMM process have also been discussed. Influences of different process parameters on various responses of USMM have been discussed in this chapter.

2.2 Fundamentals of Ultrasonic Machining (USM) Process

Ultrasonic Machining (USM) is a mechanical type non-conventional machining method used for machining both electrically conductive and non conductive materials, which have low ductility and a high hardness above 40 HRC such as ceramics, inorganic glasses, and quartz etc. The ultrasonic frequency of the tool is above 20 kHz with amplitude of 8–30 μm. Water based abrasive particles are supplied through the machining zone by a recirculation pump. The tool is pressed on the workpiece with fixed static load and the workpiece material is nonstop hammered by numerous abrasive particles with high kinetic energy imparted by the vibrated tool. Flushing of abrasive slurry refreshes the machining area and also removes the debris from the machining zone.

2.2.1 Background of USM

In 1927, R.W. Wood and A.L Loomis observed the drilling and cutting action with the help of ultrasonic vibration. After a long interval L. Balamuth proposed ultrasonic machining in 1942, while he was investigated the dispersion of solid in liquid by means of a magnetostrictively vibrating nickel tube. Several types of ultrasonic machine tool have been developed in recent year. The first communication on equipment and techniques for ultrasonic cutting appeared in 1953–54 [5]. Originally USM used to be a finishing operation for the component processed by the electro spark machines. However, this use becomes less important because of the development in electric discharge machining. But then with the boom in solid state electronics, the machining of electrically non-conducting, semi conductive, and brittle material become more and more important. For this reason, ultrasonic machining again gained importance and prominence. Through research and development on ultrasonic machining technology, different types of ultrasonic machining systems have been developed.

2.2.2 Process Development of USM

The material removal mechanism of ultrasonic machining process is purely mechanical. Figure 2.1 shows the various process equipments of ultrasonic machining process. In this process water based abrasive particles are presented to the machining zone.

Workpiece material does not depend on electrically conductive or chemically reactive in this machining process. Cutting, drilling and engraving is easily done by this machining method. The range of vibration of the tool is of 18–40 kHz and amplitude of 10–50 μm. Water based abrasive slurry is continuously supplied into the gap between the bottom area of the tool and the top surface of stationary workpiece. The tool is pressed with a static pressure depending on the size of the tool tip. The abrasive particles are hammered by the tool into the top surface of the workpiece, and as a result the particles abrade the workpiece material into a

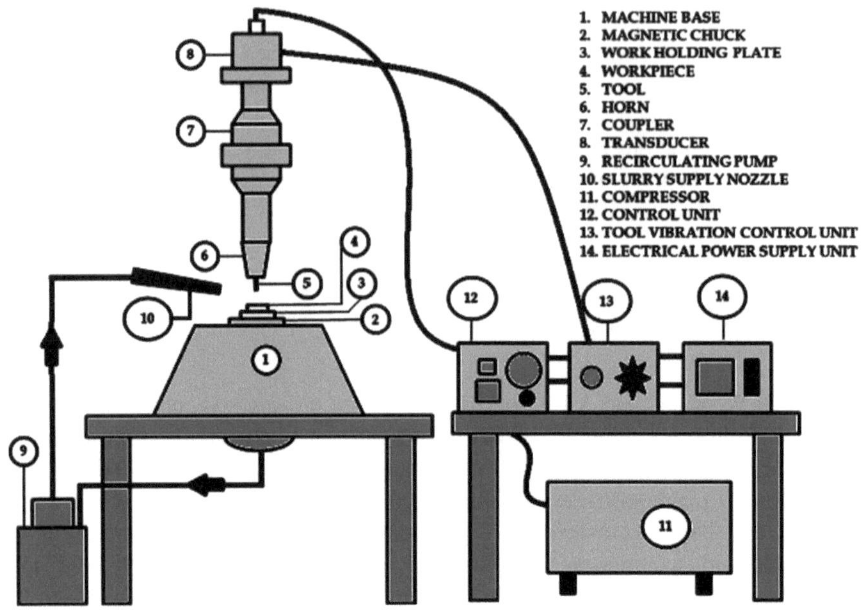

Fig. 2.1 Schematic diagram of ultrasonic machining set-up with process equipments

conjugate image of the tool form. Due to the fact USM is not restricted to the manufacture of circular holes. The tool can be prepared to the required shape, and therefore extremely complex shapes can be created on hard and brittle materials. Any damaging or thermal effects on the metallic structure of the workpiece does not found in this machining process.

2.2.3 Types of Machining Operation

Depending on the machining condition different types of ultrasonic machining is shown. The different types of machining configuration are discussed here.

i. **Stationary ultrasonic machining (USM)** is a non traditional machining process that removes material from the workpiece surface through high frequency with low amplitude vibrations of a tool against the material surface in the presence of water based abrasive particles. The low-frequency electrical signal is applied to the transducer for converting the electrical energy into high-frequency (above 20 kHz) mechanical vibration. This mechanical vibration is transmitted to the horn and tool assembly. The tool movements are vertically or orthogonal to the top of the workpiece surface. The abrasive grains are mixed with water is distributed across the tip of the tool. The abrasive particles are mixed with water and are hammered by the tool into the workpiece surface. Then the material is removed from the workpiece and the conjugate images of the tool form are produced [6].

ii. **Rotary ultrasonic machining (RUM)** is a hybrid machining process that combines the rotating diamond grinding tool with ultrasonic machining (USM). Higher material removal rates (MRR) are found using rotary ultrasonic machining. In this machine set up, a rotary core drill with metal bonded diamond abrasives tool is ultrasonically vibrated in the axial direction and the tool is fed toward the workpiece at a constant pressure. Coolant is pumped through the machining zone to prevent jamming of the tool and keeps it cool. RUM provides a fast, high-quality material removal process for a variety of glass and ceramic applications using abrasives bonded tools with combining rotation and vibration [7].

iii. USM combined with electric discharge machining [8].

iv. Ultrasonic supported cutting or grinding. Ultrasonic assisted turning is the most familiar process and is claimed to reduce machining time, residual stresses of the working zone and improve surface quality and tool life compared to conventional turning [9, 10].

2.3 Fundamentals of Ultrasonic Micro Machining (USMM) Process

The ultrasonic micro machining (USMM) process is copied from macro ultrasonic machining (USM) for micro machining on any hard and fragile materials. The tool frequency of USMM is above 20 kHz and lower amplitude of 0.5–5 μm. The abrasive particles size of USMM is few microns of 0.5–15 μm. The material removal mechanism is same as conventional USM.

2.3.1 Background of USMM

Masuzawa of Tokyo University in the mid 1990s was first attempted of downsizing macro-USM for micromachining [11]. The ultrasonic micro machining is used for producing micro feature on any hard and brittle materials such as any bio ceramic materials, glass and silicon etc. [12]. Ultrasonic micro machining has similar process parameters as macro-USM. However, the micromachining process requires a micro-sized tool with smaller amplitude, and micro-sized abrasive particles [13].

Table 2.1 Comparison of ultrasonic micro machining (USMM) and macro USM

Parameters	Ultrasonic micro machining (USMM)	MACRO-USM
Vibration frequency, kHz	Usually >20	Usually >20
Vibrated component	Tool or workpiece	Tool
Amplitude, μm	0.5–5 μm	8–30 μm
Abrasive particle size, μm	0.5–15 μm	50–300 μm
Static load	Gram force	Kilogram force
Tool diameter	<500 μm	>500 μm
Thickness of workpiece	maximum 3 mm	Maximum 50 mm
Tool material	Tungsten carbide (preferred)	Stainless steel (preferred)

2.3.2 Process Development of USMM

Ultrasonic micro machining (USMM) is a very essential micromachining method to generate micro features with high aspect ratio. Due to technological development, lot of improvements were done in the macro USM. Then the macro USM is converted into micro USM. A comparison of the Ultrasonic micro machining (USMM) and macro USM parameters are presented in Table 2.1.

2.3.3 Types of Ultrasonic Micro Machining Operation

Ultrasonic Vibration can be provided either in tool or in workpiece. There are three types of machining setup.

i. **μ-USM setup with Tool vibration**: Sun et al. developed μ-USM setup with provision of tool making using Electric Discharge Machining (EDM) and Wire electro Discharge Grinding (WEDG). In this μ-USM setup the tool is vibrated with ultrasonic vibration. The WEDG unit is permanently fixed on the machine body for grinding fine electrodes. Four axes X, Y, Z and C are numerically controlled by the machine. The μ-USM unit consists of a high frequency oscillation generator, transducer and horn-tool combination. Computer control system is used to control the operation of μ-USM setup [14]. Figure 2.2 shows the μ-USM setup with tool vibration.
ii. **μ-USM setup with Workpiece vibration**: The μ-USM setup with Workpiece vibration as shown in Fig. 2.3. In this setup a micro tool is holed properly and also rotated by a DC motor to move in X, Y and Z directions. Micro tools are prepared by wire electrical discharge grinding (WEDG). Due to sensitiveness, bending, vibration, and breakage of the micro tool, the controlled and limited

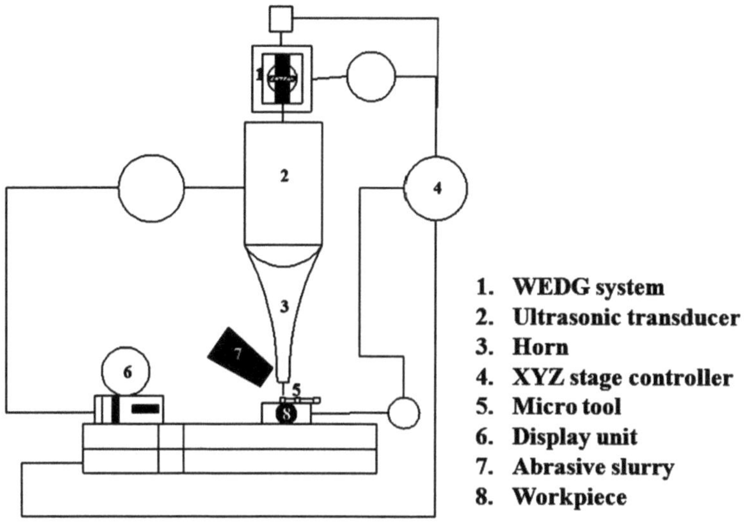

Fig. 2.2 Schematic diagram of μ-USM setup with tool vibration

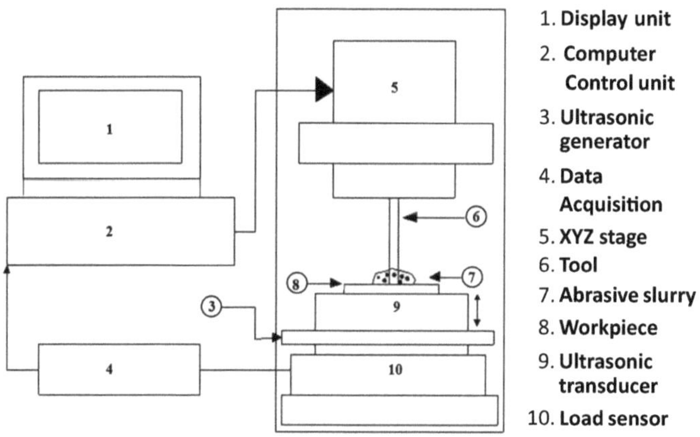

Fig. 2.3 Schematic diagram of μ-USM setup with workpiece vibration

contact force between tool and workpiece is required during machining. A close-loop control system with force feedback from an electronic balance is implemented in this machining set up, which is positioned near the vibrator. The contact force between the tool and workpiece is regulated by this system [15].

iii. **μ-RUSM setup with Workpiece vibration**: Micro rotary ultrasonic machining (MRUM) experimental set up is discussed here. Figure 2.4 exhibits the schematic diagram of micro rotary ultrasonic machining.

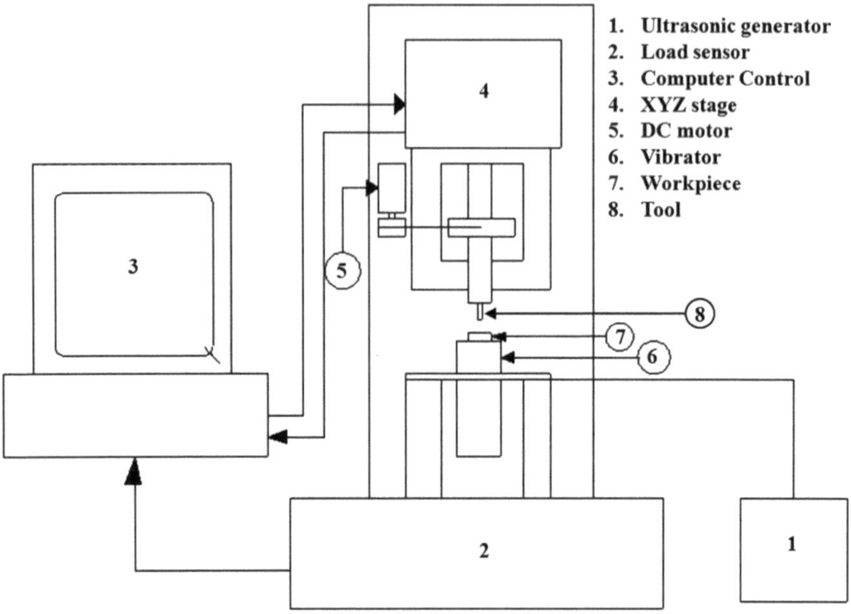

Fig. 2.4 Schematic diagram of micro rotary ultrasonic machining

The basic component of the micro rotary ultrasonic machine system is ultrasonic vibration system (transducer and generator), positioning system (XYZ-stages), cutting force feedback sensor, system controller, machine spindle, tool holder and work piece holder. The system is an assembly of a piezoelectric ultrasonic transducer. A spindle for rotating tool and position of tool is controlled in X, Y and Z axis by a precision motion controller with 25 nm resolution. The work piece is vibrated ultrasonically at 39.5 kHz by mounting it on the free end of the transducer. The abrasive slurry mixed with water is injected into the machining gap between the work piece and tool [16, 17].

2.4 Principle of Material Removal in Ultrasonic Micro Machining (USMM) Process

The mechanism of material removal of ultrasonic micro machining (USMM) process is classified in two parts, one is ductile mode and another is brittle mode. Elastic or plastic deformation in the surface of the workpiece is only responsible for ductile mode. On the other hand, the mode of brittle mechanism is found due to brittle fracture on the workpiece surface. It is extensively recognized that the transition mode from ductile mode to brittle mode absolutely exists in the material removal mechanism of brittle materials, if the depth of the cut is reduced [18].

Zarepour et al. investigated the mechanism of material removal and the relationship between process parameters. They also investigated the analytical mode of material removal mechanism in ultrasonic micro machining (USMM) process on silicon material [19]. Yu et al. [20] studied the debris accumulation and reported an explanation of the influences of abrasive particles on the surface roughness. Guodong Lia et al. investigated the material removal mechanism on quartz crystals. When the roughness value is very less, the machined surface by ultrasonic micro machining (USMM) process is comparatively smooth. This was one of the proofs of the existing of ductile machining [21]. Zarepour et al. [22] presented a new method for the static force measuring and the workpiece clamping for the ultrasonic micro machining (USMM) process. The proposed control system and force measurement provide a compact arrangement and prevent the undesirable effects of vibration of the horn.

In the ultrasonic micro machining (USMM) process, abrasive particles with random shapes and sizes are supplied in the machining gap between the micro tool and the workpiece. At each vibration cycle, each abrasive particle is interacted with the workpiece in one of the following ways [23]:

i. Appling direct impact force on the workpiece surface by the abrasive particles: this is common when the abrasive grain size is larger than the machining gap.
ii. Impact of free-moving abrasive particles on the workpiece surface: this is common when the abrasive grain size is smaller than the machining gap.
iii. Cavitations effect erosion: the high frequency vibration power generates a high frequency mechanical pressure in the slurry medium causing the abrasive particles to impact the workpiece surface.

2.4.1 Mechanism of Material Removal

In USMM, the tool vibrates above 20 kHz, and water based abrasive slurry is pumped to the small gap between the workpiece and the tool [2]. The removal of material from the workpiece by USMM process is caused of micro-cracks by the brittle fracturing of the workpiece. Understand crack generation in USMM in order to improve machining efficiency and precision is investigated. The impact of a single abrasive grain during USMM process can be very useful for understanding crack generation of the workpiece. The main activities for removing material from the workpiece have been described by some researchers as follows:

i. The hammering effect by abrasive particle is shown when the tool and workpiece are in contact.
ii. The impact action is also shown by a free-moving abrasive particle in the working gap.
iii. The cavitations erosion phenomenon is shown by abrasive particle in slurry concentration [24–26].

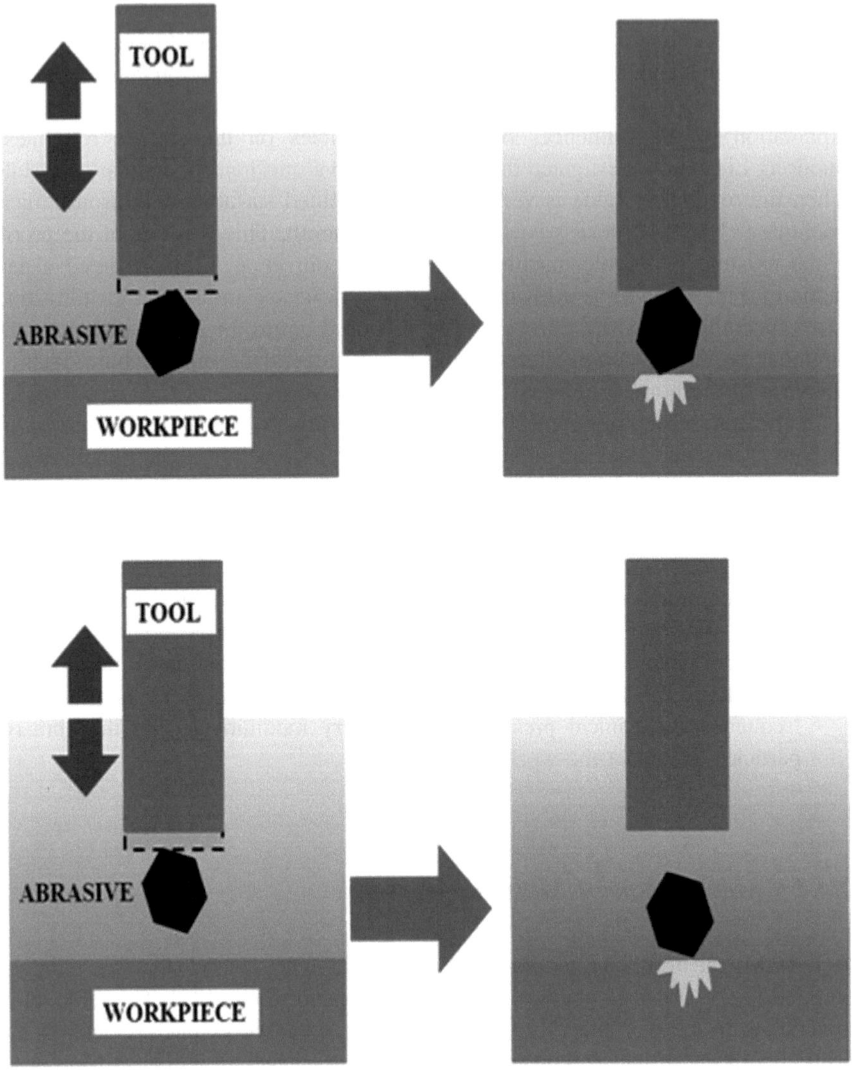

Fig. 2.5 Schematic diagram of material removal in MUSM: **a** Hammering action, **b** Impacting action

Generally, cavitations erosion is not significant for regular material removal mechanism of USMM [27]. The whole mechanisms are schematically shown in Fig. 2.5. The first two mechanisms are mainly responsible for material removal mechanism due to micro cracks, which are generated due to the rapid striking of the abrasive particles. The hammering action is broadly accepted as a major factor in crack generation.

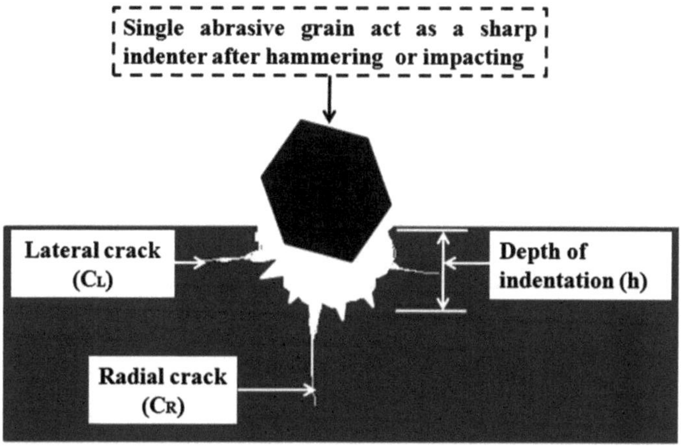

Fig. 2.6 Deformations and fracture of brittle material due to indentation

2.4.1.1 Formation of Cracks in Brittle Materials Due to Indentation

The material removes from the workpiece in the ultrasonic micro machining process is very much depended by Initiation and propagation of median as well as lateral cracks. Figure 2.6 shows that initiation of cracks and localized deformation of a brittle material. This is generated by large or a small abrasive grain, which acts as an indenter [28–30].

The abrasive particles act as indenter on the workpiece surface. The fracture is started on the surface by impression of an indenter in the machining area. Many authors have investigated in this area. It has been found that at first the sharp point of the indenter produces an inelastic deformation zone during machining. Then deformation is induced and suddenly develops into a small crack or median crack. The median crack is increased progressively by increasing the load. The median crack begins to close during unloading and inducing the formation of lateral vents. The lateral vents are carried on their extension toward the workpiece surface upon complete unloading and may lead to chipping. The sequence of crack procedures that occur during the machining has been shown in Fig. 2.7.

+P means increasing load; −P means decreasing load. The sequence of vent crack propagation is summarized as follows:

i. An inelastic deformation zone is produced by the sharp indenter.
ii. After that the deformation flow suddenly develops into a little crack. This is known as median crack.
iii. Next the median cracks are developed gradually with increasing the load.

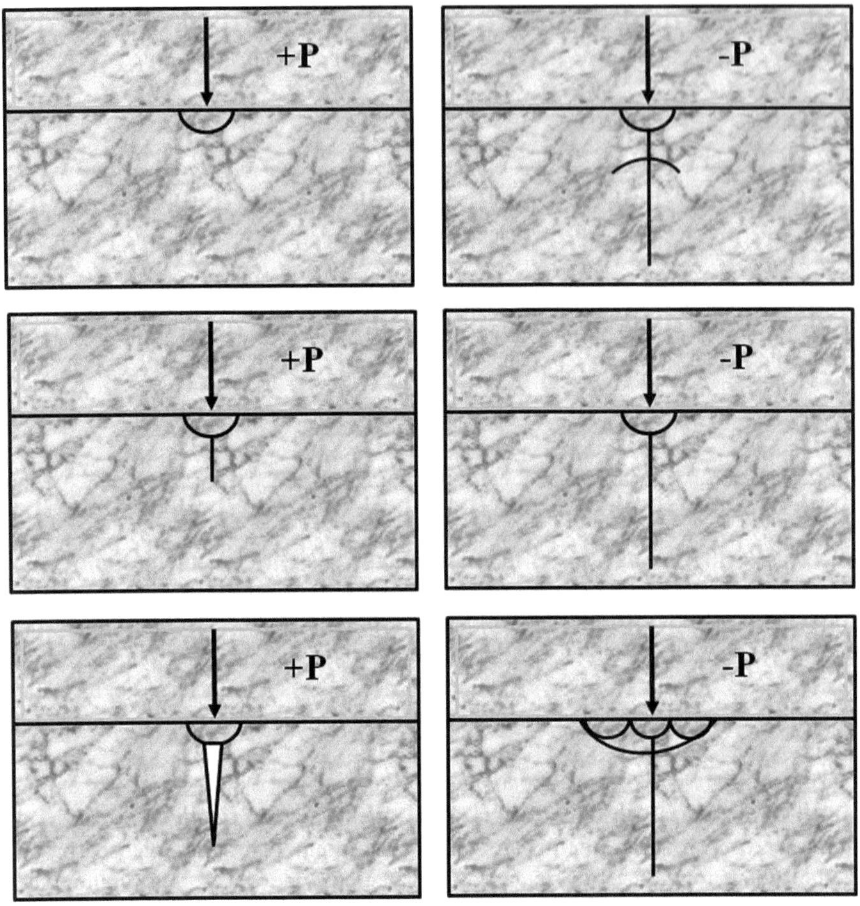

Fig. 2.7 Representation of vent-crack development under point indentation

iv. The median crack begins to close upon unloading.
v. The lateral vents are carried on their extension toward the workpiece surface upon complete unloading and may lead to chipping.
vi. The propagation of these median and lateral cracks lead eventually to chipping of the brittle material.

A critical load P_c for the beginning of a median crack is given by [31]:

$$P_c = \alpha \frac{K^4}{H_v^3} \tag{2.1}$$

where

α is a dimensionless feature which is related to the indenter geometry;
K is the fracture toughness of the workpiece; and
H_V is the vickers diamond hardness of the workpiece.

For the size of the median or radial crack (C_R) and lateral crack (C_L), the following equations have been derived [32]:

$$C_R^m = \xi_1 P \tag{2.2}$$

$$C_L = \xi_2 \left(\frac{P}{K}\right)^{3/4} \tag{2.3}$$

where, m = 1−1.5,

P is the applied load,
ξ_1 and ξ_2 are proportionality constants [33].

The depth of indentation is established by equating the mean static force. This mean force comes from impacting the tool on the grains. Assume the number of abrasive particles in the gap between tool and workpiece are inversely proportional to the square of the diameter of each of the abrasive particles. The following expression for machining depth h is described as follows [34]:

$$h = \left[\frac{8FAd}{\Pi K_1 H_V C(1+q)}\right]^{1/2} \tag{2.4}$$

where

F is the applied static force,
A is the vibration amplitude of the tool,
H_V is the hardness of the workpiece,
q is the hardness ratio of the workpiece to that of the tool,
d is average abrasive grain diameter,
C is the abrasive slurry concentration and
K_1 is proportionality constant.

2.4.2 Models on Material Removal Mechanism

Based on previous research work various models on material removal mechanism along with assumptions taken for developing the models are presented as follows:

Name of investigators	Mechanism of material removal	Assumptions
Shaw [34]	At first direct hammering of abrasive particle. After that impacting by free moving particles.	i. All abrasive grains are the same size, very stiff, and sphere-shaped. Impact forces of all particles are identical. ii. The removal of material is comparative to volume of material is removed for each impact and it also depends on number of particles impacting per cycle and frequency of impacting. iii. Depth of cut for penetrating the material is inversely proportional to hardness and flow stress of work material. iv. A number of active grains are inversely proportional to square off the mean diameter of all grains under the tool face area.
Miller [35]	Material is removed by chipping plastically deformed and work hardened material. MRR also depends upon work hardening while in brittle material on size and rate of chip for ductile material formation.	i. All abrasive particles are same size and cubical form. Plastic deformation is directly proportional to the applied stress. ii. Plastic flow stress equals burger vector times of shear modulus. iii. Cross sectional area of the cut does not vary in the machining time. iv. Viscous effects in water of the abrasive slurry are almost negligible.
Rozenberg et al. [36]	Brittle fracture.	All abrasive particles are irregular shape and incompressible but can be considered as spherical shaped having projections whose radius of curvature are proportional to the average dimensions of particle. From the experimental proof, the numerical distribution of each abrasive grain size d is given by: $\varphi(d) = 1.095\frac{N}{d_m}\left[1 - \left(\frac{d}{d_m} - 1\right)^2\right]^3$ where N is number of active abrasive grains, d_m is the mean diameter of grains.
Cook [37]	Hemispherical indentation fracture.	i. All abrasive grains are uniform and spherical in form. ii. The tool and all abrasive particles are rigid. iii. Viscous effects of the slurry concentration are negligible. iv. A linear relationship between part of active grits and ratio of indentation depth of radius of the each abrasive grit has been assumed.
Kainth et al. [38]	Indentation fracture due to direct hammering action.	i. All abrasive grains are spherical in shape and followed by Rozenberg's size distribution functions to take into account particle size in homogeneity. ii. Motion of tool remains sinusoidal under loaded conditions.
Nair and Ghosh [39]	Brittle fracture.	i. All abrasive grains are rigid and spherical in shape. ii. No consideration for MRR for impacting the abrasive particles, cavitations or chemical action of abrasive slurry. iii. The movement of tool tip is simple harmonic motion.

(continued)

(continued)

Name of investigators	Mechanism of material removal	Assumptions
Rajurkar et al. [40]	Combined effect of impact indentation and fracture phenomenon.	i. Work-piece is considered as a semi-infinite solid. ii. The movement of abrasive grit is perpendicular to the free surface during machining. iii. Speed of abrasive is same as that of vibrating tool.
Lee and Chan [41]	Brittle fracture.	i. Pre-existing defect is considered in the workpiece material for the initiation of median or lateral cracks. ii. Median or lateral crack size is related to pseudo pressure between tool and work-piece.
Wiercigroch et al. [42]	Micro-cracking due to impacts of grains.	i. MRR depends on the magnitude of impact force and its frequency. ii. Diamond grain is equally distributed over the working part of the tool with a uniform grain size. iii. Ultrasonic amplitude of vibration, frequency and tool geometry remain constant.
Nath et al. [43]	The hole entrance chipping, lateral gap.	i. The effects of material removal phenomena from the workpiece on the hole integrity such as entrance chipping, wall roughness and subsurface damage are considered. ii. Material removal mechanism takes place in the gap between the outer periphery of tool and the hole wall (called 'lateral gap'). iii. The radial and the lateral cracks are founded due to adjacent abrasives, which are under the tool face. It extends towards radial direction of the hole resulting in entrance chipping. iv. The angle penetration and the rolling actions of the each abrasive particle, which are at the edge of the cutting tool. It contributes the entrance chipping. After that the lateral gap of the sliding and the rolling mechanisms by the larger abrasive particles take part of the material removal. v. Unfavourably micro-cracks in the radial direction are produced and cause subsurface damages, which are ultimately responsible for rough surface. vi. The size of micro-cracks in brittle materials is depended by abrasive grain size. vii. It is realized that such nature of the hole integrity during USM can only be minimized by using smaller abrasive grain size, but cannot fully be reduced.
Ichida et al. [44]	liquid cavitations, cavitations erosion.	i. Non-contact ultrasonic abrasive machining (NUAM) method is introduced. ii. Abrasive particles are excited by ultrasonic energy. iii. Material is removed by the erosion due to liquid cavitations collapse pressure or impact force. iv. Material removal based on colliding or sliding of the abrasive grains, which are accelerated by the impact force due to cavitations collapse on the workpiece surface.

2.5 Basic Element of Ultrasonic Micro Machining Set up

The essential part of ultrasonic micro machining set up consists of ultrasonic power supply, oscillating system, horn, coupler, the mechanism of tool feeding, and the abrasive slurry supply system unit. Figure 2.8 shows the schematic diagram of ultrasonic micro machining set up.

2.5.1 *The Ultrasonic Power Supply*

The power supply for an ultrasonic machine tool is characterized as a sine wave generator that supplies the frequency and power of the generated signal. The output power supply with internal or external power control is variable. Low frequency (50 Hz) electrical power is converted high frequency (greater than 20 kHz) by this. The power supply is capable of automatic frequency control and automatic load compensation at optimum efficiency. It also provides constant output amplitude at required setting during the operating cycle for meeting the different energy requirement. One overload monitor system is incorporated to the power supply for defending the system from any failure conditions. If any overload condition takes place, the overload indicator lamp is indicated on monitor and the ultrasonic frequency is closed for balancing the cycle. After that the monitor is automatically come to the reset position itself for stopping the operation until the fault condition is

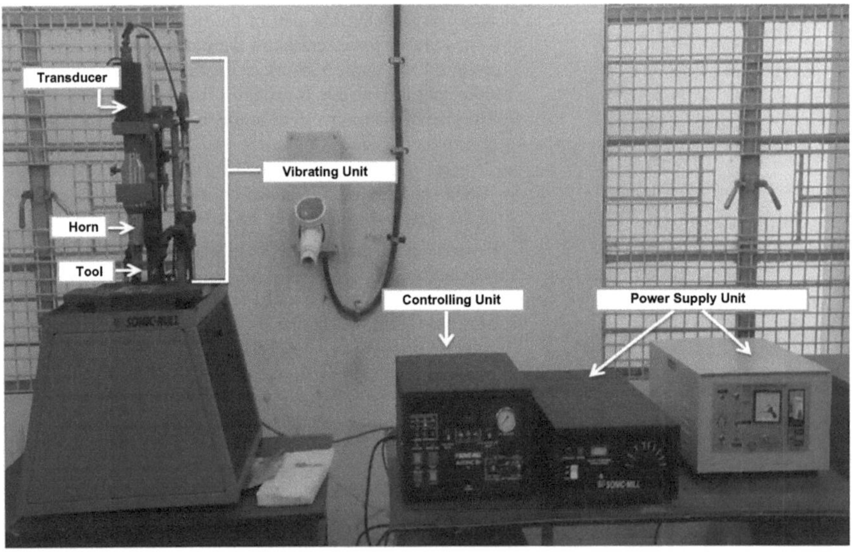

Fig. 2.8 Ultrasonic micro machining setup

eliminated. The monitor is reactivated on every operating cycle for avoiding the damage to the transducer.

2.5.2 *Oscillating System*

Ultrasonic transducer is the main part of this machining setup. It generates the ultrasonic frequency of vibration above 20 kHz and range of the amplitude of 0.8–5 μm for USMM. The electrical supply voltage provides 50 Hz electrical energy, which is also applied to the transducer element. The converter transforms the high frequency electrical vibrations into high frequency mechanical vibrations. Lead zirconate is the heart of the converter. This is actually titanate electrostrictive element. When alternating voltage is applied, it expands and contracts at the frequency of the voltage. This electrostrictive converter is highly efficient for saving energy.

i. **Magnetostrictive Transducer**
   Joule was first discovered the magnetostriction effect in 1874. According to this effect, ferromagnetic metals and alloys is changed in their length in the presence of an applied magnetic field. The ferromagnetic material is only responsible for positive or negative deformation. One coil is wrapped around a stack made of magnetostrive material (iron–nickel alloy) and electric signal of ultrasonic frequency is fed into this coil. This stack is fully laminated for minimizing eddy current and hysteresis losses; moreover, it must be cooled to dissipate the generated heat. The stack expands and contracts at the same frequency with the help of the alternating magnetic field produced by the AC generator. The maximum magnetostriction effect is achieved using the high frequency AC current, which must be superimposed on appropriate DC pre-magnetizing current. That must be exactly adjusted for getting optimum or working point. When this point is accurately adjusted, the maximum magnetostriction effect (maximum oscillating amplitude) is provided and it also prevents the frequency doubling phenomenon.
   If the frequency of the ac signal in the magnetic field is tuned by same value of the natural frequency of the transducer and the whole oscillating system, so it acts at mechanical resonance. This time oscillation amplitude becomes moderately large and the exciting power reaches its smallest value.
ii. **Piezoelectric Transducer**
   Piezoelectric transducer works with piezoelectric effect. When an electric current is passed through the piezoelectric material, then it expands. But when the current is removed the piezoelectric material reaches its original size. This phenomenon actually is called piezoelectric effect. The voltage measuring instrument shows the electric voltage produced by piezoelectric transducer, which can be used to measure forces. The main drawback of magnetostrictive transducers is the high power loss ($\eta = 55\%$). The power loss is converted into

heat. For this reason piezoelectric transducers are used and the efficiency is greater than 90%, even at higher frequencies (f = 25–40 kHz). Piezoelectric transducers use crystals like quartz that undergoes the dimensional changing proportional to the applied voltage.

2.5.3 Horn

The horn is also called mechanical amplifiers or concentrators. The functions of the horn describes as follows:

i. The mechanical energy is transmitted to the tool by the horn.
ii. The horn amplifies the amplitudes of vibration.
iii. The horn concentrates the all power on the small machining zone.

The main acoustic horn head is supplied by the company as an essential part with the machine. The required tool is fixed by silver brazing to the free end of the by threading. The difference of amplitude of vibration between the primary horn and secondary acoustic horn is small enough.

2.5.4 Coupler

The coupler attaches between the converter or transducer and horn. This allows for clamping of the converter, coupler, and horn assembly and provides amplitude choices for various applications.

2.5.5 The Mechanism of Tool Feeding

The mechanism of tool feeding should describe as follows:

i. The tool slowly comes to the workpiece.
ii. Sufficient static pressure is provided during machining.
iii. The pressure is decreased before the end of cut to remove fractures of the lower part of the workpiece.
iv. Overrun a little distance for ensuring the desired hole size at the end surface of the workpiece.
v. The tool is took out upward very quickly after machining.

The static pressure is zero, when the tool and the workpiece are not contact each other. When the tool is first making contact with the workpiece during machining, the spring of the machine spindle is expanded for giving the static pressure. The dial gauge indicates the static force. The tool displacement is indicated by the dial gauge.

2.5.6 *The Abrasive Slurry Supply System Unit*

Several abrasives are available in various sizes for ultrasonic machining. The criteria for selection of an abrasive for a particular application include hardness, size of the abrasive particle, cost of the particle and working life. In order of hardness, aluminium oxide (Al_2O_3), silicon carbide (SiC) and boron carbide (B_4C) are the most commonly used abrasives. The abrasive particles are mixed with water with proper ratio to from the slurry. Average particle size of 3–10 μm is selected and the abrasive slurry is supplied by a recirculating pump into the machining zone. The abrasive used for an application should be harder than the material being machined otherwise the usable life time of the abrasive will be substantially shortened. Boron carbide is selected when machining the hardest work piece materials or when the highest material removal rates are desired. Although the cost is five to ten times greater than the next hardest abrasive, silicon carbide, the usable life of boron carbide is 200 machine operating hours before cutting effectiveness is lost and disposal is necessary. This compares with a usable life time of approximately 60 h for silicon carbide. The combination of high removal rates and extended life time justify the higher cost of boron carbide. The abrasive particles size influences the material removal rate and surface finish. Abrasive for USM are generally available in grit sizes ranging from 240 to 800 while the coarser grit exhibit the highest removal rates, they also result in the roughest surface finish and are therefore, used only for roughing operation, conversely, 800 grit abrasives will result in fine surface finishes but at a drastic reduction in metal removal rate. The most popular general purpose abrasive is used, based on the above considerations is 320 grit of boron carbide. The common abrasive slurry concentration is 50% by weight; however, the concentration can vary from 30 to 60%. The thinner mixtures are used to promote efficient flow when drilling deep holes or when forming complex cavities. The abrasive particles are stored in a reservoir at the USM machine and pumped to the tool work piece interface by recirculation pumps at rates up to 26.5 lit/min. higher power ultrasonic machine require the addition of a light-duty cooling system to remove waste heat from the abrasive slurry.

2.6 Design and Developments of Microtools for USMM

In ultrasonic micro machining (USMM), the shape and dimension of the final products depends on the developed tool. The fabrication of micro tool is a really big challenge. It is extremely not easy to hold the micro tool properly with good accuracy of the job due the application of vibrations at the tool end. One method was introduced to overcome this problem. The tool is prepared on the machining time in this method. At first the macro size tool was fixed to the tool head. Wire electro discharge grinding (WEDG) method was applied to fabricate micro tool and machining was done on the same machine tool. Using wire electro discharge grinding (WEDG) method micro tool of less than 20 μm can be achieved [3]. Another method was introduced in 1999, in which workpiece was vibrated ultrasonically. In this method, the better tool holding is possible and high precision can be achieved [4]. Figure 2.9 exhibits the micro tool developed by WEDG method.

Tool wear also occurs during USMM process. To reduce tool wear in USMM, micro tool should be fabricated from stainless steel, brass and mild steel etc.

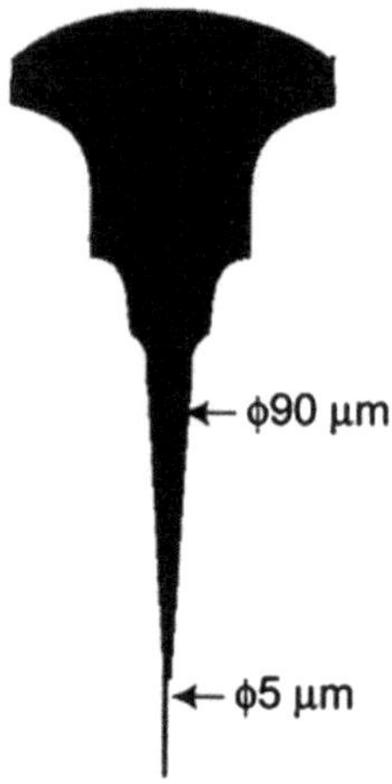

Fig. 2.9 Micro tool developed by WEDG method [4]

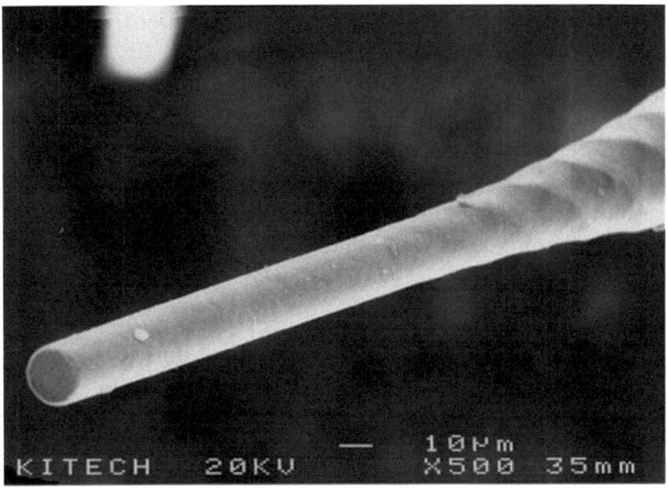

Fig. 2.10 Image of micro tool [47]

More tool wear was found using very hard tool material due to material removal from tool by abrasive particle in the machining zone [45]. The tool material should have superior elastic strength, high wear resistance most favorable values of hardness and toughness, high fatigue strength properties for the application of any hard and brittle material [46]. Tungsten carbide (WC) and sintered diamond (SD) alloy tools were introduced and up to 5 μm micro hole was produced by this tool with good accuracy. With the help of SD tool, micro tool of 21 μm diameter and 150 μm depth on soda glass was generated and the aspect ratio of 7 could be achieved [4, 19]. Figure 2.10 shows micro tool for USMM.

2.7 Parametric Influences of Various Responses of Ultrasonic Micro Machining (USMM)

Very few research works have focussed on the various responses of ultrasonic micro machining such as material removal rate (MRR), tool wear rate, geometrical accuracy and quality of the machined surface etc. The main objective of ultrasonic micro machining (USMM) is to fabricate a micro feature with desired quality of the surface, negligible surface damage, good geometrical accuracy, and satisfactory material removal rate very economically. The possible process parameters in ultrasonic micro machining (USMM) is illustrated by cause effect diagram as shown in Fig. 2.11.

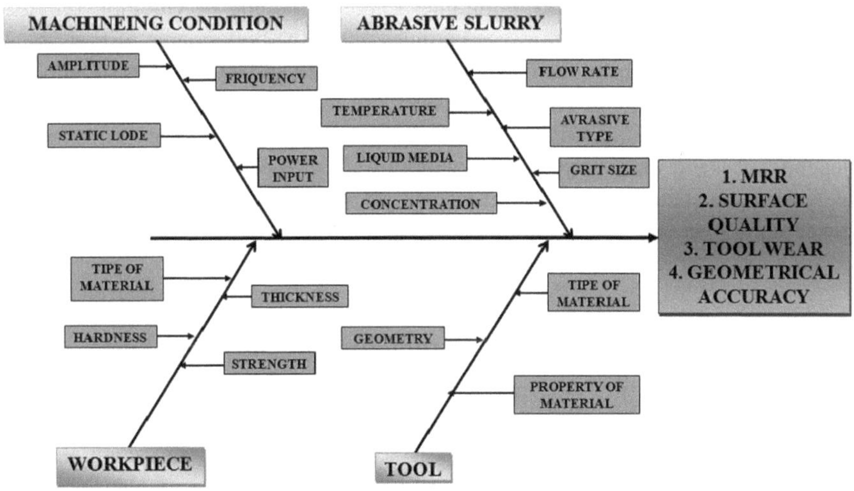

Fig. 2.11 Cause effect diagram for process parameter selection of USMM

2.7.1 *Influences of Process Parameters on Material Removal Rate*

The material removal rate significantly depends on the type and size of abrasive particle in USMM. More material removal rate is found when the average grain diameter of abrasives particle and hardness of the abrasive particles are increased [3]. Extensively more materials are removed from the workpiece using coarser abrasive particles. So that, material removal rate is increased [47]. The material removal rate also depends on the average applied static load. However, MRR is decreased beyond a certain value with increasing of the average static load. The abrasive particles in the machining region are influenced by the static load. The abrasive particles are striking and removing the more material from the workpiece using high static load. As a result, MRR is increased [6, 48]. The vibration amplitude is one of the major parameter. The material removal rate is increased with increasing the amplitude of vibration [3].

2.7.2 *Influences of Process Parameters on Tool Wear*

The abrasive particles hit the workpiece and tool also into the machining zone during machining. So that, the tool tendency to erode the materials due to abrasion of those abrasive particles. Tool wear is an significant response of USMM process. The more tool wear is found when coarser abrasive particles are used [3]. Tool wear also depends on the vibration amplitude. Large amplitude of vibration provides more kinetic energy of the abrasives particles under the bottom area of tool tip. At this time the material from the tool tip also is removed and tool wear is shown [49]. Due to extreme tool wear, the stainless steel tool is not suitable in USMM process. More tool wear is found, when the tool diameter is small [16]. The tool wear rate depends on static load, amplitude of vibration also. The tool wear increases with increasing the static load and amplitude of vibration. For achieving low tool wear lower static load and lower amplitude of vibration are used [3].

2.7.3 *Influences of Process Parameters on Surface Finish*

Surface finish depends on the type and size of abrasive particle, concentration of abrasive slurry. Good surface finished was achieved using fine and small abrasive particles.

Egashira et al. [3] has investigated that most unpleasant surface finish was found using tungsten carbide (WC) tool as compared to titanium carbide (TiC) tool. Less material is removed when fine abrasive particle size of less than 10 μm is used and as a result the smooth surface is generated [6]. The static load and vibration amplitude are the important process parameters. The Ra value of the surface increases with increasing the static load and vibration amplitude due to more material removal from the workpiece [48].

2.8 Development of Micro Feature Using USMM

Ultrasonic micro machining has faced a problem which is proper fixing the micro tools with high precision to the horn in this set up. One method was proposed for USMM set up which was fabricating the micro tools by wire electro discharge grinding (WEDG) procedure. Appling this method, micro holes on a silicon plate and quartz of 20 μm in diameter and 50 μm in depth could be achieved [3, 4]. Figure 2.12 illustrates a micro-hole machining in Al_2O_3 ceramics, with an aspect ratios of 10. Figures 2.13 and 2.14 shows the square, triangular, circular micro holes on silicon and glass. Drilling multiple holes can be made by USMM process. Figure 2.15 shows the array of microholes machined by USMM process.

Figure 2.16 shows micro cavity with 3D view, which was effectively fabricated on silicon with a cylindrical-shaped micro tungsten tool [48]. Figure 2.17 shows that machining of 48 holes with a single SD tool on Silicon, with 22 μm hole diameter, and 20 μm tool diameter and 0.8 μm of amplitude [4].

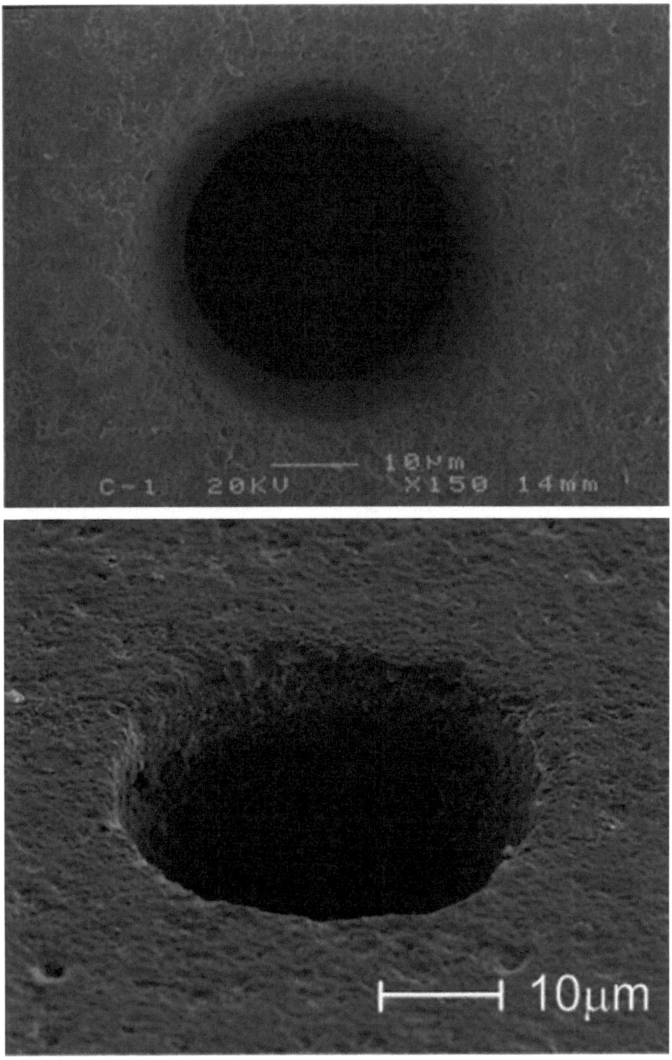

Fig. 2.12 An example of micro-hole machining in Al_2O_3 ceramics, with an aspect ratios of 10 [47]

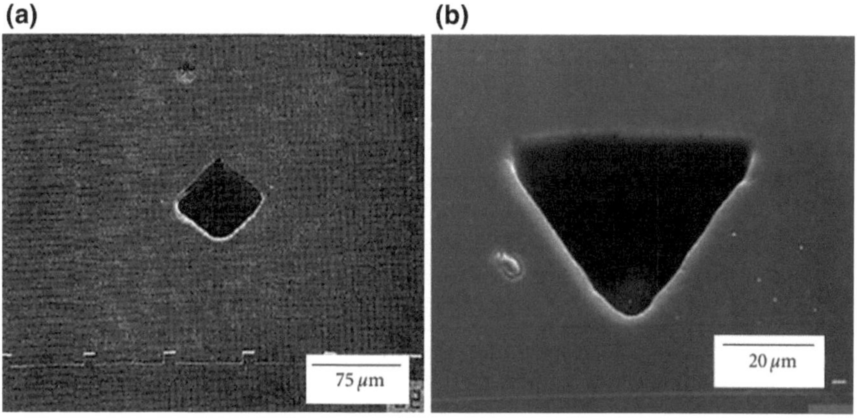

Fig. 2.13 Square and triangular micro hole on silicon [3]

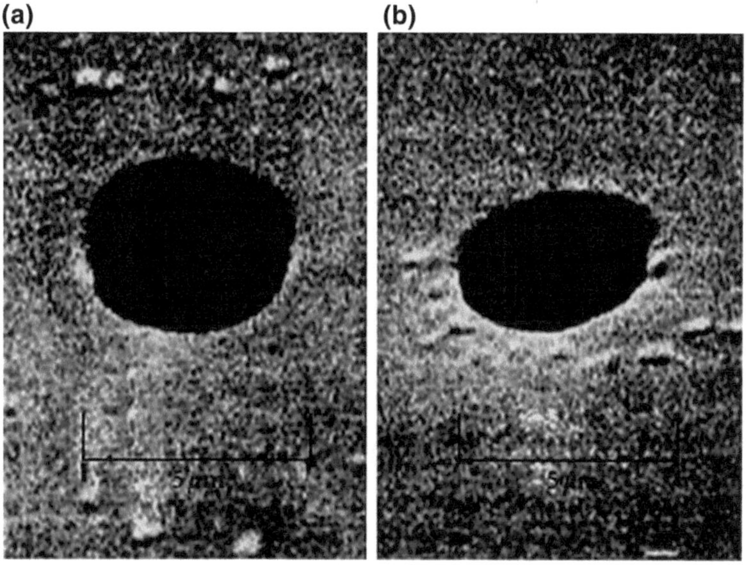

Fig. 2.14 Micro holes on silicon and glass [4]

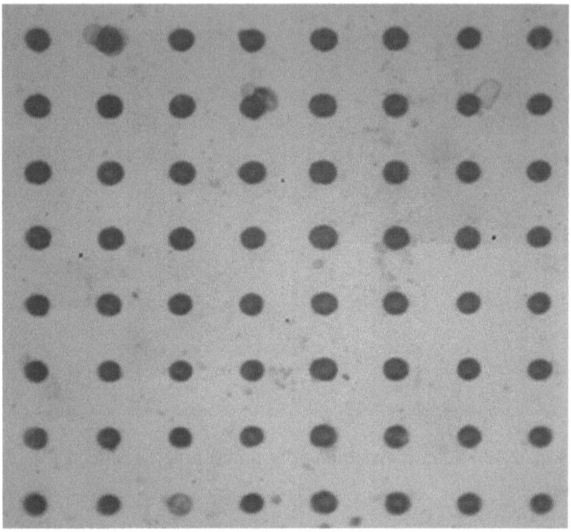

Fig. 2.15 Microholes 20 μm in diameter [50]

Fig. 2.16 A typical 3D cavity [48]

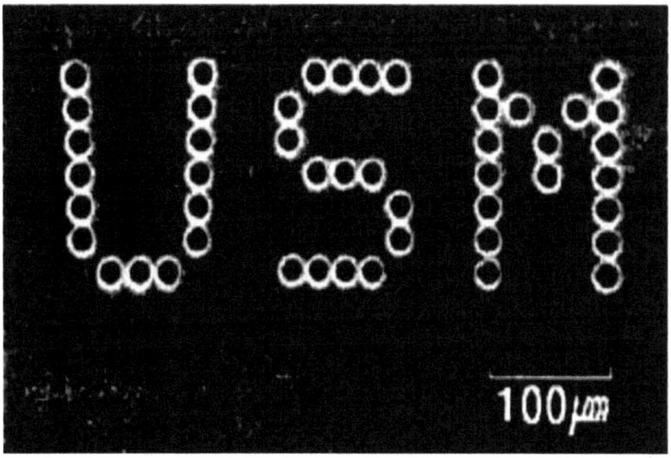

Fig. 2.17 Machining of 48 holes with a single SD tool [4]

2.9 Advantages, Limitations, and Applications of USMM

Advantages of ultrasonic micro machining process are discussed here:

i. The tool and workpiece are not contact together directly and the slurry is used during machining. So that, it is a wet cutting process. No thermal damage is found on the workpiece.
ii. The surfaces produced are free from stress and damages.
iii. It can produce good surface finish and structural integrity.
iv. It can produce burr-less surface.
v. It can machine any hard and brittle material regardless of conductivity.
vi. It does not produce electric, thermal or chemical defects at the surface.
vii. It can drill circular or non-circular holes on very hard and brittle materials, such as glass, silicon, quartz crystal, sapphire, nitride, ferrite and optics.
viii. It produces no thermal stress because of its non-thermal nature.
ix. Micro-holes drilling (hole diameter 5–21 µm) is possible by USMM process.

Limitations of ultrasonic micro machining process:

i. Material removal rate (MRR) is very poor in this machining.
ii. Tool wear is found.
iii. Depth of hole is limited.

Applications of ultrasonic micro machining process:

i. This process is very much applicable on hard and brittle materials like as alloys, glass, fiber material, ceramics, carbides etc.
ii. Engine parts can be machined by this process.
iii. Drilling micro holes in borosilicate glass for the sensors can be possible.

iv. Drilling very fine holes can be possible on shafts and gears which are used in helicopter power transmission.
v. It can be used for producing round, square, irregular shaped holes and surface impressions.
vi. Round through holes with tight-tolerance for processing semiconductor's equipment can be possible.
vii. Micro machining of micro-structured glass wafers is fabricated by USMM process for micro electromechanical systems (MEMS) applications.
viii. High-aspect ratio, 25:1 can be possible in glass and advanced material.
ix. USMM process can be applied in aerospace, tool/mold/machine construction, pump and valve industry, time piece industry/precision mechanics etc.

2.10 Scope of Advanced Research on USMM

Process development for ultrasonic micro machining is one of the most significant issues. In ultrasonic micro machining most difficult task is understanding of the material removal mechanism. Few discussions are available till now but further analysis of the material removal mechanism is necessary for this process. Micro tool fabrication is one another aspect of this process. Preparation of micro tools, handling of micro tools and fixing the micro tools with the tool holder are also the challenging areas of research. The accuracy of micromachining processes depend on these condition of the micro tools. In micromachining, there is reduction in size of tool dimension, abrasive dimension and vibration amplitude etc.

Environmental aspects are one of the essential topics which have hardly been reported. It is one of the important issues of modern manufacturing process.

The fabrication of components for micro fluidic application using USMM process is one of the exploring areas. There is a scope of research in the area of fabrication of micro channels on ceramics, silicon and glass for micro heat exchanger and sensors applications by USMM process.

2.11 Summary

In this book chapter the important issues regarding different aspects of ultrasonic micro machining (USMM) have been discussed. This chapter also focuses on the process development of USMM, working principle and limitations of USMM. Different types of USMM processes, micro tooling for USMM process, material removal mechanism and influences of process parameters have been discussed. Geometrical accuracy and capabilities of the USMM process have been reviewed.

Ultrasonic micromachining (USMM) is well-established process to machine hard and brittle materials (like alumina, zirconia, silicon and glass etc.) The material removal mechanisms in ultrasonic micromachining (USMM) are mechanical abrasion and micro chipping by the abrasive particles against the workpiece surface, cavitations erosion by ultrasonic vibration. In USMM process higher abrasive grain sizes and higher concentrations of abrasive slurry provides higher MRR but poor surface finish. In USMM process good surface finish can be achieved using finer abrasives particles.

References

1. S.G. Amin, M.H.M. Ahmed, H.A. Youssef, "Computer-aided design of acoustic horns for ultrasonic machining using finite-element analysis", Journal of Materials Processing Technology 55 (1995) 254–260.
2. Thoe, T., D. Aspinwall, Wise M. Review on ultrasonic machining. International Journal of Machine Tools and Manufacture, 1998. 38(4): p. 239–255.
3. Egashira K.; Masuzawa T.; Fujino M. and Sun X. Q. "Application of USM to micromachining by on-the-machine tool fabrication," International Journal of Electrical Machining, no. 2, pp. 31–36, 1997.
4. Egashira K. and Masuzawa T. "Microultrasonic machining by the application of workpiece vibration," CIRP Annals—Manufacturing Technology, vol. 48, no. 1, pp. 131–134, 1999.
5. Kremer D. Saleh S. M. Ghabrial S. R. and Moisan A. "The State of the Art of Ultrasonic Machining," CIRP Annals—Manufacturing Technology, vol. 30, no. 1, pp. 107–110, 1981.
6. Barash M.M. and Watanapongse D. (1970), "On the effect of ambient pressure on the rate of material removal in ultrasonic machining", International Journal of Mechanical Sciences, Vol. 12, pp. 775–779.
7. Li Z.C., Jiaoa Y., Deinesa T.W., Pei Z.J.(2005), "Rotary ultrasonic machining of ceramics matrix composites: feasibility study and designed experiments", International Journal of Machine Tools and Manufacture, Vol. 45, No. 12–13, pp. 1402–1411.
8. Jianxin, D. and Taichiu, L (2000), "Surface integrity in electro-discharge machining, ultrasonic machining and diamond saw cutting of ceramic composites", Ceramic International, Vol. 26, No. 8, pp. 825–830.
9. Balamuth L.A. (1966), "Ultrasonic assistance to conventional metal removal", Ultrasonics, Vol. 4, pp. 125–130.
10. Babitsky V.I., Mitrofanov A.V., Silverschmidt V.V. (2004), "Ultrasonically assisted turning of aviation materials: simulations and experimental study", Ultrasonics, Vol. 42, pp. 81–86.
11. X. Q. Sun, T. Masuzawa, and M. Fujino, "Micro ultrasonic machining and self-aligned multilayer machining/assembly technologies for 3D micromachines," in Proceedings of the IEEE Micro Electro Mechanical Systems (MEMS '96), pp. 312–317, 1996.
12. Ghahramani B. and Z. Y. Wang Z. Y., "Precision ultrasonic machining process: a case study of stress analysis of ceramic (Al_2O_3)," International Journal of Machine Tools and Manufacture, vol. 41, no. 8, pp. 1189–1208, 2001.
13. Masuzawa T. and Tönshoff H. K. "Three-dimensional micromachining by machine tools," CIRP Annals —Manufacturing Technology, vol. 46, no. 2, pp. 621–628, 1997.
14. Sun, X.Q., Masuzawa, T, Fujino, M., Micro ultrasonic machining and self-aligned multilayer ma-chining/assembly technologies for 3D micromachines, Proc. IEEE, 1996, pp. 312–317.
15. Yu, Z.Y., Rajurkar, K.P., and Tandon, A. Study of 3D Micro-Ultrasonic Machining. Journal of Manufacturing Science and Engineering, 2004, 126(4), 727–732.
16. Sarwade, A., Sundaram, M. M., & Rajurkar, K. P. (2010). Investigation of micro hole drilling in bovine rib using micro rotary ultrasonic machining. In 16th International Symposium on Electromachining, ISEM 2010. (pp. 411–416). Shanghai Jiaotong University Press.

17. Sarwade, A., Sundaram, M. M., & Rajurkar, K. P. (2010). Micro rotary ultrasonic machining: Effect of machining parameters on material removal rate. In Transactions of the North American Manufacturing Research Institution of SME. (Vol. 38, pp. 113–120).
18. Shimada, S., et al., Brittle-Ductile Transition Phenomena in Microindentation and Micromachining. CIRP Annals—Manufacturing Technology, 1995. 44(1): p. 523–526.
19. Zarepour, H., Yeo. S.H. (2012) Predictive modeling of material removal modes in micro ultrasonic machining. International Journal of Machine Tools and Manufacture, vol. 62, pp. 13–23.
20. Yu, Z., X. Hu, K.P. Rajurkar. Influence of Debris Accumulation on Material Removal and Surface Roughness in Micro Ultrasonic Machining of Silicon. CIRP Annals—Manufacturing Technology, 2006. 55(1): p. 201–204.
21. Lia G.; Yu Z.; Song J.; Li C., Li J. and Wataru Natsu W., "Material Removal Modes of Quartz Crystals by Micro USM", Procedia CIRP 42 (2016) 842–846.
22. H. Zarepour & S. H. Yeo & P. C. Tan, E. Aligiri, "A new approach for force measurement and workpiece clamping in micro-ultrasonic machining", Int J Adv Manuf Technol (2011) 53:517–522.
23. K. P. Rajurkar, and W. M. Wang, "Nontraditional Machining," CRC Handbook of Mechanical Engineering-Ch.13, 1997, pp. 29–34.
24. W. H. Fan, C. L. Chao, W. C. Chou, T. T. Chen, and C. W. Chao, "Study on the surface integrity of micro-ultrasonic machined glass-ceramic material," Key Engineering Materials, Vol.407–408, pp. 731–734, 2009.
25. M. Komaraiah and P. Narasimha Reddy, "A study on the influence of workpiece properties in ultrasonic machining," Int. J. of Machine Tools and Manufacture, Vol.33, No.3, pp. 495–505, 1993.
26. Soundararajan V. and Radhakrishnan, V., "An experimental investigation on the basic mechanisms involved in ultrasonic machining," Int. J. of Machine Tool Design and Research, Vol.26, No.3, pp. 307–321, 1986.
27. Y. Ichida, R. Sato, Y. Morimoto, and K. Kobayashi, "Material removal mechanisms in non-contact ultrasonic abrasive machining," Wear, Vol.258, No.1–4, pp. 107–114, 2005.
28. D. Kremer, S. M. Saleh, S. R. Ghabrial, and A. Moisan, "The state of the art of ultrasonic machining," CIRP Annals—Manufacturing Technology, Vol.30, No.1, pp. 107–110, 1981.
29. A.G. Evans, Fracture mechanics determinations, in: Fracture Mechanics of Ceramics, vol. 1, Plenum, New York, 1974.
30. D. Charkaborty, J. Mukeryi, Indentation induced cracks in hot pressed Si3N4, Indian J. Technol. 20 (1982) 361–365.
31. A.G. Evans, D.B. Marshall, in: D.A. Rigney (Ed.), Fundamentals of Friction and Wear of Materials, ASME, 1981, pp. 439–440.
32. A.G. Evans, T.R. Wilshaw, Quasi-static solid particle damage in brittle solids—I. Observations, analysis and implications, Acta Metallurgica, vol. 24, Pergamon Press, 1976, pp. 939–956, printed in Great Britain.
33. M. Komaraiah, P.N. Reddy, A study on the influence of workpiece properties in ultrasonic machining, Int. J. Mach. Tools Manuf. 33 (3) (1993) 495–505.
34. Shaw M.C. (1956), "Ultrasonic grinding", Annals of CIRP, Vol. 5, pp. 25–53.
35. Miller G.E. (1957), "Special theory of ultrasonic machining", Journal of Applied Physics, Vol. 28, No. 2, pp. 149–156.
36. Rozenberg, L.D.; Kazantsev, V.F.; Makarov, L.O. and Yakhimovich D. F. (1964), "Ultrasonic Cutting", Consultant Bureau, New York, pp. 97–102.
37. Cook N.H. (1966), "Manufacturing analysis", Addison-Wesley, New York, pp. 133–148.
38. Kainth G.S.; Nandy A. and Singh K. (1979), "On the mechanisms of material removal in ultrasonic machining" International Journal of Machine Tool Design and Research, Vol. 19, pp. 33–41.
39. Nair E.V. and Ghosh A. (1985), "A fundamental approach to the study of mechanics of ultrasonic machining", Int. Journal of Prod. Research, Vol. 23, pp. 731–753.

40. Rajurkar K.P.; Wang Z.Y. and Kuppattan A. (1999), "Micro removal of ceramic material (Al_2O_3) in the precision ultrasonic machining", Precision Engineering, Vol. 23, No.2, pp 73–78.
41. Lee T.C. and Chan C.W. (1997), "Mechanism of the ultrasonic machining of ceramic composites", Journal of Materials Processing Technology, Vol. 71, pp. 195–201.
42. Wiercigroch M., Neilson R.D., Player M.A. (1999), "Material removal rate prediction for ultrasonic drilling of hard materials using an impact oscillator approach", Physics Letters, Vol. 259, pp. 91–96.
43. Nath C.; Lim G. C.; and Zheng H.Y. (2012) "Influence of the material removal mechanisms on hole integrity in ultrasonic machining of structural ceramics", Ultrasonics, Vol.52, pp. 605–613.
44. Ichida, Y.; Sato, R.; Y. Morimoto, Y. and Kobayashi, K. (2005) "Material removal mechanisms in non-contact ultrasonic abrasive machining", Wear, Vol. 258, pp. 107–114.
45. Benedict G.F. (1987), Nontraditional Manufacturing Processes, new York, Marcel Decker Inc. pp. 67–83.
46. Kennedy, D.K. and Grieve, R.J. (1975), "Ultrasonic machining—A review", The Production Engineering, Vol. 54, pp. 481.
47. Lee B.J., Kim K. E. (2009), "Characteristics of micro-hole machining of Al_2O_3 ceramics by ultrasonic longitudinal vibration", Journal of ceramics processing research, Vol. 10, No. 4, pp. 482–490.
48. Yu Z. Y.; Rajurkar K.P.; and Tandon A. (2004), "Study of 3D micro-ultrasonic machining," Journal of Manufacturing Science and Engineering, Transactions of the ASME, vol. 126, no. 4, pp. 727–732.
49. Hocheng H.; Kuo K.L. and Lin J.T. (1999), Machineability of zirconia ceramics in ultrasonic drilling. Mater. Manuf. Process. vol. 14, no. 5, pp. 713–724.
50. Egashira, K.; Taniguchi, T.; Tsuchiya, H. and Miyazaki, M. "Micro ultrasonic machining using multi tools," in Proceedings of the 7th International Conference on Progress Machining Technology (ICPMT '04), pp. 297–301, December 2004.

Chapter 3
Electrical Discharge Micro-hole Machining Process of Ti–6Al–4V: Improvement of Accuracy and Performance

Golam Kibria, I. Shivakoti, B.B. Pradhan and B. Bhattacharyya

Abstract Micro-electrical discharge machining (Micro-EDM) has become one of the promising micromachining processes utilizing which high accurate intricate micro-features can be machined efficiently in shop floor. In this chapter, an overview of micro-EDM process and its capabilities is presented for obtaining different desired shape/profile utilizing various machining techniques. The chapter also deals with differences between EDM and micro-EDM, details of system components and micro-EDM process parameters. The significant performance characteristics of micro-EDM process are also discussed. For improving the machining rate as well as for producing high accurate micro-features in different engineering materials, experimental investigation of micro-hole drilling process on Ti–6Al–4V material is carried out implementing several innovative machining strategies such as comparative study of employing kerosene and de-ionized water as dielectrics, the effects of mixing of boron carbide additive in kerosene and de-ionized water, effects of polarity changing between the electrode and effects of rotating the micro-tool. Detailed parametric analysis is carried out to explore the effects of process parameters utilizing these novel machining strategies. Optical and SEM micrographs taken at different parametric combinations have also been analyzed.

G. Kibria (✉)
Mechanical Engineering Department, Aliah University, Kolkata 700156, India
e-mail: prince_me16@rediffmail.com

I. Shivakoti · B.B. Pradhan
Mechanical Engineering Department, Sikkim Manipal Institute of Technology, Sikkim 737132, India
e-mail: ishwar.siwa@gmail.com

B.B. Pradhan
e-mail: bbpradhan1@rediffmail.com

B. Bhattacharyya
Production Engineering Department, Jadavpur University, Kolkata 700032, India
e-mail: bb13@rediffmail.com

G. Kibria et al. (eds.), *Non-traditional Micromachining Processes*,
Materials Forming, Machining and Tribology,
DOI 10.1007/978-3-319-52009-4_3

3.1 Introduction

In the last several years, it has been seen that the demand of micromachining technologies has increased in every shop floor and in wide range of industry segment from biomedical appliances to aerospace products or parts and automotive world. Technocrats and manufacturing engineers are always seeking the challenges to develop new micro manufacturing techniques to manufacture products from hard-to-machine materials to meet the requirement of product miniaturization considering numerous technical challenges. Therefore, to support the growth of product miniaturization and to manufacture micro components effectively and efficiently, it is the responsibility of manufacturing researchers to develop appropriate micro manufacturing processes which have the ability to meet the above-mentioned challenges [1]. Micro-electrical discharge machining (μ-EDM) is an cost-effective and widely utilized micromachining process for machining materials which are difficult-to-machine in conventional methods. The inaccuracies due to vibration of tool or workpiece can be eliminated in this process because the machining method is non-contact and thermo-electric type [2]. Material erosion using controlled spark discharges was first reported by B.R. and N.I. Lazarenko in 1940s in the Union Soviet Socialist Republics (USSR). After that, significant progress and developments of EDM technologies has been carried out by researchers worldwide to control the discharge phenomena, develop reliable machine tools, implementing adaptive control mechanism, employ computer numerical control (CNC), increase accuracy of geometrical features, high precision micro-tool fabrication, tool wear compensations, etc.

3.2 Brief Overview of EDM and Micro-EDM

In Electrical Discharge Machining (EDM) process, controlled spatially and temporally separated pulsed discharges is created between the tool and workpiece in a very narrow inter electrode gap (IEG). The material removal from electrically conductive material is irrespective of the thermal, physical, chemical and mechanical properties [3]. During discharge between the tool tip and workpiece surface, the machining zone must be immerged in dielectric fluid namely kerosene, EDM oil, de-ionized water and paraffin oil etc. The schematic view of basic EDM principle is depicted in Fig. 3.1. The discharge is produced at the location of IEG where there is smallest gap underneath the tool tip surface. The succession phenomena which occur during a typical discharge between the electrodes i.e. at IEG is shown in Fig. 3.2. When a pulsed DC is applied between two electrodes kept at very small electrode gap, strong electrical field is generated. Due to the electromagnetic field, the microscopic contaminants suspended in the dielectric fluid

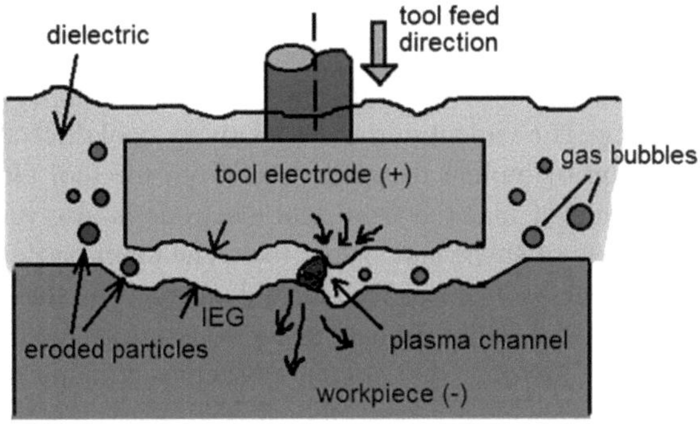

Fig. 3.1 Schematic representation of EDM process

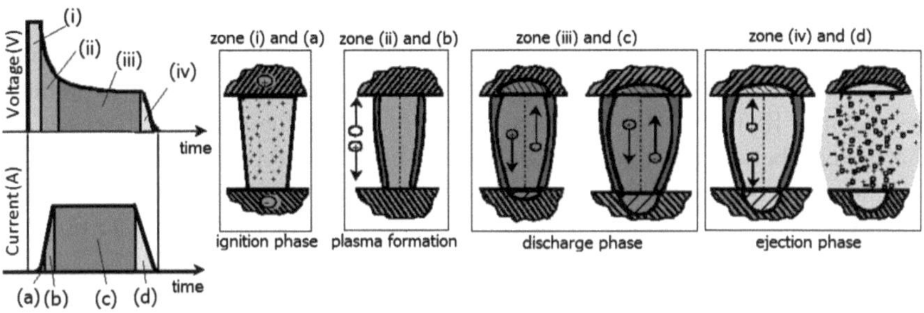

Fig. 3.2 Representation of typical voltage and current trends with sequential phenomena occurring due to single discharge

initiate to drift and they align at the strongest point of the field. These contaminants along with other particles construct the conductive bridge across the IEG, typical spark gap distance varies from 10 to 100 μm. As voltage between the electrode and workpiece increases at the beginning of the pulse, the surface temperature of the workpiece material increases. Some amount of the dielectric fluid and charged particles of the conductive bridge vaporizes and ionizes thereby forming a plasma channel. When the potential difference across the spark gap sharply falls, voltage breakdown occurs. At this time, the plasma channel starts to conduct the applied current whose magnitude rises instantaneously. The abrupt increase in current creates instantaneous increase in localized temperature and pressure in the plasma channel. During discharge between the tool and workpiece, when the local temperature rises beyond the melting point of work material, material melted and vaporized at the location of discharge and small tiny debri particles are ejected out from material's surface creating several number of craters [4]. The gaseous bubbles in the plasma channel expand outward radially from the point of its origin.

At the end of the discharge, the supply of electrical pulse is terminated. This sudden termination of the pulsed power results in collapse of plasma channel and consequently the vapour bubble under the influence of pressure imposed by dielectric fluid from surrounding. The violent in rush of relatively cool dielectric fluid creates an explosive expulsion of molten materials both from the tool electrode. Due to discharge, small amount of material from tool electrode is also removed creating geometrical inaccuracies of shape and size of tool. The debri particles are removed from the machining zone by fresh dielectric fluid supplied by the flushing jet.

The principle of material removal phenomena from workpiece in micro-EDM is same as EDM process. However, the differences such as utilizing micro sized tool, the amount of discharge energy and the X-Y-Z axes resolution make the micro-EDM process more precise and reliable and capable of micro features generation [1, 5]. In micro-EDM, very high frequency pulses (>200 Hz), small discharge energies (10^{-6}–10^{-7} J) and applied potential difference (40–100 V) between the electrodes are applied to achieve high accuracy features and surface finish (roughness as much as 0.1 μm) [6]. Utilizing improved pulse generator and precise servo feed system, the micro-tool electrode can be moved at micron rate to maintain the required IEG and also to retract the micro tool if the servo feed senses any short-circuit between the electrodes. In micro-EDM, the type of dielectric flushing and the value of flushing pressure are important parameters as during machining, micro-tool may vibrates at high dielectric pressure and deteriorate the accuracy of micro features to be produced.

Micro-EDM is not only the scaled down version of EDM process, but also there are significant differences of these two processes in terms of tool electrode type, shape and dimensions, range of discharge energy, resolution of moving axis, inter electrode gap control, type of dielectric flushing for debris removal, etc. The radius of plasma channel in micro-EDM is comparable with tool dimension as micro-EDM uses micro-tool electrode. However, in EDM, size of plasma channel is much less compared to tool size [7]. As in micro-EDM, material removal per discharge (unit removal, UR) should be less; therefore, the applied discharge energy is limited. On the other hand, in EDM, unit removal is much higher. Further, to avoid the micro-tool to be rupture or burnt due to excessive discharge energy, the maximum applied energy in micro-EDM is limited. As in micro-EDM, small sized (in micron dimensions) tool electrode is used and it has low electrode stiffness, thus, during micro machining operation in micro-EDM, micro-tool may vibrate due to flushing pressure and geometrical accuracy of micro features to be produced is deteriorated. Therefore, the range of dielectric flushing pressure is kept in lower settings. This problem is absent in EDM process. Further, in micro-EDM, the amount of thermal load on micro-sized tool electrode is high and thus, the amount of tool wear also high compared to conventional EDM. Thus, various tool wear compensation strategies are considered for generating precise and accurate micro features. Due to very low IEG in micro-EDM, short circuiting occurs frequently, which renders machined surface uneven. Moreover, when the debris are ejecting out from work material, secondary sparking occurs and geometry of micro-tool get degenerated. These drawbacks are very less in conventional EDM.

3.3 Micro-EDM System Details

The micro EDM system typically consist of several sub-components such as servo system, control unit, positioning system, dielectric circulating unit and the working chamber. Figure 3.3 shows the schematic diagram of micro EDM system with sub-systems. The servo system guides the micro electrode during the machining and the feed movement of micro-electrode is controlled by the servo control unit. The workpiece is firmly held in the working table and the positioning system control the movement of work chamber in the X and Y axis. The constant feed of the micro-tool is given by servo feed mechanism in Z axis direction. The dielectric circulating system is another important component in micro-EDM in which continuous flow of fresh dielectric fluid is supplied to the machining chamber. The servo control unit is one of the essential components of micro EDM as it controls the movement of micro-tool and also it maintains the inter-electrode gap between the micro-tool and the workpiece in micron range. In order to generate dimensional accurate features, it is very important to keep constant inter electrode gap between the electrodes [8]. The dielectric system serves an important role in micro-EDM as the dielectric possesses a crucial role during machining. The dielectric tank contains a dielectric which is pumped to the machining chamber. The pressure regulator controls the flushing pressure of the dielectric jet onto the machining chambcr. The dielectric is continuously circulated for efficient machining and the used dielectric is passed through the micro filters which captured the debris particle generated during machining. The dielectric jet removes the debris from work material and also it serves as insulation between the micro tool and the work material. Further, the

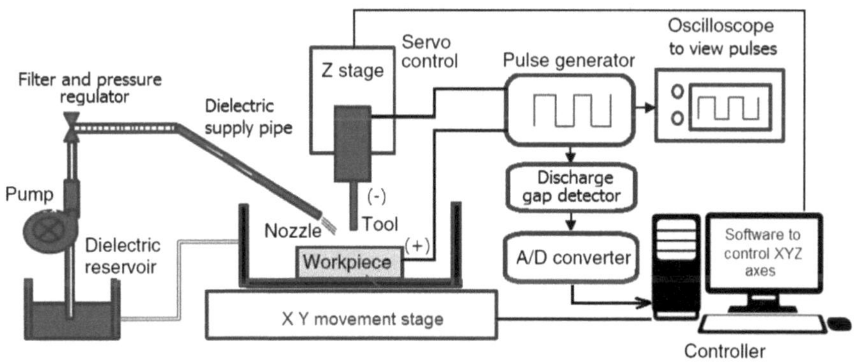

Fig. 3.3 Schematic diagram of a typical micro-EDM system showing different sub-systems

dielectric is required for the ionization and act as a medium to cool the micro-tool and the workpiece.

3.4 Pulse Generators for Micro-EDM

The series of pulse has been generated for micro EDM with the help of pulse generator. The various type of pulse generator has been used for producing series of pulse for micro-EDM process. In RC type capacitor, the charging time is more and due to this, it cannot produce high frequency pulses and it also produces thermal damage of the wok material. Sometimes, discharge current flows through the previous plasma channel without recharging the series of capacitors [9]. Since the capacitor is absent in the transistor type pulse generator as a result high discharge frequency pulses is produced which enhances the material removal rate. Moreover, the transistor type pulse generator is controlled easily as compared to RC type generator. The transistor type pulse generator cannot be used when nanoseconds pulses are required for machining as it is not capable of producing nanoseconds pulses. To reduce the delay time of transistor-type pulse generator, the transistor-type isopulse generator was developed and successfully employed for micro-EDM operation [10]. For rough and semi finish machining, a field effect transistor is used to cut off the discharge current. Instead of observing the gap voltage, the pulse current is observed for detecting the discharge. As the current sensor provide an output less than 5 V and also it acts as input to pulse control circuit, it eliminates the voltage attenuation circuit. In this way, the delay time is shortened to significant amount and ultimately, about 80 ns pulse duration was achieved. As in RC-type pulse generator, stray capacitance determines the minimum discharge energy per pulse, therefore, it act as the limitation for generating micro-features on workpiece [11]. Due to difficulty in eliminating the stray capacitance, crater diameter less than 2 μm cannot be achieved [12]. To avoid these problems, a capacity coupling based pulse generator was developed. In this generator, the effect of the stray capacitance can be eliminated as electric feeding is done without touching the micro-tool electrode. Thus, discharge crater dimension of nanometer domain is realized.

3.5 Control Parameters of Micro-EDM

The micro EDM process parameters have significant effect on various process performances. The proper selection of control parameters results in better machining performance. Figure 3.4 shows the fish-bone diagram showing all process parameters related to machining parameters, work material, tool-electrode, polarity, flushing pressure, etc. The micro EDM control parameters are categorized as follows: (a) electrical (b) non-electrical and (c) gap and motion control parameters. Brief discussions on these process parameters are given hereunder.

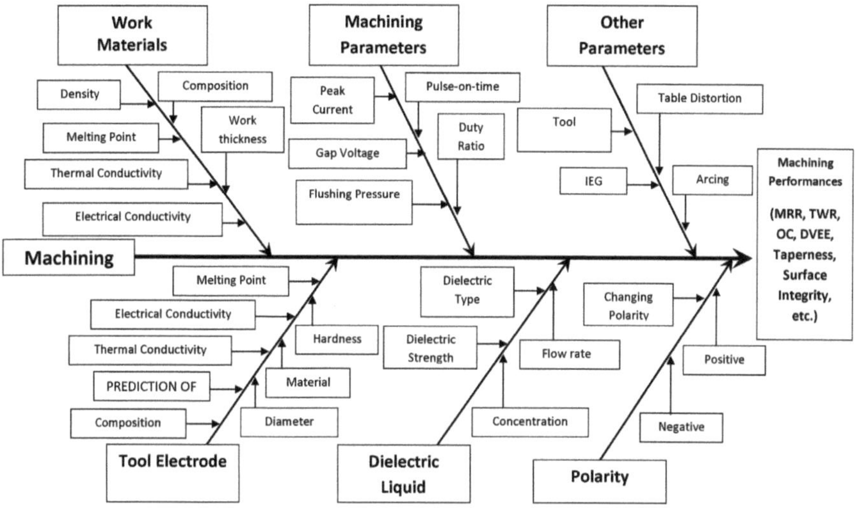

Fig. 3.4 Representation of process parameters in micro-EDM using fish-bone structure

3.5.1 Electrical Process Parameters

(a) **Discharge energy**: In micro-EDM, the most significant process parameter is discharge energy. This parameter is a collection of other operating parameters which are related to the energy of the discharge created between the electrodes at inter electrode gap. For different type of pulse generator, the calculation for discharge energy is different. The material removal rate is directly related to discharge energy during micro-EDM operation. On the contrary, tool wear ratio also increases, which deteriorate the surface finish and accuracy of micro features generated.

(b) **Gap voltage**: Gap voltage is the voltage in the gap between the two electrodes. The total energy of the spark is determined by applied voltage. Depending upon the setting of voltage, the IEG is set by servo control. Larger value of IEG improves the flushing of debris from machining zone and makes the next discharge stable and ultimately improves material removal rate. However, the surface finish deteriorated due to large size of crater dimension at high voltage condition. The voltage of IEG at which discharge occurs between the micro tool and workpiece is known as discharge voltage. The discharge voltage mainly depends upon the breakdown strength of the dielectric and IEG.

(c) **Peak current**: The average current is the average of amperage in spark gap measured over a complete cycle. This is read on the ammeter during the process. The theoretical average current can be measured by multiplying the duty cycle and the peak current i.e. maximum current available for each pulse from the power supply. The amount of energy/power which is used for discharge is mainly determined by peak current. Higher value to peak current signifies better machining efficiency in terms of material removal rate. At the same time, the surface finish is deteriorated and tool wear ratio increases.

(d) **Duty factor**: Duty factor is the ratio between the pulse duration to the total cycle time. Mathematically, it is evaluated by using Eq. (3.1)

$$Duty\,factor\ (DF)\% = \frac{Pulse\ duration}{Cycle\ time} \times 100\% \tag{3.1}$$

If the duty factor is high, the flushing time is very less and this might lead to the short circuit condition and a small duty factor indicates a high pulse off time and low machining rate.

(e) **Pulse duration**: Pulse duration or pulse-on-time is the time interval in which the applied current is flowing through the IEG of two electrodes. In this time period, breakdown of dielectric occurred and removal of material from workpiece surface takes place. Large value of pulse duration means higher material removal rate. Broader and deeper craters are achieved at longer the pulse duration setting and consequently, rough machined surface is attained. On the contrary, smaller craters which are obtained at low value of pulse duration provides smoother surface finish.

(f) **Polarity**: Polarity refers to the electrical conditions determining the direction of the current flow relative to the electrode. The polarity condition of electrodes is of two type, (i) straight polarity and (ii) reverse polarity. Straight polarity is that condition when the micro tool is connected to cathode (−), whereas, reverse polarity is that condition in which tool electrode is connected to anode (+) and workpiece to cathode (−). For achieving high material removal rate from workpiece, tool electrode is used as cathode and workpiece as anode. Depending on the application, some electrode/work material combinations provide better results when the polarity is changed. Generally for graphite electrode, a positive polarity gives better wear condition and negative polarity gives better machining speed.

(g) **Pulse frequency**: It is the measure of number of cycle per second. Larger value of pulse frequency decreases the pulse duration which results in minimum thermal damage in the work piece during machining. However, at high value of pulse frequency, the surface finish of the machined surface is improved.

3.5.2 Nonelectrical Process Parameters

(a) **Micro-tool electrode**: Tool electrode is an important part for achieving effective and efficient machining condition. The thermal properties of tool electrode material play a significant role during micro-EDM as it is a thermal process. Materials of higher melting, boiling points and heat conductivity are used to fabricate the micro-tool for micro-EDM process [13]. There are several criteria to choose proper tool electrode materials such as (i) machinability, (ii) electrical and thermal conductivity, (iii) density, (iv) Hardness and toughness, (v) cost and availability, (vi) material removal rate and wear ratio, etc.

(b) **Work materials**: The micro EDM can only machine electrically conductive materials. Moreover, for efficient machining, the thermal properties of the material such as thermal conductivity, specific heat, melting point are the important aspects need to be considered while selecting the work piece material [14].

(c) **Dielectric fluids**: The dielectric fluid has significant role during micro-EDM possess as without it, it is no longer possible to generate efficient discharge between the micro-tool tip and workpiece surface. The quality of surface finish and geometrical accuracy of machined parts depend on several properties of dielectric such as viscosity, dielectric strength, cooling capability, chemical compositions, etc. For safe machining operation and stable sparking condition, the dielectric strength and flash point temperature of the dielectric fluid should be higher. Furthermore, low value of viscosity and specific gravity are another two desirable properties of dielectric fluid. These properties significantly affect the machining efficiency and consequently improves material removal rate, lowers tool wear rate and enhances the surface finish of machined features.

3.5.3 Gap Control and Motion Parameters

Electrode rotation: The rotation of micro-tool about its axis during the micro-EDM enhances the machining performance as the rotation of micro-tool enhances the flushing action of the debris formed during micro-EDM at inter electrode gap. Due to rotation of micro-tool, a tangential force is produced at small IEG and this leads to effective and efficient discharge by smooth removal of debris through small gap of micro-tool surface and micro-hole wall [15]. Moreover, electrode wear ratio also reduces due to micro-tool rotation.

Tool geometry and shape: The micro-feature generated in the workpiece during micro-EDM is the replication of the micro-tool geometry. Moreover, the geometry of micro-tool has significant effect on tool wear ratio. Depending upon the requirement, the shape of micro-electrode can be circular, rectangular, cylindrical etc. Vibration assisted micro-EDM has significant effect on machining rate and taper during micro-hole drilling [16].

Servo feed: For properly maintaining the discharge gap width and to avoid arcing and short-circuiting between the micro-tool and the workpiece, the servo feed control system play vital role during micro-EDM process. As soon as the value of average gap voltage approach more than preset threshold voltage of pulse generator, the feed rate of servo increases and compensating the discharge gap between the electrodes and vice versa [3].

Tool and workpiece vibration: The performance of the micro-EDM has been improved by incorporating the vibration of micro-tool or workpiece at certain frequency and amplitude. During vibration of micro-tool or workpiece, the forward and backward motion of tool or workpiece changes the discharge gap and consequently, dielectric fluid pressure in the IEG also changes constantly. When the micro-tool is advancing towards the workpiece, the dielectric fluid is forced out from the machining zone. Then after, when the micro-tool move away from machine zone, fresh dielectric is taken by discharge gap and thus, overall flushing efficiency increases.

Flushing techniques: Dielectric flushing has important role for removing the debris from machining zone and consequently, it enables stable discharge condition by supplying fresh dielectric fluid in the gap. In general, there are mainly two types of flushing, pressure flushing and suction flushing. Depending upon the type of flushing, the amount of flushing pressure is provided. In micro-EDM operation, for effective generation of high aspect ratio micro-features such as micro-holes, jet flushing is more effectively used. In other cases, side flushing is commonly used. If the jet flushing is provided from one direction, there may be chances to accumulate the debris in the downstream, which creates irregular gap width and as a result, accuracy of micro-feature is deteriorated [17]. To avoid this, jet flushing from both side and sweeping type flushing also sometime recommended. In Fig. 3.5, different types of flushing for micro-EDM are illustrated.

Flushing pressure: For quick removal of the debris from the IEG during micro-EDM, it is more important to flush out debris particles from very narrow discharge gap. Higher value of flushing pressure is preferable for effective debris removal, stable machining and high aspect ratio micro-feature generation. However, as stiffness of micro-tool is low, high flushing pressure may deteriorate dimensional accuracy due to micro-tool vibration or deflection.

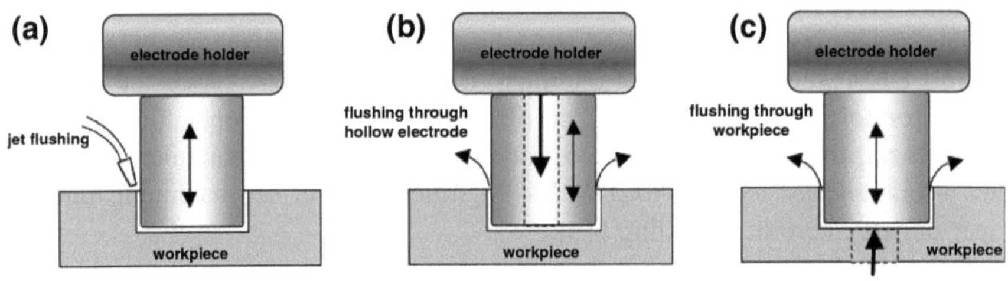

Fig. 3.5 Schematic of **a** jet, **b** flushing through tube electrode and **c** flushing through workpiece in micro-EDM process

3.6 Micro-EDM Performance Measurements

Material removal rate: The amount of material removed from the workpiece per unit time is known as material removal rate (MRR). It is evaluated in terms of volume of material removed from the workpiece or can be evaluated by differentiating the weight of the workpiece taken before and after machining. It is calculated using Eq. (3.2)

$$\text{Material removal rate }(MRR) = \frac{\text{Weight of workpiece before machining} - \text{Weight of workpiece after machining}}{\text{Machining time}} \tag{3.2}$$

The higher material removal depicts the increases the productivity and hence, the material removal is always consider to be the higher the better type. Higher MRR can be achieved at high discharge voltage, peak current, pulse duration and duty cycle. However, other desirable process performances such as tool wear ratio, surface finish and dimensional accuracy is also important aspects and taken into account in micro-EDM.

Tool wear rate: The amount of material removed from the micro-tool during machining is known as tool wear rate (TWR). The high electrode wear is not desirable in micro-EDM as it changes the geometry of the tool which reduces quality of the machined surface and inaccurate geometric features are achieved onto the machined surface. The electrode wear is calculated by differentiating the weight of the micro-tool before and after machining and is calculated by Eq. (3.3).

$$\text{Tool wear rate }(TWR) = \frac{\text{Weight of microtool before machining} - \text{Weight of microtool after machining}}{\text{Machining time}} \tag{3.3}$$

Surface roughness: Surface roughness of the machined features mainly depends on the crater size (diameter and depth) that is formed by each discharge. In addition, if the dielectric circulation in the discharge gap is not efficient, then some molten material from tool and workpiece resolidifies on the micro-feature surface and makes the surface rough. Thus, effective flushing technique has significant effect on surface finish of the features. Crater dimensions also largely depend on pulse energy of discharge and other process parameters such as peak current, pulse frequency and pulse duration [18]. Required amount of flushing pressure can decrease the roughness of machined surface. Moreover, the properties of material of the micro-tool and the workpiece have considerable effects on surface finish.

Overcut: Overcut is the difference in diameter of the micro-hole at the entrance and the diameter of the tip of the cylindrical shaped micro electrode. It produces inaccuracy in the dimension of the machined features. For micro-hole drilling process in micro-EDM, overcut is calculated using Eq. (3.4). The overcut occurs mainly due to secondary sparking during material removal from sidewall of the micro-feature surface while the debris try to ejecting out by flushing pressure from the machining zone.

$$Overcut\ (OC) = \frac{Diameter\ of\ entry\ hole - diameter\ of\ microtool\ tip}{2} \tag{3.4}$$

Diameteral variance at entry and exit hole: This performance criterion is related to through micro-hole drilling in micro-EDM. The diameteral variance of entry and exit hole is measured by differentiating the micro-hole diameter at entry and at exit side of from end to end hole on workpiece. During micro-drilling operation of high aspect ratio hole, if the secondary sparking occurs for a long time, then large amount of diameteral difference is realized. Dimeteral variance of entry and exit hole is affected by a number of process parameters such as peak current, pulse duration, duty cycle and flushing pressure.

Circularity: Circularity is the measure of roundness of micro-hole. The circularity of micro-hole is calculated by Eq. (3.5). Due to high flushing pressure or non-uniform discharge condition, the degree of roundness of micro-hole deteriorated and further reduces the circularity of the micro-hole.

$$Circularity = 4\pi \frac{Area\ of\ microhole}{[Perimeter\ of\ microhole]^2} \tag{3.5}$$

Machining time: The time during which the machining is performed in order to produce micro-features on the workpiece. Due to inefficient discharge and various unwanted phenomena, machining time increases for producing particular micro-features on workpiece surface.

3.7 Varieties of Micro-EDM Processes

Depending upon the tool-work configuration and relative motion between the micro-tool and workpiece to be machined, there are several varieties of micro-EDM process. These machining varieties have been developed based upon the requirements of product intricacy and features.

3.7.1 Micro-EDM Drilling

The process in which deep micro-sized hole is generated by EDM process is termed as micro-EDM drilling. In conventional EDM process, both tube and solid electrodes can be used as tool material. However, for micro-EDM, as the geometrical

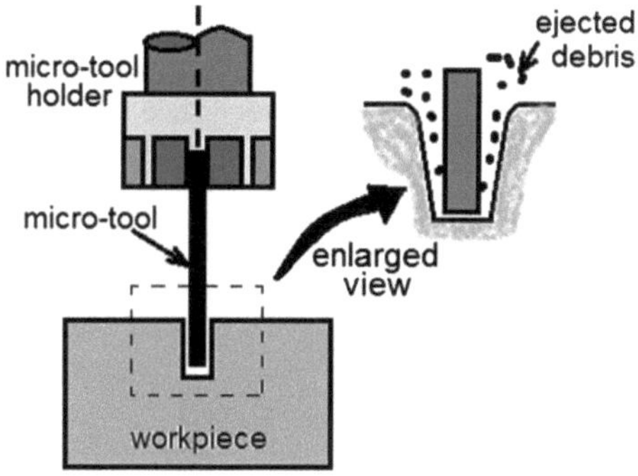

Fig. 3.6 Schematic representation of micro-EDM drilling process with enlarged view of machining zone

dimensions of tool electrode is in micron range, therefore, only solid micro-rods can be used. For efficient removal of eroded particles from the machining zone, it is recommended to employ jet flushing in micro-EDM with low value of flushing pressure to avoid vibration of micro-tool electrode. In Fig. 3.6, the schematic of micro-EDM drilling process is depicted. When the debris particles try to eject out from the IEG, due to secondary sparking between micro-tool and micro-hole walls, material removal is more at entry side of hole and this result in taper in micro-hole geometry. For high precision micro-hole drilling in EDM, sometimes, micro-tool electrode is fabricated in the machine itself by wire-electrical discharge grinding (wire-EDG) process to avoid tool deflection or breakage [19]. Different types of micro-features (irregular, curved, inclined, tapered, etc) can be produced using different alignment of micro-tool as well as tool feed mechanisms. Cooling holes in turbine blades, nozzles for fuel injection system, parts for manufacturing of micro-turbines and surgical instruments are the typical examples of micro-EDM drilling.

3.7.2 Micro Wire-EDM

To cut conductive hard-to-machine materials like Ti-alloy, Ni-alloy, HSS, etc, micro-wire EDM is employed. Micro-wire which is continuously travelling through the upper and lower wire guide, is used here as tool electrode. According to the profile of intricate cut, the machine axes moves in the direction of programmed path and material is removed from workpiece by series of electrical discharges between the micro-wire and workpiece as shown in Fig. 3.7. Since there are additional axes U and V, therefore, taper cutting is possible efficiently by keeping the wire at some angle with vertical by positioning the upper wire guide in respect of lower one [20].

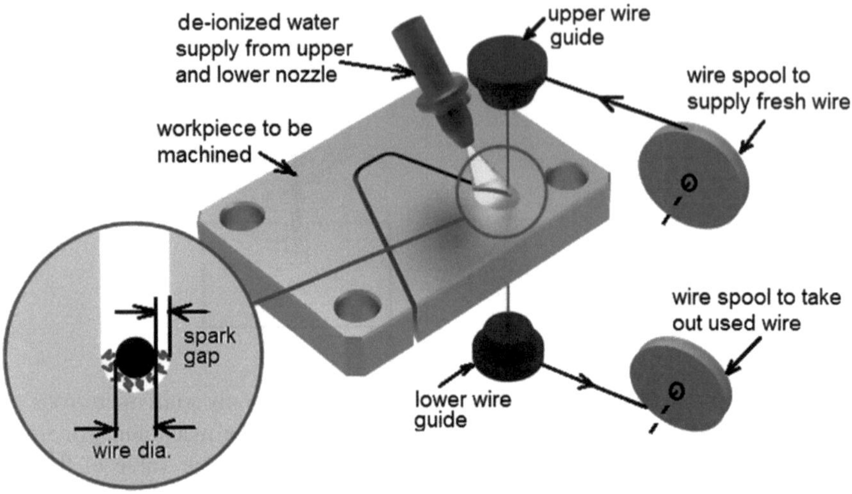

Fig. 3.7 Schematic view of configuration of micro-wire EDM process

Fig. 3.8 Schematic of micro-EDM milling operation

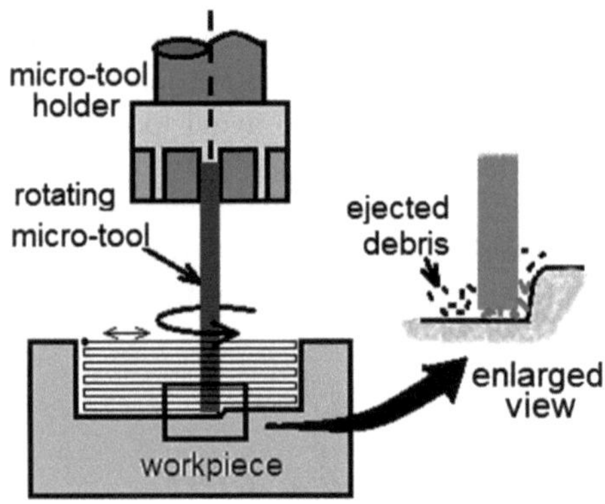

3.7.3 *Micro-EDM Milling*

To produce high precision complex geometries, micro-EDM milling is used where the elimination of complex die-sinking tool required for obtaining such profile is possible. The process utilizes cylindrical solid micro-tool electrode to fabricate desired three-dimensional complex structure by scanning the programmed path layer-by-layer using CNC software as shown by schematic representation in Fig. 3.8. Here, the rotation of micro electrode in its axis is a requirement while scanning X-Y planes. Tool length compensation is very much important factor in this process because during layer-by-layer scanning, due to wear of micro-tool, tool length gradually reduced.

3.7.4 Dry and Near-Dry Micro-EDM

In conventional EDM and general micro-EDM processes, mainly hydrocarbon dielectric fluid is used. However, during machining with hydrocarbon oil, harmful vapors are produced which make the machine environment toxic. Thus, to reduce the environmental pollution and also reduce the cost of waste management, dry and near-dry micro-EDM process has been developed. Pure oxygen gas or air is used as dielectric in dry micro-EDM process [21]. However, in near-dry micro-EDM process, mixture of gas and liquid is used as dielectric [22]. The schematic view of dry and near-dry micro-EDM process is depicted in Fig. 3.9. Tube type electrode is used as tool electrode in dry micro-EDM in which air or gas in supplied to the IEG to act as dielectric which ultimately act as cooling medium of micro-tool and workpiece and remove the molten and vaporised material from narrow IEG.

3.7.5 Planetary or Orbital Micro-EDM

In micro-EDM drilling, especially for producing high aspect ratio micro-holes, a common problem is the debris accumulation at machining zone. Due to this, secondary sparking and short-circuiting occur. To avoid this problem, a relative motion (orbital or planetary type) is provided between the micro-tool electrode and workpiece to be machined. The schematic representation of orbital or planetary micro-EDM process is shown in Fig. 3.10. The planetary motion of micro-electrode provides space for dielectric fluid circulation and thus, it helps to reduce the concentration of debris in IEG. It further helps to improve material removal rate, increase the degree of accuracy of micro-structure and reduce tool wear ratio. One of the important benefits of planetary micro-EDM is that in this process, micro-holes of different diameters can be produced by just changing orbital radius and using a single micro-electrode [23].

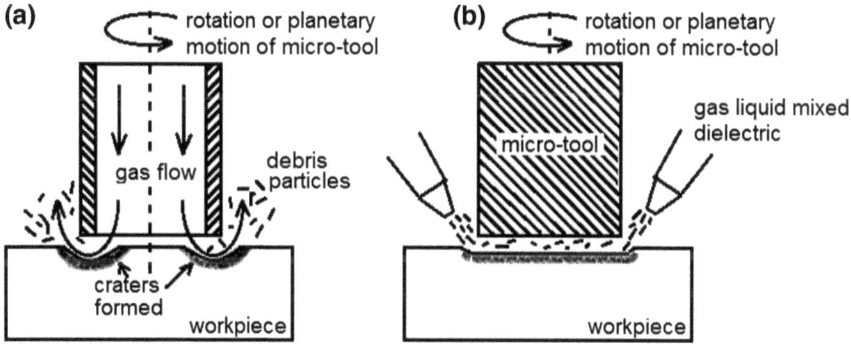

Fig. 3.9 Schematic of **a** dry micro-EDM and **b** near-dry micro-EDM

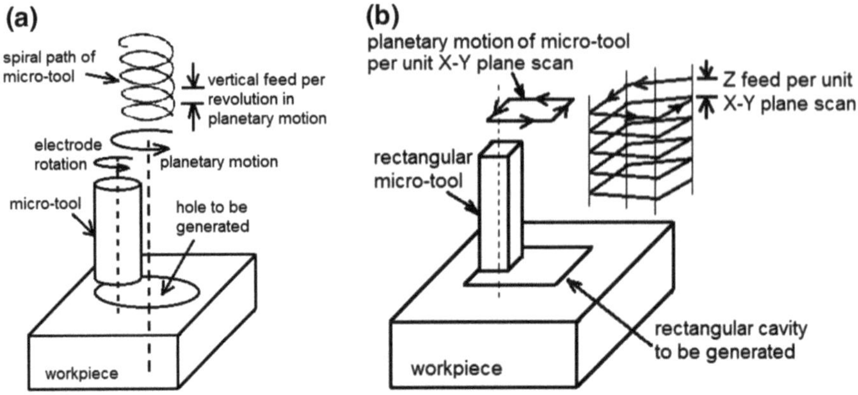

Fig. 3.10 Schematic representation of planetary micro-EDM for generation of **a** circular hole and **b** non-circular hole

3.7.6 Reverse Micro-EDM

For fabrication of high aspect ratio micro-tool electrode, reverse micro-EDM is employed. The process comprises of two steps. In first step, micro-EDM drilling operation is carried out on low thickness material using normal polarity setting. In next step, the material from which high aspect ratio micro-tool electrode is to be fabricated is attached with z axis of machine and already fabricated micro-hole is clamped with machining chamber. The polarity of electrodes is reversed and material erosion takes place from workpiece material as shown in Fig. 3.11. For generating array micro-electrodes on a single material, reverse micro-EDM is an excellent method. The number of micro-electrodes and gap between the electrodes depends upon the number of micro-holes on sacrificial electrode plate.

3.7.7 Micro Electro-discharge Grinding

For fabricating micro-electrode on the machine itself from an electrode material which is larger than required dimensions, micro electro-discharge grinding (micro-EDG) process is employed. The process is classified as (i) block micro-EDG, (ii) moving block micro-EDG, and (iii) micro-wire EDG. In the first type, a sacrificial block (rectangular) with perfectly vertical is used and it should have high degree of wear resistance. If the degree of alignment of the sacrificial block reduces, the geometrical accuracy of fabricated micro-tool by micro-EDG is reduced. Due to wear from sacrificial block, the fabricated micro-electrode diameter is not achieved as straight and wear compensation prediction is difficult to calculate. To overcome this, on-machine dimension measuring instrument and high resolution camera are installed for instantaneous measurement of micro-tool geometries.

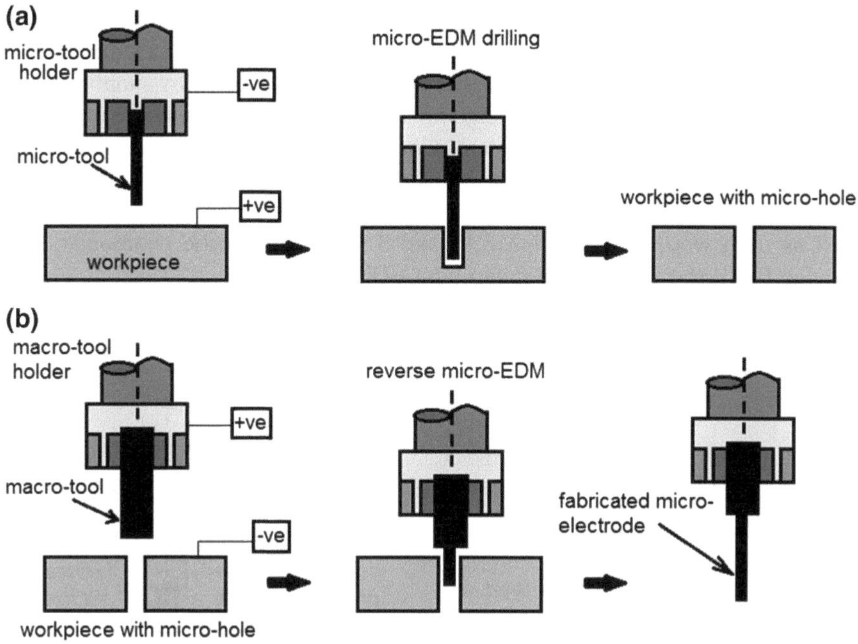

Fig. 3.11 Sequential representation of reverse micro-EDM, **a** micro-hole generation in normal micro-EDM and then **b** fabrication of high aspect ratio micro-electrode using micro-hole

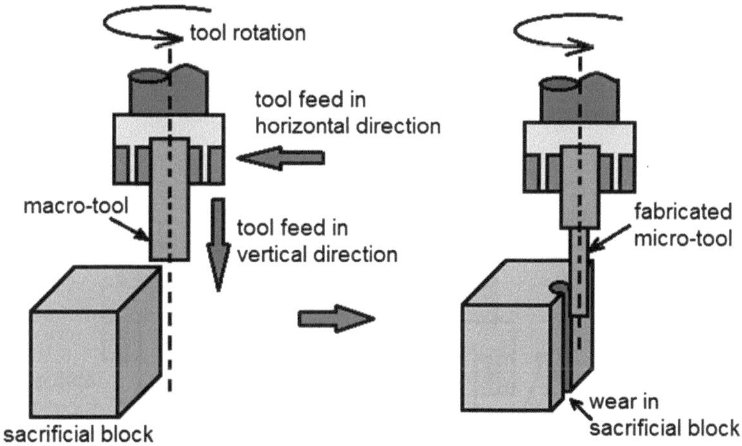

Fig. 3.12 Methodology of stationary block micro-EDG process

The schematic view of block micro-EDG process is shown in Fig. 3.12. During block micro-EDG process, cylindrical material is considered as workpiece and the rectangular block is considered as electrode. During sparking between block surface and rotating workpiece, it is very important factor to apply dielectric jet properly to avoid deflection of material as well as uniformity of micro-tool geometry.

To avoid tapering of micro-electrode in block micro-EDG process, a new and novel micro-EDG process is developed. In moving block micro-EDG, dimensional accurate and high aspect ratio micro-electrode can be fabricated by simply moving the sacrificial block in a direction perpendicular to cylindrical workpiece in addition to rotary and feed movement. In micro-wire EDG process, the workpiece is rotated at vertical axis and a travelling wire is fed towards the rotating workpiece using X and Y axes controller. Due to discharge, material is eroded from travelling wire as well as from workpiece. In Figs. 3.13 and 3.14, the schematic representation of moving block micro-EDG and micro-wire EDG are viewed.

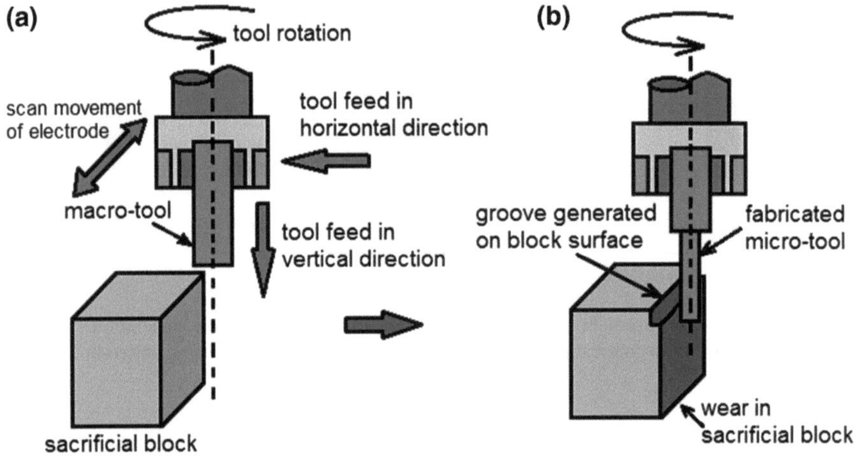

Fig. 3.13 Schematic illustration of moving block micro-EDG, **a** showing the movement of block and macro-tool, **b** fabricated high aspect ratio micro-tool

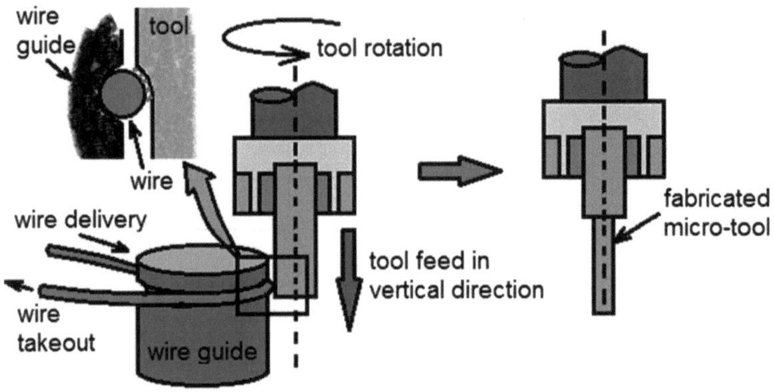

Fig. 3.14 Representation of micro-wire EDG process

3.8 Applications of Titanium and Its Alloys

Titanium alloy was developed in the early 1950s for defence and aeronautic applications due to its very high strength-to-weight ratio. The unique combination of high strength, low weight, and excellent corrosion resistant property of titanium alloy has made it suitable for a wide variety of industrial applications. Its corrosion resistant results from a very thin (−10 nm), stable, continuous oxide layer that regenerates instantaneously if any oxygen or moisture is present. Commercially Pure (CP) Titanium is used primarily for its corrosion resistance. Further titanium alloy can withstand pitting, crevice and cavitations, corrosion, erosion, and stress corrosion cracking in salt water, marine atmospheres, and a broad range of acids, alkalis, and industrial chemicals. This alpha-beta alloy is the workhorse alloy of the titanium industry. The alloy is fully heat-treatable in section sizes up to one inch and is used up to approximately 400 °C. Since it is the most commonly used alloy, over 70% of all alloy grades melted are sub- grade of Ti–6Al–4V. Its uses arc not only confined and concentrated in aerospace engine and airframe components but it is widely uses in major non-aerospace applications in the marine offshore, power generation industries and biomedical applications also. Ti–6Al–4V is the most widely used titanium alloy, accounting for more than half of all titanium tonnage worldwide.

Various excellent properties possessed by titanium super alloys have led to a wide and diversified range of successful applications in medical science as well as in automotive, aerospace, chemical plants, and other precision engineering fields. The wide variety of applications of this exotic material in almost all the precision engineering fields demand the huge need of micro-machining of Ti–6Al–4V. Thus, in order to meet up the huge market demand of micro-products manufactured from Commercially Pure (CP) Titanium and its alloy necessitated the need of micro-machining using micro-EDM technology utilizing different novel machining strategies.

3.9 Brief Background of Machining Ti–6Al–4V in Micro-EDM

In last 15 years, a number of research and development activities have been carried out around the globe for improving the machining performance as well as dimensional accuracy of machined components in micro-EDM process. Moreover, many researchers have utilized various statistical tools for finding out influencing process parameters and optimization of the process for obtaining parametric condition to achieve high performance machining and accurate geometrical features. Micro-hole electro discharge machining has been carried out by Pradhan et al. on Ti–6Al–4V alloy using 500 μm diameter brass micro-electrodes [24]. Four process parameters such as peak current, pulse-on time, flushing pressure and duty ratio were considered as varying parameters to investigate the process criteria i.e. material removal rate, tool wear rate and overcut. The most influencing process parameters were achieved as peak current and pulse-on time. Flushing pressure and duty factor have no significant effect on material removal rate and tool wear rate. However, overcut of micro-hole generated is mostly affected by peak current and pulse-on-time. Using Response Surface Methodology (RSM) statistical tool, Pradhan and Bhattacharyya investigated the parametric effect on machining performance as well as optimized the process to achieve high MRR, low TWR as well as least overcut during machining of Ti–6Al–4V material [25]. Combined RSM-ANN based mathematic models were developed to correlate machining performances and process parameters. Silicon carbide (SiC) powder additive mixed dielectric was used to investigate and analysis of micro-EDM of Ti–6Al–4V alloy by Ali et al. and the authors concluded that by employing SiC mixed dielectric, material removal rate is enhanced when compared to machining with conventional dielectric [26]. The experimental results revealed that optimization for MRR of 7.31 μg/min is achieved at powder concentration of 24.75 g/L and discharge energy of 56.77 μJ.

Optimization of MRR, TWR and overcut during micro-EDM of Ti–6Al–4V alloy considering voltage, pulse frequency, current and pulse width is achieved by Meena and Azad utilizing combined approach of Taguchi Method and Grey relational analysis (GRA) [27]. Analysis of variance (ANOVA) of the test results have been performed to in order to estimate the predictive accuracy of the developed models and to determine the relative significance of considered process parameters. Hole sinking electrical discharge micromachining (HS-EDMM) process of Ti–6Al–4V thin sheet was carried out by Porwal et al. to develop predictive integrated model (ANN-GRA-PCA) using single hidden layer BPNN [28]. The authors used GRA coupled with PCA hybrid optimization strategy for achieving optimal MRR, TWR and taper of micro-hole at parametric setting of gap voltage of 140 V and capacitance of 100 nF. Experimental investigation and parametric study was carried out by Tiwary et al. during micro-hole machining in micro-EDM of Ti–6Al–4V material based using 1 mm thick titanium alloy material as workpiece and brass electrode of diameter 300 μm as micro-tool on RSM approach considering pulse on time, peak current, gap voltage and flushing pressure as process parameters [29]. Using various surface plots, influences of the process parameters on MRR, TWR, overcut and taper of micro-hole were studied. A combined approach of RSM and fuzzy-TOPSIS method is used to find out optimal parametric combination. The effect of gap voltage, capacitance, rotational speed of electrode and feed rate on MRR during micro-EDM milling operation of Ti–6Al–4V material was investigated by Kuriachen and Mathew [30]. Based on RSM-Box Behnken experimental design, quadratic regression model for MRR was developed and from the test results, it was seen that capacitance and rotational speed of electrode has direct effect on MRR. Study and analysis on the influence of micro-EDM parameters on MRR, TWR, machining time and quality of micro-hole was carried out by Plaza et al. during machining of Ti–6Al–4V [31]. The authors used a new strategy to use helical-shaped micro-tool electrodes. The influences of helix angle and flute depth of the helical micro-tool on process performances have also been studied. Blind and through micro-holes and micro-slots were machined on brass and Ti–6Al–4V materials by Moses and Jahan [32]. Dimensional accuracy, surface finish, and profile accuracy of holes and slots were measured for analysis. Single through micro-hole, single blind hole, letter H blind, three blind slots and three through slots were machined also. The influence of various process parameters such as pulse-on-time, peak current, gap voltage and flushing pressure on MRR, TWR, overcut and taper of micro-hole during machining of Ti–6Al–4V were studied by Tiwary et al. and empirical models were developed for to correlate the process parameters and machining performances [33]. Multi performance optimization was achieved as MRR of 0.0777 mg/min, TWR of 0.0088 mg/min, OC of 0.0765 mm and taper of 0.0013 at parametric setting of pulse-on-time of 1 μs, peak current of 2.5 A, gap voltage of 50 V, and flushing pressure of 0.20 kgf/cm^2.

Kuriachen and Mathew carried out investigation to machine Ti–6Al–4V with tungsten carbide electrode employing SiC micro particle suspended dielectric during micro-EDM milling operation [34]. The effects of various process parameters such as voltage, capacitance and powder concentration on MRR and TWR was studied. The recommended process parametric setting was powder concentration of 5 g/L, capacitance of 0.1 μF and voltage of 115 V for achieving high material removal and low tool wear rate. Predictive thermal model was developed by B. Kuriachen et al. for simulation of single-spark micro electric discharge machining [35]. The crater geometry and temperature distribution in the workpiece at various process parametric setting were predicted using Gaussian distribution of heat flux, percentage distribution of energy among the workpiece, tool electrode and dielectric. Mathematical model which predicts the radius of the single-spark during micro-EDM process of Ti–6Al–4V was developed by Kuriachen and Mathew [36]. The authors concluded that the spark radius increases proportionally with in capacitance except in the higher energy levels where double sparking phenomenon was observed.

It is observed from the literature review that most of the research investigation and analysis is on investigating the effects of process parameters on machining performances such as MRR, TWR, overcut, surface roughness, etc. Furthermore, optimal parametric combination was also found utilizing several statistical techniques as well as combined approach of different predictive tools for achieving high machining rate, less tool wear rate and good geometrical micro-structures. However, there are many issues which remained unsolved in micro-EDM. Such issues are improving material removal rate, methodology for compensating the micro-tool wear, improving the stability of discharge, improving the accuracy of micro-feature implementing innovative ideas, etc. To solve these important issues and to improve overall micro-EDM efficiency, several new micro-EDM machining strategies have been developed and researchers already started to utilize these innovative strategies during micro-EDM of Ti–6Al–4V material. In the following sections, experimental investigation and analysis of micro-EDM process of Ti–6Al–4V have been carried out extensively utilizing some innovative machining strategies such as ultrasonic vibration assisted micro-ED machining, utilization of non-hydrocarbon oil as dielectric, reversing the polarity between electrodes, rotating the micro-tool electrode, utilizing powder mixed dielectric fluid, etc.

3.10 Innovative Machining Strategies for Improving Micro-EDM

To achieve effective machining performance of micro-EDM for real time utilization and also for obtaining high productivity with a higher degree of desired accuracy and surface integrity, manufacturer engineers and research scientists are always engaged to develop innovative machining techniques by considering various innovative machining strategies during micro-EDM of various difficult-to-machine materials. In this section, research investigation and analysis in the direction of implementing such novel machining strategies during micro-EDM of Ti–6Al–4V superalloy is discussed.

3.10.1 Ultrasonic Vibration Assisted Micro-EDM

The use of ultrasonic vibration during micro-EDM shows effectiveness for improving machining time as well as geometrical accuracy of fabricated features of micro components. During ultrasonic assisted micro-EDM, micro-tool or dielectric fluid is vibrated ultrasonically. In micro-EDM, the debris particles which are generated are difficult to remove from very small IEG. This phenomenon causes the chance of short circuit between electrodes and creates the non-uniform discharge during micro-EDM. With the implementation of ultrasonic vibration, the dielectric can be circulated even in the narrow gap as the assisted ultrasonic vibration produces a pumping effect and enhances the circulation of dielectric at inter electrode gap which reduces the machining time, reduces the amount of electrode wear, improves the material removal rate and reduces the chances of micro-cracks generation on machined surface [37–39]. The ultrasonic vibration assisted micro-EDM helps to produce high aspect ratio micro-holes efficiently compared to normal micro-EDM without ultrasonic vibration. Moreover, the material removal rate increases significantly with less electrode wear ratio and quality of machined surface and dimensional accuracy has been improved [39]. The ultrasonic vibration of work material in micro-EDM shows a drastic improvement. The machining efficiency has been increased up to 8 times when compared to traditional micro-EDM process during machining of stainless steel of thickness 0.5 mm and up to 60 times during the machining of Nitinol [40, 41]. Figure 3.15 shows the ultrasonic vibration assisted micro-EDM process along with different tool tip positions and IEG at different positions of vibration. The powder mixed dielectric always contributes on the performance of micro-EDM process. During vibration assisted powder mixed

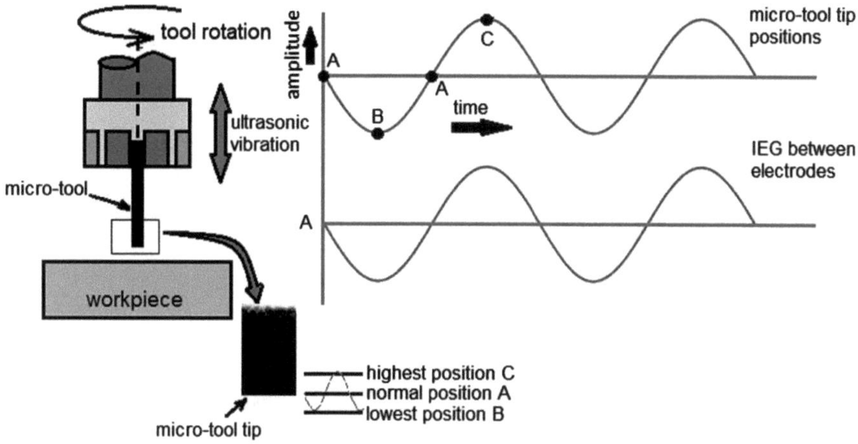

Fig. 3.15 Schematic of ultrasonic vibration assisted micro-EDM process with amplitude of vibration

micro-EDM, the adhesion of powder or abrasives in the narrow gap between the micro tool and the work material occurs. Therefore, implementing ultrasonic vibration during powder mixed micro-EDM, MRR increases significantly as it enhances the dielectric circulation in the machining zone [42]. The drawbacks due to arcing and short circuits in micro-EDM can be reduced substantially when assisted with ultrasonic vibration as the number of normal pulses and the average pulse energy increases [43, 44]. The machining time of micro-EDM can be reduced significantly with increase in vibration frequency at constant amplitude which significantly reduces the start-up process. The vibration between the micro-tool and the work material obstructs the arc which ends the arcing state [45]. Thus, the total duration of arcing event reduces and decreases with increase in vibration frequency. Figure 3.16 shows the comparison of quality of blind micro holes produced with and without ultrasonic vibration (frequency of 6 kHz and amplitude of 3 μm) [46]. The comparison shows the improvement in the dimensional accuracy by 10.5 μm. Furthermore, the micro-hole produced without vibration is larger (diameter is 201.5 μm) than micro-hole generated with ultrasonic vibration (diameter is 191 μm) and this result indicates that the electrode wear is less in ultrasonic assisted micro-EDM process. The recast layer formation has been always a drawback during micro-EDM process. Due to better dielectric circulation in the machining zone, the debris removal from narrow gap due to ultrasonic vibration is improved and results in less re-solidification or recast layer [37]. Moreover, this leads to thinner heat-affected zone produced around the machined cavities. The ultrasonic vibration assisted micro-EDM machining not only shows the improvement in the process performance when used with conventional and powder mixed dielectric, but it also shows the tremendous contribution during dry EDM/micro-EDM. The improvement in MRR has been reported when compared to gas based EDM and liquid

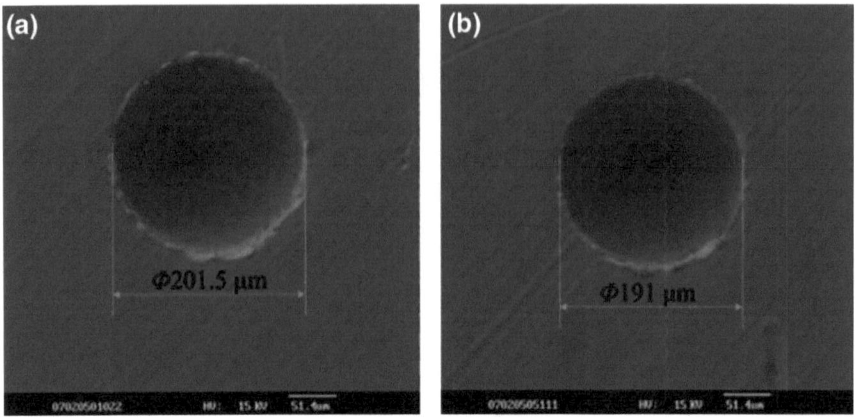

Fig. 3.16 SEM images of blind holes generated **a** without ultrasonic vibration and **b** with vibration [46]

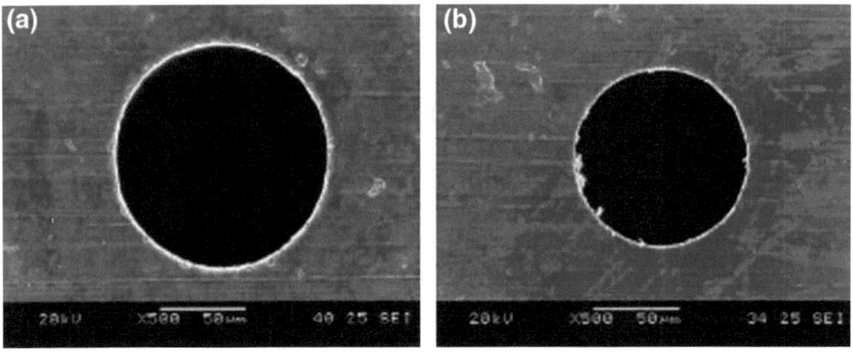

Fig. 3.17 SEM images of micro-hole of **a** entrance and **b** exit side machined with ultrasonic vibration micro-EDM [48]

based EDM without vibration [47]. The reduction in taperness has been achieved when micro-hole is machined with the aid of ultrasonic vibration. Figure 3.17 shows SEM images of micro-hole of entrance (diameter of 122 μm) and exit (diameter of 106 μm) side when machined with ultrasonic assisted micro-EDM [48]. The difference in the diameter (16 μm) is due to secondary discharge and wear of micro-tool.

3.10.2 Utilization of Non-hydrocarbon Dielectrics

In most of the conventional EDM systems, hydrocarbon oil kerosene is used. However, during machining with kerosene dielectric, it creates several drawbacks such as degradation of dielectric properties, pollution of air, and adhesion of carbon particles on the work surface etc. Due to these, unstable machining operation takes place and results in inefficient performance during micro-EDM. Thus, it is the task of manufacturer engineers and research scientists to find out alternate dielectric which can overcome the above-mentioned drawbacks. De-ionized water is one of the non-hydrocarbon dielectric which can efficiently be used during micro-EDM process. The use of de-ionized water as dielectric does not produce any toxic vapours such as CO and CH_4 as liberated in case of kerosene oil [49]. Compared to kerosene, de-ionized water keeps the machining environment clean and safe. Moreover, de-ionized water supplies oxygen in the machining zone and this promotes stable discharges in IEG. Furthermore, the high degree of fluidity and high cooling rate properties of de-ionized water encourages the debris particles to flush out quickly from IEG and cool the discharge zone quickly and prepares for next discharges.

From the exhaustive review of micro-EDM of Ti–6Al–4V, it is found that very few research investigation were performed on implementing de-ionized water dielectric during micro-hole machining. Further, comparative study on the performance of pure kerosene and pure de-ionized water has not been performed during machining of Ti–6Al–4V alloy. Keeping in view of the requirements and developmental research issue for improving the micro-EDM performance measures, a well-planned research methodology has been designed to investigate the influence of various dielectrics such as kerosene and de-ionized water on various micro-EDM performances such as material removal rate, tool wear rate, overcut, diameteral variance at entry and exit hole (DVEE) and machined surface topography through various test results and scanning electron microscope (SEM) micrographs.

(a) **Experimental details and machining conditions**

The experiments are performed on a traditional die sinking EDM (model: series 2000, EMS-5535-R50, ZNC EDM machine, Manufacturer: Electronica Machine Tools Pvt. Ltd., Pune, India). This EDM set up consists of (i) Spark generator unit, (ii) Z axis unit with servo feed mechanism, (iii) X-Y table unit and (iv) Dielectric pumping and filtering unit. Figure 3.18a, b shows the photographic view of the main components of micro-EDM set up used in this experimental study. When machining experiments were carried out using de-ionized water, a separate dielectric chamber with separate pump, and pressure-regulating valve and filter were used to circulate the dielectrics without affecting dielectric supply system of the main machine. Figure 3.18c shows the developed external dielectric supply system to circulate dielectrics during micro-EDM experiments. Through micro-holes were machined on Ti–6Al–4V plates of size 13 × 15 × 1 mm. Cylindrical shaped tungsten electrodes with flat front of diameter 300 μm were used as tool. Tables 3.1 and 3.2 show the chemical compositions and physical as well as mechanical properties of titanium alloy (Ti–6Al–4V), respectively. Electrical resistivity of boron carbide is in the range of 0.1–10 Ω-cm. As the micro-EDM performances are affected mostly by peak current (Ip) and

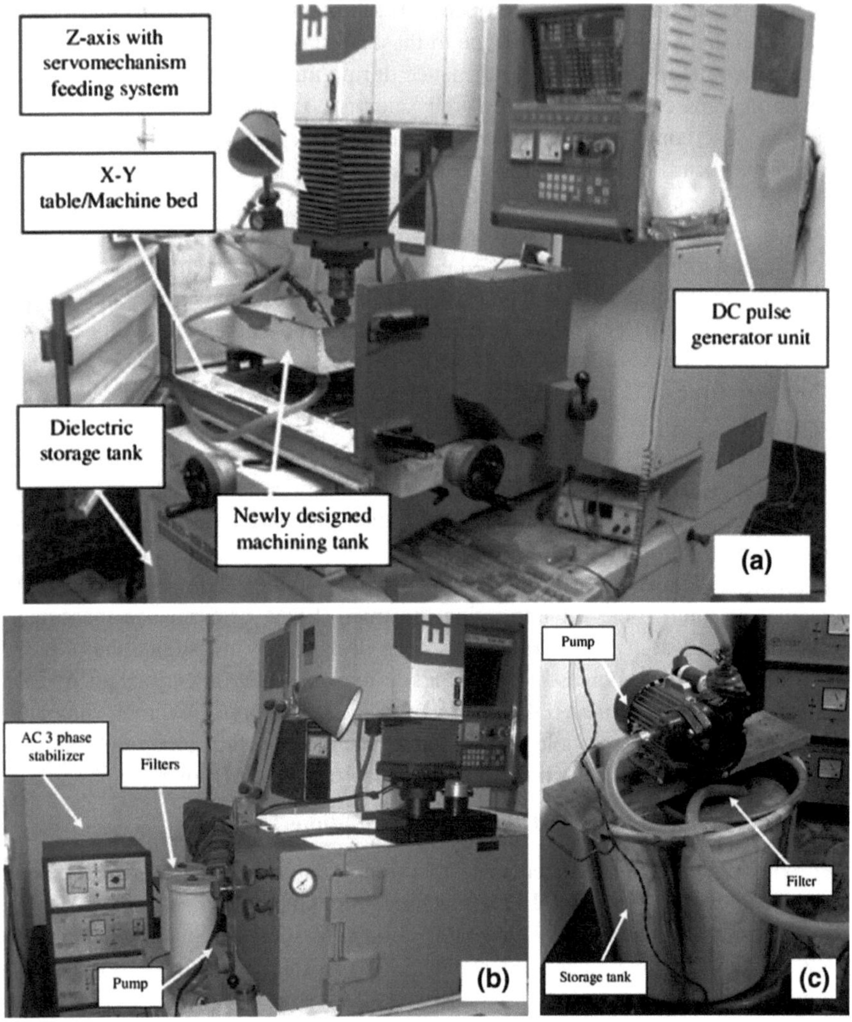

Fig. 3.18 Photographic views of **a** and **b** EMS-5535-R50, ZNC EDM machine **c** external dielectric supply system for circulating de-ionized water

pulse-on-time (Ton) during machining of Ti–6Al–4V alloy [24, 50], so influence of these two predominant process parameters were considered as the varying parameters keeping other process parameters like flushing pressure (Pr), duty factor (t) as constant during machining with each dielectrics.

In this research investigation, experimentations have been carried out at various peak current settings i.e. 0.5, 1, 1.5, 2 A employing kerosene and deionized water. The details of the machining conditions and other parameter details are enlisted in Table 3.3. In this research study, micro-EDM characteristics such as material removal rate (MRR), tool wear rate (TWR), overcut (OC) and diameteral variance

Table 3.1 Compositions of Ti–6Al–4V superalloy

Element	Percentage
Aluminium	5.5–6.75
Carbon	≤ 0.10
Iron	≤ 0.50
Hydrogen	≤ 0.015
Nitrogen	≤ 0.05
Oxygen	≤ 0.45
Other	<0.4
Vanadium	3.5–4.5
Titanium	Balance

Table 3.2 Physical and mechanical properties of Ti–6Al–4V

Property	Typical value
Density (g/cm^3)	4.42
Melting range (°C±15 °C)	1649
Specific heat (J/kg °C)	560
Thermal conductivity (W/m-K)	7.2
Tensile strength (Mpa)	897
Elastic modulus (GPa)	114
Hardness Rockwell C	36

in entry and exit (DVEE) were considered as the machining characteristics. Material removal rate was calculated using Eq. (3.2). Similarly, tool wear rate is calculated using Eq. (3.3). A high precision weighing machine (Manufacturer: Mettler Toledo, Switzerland, Least measurable weight = 0.01 mg) is used to measure various weights of workpiece and micro-tools. The overcut of micro-hole machined is measured using Eq. (3.4). A high precision measuring microscope (Manufacturer: OLYMPUS, Japan, Model: STM6, minimum measurable dimension = 0.5 μm) is used to measure all the dimensions of micro-hole diameters as well as of micro-tool. Experimentation were conducted at various micro-EDM parametric combinations, measured and calculated the performance criteria. The discussion and analysis of various test results are described in the following sections.

(b) **Results and discussion**

In this section, a detailed comparison of various machining performance characteristics has been performed for the investigation of influence of dielectric liquid with kerosene and de-ionized water in micro-EDM for micro machining of titanium alloy (Ti–6Al–4V). Figure 3.19 depict the comparison of the material removal rate (MRR) using two different dielectrics such as kerosene and de-ionized water varying peak current (Ip) and pulse-on-time (Ton). The material removal rate (MRR) is much more using de-ionized water than the kerosene throughout the considered range of pulse duration and increase of peak current. When machining is done with kerosene, as the dielectric fluid is kerosene which is a chemical compound of carbon and hydrogen, decomposes and produces a layer of titanium

Table 3.3 Experimental condition for through micro-hole machining on Ti–6Al–4V alloy

Condition	Description
Workpiece material (anode)	Ti–6Al–4V plate of size 13 mm × 15 mm and thickness of 1 mm
Tool electrode (cathode)	Solid tungsten micro-tool, diameter of 300 μm
Dielectric fluids	Kerosene, de-ionized water
Peak current (A)	0.5, 1, 1.5, 2
Pulse-on-time (μs)	1, 2, 5, 10, 20
Duty factor (%)	95
Flushing pressure (kgf/cm^2)	0.5
Resistivity of pure de-ionized water (megohm-cm)	4.2

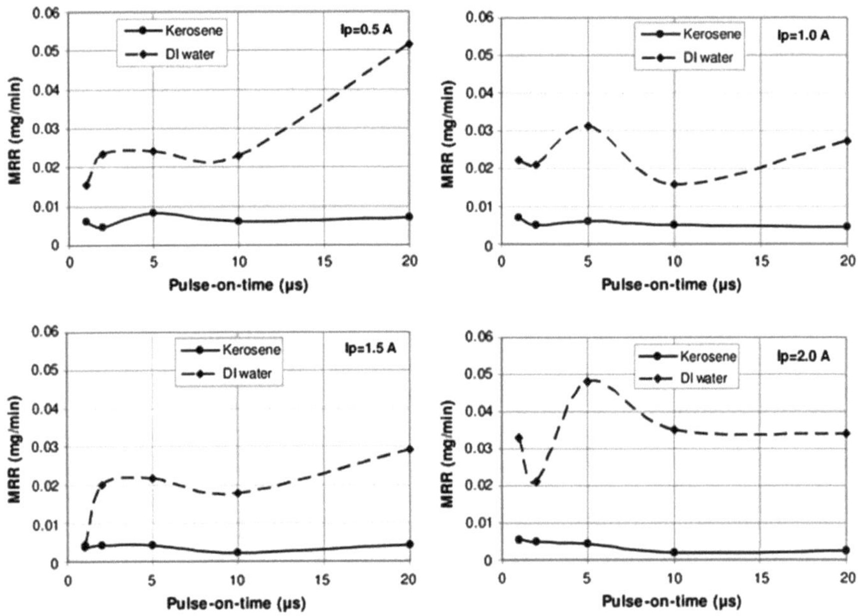

Fig. 3.19 Influence of different dielectrics on the material removal rate (MRR) with varying pulse-on-time (Ton)

carbide (TiC) on the workpiece surface. But, when using de-ionized water, the water decomposes and a layer of titanium oxide (TiO_2) is produced on the machined surface. Since, TiC has a higher melting temperature (3150 °C) than that of TiO_2 (1750 °C), a large discharge energy is required for improving the material removal rate using kerosene. Also, the size of debris formed during machining with kerosene is less compared to machining with de-ionized water, thus improving the material removal rate.

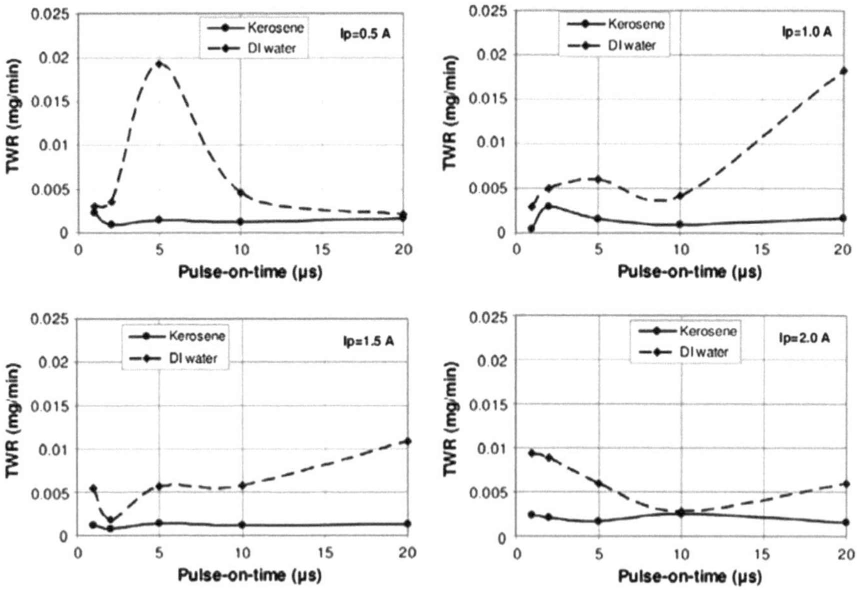

Fig. 3.20 Influence of different dielectrics on the tool wear rate (TWR) with varying pulse-on-time (Ton)

Figure 3.20 shows the comparison of the tool wear rate (TWR) using two different dielectrics such as kerosene and de-ionized water varying peak current (Ip) and pulse-on-time (Ton). It is seen from the figure that the tool wear rate (TWR) is high using de-ionized water compared to machining with kerosene. When using kerosene as the dielectric, it decomposes in the high discharge energy and produces carbon particles that stuck or adhere to the surface of the electrode. These carbon particles restrict the rapid wear of the tool. So, the tool wear rate (TWR) is less while using kerosene as dielectric. On the other hand, when using deionized water, no carbon adhere to the tool electrode surface, thus TWR is higher enough with de-ionized water.

Figure 3.21 shows the comparative study of the overcut (OC) using two different dielectrics i.e. kerosene and de-ionized water with varying peak current (Ip) and pulse-on-time (Ton). From the figures, it is clear that the overcut of the machined micro-holes is larger when using de-ionized water for the pulse duration of 1 and 2 μs when machining is done by varying peak current. But at higher pulse duration the overcut of the micro-holes is larger when using kerosene as the dielectric. When deionized water is used, it releases oxygen decomposed from water. This oxygen influences the machining stability and helps to form more debris. These debris particles ejected through the short gap of tool surface and micro-hole walls. Thus increasing secondary sparking, resulting in higher overcut compared to kerosene. But at higher pulse duration, the machining stability and efficiency increases due to more pulses per cycle, resulting higher overcut with kerosene compared to

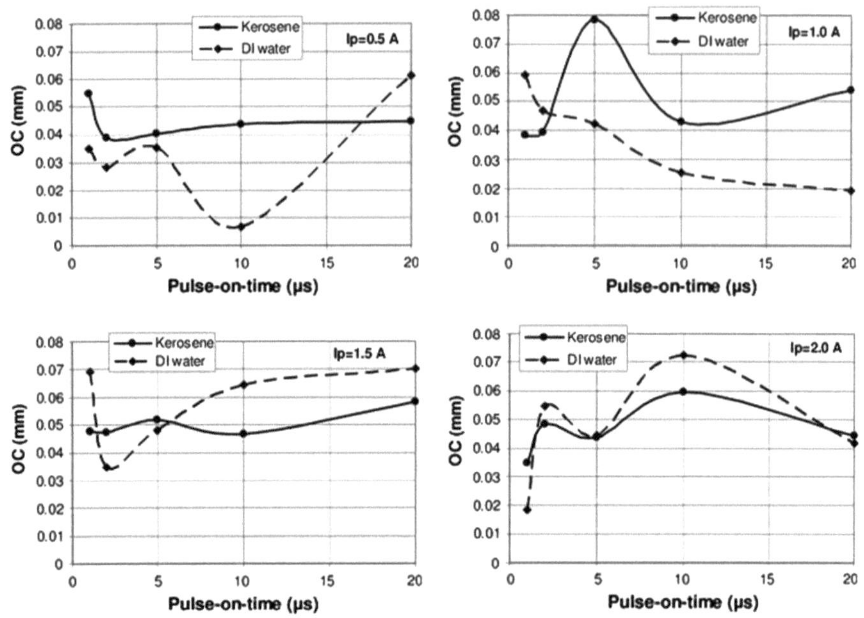

Fig. 3.21 Influence of different dielectrics on the overcut (OC) with varying pulse-on-time (Ton)

de-ionized water. When the overcut is analyzed with the increase of peak current it is found that a high peak current results higher overcut with de-ionized water compared to kerosene.

Figure 3.22 shows the comparison of the diameteral variance in entry and exit (DVEE) using two different dielectrics i.e. kerosene and de-ionized water with varying peak current (Ip) and pulse-on-time (Ton), respectively. It can be observed from the figures that DVEE of the micro-holes increases at low discharge duration when varying peak current was employed using de-ionized water as dielectric fluid. But further increase of pulse duration results the decrease of DVEE of the holes with de-ionized water. A straight through micro-hole can be generated at pulse duration of 5 μs and peak current of 1.5 A. On the other hand, DVEE is lower at peak current 0.5 and 1 A employing deionized water compared to kerosene. But, as the peak current increases, the diameter variance increases with de-ionized water. Thus, a straight through micro-hole is not achieved.

After machining of micro-holes utilizing various type of dielectric fluids, the workpiece were carefully polished, cleaned, and etched with a solution of 2.5 ml of HF acid (40%), 5 ml concentration of HNO_3 and 42.5 ml of de-ionized water for examining the surface topography of micro-holes as well as the recast layer formed on the machined micro-hole surfaces with the aid of optical and SEM micrographs. Figure 3.23 shows some optical micrographs of machined micro-holes that were taken using a 10X zoom lens in a precision optical microscope for both pure dielectrics i.e. kerosene and de-ionized water at the machining condition of 1 A

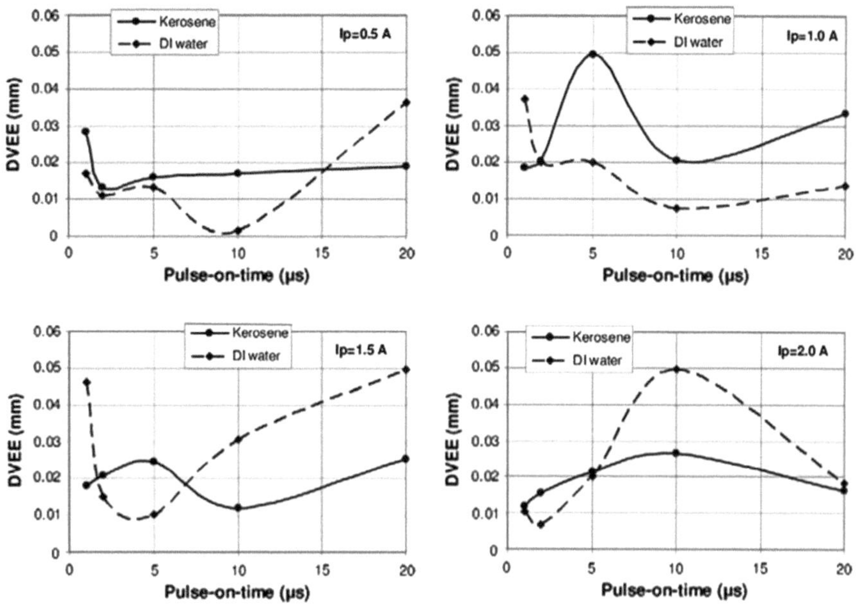

Fig. 3.22 Influence of different dielectrics on the diameter variance between entry and exit (DVEE) with varying pulse-on-time (Ton)

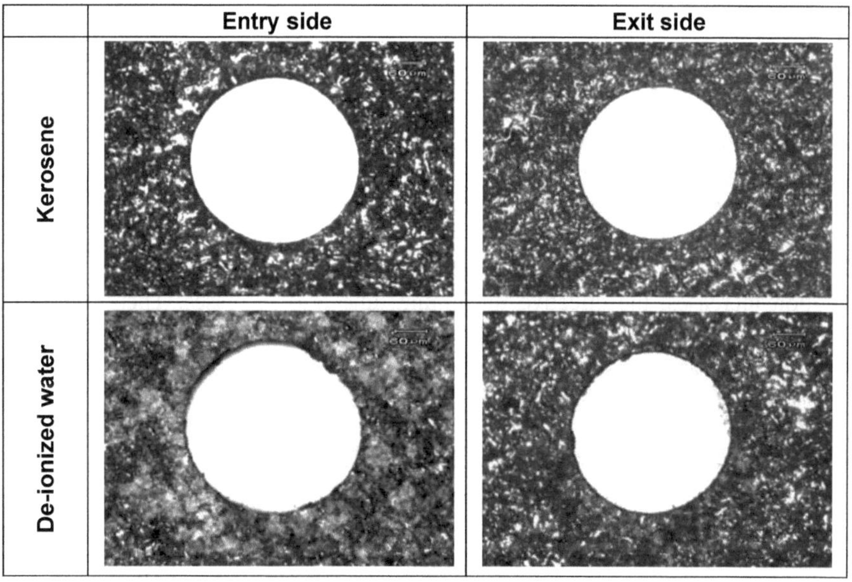

Fig. 3.23 Optical photographs of machined micro-holes with different dielectrics at 1 A/2 μs [51]

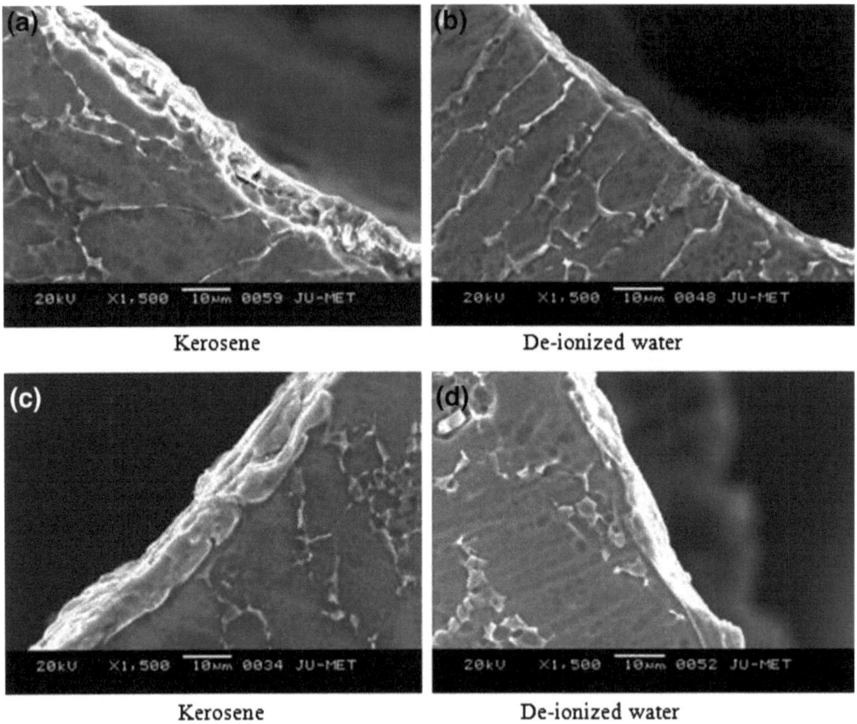

Fig. 3.24 SEM micrographs of white layer formed on micro-hole edge **a** and **b** machined at 0.5 A/1 μs and **c**, **d** machined at 1.5 A/10 μs [51]

peak current and 2 μs pulse duration. It is clear from these figures that kerosene dielectric result in improved quality micro-hole than de-ionized water. In Fig. 3.24, SEM micrographs of the micro-hole's edges are shown to examine the white/recast layer formation using kerosene and de-ionized water. The thickness of the white layer is much lower using de-ionized water compared to kerosene. Moreover, with increase of the pulse-on-time, the thickness of the white layer increases. As pulse duration increases, the effective machining time also increases. Therefore, more debris is generated and this debris adheres to the micro-hole surface and resolidified as deionized water has high cooling rate than kerosene.

3.10.3 Abrasive Mixed Dielectric in Micro-EDM

In micro-EDM, debris in the inter electrode gap facilitates the ignition process and further increases the gap size and overall flushing conditions [52]. Absence of debris particles in the gap can results arcing between the electrodes and it further results in lack of precise feeding mechanism. However, excess debris leads to uneven discharge and short-circuiting. Some debris particles in the machining gap provide better discharge transitivity, gap size, breakdown strength, and deionization [53]. From the past research, it is revealed that most of the research has been performed on EDM employing powder mixed dielectrics. However, no research was reported to use powder mixed dielectric during micro-EDM of Ti–6Al–4V. Thus, in the present experimental study, detailed analysis has been carried out to investigate comparatively the influence of mixing boron carbide powder in kerosene and de-ionized water during micro-hole machining in micro-EDM on Ti–6Al–4V material. This B_4C powder has some excellent physical and chemical properties such as high chemical resistance and hardness, excellent wear and abrasion resistant etc. These exceptional characteristics of boron carbide may provide effective and efficient discharge conditions at the machining zone and also enhancement in above-mentioned machining performances. Electrical resistivity of boron carbide is in the range of 0.1–10 Ω cm. As B_4C abrasive lie in a transition zone between good conductors and isolators, the potential difference as well as the plasma channel produced in the micron sized inter electrode gap can make the abrasive to conduct thermoelectric power in the machining zone [54, 55]. Boron carbide is characterized by a relatively wide gap in its forbidden band, a low thermal conductivity, and a high thermoelectric power. These properties make it a potentially useful material for high-temperature thermoelectric energy conversion compared to silicon carbide as well as tungsten carbide abrasives.

(a) **Experimental details and machining conditions**

As micro-EDM process uses discharge energy in the range of 5–150 μJ, so there is a little chance to melt and evaporate the B_4C particles due to discharge in the inter electrode gap. It was observed in the optical measuring microscope that the powder particles comprise of mixture of different shapes and sizes ranging from 8 to 20 μm. When, Boron carbide (B_4C) powder-mixed de-ionized water was applied in the machining zone, most of the particles, which are more than 10 μm in size, get accumulated at the base of machining zone because of self-weight of particles even though a motor driven stirrer was applied to provide turbulence in powder mixed dielectrics in the machining tank. Hence, the effective average particle size that may involve in the machining phenomena at the micro-machining zone ranges from 8 to 10 μm during micro-EDM. Therefore, the average size of B_4C particles which actively take part in machining is in the range of 8–10 μm. The thermo-physical properties of boron carbide powder are enlisted in Table 3.4. B_4C powder additive

Table 3.4 Thermo-physical properties of Boron carbide (B_4C) additives

Property	Typical value
Density (g/cm^3)	2.52
Melting point (°C)	2445
Electrical conductivity (at 25 °C) (S)	140
Thermal conductivity (at 25 °C) (W/m K)	30–42
Young's modulus (GPa)	450–470
Hardness (Knoop 100 g) (kg mm^{-2})	2900–3580
Specific heat (J K^{-1} kg^{-1}) (at 25 °C)	950

of size 8–10 μm and of concentration 4 g/l was added to pure kerosene and deionized water when experimentations were performed with powder-mixed dielectrics. This particular concentration of additive in dielectrics has been selected based on past research studies on powder mixed dielectrics in EDM [56, 57]. The same micro-EDM machine was utilized for this experimental investigation. When experiments are done with powder mixed kerosene and powder mixed de-ionized water, a separate dielectric chamber with separate pump, and pressure-regulating valve and filter were used to circulate the dielectrics without affecting dielectric supply system of the main machine. When machining was done with powder-mixed dielectric, the external filter unit was removed, and a magnetic field was employed to remove the machining debris from the dielectrics. In this research study, micro-EDM characteristics such as material removal rate (MRR), tool wear rate (TWR), overcut (OC), diameteral variance in entry and exit (DVEE) were considered as the machining characteristics. The discussion and analysis of various test results are described in the following sections.

(b) **Results and discussion**

Figure 3.25 shows the comparative plots of the material removal rate (MRR) using different dielectrics such as pure kerosene, pure de-ionized water and Boron Carbide (B_4C) mixed dielectrics powder for varying pulse-on-time (Ton) at different peak currents (Ip). This figure reveals that MRR is high with de-ionized water

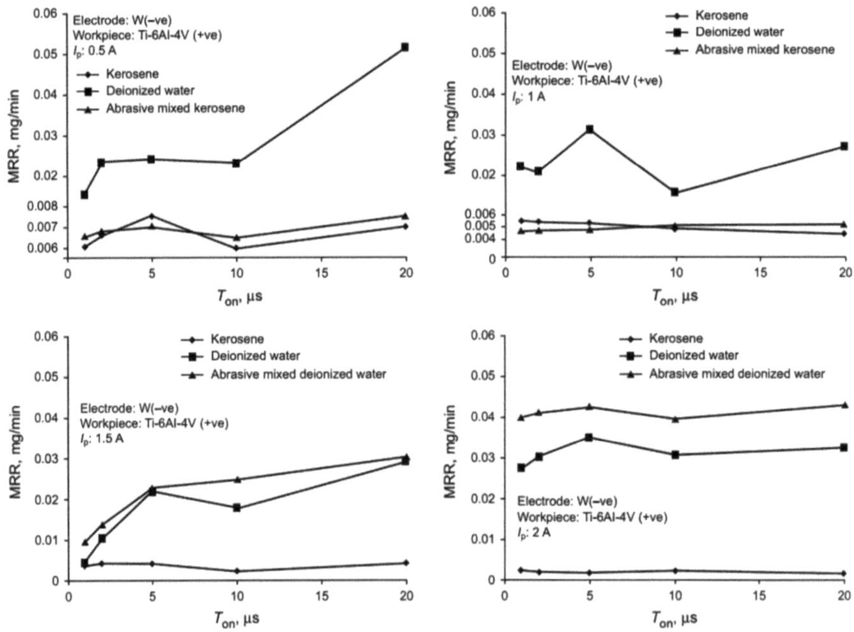

Fig. 3.25 Variation of material removal rate (MRR) with pulse duration (Ton) at various fixed peak current (Ip) for different dielectrics [51]

than kerosene for all considered settings of pulse duration and peak current during experimentation. Additionally, when machining is done by mixing B_4C powder additives in kerosene dielectric it is clearly seen that MRR increases with the increase of pulse duration at constant peak current of 1.5 and 2 A. Also the MRR with powder mixed dielectrics is larger compared to machining with pure kerosene and de-ionized water at higher pulse duration discharge settings. The increase of MRR with the increase of pulse duration using B_4C mixed kerosene is due to increase of spark discharge time i.e. longer effective machining time per pulse. The presence of boron carbide additive in kerosene further helps in uniform distribution of discharge energy and better conduction of discharge current thereby enabling better machining condition. When B_4C powder was applied to de-ionized water, it is seen that MRR is more using additive compared to pure deionized water at peak current 1.5 and 2 A. The same reason is applicable here also for the increase of MRR as of additive mixed kerosene. So, it is concluded that the addition of carbide powder particles in dielectrics prevails better machining efficiency than the pure dielectrics due to uniform distribution of discharge energy in the machining zone.

In Fig. 3.26 comparative results of tool wear rate (TWR) with pulse duration at different pulse discharge with various constant peak current are shown employing kerosene, de-ionized water and B_4C powder mixed dielectrics. This figure reveals that TWR is high using de-ionized water compared to machining with kerosene dielectric. Furthermore, it is revealed from the same figure that tool wear rate

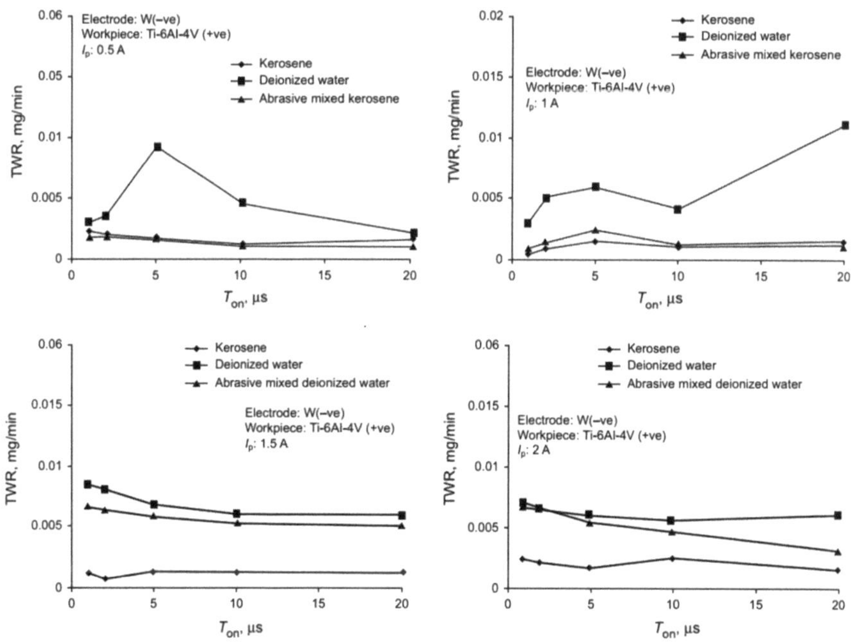

Fig. 3.26 Variation of tool wear rate (TWR) with pulse duration (Ton) at various fixed peak current (Ip) for different dielectrics [51]

associated with B_4C mixed kerosene is less compared to machining with pure kerosene at peak current of 0.5 and 1 A. When machining is done with boron carbide abrasive mixed kerosene dielectric, the tool wear is less due to the presence of more number of carbon particles evolving from the decomposition of kerosene dielectric as well as boron carbide abrasive in the machining zone. From the same figure it is observed that at fixed 1 A peak current tool wear rate is lesser than at fixed 0.5 A current setting using additive mixed kerosene. This is due to the fact that the higher discharge energy results in more decomposition of kerosene and it further generates more carbon which in turn adheres onto the tool surface preventing secondary sparking. Although higher peak current i.e. 2 A produces more discharge energy, but that results in more current density and subjects the tool electrode under large thermal stresses. The powder mixed de-ionized water results in less tool wear rate at 2 A compared to 1.5 A due to more deposition of carbon particles from B_4C additives. It is also found that machining combined with boron carbide powder mixed deionized water results in less tool wear compared to pure de-ionized water due to adhesion of carbon particles from boron carbide powder on the tool surface, which restrict tool wear to certain extent.

The comparative plots of overcut (OC) of micro-holes on Ti–6Al–4V employing kerosene, deionized water and B4C abrasive mixed with these dielectrics are shown in Fig. 3.27 when pulse duration was varied for fixed different peak current. It is observed from these figures that the overcut of the machined micro-holes is less

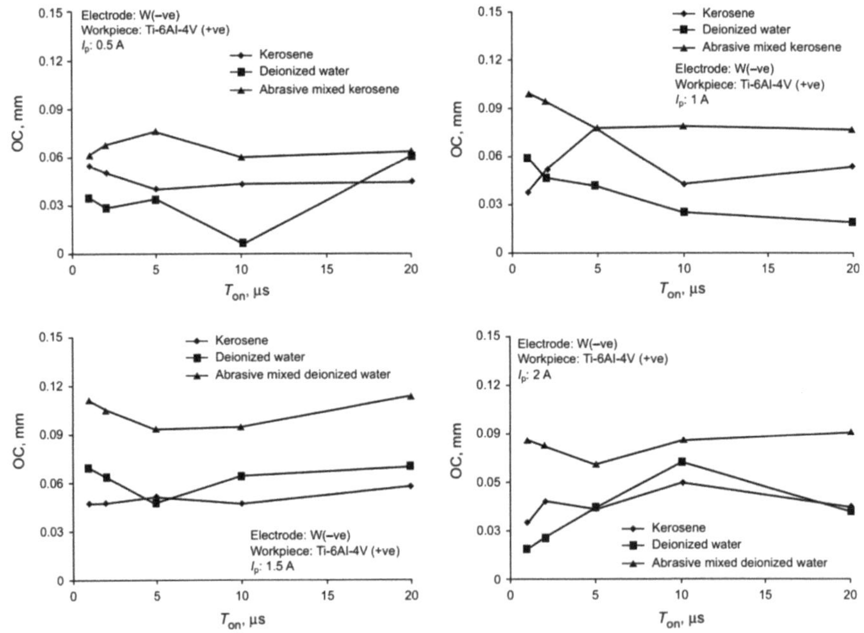

Fig. 3.27 Variation of overcut (OC) with pulse duration (Ton) at various fixed peak current (Ip) for different dielectrics [51]

when dielectric was de-ionized water for peak current setting of 0.5 and 1 A. However, at higher peak current i.e. 1.5 and 2 A, overcut is more in case of de-ionized water compared to pure kerosene dielectric. In addition, when B_4C additive was used in dielectrics, it is found that OC is larger with both powder mixed dielectrics compared to pure dielectrics. It is so because the suspended additive particles remove the molten layer from the machining zone and further reduce the formation of thick white layer, resulting larger OC. It is also revealed that OC decreases with increase in pulse duration while using B_4C suspended kerosene as dielectric. It is due to decrease of overall machining time i.e. faster machining. However, larger OC is found in case of B_4C mixed deionized water as dielectric because of secondary sparking.

Figure 3.28 shows the comparative outcomes of diameteral variance at entry and exit (DVEE) using kerosene, de-ionized water and boron carbide powder mixed with these dielectrics. It is found from this figure that DVEE of the micro-holes is lower employing de-ionized water compared to kerosene as dielectric fluid at lower peak current i.e. 0.5 and 1 A. However, at higher peak current i.e. 1.5 and 2 A, DVEE is larger using de-ionized water. It is also found that boron carbide powder mixed kerosene results in large DVEE compared to pure kerosene at low peak current of 0.5 and 1 A. As the machining progresses, the additive boron carbide particles creates more carbon adhesion on the work surface, that further results in

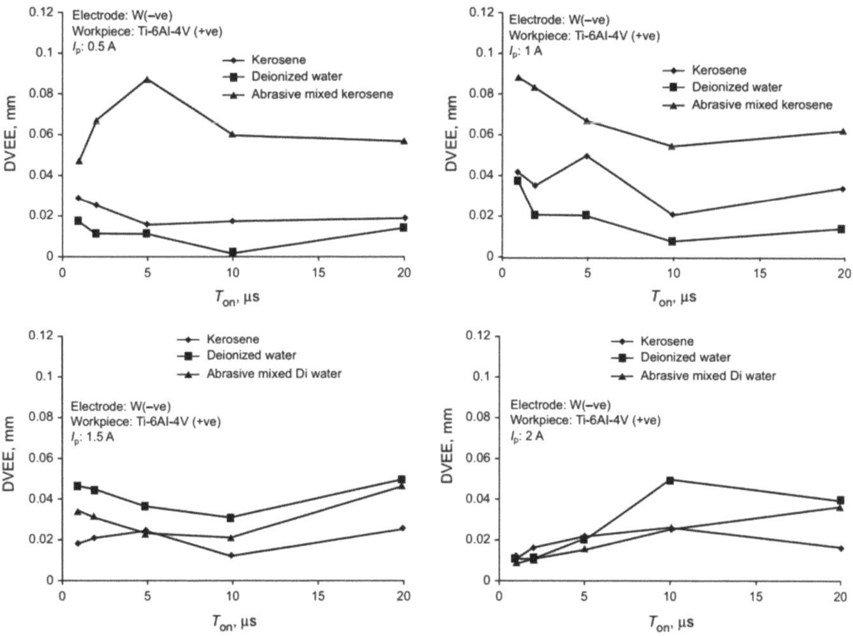

Fig. 3.28 Variation of diameteral variance at entry and exit (DVEE) with pulse duration (Ton) at various fixed peak current (Ip) for different dielectrics [51]

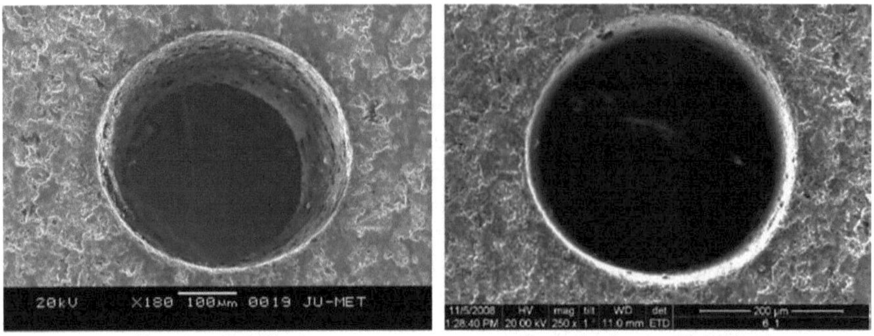

Kerosene (1 A/5 μs) De-ionized water (2 A/10 μs)

Fig. 3.29 SEM micrographs of machined micro-holes using powder mixed dielectrics [51]

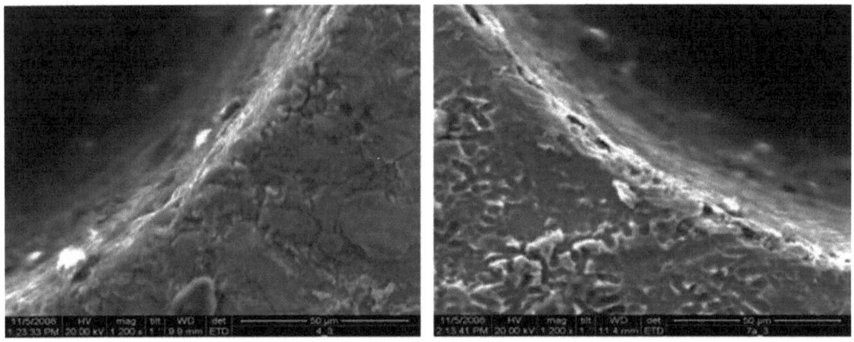

Kerosene (0.5 A/1 μs) De-ionized water (1.5 A/10 μs)

Fig. 3.30 SEM micrographs of white layer of machined micro-hole's edges using powder mixed dielectrics [51]

lower material removal at exit side of micro-hole and greater variance in entry and exit diameters. But, when B_4C additive mixed de-ionized water is used at higher peak current of 1.5 and 2 A, the powder particles help in uniform distribution of discharge energy which in turn leads to better dimensional accuracy micro-holes compared to pure de-ionized water.

Figure 3.29 shows SEM micrographs of the inner surface of machined micro-hole using powder mixed kerosene and de-ionized water at parametric combinations of 1 A/5 μs and 2 A/10 μs of peak current and pulse-on-time. It is revealed from these micrographs that with more discharge energy, inaccurate micro-hole is generated using powder mixed de-ionized water due to more secondary sparking phenomena. However, smooth inner surface is produced using powder mixed de-ionized water than powder mixed kerosene. In Fig. 3.30, SEM micrographs of micro-hole's edge are viewed for examining the recast/white layer formation during micro-hole machining. It is revealed from these figures that the recast layer formed on the edges is very less than using pure dielectrics. It is due to the fact that the additive particles help to remove the molten debris and restrict to form thick white layer on machined micro-hole edges.

3.10.4 Rotation of Micro-tool Electrode

During micro-EDM process, the machining performances can be enhanced by implementing micro-tool electrode rotation about its axis. While rotating the micro-tool, flushing out of debris particles from very narrow IEG occurs quickly due to generation of tangential force in the machining zone. This leads to effective and efficient discharge between the electrodes. In this section, the effects of peak current (Ip), pulse-on-time (Ton) and rotational speed of the micro-tool electrode are explored during micro-hole machining in micro-EDM process.

(a) **Experimental method and machining conditions**

The experiments were conducted using the same ZNC R50 EDM machine to investigate the effects of rotation of electrode in respect of material removal rate (MRR), tool wear rate (TWR), overcut (OC) and DVEE on Ti–6Al–4V workpiece of 1 mm thickness with a brass electrode of 300 μm in diameter. During machining, kerosene is used as dielectric fluid. For rotating the tool electrode, tool rotational attachment has been used in which the rotational speed can be varied between 1 and 300 rpm in a resolution of 2 rpm. The range of peak current (Ip) and pulse-on-time (Ton) selected were 0.5–2 A and 1–20 μs respectively. The experiments were conducted in two stages: (i) varying only peak current from 0.5 to 2 A keeping pulse-on-time, duty factor and flushing pressure as constant at 10 μs, 95%, 0.5 kgf cm^{-2} respectively and (ii) varying only pulse-on-time from 1 to 20 μs keeping peak current, duty factor, flushing pressure constant at 1 A, 95% and 0.5 kgf cm^{-2} respectively with stationary and rotating electrode with a rotational speed of 150 rpm in each case. The peak current has been fixed at 1 A because for micro-machining, very low current density is not sufficient to melt and vaporize the work material and very high current density leads to higher TWR and larger thermal damage of the workpiece surface. On the other hand, pulse-on-time is fixed at 10 μs because short duration is more beneficial for micromachining as it helps in reducing the tool wear. A moderate rotational speed of the electrode has been selected i.e. 150 rpm after conducting trail runs of experiment, because this helps in easy removal of debris from the machining zone and keeps the tool wear at minimum.

(b) **Results and discussion**

Utilizing the micro-tool electrode rotating facility with developed tool holder, the experiments were conducted to study the effects of micro-EDM parameters on process criteria namely material removal rate (MRR), tool wear rate (TWR), overcut (OC) and diameteral variation at entry and exit (DVEE) of the machined micro-holes. The variations of MRR with peak current (Ip) and pulse-on-time

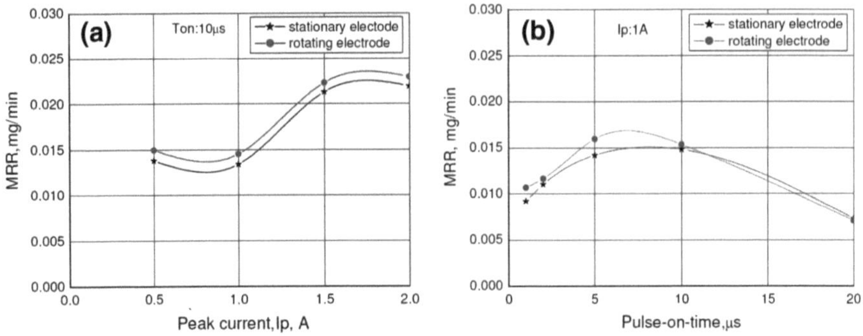

Fig. 3.31 Variation of MRR with **a** peak current and **b** pulse-on-time [58]

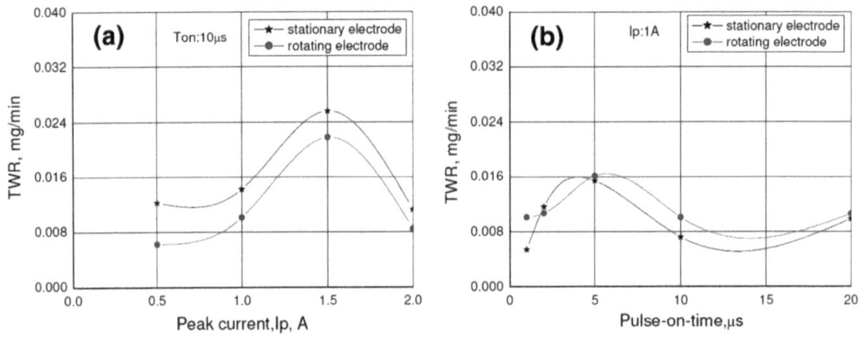

Fig. 3.32 Variation of TWR with **a** peak current and **b** pulse-on-time [58]

(Ton) with stationary and rotating electrode are shown in Fig. 3.31a, b, respectively. It is observed from the graph that MRR increases with increase in peak current as was expected for both stationary and rotating electrodes. Since higher peak current leads to higher discharge energy, so it results in increase of MRR. It is also observed from the same figure that with rotating electrode higher MRR is achieved. The higher MRR with rotating electrode may be attributed to better removal of sludge and carbonized particles from the machining zone due to centrifugal force of rotation. This improved sludge removal due to rotating effect of electrode helps in exposing the actual machining surface, which in turn improves the overall machining condition leading to higher MRR. Thus, for maximum MRR from within the considered range of parametric setting, the best parametric combination in the present case study is 2 A/10 μs/0.5 kgf cm^{-2}/95%/150 rpm.

Figure 3.32a, b respectively show the variations of TWR with peak current (Ip) and pulse-on-time (Ton) with stationary and rotating electrode. It can be observed from the graph that TWR increases as the peak current increases from 0.5 to 1.5 A. However, TWR is observed to decrease from 1.5 to 2 A. The increase in TWR with the increase in peak current can be attributed to the increase in discharge

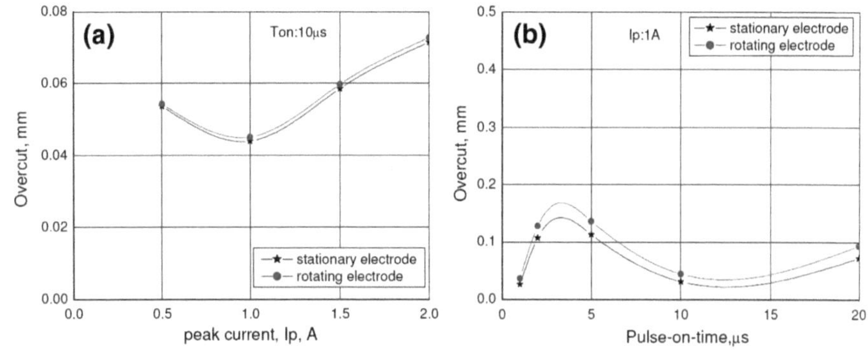

Fig. 3.33 Variation of OC with **a** peak current and **b** pulse-on-time

energy and the rotational effect of the electrode. Further increase in peak current, increases machining efficiency and decreases the tool wear because the tool electrode is subjected to high-energy electric field for shorter duration. Thus, this figure clearly indicates that for micro-hole machining with low TWR, smaller peak current is suitable. Owing to rotational effect, the magnitude of TWR further decreases which is evident from the figure under consideration. This may be attributed to better removal of debris due to the tangential force of rotation. Thus, for least TWR, the best parametric combination within the considered range of parametric settings is 1 A/10 μs/0.5 kgf cm^{-2}/95%/150 rpm.

Figure 3.33a, b respectively show the variations of OC with peak current (Ip) and pulse-on-time (Ton) with stationary and rotating electrode. It is observed from the figure that overcut decreases with the increase in peak current in the range 0.5–1 A which may be attributed to increase in discharge energy with the increase in peak current and enabling faster machining and thereby reducing effective machining time. The reduction in machining time means less exposure of tool electrode to discharge energy which is responsible for tool wear. However, peak current beyond 1 A, overcut has found to increase monotonically. Increase in peak current above 1 A results in larger discharge energy which causes larger MRR resulting in larger overcut. Thus the most suitable parametric combination for least overcut from within the considered range of parametric settings is 1 A/10 μs/95% duty factor/0.5 kgf cm^{-2}/150 rpm.

The variations of DVEE with peak current (Ip) and pulse-on-time (Ton) with stationary and rotating electrode are shown in Fig. 3.34a, b, respectively. It is observed from the same graph that with the increase in peak current, the DVEE decreases. Furthermore, the magnitude of DVEE with rotating electrode is less than that of stationary electrode throughout the considered range of peak current. As the depth of the micro-hole increases, the sparking point shifts radially inward. When the peak current is increased, the thermal energy density increases at the pointed tip of the micro-tools. This high density discharge energy rapidly melts and vapourizes the sharp micro-tool tips and subsequently making the micro-tool end broader

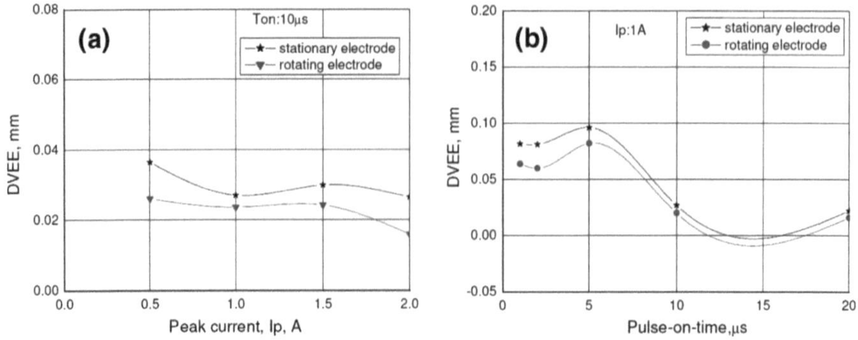

Fig. 3.34 Variation of DVEE with **a** peak current and **b** pulse-on-time

which finally helps in achieving a straight through micro-hole by decreasing DVEE. Further, the rotational speed of the tool helps to remove the sludge and debris efficiently from the machining zone and also reduces the chances of secondary discharge sparking thereby providing stable machining condition for achieving micro-hole with less DVEE. The lowest DVEE achieved is with the parametric combination of 2 A/10 μs/95%/0.5 kgf cm^{-2}/150 rpm.

The improvement in micro-hole geometry with the increase in peak current coupled with rotation of tool electrode may be attributed to better flushing of debris due to rotation of electrode, uniform tool wear, and evenly distribution of discharge energy. Figure 3.35a shows the optical micrographs of micro-hole's diameters at entry and exit machined with parametric setting of 1 A/5 μs/95% duty factor/0.5 kgf/cm^2 i.e. with stationary micro-tool and Fig. 3.35b shows the optical micrographs of micro-hole's diameters at entry and exit machined with parametric setting of 1 A/5 μs/95% duty factor/0.5 kgf/cm^2/150 rpm i.e. with rotating micro-tool. It can be observed from these two micrographs that with rotating tool electrode, straight-through micro-hole can be fabricated.

3.10.5 Reversing Polarity of Electrodes

During micro-EDM process, especially, for micro-drilling operation, the amount of carbon deposition and machining debris increases as micro-hole depth increases and its removal becomes very difficult. Therefore, with the increase in micro-hole depth it becomes necessary to increase the rate of change of polarity in order to facilitate the removal of deposited carbon and debris from the machining zone. In this section, a novel approach has been considered to change polarity of electrodes in reduced time domain. The experimental condition has been designed in a novel way in which the polarity during micro-EDM machining has been changed in a designed fashion in exponential time domain to improve the machining condition, debris removal, machining efficiency to achieve better geometrical accuracy of the

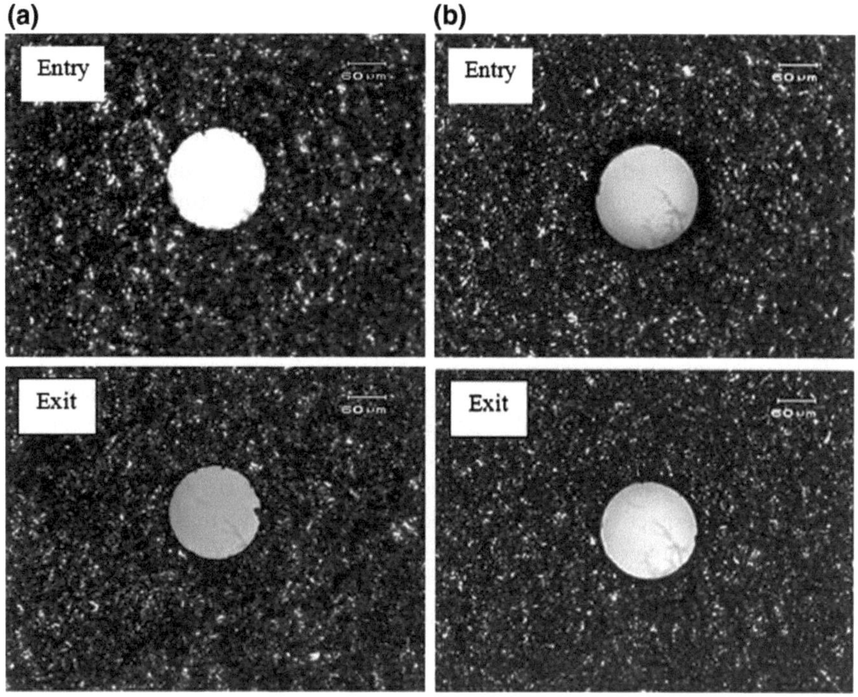

Fig. 3.35 Optical micrographs of micro-holes with **a** stationary and **b** rotating tool electrode machined at 1 A/5 μs/95% duty factor/0.5 kgf/cm^2

micro-hole. In constant polarity machining condition, the job was positive and tool electrode was negative i.e. normal polarity. But, in changing polarity machining condition, the polarity of the job and the tool electrode has been changed in exponential time domain.

(a) **Experimental method and machining conditions**

The machining begins with normal polarity for the first 10 min, then the polarity is changed for next 3 s and again the polarity is switched over to normal polarity and the machining continues for another 9 min before the second change. The time duration of machining with normal polarity reduces after every change and this continues till the through micro-hole is produced on the workpiece. However, the machining with reverse polarity is kept constant at 3 s in each change. The time chart has been prepared after conducting several trail experiments to find out the time required to machine the through micro-hole in 1 mm thick Ti–6Al–4V alloy sheet with 300 μm diameter brass tube electrode. The dielectric fluid used is kerosene. It is learnt from past literatures and experimental investigations in micro-EDM that peak current and pulse-on-time are the most influential parameters. Therefore, these dominating parameters have been selected as process parameters in the present micro-EDM experimentation. To study the effects of pulse-on-time

(Ton) and peak current (Ip), the experimental planning has been carried out firstly by only varying the peak current from 0.5 to 2 A keeping pulse-on-time (Ton), duty factor (t) and flushing pressure (Pr) constant at 10 μs, 95% and 0.5 kgf/cm^2 respectively and secondly by only varying the pulse-on-time (Ton) from 1 to 20 μs keeping peak current (Ip), duty factor (t) and flushing pressure (Pr) constant at 1 A, 95% and 0.5 kgf/cm^2 respectively. The experiments were conducted on the same ZNC R50 EDM machine. Experiment at each parametric setting was conducted three times and the average of the three were considered for calculating material removal rate (MRR), tool wear rate (TWR), overcut (OC) and diameteral variation at entry and exit (DVEE) of the micro-hole.

(b) **Results and discussion**

The variations of MRR with peak current and pulse-on-time with constant and changing polarity, keeping all other process parameters constant i.e. pulse-on-time at 10 μs in case of varying peak current and peak current at 1 A in case of varying pulse-on-time, duty factor at 95%, and flushing pressure at 0.5 kgf/cm^2, shown in Fig. 3.36a, b, respectively. It is noticed from Fig. 3.36a that MRR increases monotonically in both cases with the increase in peak current from 0.5 to 1.5 A but for changing polarity it decreases as peak current increases from 1.5 to 2 A. The magnitude of MRR in both cases is almost equal in the considered peak current range. The low MRR at smaller peak current could be due to lower discharge energy when machining in both constant and changing polarity. However, MRR at changing polarity is low compared with constant polarity as shown Fig. 3.36a due to the change in the position of maximum liberation of heat energy due to sparking. The increase in MRR with increasing peak current is attributed to larger discharge energy. Figure 3.36b shows the variation of MRR with pulse-on-time. It is observed from this figure that the MRR variation is almost opposite in nature for changing and constant polarity at lower pulse-on-time values between 1 and 5 μs. MRR decreases from 1 to 2 μs for constant polarity but in the same pulse-on-time range it increases for changing polarity. In the range from 2 to 5 μs, MRR increases and decreases sharply for constant and changing polarity respectively. However,

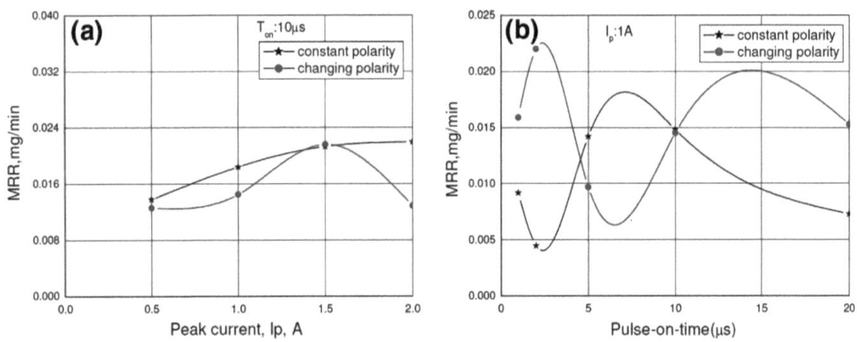

Fig. 3.36 Variation of MRR with **a** peak current and **b** pulse-on-time [59]

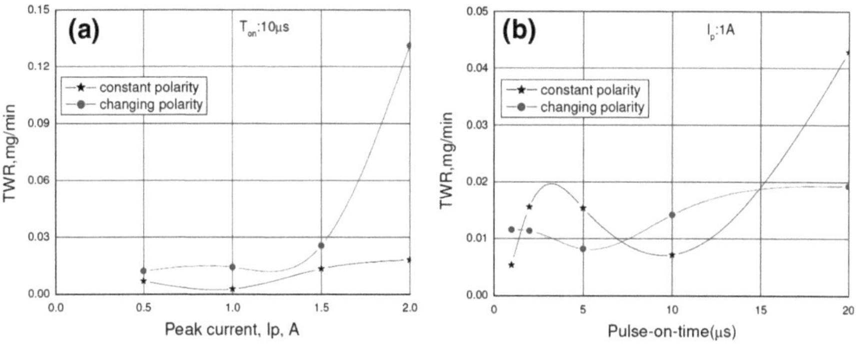

Fig. 3.37 Variation of TWR with **a** peak current and **b** pulse-on-time [59]

from 5 to 10 µs, MRR increases gradually in both cases. Furthermore, in the pulse-on-time range of 10–20 µs, MRR increases slowly for constant polarity whereas it decreases for changing polarity. It can also be observed that maximum MRR is obtained at the lower range of pulse-on-time, i.e. 1–5 µs. In the case of the changing polarity approach to machining, owing to change in polarity, the position of maximum liberation of heat energy due to sparking is changed and at the same time the number of sparking per cycle is increased for lower pulse-on-time. Thus the total heat energy generated in the discharge phenomenon is increased during the parametric setting and a high MRR is obtained.

Figure 3.37a, b show the variation of TWR with peak current and pulse-on-time respectively. It is observed from Fig. 3.37a, as expected, that TWR increases proportionally with the increase in peak current from 0.5 to 2 A at fixed pulse-on-time of 10 µs in both constant and changing polarity. The increase in TWR with the increase in peak current is due to the increase in discharge energy. The increase in magnitude of discharge energy rapidly deteriorates the tool geometry as the thermal energy is concentrated in a very small area (the size of the electrode in this case). However, it can be seen from Fig. 3.37a that TWR is greater with changing polarity; this is exactly opposite to the expectation outlined above, where it was thought that tool wear should be less in this case. This might have occurred owing to the removal of carbon deposition from the tool during normal polarity. However, when pulse-on-time was varied keeping peak current, duty factor, and flushing pressure constant as shown in Fig. 3.37b, TWR has decreased significantly with changing polarity in comparison to constant polarity. This indicates that pulse-on-time is the more critical factor in tool wear than peak current. Therefore, for minimum TWR it is better to use changing polarity with smaller pulse-on-time (1–10 µs) and low peak current of 1 A as this yields higher MRR and low TWR. This observation reveals that shorter pulse-on-time, which means a higher frequency of sparking, and low peak current are suitable for micro-EDM.

Figure 3.38a, b show the variation of OC with peak current and pulse-on-time, respectively. It is observed from Fig. 3.38a that OC decreases sharply from 0.5 to 1 A for both constant and changing polarity and thereafter increases monotonically

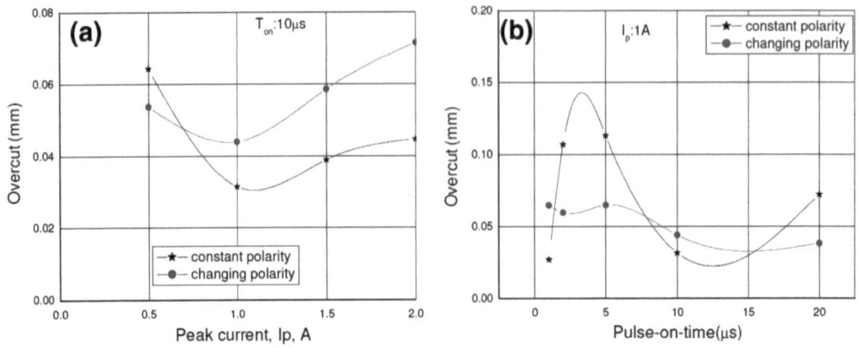

Fig. 3.38 Variation of OC with **a** peak current and **b** pulse-on-time [59]

with increasing peak current in both machining conditions. This clearly indicates that the optimum peak current setting is 1 A at the present level of process parameters range. If peak current is low, the discharge energy is also low, which means a longer machining time exposing the sidewall of the hole to secondary sparking resulting in larger OC. However, the increase in OC at higher peak current is due to higher discharge energy and larger debris concentration in the gap because of higher MRR. Also, it can be observed from Fig. 3.38b that OC increases steeply from 1 to 5 μs, decreases sharply from 5 to 10 μs, and again increases at 10–20 μs for constant polarity. Thus the OC fluctuates with the change in pulse-on-time for constant polarity. However, for changing polarity, OC is found to decrease gradually for pulse-on-time from 1 to 20 μs, thereby suggesting the benefit of using the changing polarity technique for achieving lower overcut in the machining of micro-holes, hence increasing the geometrical accuracy of the machined micro-hole in Ti–6Al–4V material. Further, it is observed throughout the range of pulse-on-time from 1 to 20 μs that the magnitude of OC is far lower with changing polarity than with constant polarity, which is an indicator in itself that changing polarity yields low OC and results in improvement of micro-hole geometry. This is due to the fact that the carbon particles, which are the by-product of the micro-EDM process, are deposited on the surface of the tool electrode and this helps in preventing secondary sparking. Thus only the end face or bottom face is exposed for sparking, thereby reducing OC and resulting in straight-through micro-hole generation, thus improving the accuracy of micro-hole machining.

It is observed from Fig. 3.39a, b that DVEE decreases sharply with the increase in peak current and pulse-on-time. The lowest DVEE is found at 1 A and 10 μs, which can be seen in the figures for both constant and changing polarity machining conditions. Also it is observed that DVEE is less with changing polarity than with constant polarity throughout the peak current and pulse-on-time ranges considered in the experiments. It can be concluded that with the novel polarity changing technique, a straight micro-hole is possible. As the depth of the micro-hole increases, the sparking points shift radially inward. When the polarity is changed the pointed tip of the micro-tool wears off uniformly, making the tool end broader and helping in machining a straight micro-hole, resulting in the decrease in DVEE.

Optical micrographs of the micro-holes (entry and exit diameter) machined with constant and changing polarity at 1 A/20 μs/95% duty factor/0.5 kgf/cm^2 are shown in Fig. 3.40a, b, respectively. The DVEE at parametric combination of 1 A/10 μs/95% duty factor/0.5 kg cm^{-2} for the constant polarity machining condition is 0.0270 mm, and for changing polarity the value is 0.0173 mm.

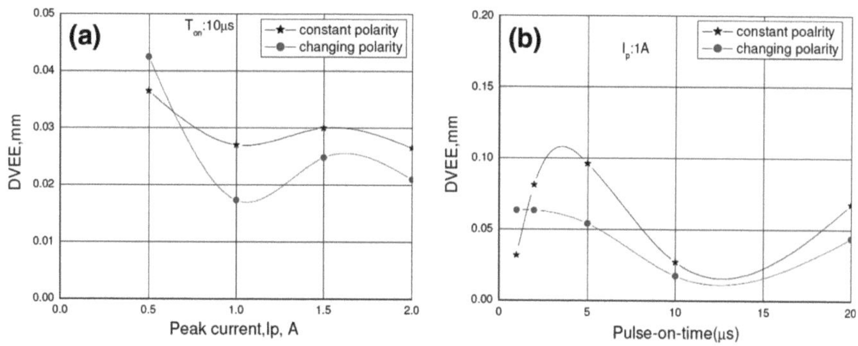

Fig. 3.39 Variation of DVEE with **a** peak current and **b** pulse-on-time [59]

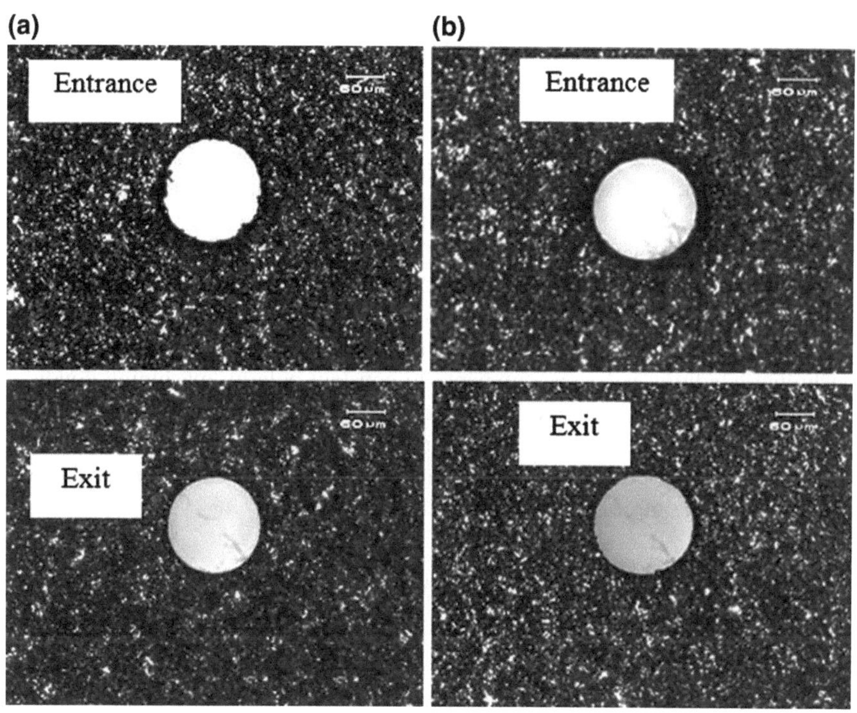

Fig. 3.40 Optical micrographs of micro-holes with **a** constant and **b** changing polarity machined at 1 A/20 μs/95% duty factor/0.5 kgf/cm^2

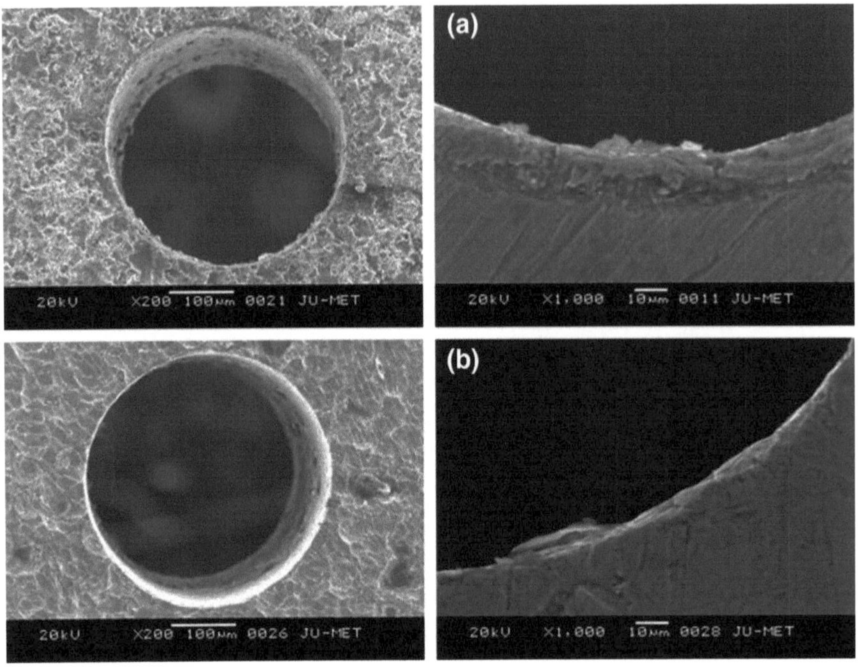

Fig. 3.41 SEM micrographs of micro-holes with **a** constant and **b** changing polarity machined at 1 A/20 μs/95% duty factor/0.5 kgf/cm^2

The comparison of these data and the micrographs clearly indicate that DVEE and geometrical shape are better when using the changing polarity condition for the micromachining of micro-holes by the micro-EDM. The SEM micrographs of machined micro-holes at parametric settings of 1 A/20 μs/95% duty factor/0.5 kg cm^{-2} with constant and changing polarity techniques are shown in Fig. 3.41a, b, respectively. It can be observed from these figures that the thickness of recast layer formed on micro-hole surface with polarity changing technique machining condition is less as compared to constant polarity machining condition corroborating the fact that the surface quality of the machined micro-hole has improved with this new machining technique. This observation clearly indicates that with changing polarity technique, a better machining condition is achieved during micro-hole machining by micro-EDM process.

3.11 Conclusions

In this chapter, a brief introduction of micro-EDM process and the implementation of innovative machining strategies for improving the process performances during machining was presented. Moreover, the principle of normal EDM and then micro-EDM were described with brief details of sub-systems and significant process parameters with performance measures. Detailed research and investigation was performed during micro-EDM of Ti–6Al–4V alloy implementing various machining strategies such as ultrasonic vibration assisted micro-EDM, parametric influence and comparative study of process criteria employing two different dielectrics i.e. kerosene and de-ionized water, use of boron carbide mixed kerosene and de-ionized water as dielectrics, providing rotation of micro-tool electrode during micro-hole generation and reversing the polarity of electrodes. The experimental results show that these innovative machining strategies have great influences for improving the machining rate and geometrical accuracy in terms of material removal rate, tool wear rate, overcut and diameteral variance of entry and exit of micro-hole. Furthermore, the process performances also greatly depend on significant process parameters such as peak current, pulse-on-time and flushing pressure. As micro-EDM process is slow process, therefore, novel machining strategies mentioned above must be implemented to improve the process performances to improve the accuracy of micro-structure, surface finish and efficiency. Furthermore, innovative hybrid micro-machining processes can also be developed for micro-EDM to machine hard-to-machine materials like Ti–6Al–4V alloy for increasing the yield of micro-manufacturing.

Acknowledgements The authors acknowledge the financial support and assistance provided by CAS Ph-IV programme of Production Engineering Department of Jadavpur University under University Grants Commission (UGC), New Delhi, India.

References

1. Masuzawa T (2000) State of the Art of Micromachining. Annals of the CIRP 49(2):473-488
2. Ho KH, Newman ST (2003) State of the art electrical discharge machining (EDM). International Journal of Machine Tools & Manufacture 43(13):1287-1300
3. Kunieda M, Lauwers B, Rajurkar KP, Schumacher BM (2005) Advancing EDM through Fundamental Insight into the Process. CIRP Annals - Manufacturing Technology 54(2):64–87
4. König W, Klocke F (1997) Fertigungsverfahren - 3: Abtragen Und Generieren, Vol.3. Springer, Berlin
5. Pham DT, Dimov S, Bigot S, Ivanov A, Popov KB (2004) Micro-EDM—recent developments and research issues. Journal of Materials Processing Technology 149(1–3): 50–57
6. Gentili E, Tabaglio L, Aggogeri F (2005), Review on micromachining techniques. Courses Lectures, CISM International Centre for Mechanical Science 486:387-396
7. Katz Z, Tibbles CJ (2005) Analysis of Micro-scale EDM process. International Journal of Advanced Manufacturing Technology 25(9):923-928
8. Rajurkar KP, Levy G, Malshe A, Sundaram MM, McGeough J, Hu X, Resnick R, DeSilva A (2006) Micro and nano machining by electro-physical and chemical processes. Annals of the CIRP 55(2):643–666
9. Masuzawa T, Sata T, Kinoshita N (1971) The occurring mechanism of the continuous arc in micro-energy EDM by RC circuit. Journal of Japan Society of Electrical-Machining Engineers 5(9):35–52

10. Han F, Yamada Y, Kawakami T, Kunieda M (2004) Improvement of Machining Characteristics of Micro-EDM Using Transistor Type Isopulse Generator and Servo Feed Control. Precision Engineering 28(4):378–385
11. Masuzawa T, Fujino M (1980) Micro Pulse for EDM. In: Proceedings of the Japan Society for Precision Engineering Autumn Conference, Japan 140-142 (in Japanese)
12. Han F, Yamada Y, Kawakami T, Kunieda M (2003) Investigations on Feasibility of Sub Micrometer Order Manufacturing Using Micro-EDM. In: Proceedings of American Society of Precision Engineering (ASPE) Annual Meeting 30:551–554
13. Jahan MP, Wong YS, Rahman M (2009) A Study on the Fine-Finish Die-Sinking Micro-EDM of Tungsten Carbide Using Different Electrode Materials. Journal of Materials Processing Technology 209(8):3956–3967
14. Mahardika M, Tsujimoto T, Mitsui K (2008) A New Approach on the Determination of Ease of Machining by EDM Processes. International Journal of Machine Tools & Manufacture 48 (7-8):746–760
15. Yan BH, Huang FY, Chow HM, Tsai JY (1999) Micro-Hole Machining of Carbide by Electrical Discharge Machining. Journal of Materials Processing Technology 87(1-3):139–145
16. Hung JC, Lin JK, Yan BH, Liu HS, Ho PH (2006) Using a Helical Micro-Tool in Micro-EDM Combined with Ultrasonic Vibration for Micro-Hole Machining. Journal of Micromechanics and Microengineering 16(12):2705–2713
17. Levy GN, Ferroni B (1975) Planetary Spark Erosion - Applications and Optimization. In: Proceedings of the 16th MTDR Conference 291–297
18. Jahan MP, Rahman M, Wong YS (2011) Study on the Nano-Powder Mixed Sinking and Milling Micro-EDM of WC–Co. International Journal of Advanced Manufacturing Technology 53(1):167–180
19. Tsai YY, Masuzawa T (2004) An index to evaluate the wear resistance of the electrode in micro-EDM. Journal of Materials Processing Technology 149(1-3):304–309
20. Kunieda M, Lauwers B, Rajurkar KP, Schumacher BM (2005) Advancing EDM through Fundamental Insight into the Process. Annals of the CIRP 54(2):599–622
21. Yu Z, Jun T, Kunieda M (2004) Dry EDM of Cemented Carbide. Journal of Materials Processing Technology 149(1-3):353–357
22. Tao J, Shih AJ, Ni J (2008) Near-Dry EDM Milling of Mirror-Like Surface Finish. International Journal of Electrical Machining 13:29–33
23. Egashira K, Taniguchi T, Hanajima S (2006) Planetary EDM of Micro Holes. International Journal of Electrical Machining 11:15–18
24. Pradhan BB, Masanta M, Sarkar BR, Bhattacharyya B (2009) Investigation of electro-discharge micro-machining of titanium super alloy. International Journal of Advanced Manufacturing Technology 41:1094-1106
25. Pradhan BB, Bhattacharyya B (2009) Modelling of micro-electro discharge machining during machining of titanium alloy Ti-6Al-4V using response surface methodology and artificial neural network algorithm. Proceedings of the Institution of Mechanical Engineers Part B: Journal of Engineering Manufacture 223(6):683–693
26. Ali MY, Rahman NABA, Aris EBM (2012) Powder mixed micro electro discharge milling of titanium alloy: investigation of material removal rate. Advanced Materials Research 383–390:1759–1763
27. Meena VK, Azad MS (2012) Grey Relational Analysis of Micro-EDM Machining of Ti-6Al-4V Alloy. Materials and Manufacturing Processes 27(9):973-977
28. Porwal RK, Yadava V, Ramkumar J (2014) Modelling and multi-response optimization of hole sinking electrical discharge micromachining of titanium alloy thin sheet. Journal of Mechanical Science and Technology 28(2):653-661
29. Tiwary AP, Pradhan BB, Bhattacharyya B (2014) Application of multi-criteria decision making methods for selection of micro-EDM process parameters. Advances in Manufacturing 2(3):251-258

30. Kuriachen B, Mathew J (2014) Modeling of material removal mechanism in micro electric discharge milling of Ti-6Al-4V. Applied Mechanics and Materials 592–594:516–520
31. Plazaa S, Sancheza JA, Pereza E, Gila R, Izquierdob B, Ortegaa N, Pombob I (2014) Experimental study on micro EDM-drilling of Ti6Al4V using helical electrode. Precision Engineering 38(4):821–827
32. Moses MD, Jahan MP (2015) Micro-EDM machinability of difficult-to-cut Ti-6Al-4V against soft brass. International Journal of Advanced Manufacturing Technology 81(5):1345-1361
33. Tiwary AP, Pradhan BB, Bhattacharyya B (2015) Study on the influence of micro-EDM process parameters during machining of Ti–6Al–4V superalloy. International Journal of Advanced Manufacturing Technology 76(1–4):151–160
34. Kuriachen B, Mathew J (2016) Effect of powder mixed dielectric on material removal and surface modification in micro electric discharge machining of Ti-6Al-4V. Materials and Manufacturing Processes 31(4):439-446
35. Kuriachen B, Varghese A, Somashekhar KP, Panda S, Mathew J (2015) Three-dimensional numerical simulation of microelectric discharge machining of Ti-6Al-4V. International Journal of Advanced Manufacturing Technology 79(1):147-160
36. Kuriachen B, Mathew J (2016) Spark radius modeling of resistance-capacitance pulse discharge in micro-electric discharge machining of Ti-6Al-4V: an experimental study. International Journal of Advanced Manufacturing Technology 85(9):1983-1993
37. Kremer D, Lebrun JL, Hosari B, Moisan A (1989) Effects of ultrasonic vibrations on the performance in EDM. Annals of the CIRP 38(1):199–202
38. Ichikawa T, Natsu W (2013) Realization of micro-EDM under ultra-small discharge energy by applying ultrasonic vibration to machining fluid. The Seventeenth CIRP Conference on Electro Physical and Chemical Machining (ISEM) 6:326-331
39. Liew PJ, Yan J, Kuriyagawa T (2014) Fabrication of deep micro-holes in reaction-bonded SiC by ultrasonic cavitation assisted micro- EDM. International Journal of Machine Tools & Manufacture 76:13-20
40. Jahan MP, Rahman M, Wong YS, Fuhua L (2010) On-machine fabrication of high-aspect-ratio micro-electrodes and application in vibration-assisted micro-electro discharge drilling of tungsten carbide. Proceedings of the Institution of Mechanical Engineers Part B: Journal of Engineering Manufacture 224(5):795-814
41. Changshui G, Zhengxun L (2003) A study of ultrasonically aided micro-electrical-discharge machining by the application of work piece vibration. Journal of Materials Processing Technology 139(1-3):226-228
42. Prihandana GS, Mahardika M, Hamdi M, Wong YS, Mitsui K (2009) Effect of micro-powder suspension and ultrasonic vibration of dielectric fluid in micro-EDM processes-Taguchi approach. International Journal of Machine Tools & Manufacture 49(12-13):1035-1041
43. Ghoreishi M, Atkinson J (2002) A comparative experimental study of machining characterstics in vibratory, rotary and vibro-rotary electro- discharge machining. Journal of Materials Processing Technology 120(1-3):374-384
44. Shabgard MR, Sadizadeh B, Kakoulvand H (2009) The Effect of Ultrasonic Vibration of Workpiece in Electrical Discharge Machining of AISIH13 Tool Steel. International Journal of Mechanical, Aerospace, Industrial, Mechatronic and Manufacturing Engineering 3 (4):404-408
45. Garn R, Schubert A, Zeidler H (2011) Analysis of the effect of vibrations on the micro-EDM process at the workpiece surface. Journal of Precision Engineering 35(2):364-368
46. Tong H, Li Y, Wang Y (2008) Experimental research on vibration assisted EDM of micro-structures with non-circular cross-section. Journal of Materials Processing Technology 208(1-3):289–298
47. Zhang QH, Zhang JH, Deng JX (2002) Ultrasonic vibration electrical discharge machining in gas. Journal of Materials Processing Technology 129(1-3):135–138
48. Kim DJ, Yi SM, Lee YS, Chu CN (2006) Straight hole micro EDM with a cylindrical tool using a variable capacitance method accompanied by ultrasonic vibration. Journal of Micromechanics and Microengineering 16(5):1092–1097

49. Zhang QH, Du R, Zhang JH, Zhang QB (2006) An investigation of ultrasonic-assisted electrical discharge machining in gas. International Journal of Machine Tools & Manufacture 46(12-13):1582–1588
50. Bhattacharyya B, Gangopadhyay S, Sarkar BR (2007) Modelling and analysis of EDMed job surface integrity. Journal of Materials Processing Technology 189(1-3):169-177
51. Kibria G, Sarkar BR, Pradhan BB, Bhattacharyya B (2010) Comparative study of different dielectrics for micro-EDM performance during microhole machining of Ti-6Al-4V alloy. International Journal of Advanced Manufacturing Technology 48(5-8):557-570
52. Luo YF (1997) The dependence of interspace discharge transitivity upon the gap debris in precision electro-discharge machining. Journal of Materials Processing Technology 68 (2):127–131
53. Jeswani ML (1981) Effect of the addition of graphite powder to kerosene used as the dielectric fluid in electrical discharge machining. Wear 70(2):133–139
54. Pierson HO (1996) Handbook of Refractory Carbides and Nitrides: Properties, Characteristics, Processing and Applications, Noyes Publications, Westwood, New Jersey, USA
55. Luis CJ, Puertas I (2007) Methodology for developing technological tables used in EDM processes of conductive ceramics. Journal of Materials Processing Technology 189 (1-3):301-309
56. Ming QY, He LY (1995) Powder-suspension dielectric fluid for EDM. Journal of Materials Processing Technology 52(1):44–54
57. Tzeng YF, Lee CY (2001) Effects of powder characteristics on electro discharge machining efficiency. International Journal of Advanced Manufacturing Technology 17(8):586–592
58. Pradhan BB, Sarkar BR, Kibria G, Bhattacharyya B (2009) EDM with rotational electrode for machining micro holes in Ti-6Al-4V. Journal of Institution of Engineers (India) 89:3-8
59. Pradhan BB, Bhattacharyya B (2008) Improvement in microhole machining accuracy by polarity changing technique for microelectrode discharge machining on Ti–6Al–4V. Proceedings of the Institution of Mechanical Engineers Part B: Journal of Engineering Manufacture 222(2):163-173

Chapter 4
Advancements in Micro Wire-cut Electrical Discharge Machining

Asit Baran Puri

Abstract A WEDM process may be called as a Micro WEDM (MWEDM) process when it is used for manufacturing micro parts. The rest underlying theory is same as traditional WEDM. A micro part is machined with good dimensional accuracy and surface finish when the kerf (slot) width becomes considerably smaller compared to that obtainable in conventional WEDM. Truly speaking, the cutting tool in WEDM/MWEDM is not the wire, but it is the pulse (electrical discharge). A minimum kerf is ensured in Micro WEDM, when (i) pulse sizes are made extremely small, (ii) extremely thin wire is used ($\phi = 20$–$100\ \mu m$), and (iii) process inaccuracies along with the discharge gap are minimized. It is required for Micro WEDM to maintain the pulse energy in the order of 10^{-5}–10^{-7} J. Process inaccuracies are minimised by minimising the amplitude of wire vibration and wire lag. Hence, for an MWEDM machine, the following subsystems are modified as compared to a conventional WEDM setup: (a) the machine tool configuration is designed mainly to eliminate stray capacitance or leakage of charges from the gap, (b) the pulse generator produces discharges with pulse energy preferably in the order of 10^{-6}–10^{-7} J, (c) a pulse discrimination system is installed to avoid arc or other abnormal discharges, (d) a closed-loop controlled proper wire transportation system is needed, (e) the wire diameter should be 20–80 μm and preferably of tungsten, (f) a suitable oil dielectric is used and lastly, (g) a precise closed-loop controlled servo mechanism is used for gap and work-table feed control to operate at steps at submicron level.

A.B. Puri (✉)
Mechanical Engineering Department, National Institute of Technology Durgapur, Durgapur 713209, West Bengal, India
e-mail: abpuri2000@yahoo.co.in

G. Kibria et al. (eds.), *Non-traditional Micromachining Processes*,
Materials Forming, Machining and Tribology,
DOI 10.1007/978-3-319-52009-4_4

4.1 Introduction

Micro manufacturing and micro fabrication are synonymous while micromachining is only a subset of them. Micromachining may be defined as a process or a set of processes used to manufacture or fabricate parts or machine components whose dimensional feature(s) are most conveniently described in microns. Micromachining is done by conventional machining (also called tool based machining, TBM) and nonconventional machining whereas, surface micromachining, bulk micromachining and other high aspect ratio machining (e.g. LIGA) pertain to a completely different process technology, commonly called MEMS technology. However, micro wire-cut electrical discharge machining (Micro WEDM) process is a variant of WEDM process and WEDM process is a variant of EDM process. EDM has been a very important and versatile machining in its applicability in the current manufacturing scenario. It has applications in diverse major modes as in: (a) electrical discharge die sinking, (b) electrical discharge drilling, (c) electrical discharge grinding, and (d) wire-cut EDM. Although the traditional EDM processes are the most versatile, a few shortcomings of the process are often observed with regard to the tool in applications. In EDM, both the job and tool electrode erode up. While a part is to be produced with very fine tolerance, the tool should have even finer tolerance, which sometimes is very difficult to achieve. A tool electrode can produce only a limited number of parts, sometimes only 1 or 2 owing to the tool wear. For a definite shape, size and tolerance of a workpiece a set of new tool electrodes are required and for new jobs these tool electrodes are of no further use. Also, the consumption of labour and material in manufacturing the electrodes are relatively high. All these problems related to tool are eliminated in WEDM, since the tool electrode used here is a very thin disposable wire.

4.2 Working Principle

A WEDM process may be called as a Micro WEDM process only when it is used for manufacturing micro parts, the rest underlying theory remaining same as of WEDM. A scheme of the MWEDM setup is shown in Fig. 4.1. In WEDM, the electrode is a continuously moving electrically conductive wire. The mechanism of material removal in WEDM and MWEDM involves complex erosion effect by rapid, repetitive and discrete spark discharges between the wire tool and the job immersed in a liquid dielectric medium. A DC power supply is used to generate a very high frequency pulses. The wire is unwounded from the feeding spool by a drive unit moving at constant speed to pass through a tension mechanism and then through the workpiece. The expended wire is either rewound on a take-up spool or chopped off in small pieces to be collected in a bin. However, a fresh electrode wire is continuously presented to the machining zone. The work is connected to the positive terminal and the wire-tool is connected to the negative terminal of the

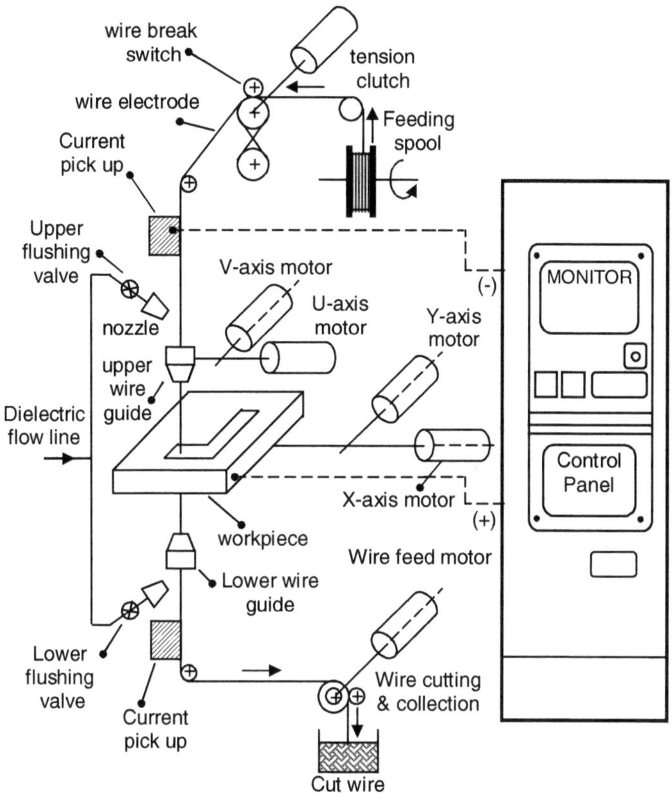

Fig. 4.1 A basic scheme of the MWEDM setup

power source. A DC power source is used as fixed polarity is desired. In MWEDM process, the electrical energy in the form of short duration electrical impulses is required to be supplied to the machining gap. The pulse energy is maintained commonly in the order of 0.1–10 μJ. Therefore, specially designed generators are employed and also different types of electrical circuits are used to meet the above requirement. Each of them has its own merits and demerits. When the circuit is switched on, a suitable voltage is generated between tool and work surface. The voltage goes on increasing and when a certain voltage (breakdown voltage) is reached, a discharge takes place. As soon as an avalanche of electron reaches the workpiece from tool, the voltage becomes almost zero between the tool and work, and requires a certain time to generate that suitable voltage to cause another spark. Also, at the end of one spark, the non-conducting medium (dielectric) helps to break the contact between the tool and work allowing a gradual increase in voltage between the terminals. The intense heat generated melts and vapourises the material in the sparking zone. The metal in the form of liquid drops is dispersed into the space surrounding the electrodes by the explosive pressure of the gaseous products in the discharge. This results in the formation of a tiny crater at the point of discharge in the workpiece. In MWEDM, the diameter of the crater may range from submicron level to a few micron. Furthermore, the particles removed from the electrodes due to the discharge fall in the liquid and cool down to form a colloidal suspension of metal. The dielectric, thus, becomes contaminated.

4.2.1 Mechanism of Sparking and Material Removal

The mechanism of sparking and material removal in WEDM and MWEDM are same. The spark gap between the electrodes at various points would vary due to the asperities present at the surfaces of tool and work as well as due to wire-tool vibration. When a suitable voltage is built up across the electrodes, a strong electrostatic field is established causing cold emission of electrons from cathode to anode at a point where least resistance is offered by the dielectric medium. These electrons start accelerating to gain sufficient velocity with which they collide with dielectric fluid molecules breaking them into electrons and positive ions. Electrons so produced also accelerate and collide with dielectric molecules to dislodge more and more electrons. Ultimately, an ionized column is established in the dielectric across the gap. The conductivity of this ionized column is very high, so it connects the tool and work, and an avalanche of electrons flow from cathode to anode. This is seen as spark. As soon as this happens, the gap between the electrodes increases at that spot and the next spark occurs at somewhere else where the spark gap is minimum. Thus, the sparks move all over the electrode surface. As a consequence of spark, the temperature generated is of the order of 8000–10,000 °C which melts and evaporates the metal in the sparking zone, i.e., both in the surfaces of tool and work. Also the dielectric fluid adjacent to the sparking zone evaporates and other gaseous products may form. The products try to expand against the inertia of the adjacent dielectric creating high compressive pressure wave. Shortly after, the discharge column collapses and gaseous products cool down. Pressure starts falling leading to cavitation (when pressure falls below its vapour pressure). So the surrounding fluid rushes towards the sparking zone and occupies the space resulting in a mechanical blast. This blast removes the molten metal from anode and cathode in the shape of metallic globules. Thus, major portion of the molten metal is dispersed into the dielectric fluid medium and forms a colloidal suspension. A small fraction of molten metal adheres to the crater surface and resolidifies upon cooling to form a thin white layer.

4.2.2 Work Material

Any electrically conductive material can be used as work material with a minimum electrical conductivity of $\kappa = 0.01$ S/cm (siemens/cm). Electrical conductivity (κ)

is the reciprocal of electrical resistivity, ρ (i.e., $\kappa = 1/\rho$). However, EDM is widely used in machining of high strength and temperature resistive (HSTR) alloys, hard carbides, hardened steels etc.

4.2.3 Dielectric Fluids

The most important requirements of a dielectric fluid are: (a) it should have high dielectric strength, i.e. it should remain non-conductive until the breakdown voltage is reached, (b) it should have high cooling ability, (c) it should deionise the spark gap quickly and completely after the discharge takes place and (d) it should possess low viscosity to assure good flow characteristics at room temperature. All these qualities are generally found in hydrocarbon oils having a viscosity of 3–6 cSt at 20–22 °C. This is the reason why kerosene is most widely used dielectric medium in die-sinking EDM process. However, a dielectric, usually, de-ionized water ($\kappa_w \sim 1$ μS/cm), is continuously forced fed to the machining zone to flush away the eroded particles from the machining zone in WEDM. De-ionized water is used as the dielectric medium because it not only affords better flushing and cooling but also creates an environment of low conductivity dielectric medium to provide a larger spark gap compared to oil dielectric. The wire electrode under tension is subjected to vibrations due to various forces (like, electrostatic, electrodynamic, hydraulic etc.) during spark discharges. That is why a larger spark gap is required to be maintained than that of traditional EDM process. It is worthy to mention that water is not at all a suitable dielectric for traditional EDM as it causes rapid electrode wear.

Although deionized water is normally used as the dielectric liquid in conventional WEDM, dielectric oil ($\kappa_w < 0.1$ μS/cm) is more preferable than water as the dielectric in micro-WEDM for several reasons. They are: (a) electrolysis corrosion of workpiece does not occur in oil dielectric, (b) the discharge craters generated in oil dielectric are smaller than those in water even under the same pulse conditions, and (c) the permittivity of water is about 36 times higher than that of oil, i.e., the capacitance of the working gap in water is about 36 times larger than that in oil [1]. This is one way of obtaining very low pulse energy content in the order of 0.1–10 μJ for MWEDM. However, in addition to the above, the tool wear is less and deionisation unit is not necessary when oil is used as dielectric.

4.2.4 Wire Electrode

In general, copper and brass are normally preferred as wire electrode material in WEDM with diameters of the wire ranging from 100–300 μm. In MWEDM, tungsten and molybdenum wire in the range of 20–100 μm are used as extremely sharp corners and high precision are the major requirements of machining. Brass

Table 4.1 Comparison of the properties of tungsten and molybdenum [2]

	Properties	Tungsten, W	Molybdenum, Mo
1	Thermal expansion	4.5 µm/(m·K) (at 25 °C)	4.8 µm/(m·K) (at 25 °C)
2	Thermal conductivity	173 W/(m·K)	138 W/(m·K)
3	Electrical resistivity	52.8 nΩ·m (at 20 °C)	53.4 nΩ·m (at 20 °C)
4	Young's modulus	411 GPa	329 GPa
5	Shear modulus	161 GPa	126 GPa
6	Bulk modulus	310 GPa	230 GPa
7	Poisson ratio	0.28	0.31
8	Mohs hardness	7.5	5.5
9	Vickers hardness	3430–4600 MPa	1400–2740 MPa
10	Brinell hardness	2000–4000 MPa	1370–2500 MPa

wire and zinc coated brass wire are also used. The resistance to erosion and tensile strength of tungsten and molybdenum are much higher than that of copper or brass wire. Hence, a very high precision machining is achieved with the least possibility of wire rupture. At the same time, tungsten and molybdenum wires are too costly. A brief comparison of the properties of tungsten and molybdenum is given in Table 4.1 [2]. It may be seen that tungsten is a few steps ahead of molybdenum so far as wire material is concerned. That is why tungsten wire is used almost universally. However, for high precision machining a very low flushing pressure is used as the wire vibrates during machining. Very often, the working gap is simply kept immersed in dielectric medium (no flushing condition).

4.2.5 *Wire Tool Failure and Its Prevention*

Among other problems, wire rupture is perhaps the most serious problem in the application of MWEDM. The wire EDM machine is used to run continuously without operator. If the electrode wire breaks during the process, rethreading of the wire is done causing the downtime of the machine. It has two drawbacks as follows: (i) it increases the overall machining time of a part or product, and (ii) the machining accuracy of the profile decreases, since a stripe (tool mark) is left on the work surface very often. This is due to the fact that it is very difficult to reset the wire at the same position where it ruptured. To make the matter worse, the wire rupture phenomenon has a more or less stochastic behaviour. However, the causes of wire rupture during machining may be listed as follows:

(a) Contact resistance (between the wire and the current pick-up): Current pick-up is a conductive ceramic object through which the electrical power is supplied into the wire. In this case, localized joule heating depending on the contact resistance between the wire and the current pick-up(s), becomes intense upon increasing working current or effective pulse frequency. Thus the surface of the wire is soiled.

(b) Improper winding system of the wire.
(c) Improper wire feed.
(d) Improper wire tension adjustment.
(e) Short circuiting: When the work-table feed rate (feeding of workpiece) is more than the cutting speed, the wire electrode may get physical contact with the workpiece. A short circuit is said to occur in this case. A high current passes through a very small segment of wire producing joule heating followed by wire rupture.
(f) Improper pulse types and frequency: The basic reason of wire rupture under this category may be called as thermal failure due to time and space concentration of spark discharges. This renders the local heating of the wire more severe and wire rupture becomes inevitable. The increase in pulse frequency leads to the same phenomenon to occur. Occurrence of arc pulses is another reason of wire rupture. Thus, the remedies to this type of wire rupture are: (i) avoiding a high pulse frequency; and (ii) avoiding occurrence of increased arc and other abnormal pulses. Wire rupture of the first type may be avoided by increasing pulse off time and to achieve a fairly stable process. The wire rupture of the second type may also be avoided by increasing pulse off time, and employing a specially designed pulse discriminator system which eliminates or suppresses the arc and other abnormal pulses by maintaining a suitable ignition delay period (t_l).

4.3 Micro WEDM Machine

In micro-WEDM, the key is to limit the energy in the discharge. Lower energy results in low MRR, high accuracy and better surface finish. To generate small energy pulses of the order of 0.1–10 μJ, a DC power supply system should be properly designed as micro energy pulse generator. For better MRR discharge frequency should be increased. This is ensured with the use of MOSFET. A MOSFET (metal oxide semiconductor field effect transistor) is a device used for amplifying or switching electronic signals. But, the biggest problem is posed by 'stray capacitance' prevailing in an R–C circuit. Any two adjacent conductors can be considered as a capacitor. If the conductors are not close together for long, they may also form a capacitor although the capacitance will be small. This (often unwanted) effect is termed 'stray capacitance'. Stray capacitance can allow signals to leak between two isolated circuits (an effect called crosstalk), and it can be a limiting factor for proper functioning of circuits at high frequency. The problem of stray capacitance is tackled with the use of granite basement, piezo-electric ceramics and other electrically nonconductive strong materials. Granite is quite suitable for micro-WEDM machine equipment because it has good insulating characteristics and has no inductance. Hence, it can be used for basement and work table for its lower heat transmission coefficient. It is insensitive to temperature, and good at absorbing vibration. Piezo-electric ceramics are good for making driving parts. It has very high displacement sensitivity and high frequency response. However, because of its limited stroke, researchers often make special mechanism to increase its force and enlarge its stroke. Apart from the above, a good MWEDM machine tool should have, (i) iso-energy and iso-frequency micro pulse generators fitted with a suitable pulse discriminating system, (ii) a wire transport system supplemented with a closed loop feedback control of precision wire tension mechanism. However, designing and developing the micro pulse generator is the task of an electrical engineer.

4.3.1 Taper Cutting System in MWEDM

While cutting a straight job, the upper guide remains stationary in u-v plane at u = 0 and v = 0 with respect to lower guide (Fig. 4.1). Actually, instead of movement of both the upper and lower guides, the worktable carrying the job blank is moved along the programmed path while the lower guide remains permanently stationary. Thus, during straight cutting of jobs, the wire tool is kept exactly vertical to x-y and u-v plane. In case of taper cutting, the upper wire guide is not kept stationary but moves along u and v axes with respect to the lower guide at a predetermined pace with the worktable movement. The lower guide is kept stationary as stated earlier. Thus, the wire does not remain vertical, embracing the generation of motion of the top slide in u and v axes. The angle so produced with the vertical axis, can be varied at various points which can be controlled by the path programming. A desired taper angle can be achieved by simultaneous control of the movement of the x-y table and the u-v auxiliary table along their respective predetermined path as designed and stored in the controller. However, taper cutting in WEDM system is performed very cautiously and at a considerably low cutting speed as compared to straight cutting due to following difficulties encountered: (i) achieving proper flushing is difficult as the wire remains tilted and flushing direction is vertical (this renders the heating of wire more severe causing the wire rupture due to the thermal load), and (ii) wide fluctuation in wire tension.

4.3.2 Trim Cutting Features in MWEDM

The trim cut (also called skim cut) is an operation where the wire electrode traces back the same path after the first cut is over (Fig. 4.2). Obviously, this is used for machining of job profile in multi-pass cuts. Multi-pass cutting of jobs is usually done in two or three cuts depending on the specific requirement. While trim cutting, very low energy pulses are applied to obtain a very less amount of material

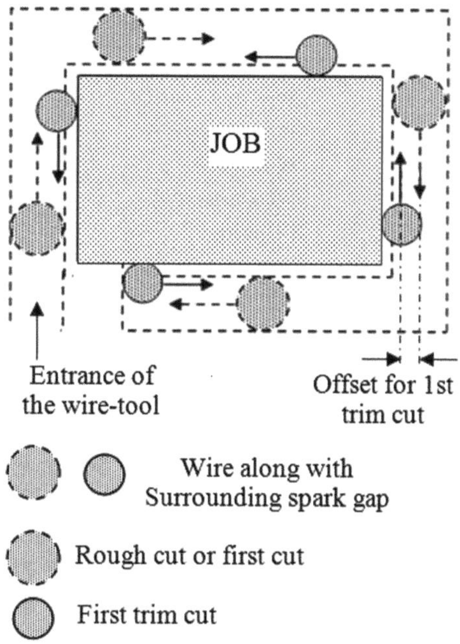

Fig. 4.2 Trim cutting in MWEDM

removal. Generally, high pressure flushing is avoided in case of trim cut to obtain better accuracy. However, the objective of using trim cuts may be listed as follows: (i) to achieve better job accuracy; (ii) to achieve improved surface finish; (iii) to reduce inaccuracies produced by minor job deformations after the first cut due to residual stresses in the job material; (iv) to reduce the bow effect (or cambering effect) on the cut job surface produced in the first cut due to wire vibration or adverse flushing conditions; (v) to improve the life of the job by reducing thermally affected layer formed on the machined surface in the first cut or previous cut.

4.3.3 Micro WEDM Process Parameters

A few important process parameters with their operating ranges in MWEDM is shown in Table 4.2.

4.4 Micro WEDM Circuits and Operating Principles

In the MWEDM process, electrical energy in the form of short duration impulses is required to be supplied in the machining gap. For this purpose several electrical circuits or even specially designed generators are employed. The circuits are

Table 4.2 Process parameters with their operating ranges in MWEDM

	Parameters	Type/value
1.	Wire tool material	W, Mo (less common)
2.	Wire diameter	20–100 μm
3.	Wire speed	20–50 mm/min
4.	Wire tension	300–600 g
5.	Dielectric	Kerosene (hydrocarbon oil) Deionised water (less common)
6.	Servo reference voltage	80–120 V
7.	Open voltage, (V_O)	100–120 V
8.	Peak current, (I_P)	a few mA to 1 A
9.	Pulse duration, (T_{ON})	0.1–64 μs
10.	Pulse off-time, (T_{OFF})	0.1–64 μs

distinguished according the way in which the voltage is built up and the pulse is controlled and also the characteristics of discharge. However, in almost all such circuits a capacitor is used for storing the electric charge before the discharge takes place across the gap. Different circuits are used for different machining conditions and requirements. The commonly used systems are:

i. R–C relaxation circuit with a constant D.C. source or Relaxation generator.
ii. High frequency electronic circuit.

However, irrespective of the circuits, the pulse energy may be determined as,

$$W_p(\text{or } E_P) = \int_{t_1}^{t_2} I(t)V(t)dt$$

4.4.1 R–C Relaxation Circuit

In micro-WEDM, a relaxation type of power supply may be used. This is because of the following advantages of the relaxation circuit. The power supply is simple, rugged and competitive for small energy for micromachining. The discharge energy (E_p) generated in the discharge circuit can be expressed as: $E_P = \frac{1}{2}(C + C')V_b^2$, where, C is the capacitance of the capacitor and C′ is the total stray capacitance in the circuit. V_o, V_b, C and C′ can be made very small. Although elimination of C′ is quite a difficult task, the machine tool and power generator in MWEDM are designed in order to eliminate stray capacitance from the system. This is of utmost importance to achieve both stability and less discharge energy simultaneously.

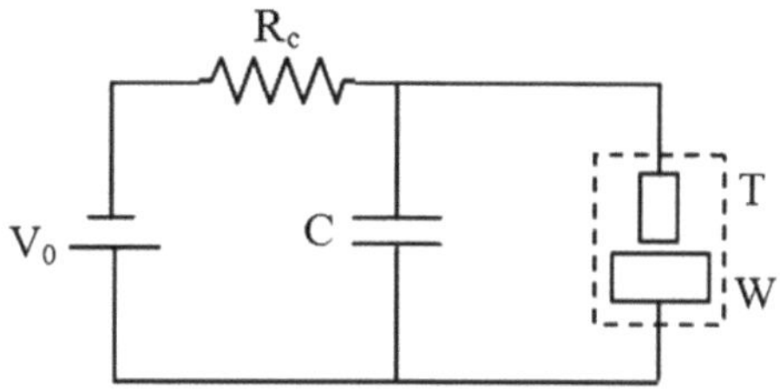

Fig. 4.3 A basic scheme of an R-C circuit

However, R–C relaxation circuit is the simplest type and it can produce sparks over a wide range of frequency (although not in higher range). It is shown in Fig. 4.3.

4.4.1.1 Current in the Charging Circuit

When the circuit is switched on, there will be a current in the charging loop while there is no current in the discharge loop and the capacitor charges itself through the resistance R_c. The voltage across the capacitor goes on increasing. At a voltage $V = V_b$, the spark occurs discharging the capacitor completely (theoretically). After the spark, the dielectric deionises, the capacitor starts itself recharging and the cycle repeats. The time taken to recharge the capacitor to the breakdown voltage (V_b) must be sufficient to allow the dielectric to deionise completely.

Let, in the charging loop, the current at any instant is I_c (i.e., instantaneous current), V_0 is the supply voltage, V_c is the voltage across the capacitor at any instant, R_c is resistance, C is the capacitance of the capacitor and t = time elapsed after switching on the circuit. The instantaneous current in the charging loop will be,

$$\begin{aligned} I_c &= \frac{V_0 - V_c}{R_c} = -C\frac{d(V_0 - V_c)}{dt} = C\frac{dV_c}{dt} \\ &\Rightarrow \frac{dV_c}{V_0 - V_c} = \frac{1}{R_cC}dt \end{aligned} \tag{4.1}$$

Using boundary conditions: (i) at $t = 0$, $V_c = 0$ and (ii) at $t = t_c$, $V_c = V_b$, (t_c = charging time of the capacitor or pulse off time, and V_b = breakdown voltage) and solving the differential Eq. (4.1), the V_c and V_b may be obtained as, $V_c = V_0\left(1 - e^{-\frac{t}{R_cC}}\right)$, and $V_b = V_0\left(1 - e^{-\frac{t_c}{R_cC}}\right)$.

Thus, $I_c = \frac{V_0 - V_c}{R_c} = \frac{V_0\, e^{-\frac{t}{R_cC}}}{R_c}$, $0 \le t \le t_c$

Therefore, the charging voltage across capacitor increases exponentially while the charging current decreases exponentially.

4.4.1.2 Current in the Discharging Circuit

Let, in the discharge loop, I_d is the current at any instant (i.e., instantaneous current), V_d is instantaneous voltage, R is the resistance. The total resistance in the discharge circuit R may be considered to be comprised of the line resistance (R_L) and the spark resistance or machine resistance (R_S). As, $R_S \gg R_L$, $R \approx R_S$. Thus,

$$\begin{aligned} I_d &= \frac{V_d}{R} = -C\frac{dV_d}{dt} = \frac{V_d}{R_s} \\ \Rightarrow \frac{dV_d}{V_d} &= -\frac{1}{R_S C}dt \end{aligned} \tag{4.2}$$

Putting the boundary conditions: At t = 0, $V_d = V_b$ and at, $t = t_d$, $V_d = V_\varepsilon$, (t_d is discharging time of the capacitor or pulse on time, and V_ε is the residual voltage across capacitor at the end of the discharge) and solving the differential Eq. (4.2), the V_d and V_ε may be obtained as,

$$V_d = V_b e^{-\frac{t}{R_S C}}, \quad \text{and} \quad V_\varepsilon = V_b e^{-\frac{t_d}{R_S C}}$$

Thus, $I_d = \frac{V_d}{R_s} = \frac{V_b\, e^{-\frac{t}{R_S C}}}{R_s}, \quad 0 \le t \le t_d$

Therefore, in the discharging circuit, the voltage will fall exponentially while the capacitor discharges itself and the discharging current will also exponentially decay, i.e., theoretically, the time taken for the current to be zero is very large. The theoretical voltage and current waveforms are shown in Fig. 4.4. However, the actual voltage and current waveforms reveal that there exists a negative voltage during the later part of the discharge. Also, the profile of the discharge voltage indicates definite existence of an inductive load in the discharging circuit. Hence,

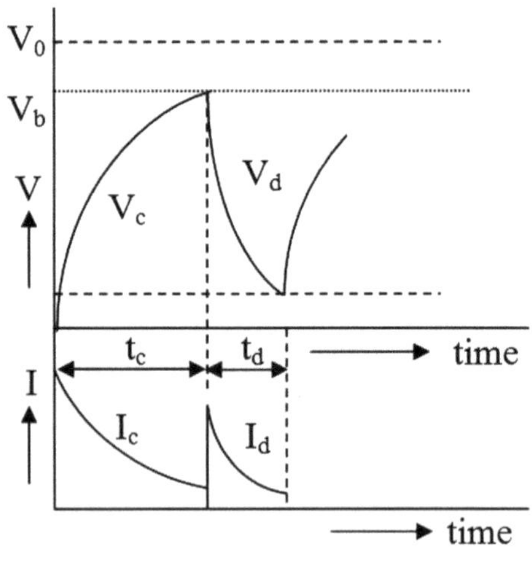

Fig. 4.4 Voltage and current waveforms (theoretical) in an R-C circuit

neither purely resistive nor purely inductive is the discharging circuit, it behaves as an R-L-C circuit.

4.4.1.3 R-L-C Discharging Circuit

Actually, $I_d = \frac{V_d - e}{R} = -C\frac{dV_d}{dt}$

Where, 'e' is the back emf generated in the discharging circuit due to inductance and $R = R_L + R_S$. Importantly, when, $e > V_d$ at the later part of the discharge (tail end), I_d becomes negative.

However, $e = L\frac{dI_d}{dt} = -LC\frac{d^2V_d}{dt^2}$

Thus,

$$\begin{aligned} V_d - e &= -RC\frac{dV_d}{dt}, \\ &\Rightarrow LC\frac{dV_d^2}{dt^2} + RC\frac{dV_d}{dt} + V_d = 0 \\ &\Rightarrow \frac{dV_d^2}{dt^2} + \frac{R}{L}\frac{dV_d}{dt} + \frac{1}{LC}V_d = 0 \end{aligned} \tag{4.3}$$

The discharge voltage will be governed by the Eq. (4.3), which is a case of damped oscillation. For purely inductive circuit (i.e., R = 0), the governing equation for 'V_d' becomes, $\frac{d^2V_d}{dt^2} + \frac{1}{LC}V_d = 0$. This is a case of simple harmonic fluctuation of voltage with a time period $\sqrt{1/LC}$. Naturally, there will be more number of reversal of sparks leading to high tool wear. Hence, this oscillation has to be damped effectively by introducing a suitable R. The total resistance in the discharge circuit R may be considered to be comprised of the line resistance (R_L) and the spark resistance (R_S). R_S is very large, almost constant and uncontrollable [3]. Hence, R_L has to be adjusted suitably.

4.4.1.4 Frequency and MRR in R-C Circuit

The pulse energy in an R-C relaxation circuit may be obtained as, $W_p = ½\, CV_b^2$ J, C = capacitance, F; V_b = breakdown voltage, V.

Thus, $t_c = R_cC\ln\left(\frac{V_0}{V_0 - V_b}\right)$, and, $t_d = R_sC\ln\left(\frac{V_b}{V_\varepsilon}\right)$

The frequency of operation, $\nu = \frac{1}{T} = \frac{1}{t_c + t_d}$.

So, $MRR = K \cdot \nu \cdot W_p = 1/2\, K\nu \cdot CV_b{}^2$; K is a constant of proportionality which depends grossly upon dielectric, type of flushing, work material properties. It may be calibrated for a certain experimental condition.

4.4.2 High Frequency Electronic Circuit

A pulse generator for micro WEDM provides iso-energy and iso-frequency pulses with low discharge energy and may be considered as 'heart' of the system. The ideal voltage and current waveforms have been shown in Fig. 4.5, where V_P is pulse peak voltage or open circuit voltage, V_m is average working voltage, V_d is discharge voltage, t_l is ignition delay time, I_P is pulse peak current, t_d is pulse discharge time, T_{ON} is pulse on time; T_{OFF} is pulse off time. Yan and Chien [4] have developed a transistor-controlled elementary power supply which minimizes the single discharge energy and also provides high frequency pulses (Fig. 4.6). The circuit is composed of a discharge circuit and a snubber circuit driven by MOSFET M_1 and control MOSFET M_2, respectively. Figure 4.7 illustrates the timing chart of the pulse control signals and associated pulse trains of MWEDM pulse generator. P_1 and P_2 are the pulse control signals of the MOSFET M_1 and M_2, respectively. When the MOSFET M_1 is turned on, the discharge circuit leads the gap to discharge by supplying a DC voltage of 110 V. After each discharge, the excessive discharge

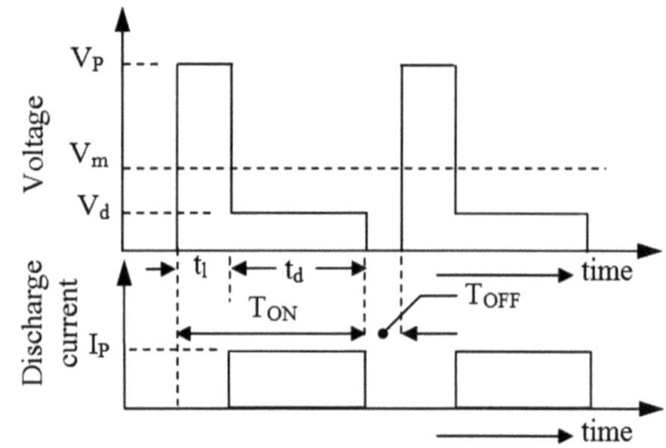

Fig. 4.5 Ideal voltage and current waveforms in a high frequency electronic circuit

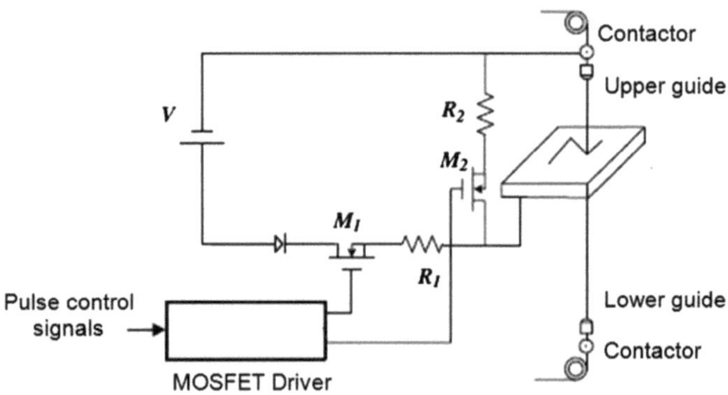

Fig. 4.6 Micro-WEDM power supply circuit with MOSFET [4]

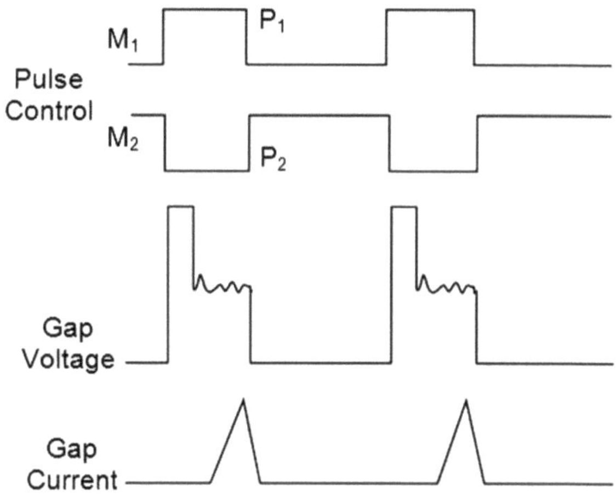

Fig. 4.7 Timing chart of the pulse control signal and associated pulse trains [4]

energy, if any, stored into the inductive part of the discharging path is directed to the resistor R_2 through the control MOSFET M_2. If the excessive energy remains in the discharging circuit and is not drained out in due time, it would result in instability of the discharging process.

4.5 Pulse Types and Pulse Discrimination

Like EDM, in Micro WEDM also, pulses are classified into four major types, namely: open, normal, arc and short pulses (Fig. 4.8). The basis of this classification is the time evolution of the gap voltage and gap current. In open pulses (also called open circuit pulses), the discharge between the wire-tool and the work piece does not take place. Hence, the energy is continuously accumulated in the electrodes for a discharge to take place. A normal pulse is an effective and the most desirable when it has a suitable ignition delay time (t_l). Ignition delay-free discharges are called arc pulses and in 'off' pulses, neither gap voltage nor gap current is detected. A short circuit occurs when metal contact between the work piece and the wire electrode takes place. The occurrence of arc discharges is due to poor flushing, gap contamination and short ionization, which depend on the instantaneous concentration of the slag in the spark gap. Also, when the pulse interval of each spark is not sufficient to flush away the spark debris and allow re-ionization of the dielectric, the arc discharges and short circuits occur.

The time between the application of the voltage between electrode and workpiece, and the start of ignition is termed as ignition delay time (t_l). It is a parameter of the WEDM process having correlation with the physical circumstances in the gap. It is inversely proportional with the electrical field strength at the spot of spark discharge. This is the reason why the average delay time is considered as an evaluation parameter for on-line process information required for adaptive control

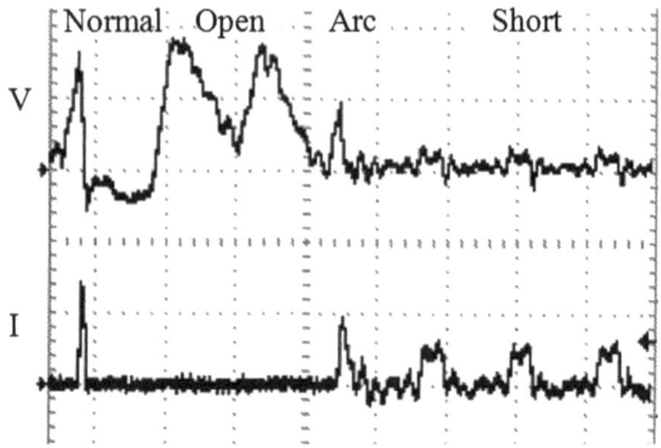

Fig. 4.8 Voltage and current waveforms of the four basic MWEDM gap states [4]

system in WEDM. The breakdown field strength as material constant, influences the delay time and a lower value of the breakdown field strength of the dielectric fluid causes a decrease of the delay time. Also, an increase in the contamination of the dielectric with the uncharged gap width results in a lower value of the delay time. It is worth mentioning here that none of these circumstances are constant not even for a stable process and a great variation of the delay time occurs for every individual pulses. However, results of various investigators have shown that generally a decreasing average ignition delay time is attended by an increased relative electrode wear and a decreased average MRR. This phenomenon had led to the idea that pulses preceded by a very short ignition delay time are not good and ineffective. But, at the same time it has also been experimentally verified that the elimination of pulses with very short ignition delay does not yield any improvement in machining performances as expected in connection with the above mention ideas [5]. However, a micro pulse generator in MWEDM should be able to generate iso-frequency and iso-energy pulses and must have a suitable pulse discriminating system incorporating the control strategy for regulating the pulse interval in real-time according to the identified gap states to avoid the occurrence of arc discharges and short circuits.

4.6 Wire Vibration in Micro WEDM

The wire vibration phenomenon is extremely complicated in nature. This is because the magnitudes and directions of various forces acting along or upon the wire are not always constant as the occurrence of sparks is highly stochastic in nature. The stochastic nature of the wire-cut electrical discharge machining (WEDM) process is also attributed to a combination of factors including fluctuation in voltage and current, decomposition and distribution of dielectric, random ionic migration, interaction of two successive discharges and presence of debris particles in machining zone. Not only that, several of the above-mentioned factors vary markedly with time and space in the discharging zone [6]. The main forces acting along or upon the wire are forces from the gas bubbles formed by the plasma of the erosion mechanism, hydraulic forces due to flushing, electrostatic force and electrodynamic force.

All these above-mentioned forces along with the axial tensile force (F) applied to the wire set the wire to vibrate when discrete sparks are generated between the electrodes. Therefore, vibration of a stretched wire supported by the two wire guides at opposite ends satisfying the standard vibration theories of vibrating strings may be mathematically modelled with the following assumptions [7, 8]: (i) The wire mass is uniformly spread along its length; (ii) The axial tension remains constant between the wire guides; (iii) The wire is static (not moving); (iv) The wire is perfectly flexible; and (v) The disturbing forces acting per unit length of the wire perpendicular to the axial force, varies as a function of time and space. A scheme of the wire vibration during cutting is shown in Fig. 4.9 [9].

Therefore, the general partial differential equation of motion (Newton's second law of motion) for a stretched vibrating string of length 'L' in a plane (along Z-axis) may be expressed [10] as follows:

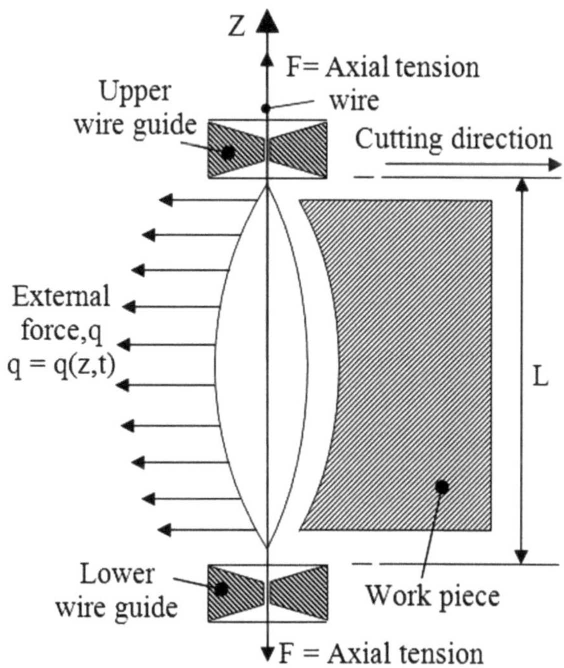

Fig. 4.9 Wire vibration during cutting [9]

$$F\frac{\partial^2 y}{\partial z^2} - \text{EI}\frac{\partial^4 y}{\partial z^4} = \rho_0\frac{\partial^2 y}{\partial t^2} + \beta\frac{\partial y}{\partial t} + q(z,t) \tag{4.4}$$

where, F is wire tension (N), y is wire deflection (m), t is time (s), ρ_0 is ρ.s (kg/m), ρ = wire mass density, kg/m^3, s = wire cross section (m^2), β = specific damping co-efficient (N-s/m^2), E = young's modulus, N/m^2, I = area moment of inertia, ($\pi d^4/64$), d = wire diameter (m), q (z, t) = specific external load or body force per unit length of the wire, N/m. The term $F\frac{\partial^2 y}{\partial z^2}$ indicates the force due to bending deformation because of the applied axial tension F. The term EI $\frac{\partial^4 y}{\partial z^4}$ represents the force required for transverse bending against flexural rigidity of the string. As the wire is perfectly flexible, as has been assumed, this term is negligibly small. The terms $\rho_0\frac{\partial^2 y}{\partial t^2}$ and $\beta\frac{\partial y}{\partial t}$ represent the acceleration force (Newton's 2nd law of motion) and damping force respectively.

It may be assumed that the workpiece is kept at the mid position of the wire-span. The wire has the length 'L' between the two wire guides. It is displaced laterally through a distance over the workpiece height 'H' upon action of the body force q (z, t) as shown in Fig. 4.10 [9]. The body force being the predominant one, pushes the wire away from the workpiece during collapsing of discharges and the

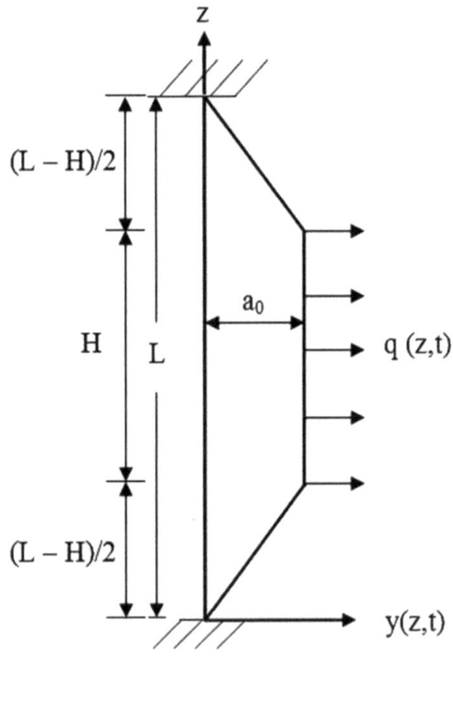

Fig. 4.10 Action of the body force before the wire is set to vibration [9]

wire deviates through a distance a_0 from its ideal initial stretched condition over the workpiece height theoretically and the same is released at t = 0.

From the geometry of the wire shape, the displacement function y (z, t) may be determined by the following initial conditions (IC):

$$\begin{aligned} \text{IC: } y(z,0) = f(z) &= \frac{2a_0 z}{L - H}; \quad \text{when, } 0 \le z \le \frac{L - H}{2} \\ &= a_0; \quad \text{when, } \frac{L-H}{2} \le z \le \frac{L+H}{2} \\ &= \frac{2a_0(L - z)}{L - H}; \quad \text{when, } \frac{L+H}{2} \le z \le L \end{aligned}$$

However, Eq. (4.4) may also be rewritten as:

$$\frac{\partial^2 y}{\partial t^2} = c^2 \frac{\partial^2 y}{\partial z^2} - K\frac{\partial y}{\partial t} - Q \tag{4.5}$$

With: $c^2 = \frac{F}{\rho_0}$, $K = \frac{\beta}{\rho_0}$ and $Q = \frac{q}{\rho_0}$, $q = q(z,t)$.

It is understood that the amplitude of vibration will go on increasing with the increase of Q and decrease of wire tension. However, as the wire tension remains unchanged during a machining process, the displacement function 'y' may be expressed as, Q = αy. 'α' is the constant of proportionality. Although it may appear apparently that this is overly simplified, the complexity lies in determining 'α'. Therefore, Eq. (4.5) is simplified as a linear homogeneous equation as stated below [9]:

$$\frac{\partial^2 y}{\partial t^2} = c^2 \frac{\partial^2 y}{\partial z^2} - K\frac{\partial y}{\partial t} - \alpha y \tag{4.6}$$

The above equation is solved by the method of separation of variables with the following initial and boundary conditions (IC and BC):

$$\begin{aligned} \text{IC : } y\ (z,0) = f\ (z); \quad \text{and, } \frac{\partial y}{\partial t}(z,0) = 0 \\ \text{BC : } y(0,t) = 0; \quad \text{and, } y(L,t) = 0 \end{aligned}$$

Thus, displacement function y = y (z, t), may be determined as,

$$y(z,t) = e^{-\frac{\beta t}{2\rho_0}} \sum_{n=1}^{\infty} \{A_n \cos(\omega_n t) + B_n \sin(\omega_n t)\} \sin\left(\frac{n\pi}{L} z\right) \tag{4.7}$$

With: $A_n = \frac{8a_0 L}{n^2\pi^2(L-H)} \sin\left(\frac{n\pi}{2}\right) \cos\left(\frac{n\pi}{2} \cdot \frac{H}{L}\right)$, $B_n = A_n \frac{\beta}{2\rho_0} \cdot \frac{1}{\omega_n}$, and

$$\omega_n = \frac{1}{2}\sqrt{\left(\frac{4n^2\pi^2F}{\rho_0L^2} - \frac{\beta^2}{\rho_0^2} + 4\alpha\right)}$$

'ω_n' is the natural angular (circular) frequency of the wire vibration.

Therefore, $A_n = 0$, for n = 2, 4, 6..., i.e., for even harmonics and for odd harmonics, i.e., (n = 1, 3, 5...), A_n possesses different values. Thus, A_n may be calculated for known values of L, H and a_0. 'a_0' may be determined for a particular machining setup experimentally. Once A_n is known, B_n may be calculated if β, ρ_0 and ω_n are known. 'ω_n' may be calculated when values of F and α are known. The values of 'α' may be determined by the calibration.

So, finally, the exact equation of motion may be determined as given by the Eq. (4.7). The term $e^{-\frac{\beta t}{2\rho_0}}$ indicates that the vibration will decay exponentially with time. $A_n\cos(\omega_n t) + B_n\sin(\omega_n t)$ may also be expressed as $\sqrt{A_n^2 + B_n^2}\cdot\sin(\omega_n t + \theta)$, where $\theta = \tan^{-1}\frac{A_n}{B_n}$. The maximum amplitude of vibration is $\sqrt{A_n^2 + B_n^2}$ and the angular or circular frequency will be ω_n. ω_1, ω_2, ω_3... are the natural frequencies for n = 1, 2, 3... respectively. The fundamental frequency of the vibration will be $\frac{\omega_n}{2\pi}$ c/s, for n = 1.

Di et al. [11] have measured the amplitude of wire vibration during MWEDM process using an indirect but accurate method. They measured the kerf width by confocal laser scanning microscopy. The vibration amplitudes were determined using known values of wire diameter, spark gap and the measured value of the kerf width for a particular experimental condition. The experiments were carried out on the self-developed micro-WEDM machine which mainly included an transistor-controlled R-C pulse generator, constant tension winding system using 30 μm tungsten wire, kerosene fluid circulation system, a gap detection system and a precise servomechanism with piezoelectric ceramics motions. In experiment, the workpiece was a 0.5 mm thick stainless steel plate, and there was 2 mm distance from workpiece to both wire guides. The experimental condition is given in Table 4.3 (condition no. I) and the corresponding vibration amplitude plot is presented in Fig. 4.11. Thus, it is seen that in a normal situation, the vibration amplitude may lie between 3 and 5 μm.

Table 4.3 Experimental condition for MWEDM [11]

Process parameters	Experimental Condition	
	I	II
Open voltage (V)	30–120	110
Discharge capacitor (pF)	400	300, 770, 1300, 1800, 2270
Charge resistance (kΩ)	1	3, 1, 0.6, 0.5, 0.33
Feeding speed (μm/min)	50	50
Reference voltage (V)	18–90	90
Wire tension (N)	1	1

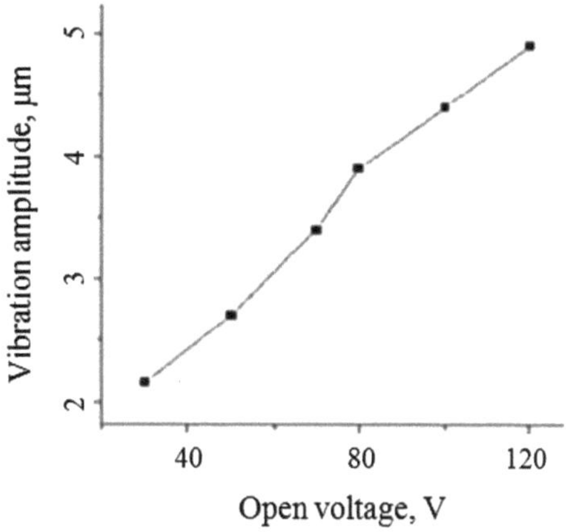

Fig. 4.11 Plot of vibration amplitude versus open voltage in MWEDM [11]

4.7 Wire Lag in Micro WEDM

The static deflection of a stretched wire supported by the two wire guides at opposite ends, is defined as wire lag in WEDM and is shown in Fig. 4.12. Thus, the wire bending or the static deflection (y) of the stretched wire of length ‘l’ in a plane (along z-axis) may be expressed as [12]:

$$F\frac{\partial^2 y}{\partial z^2} = q_0 \tag{4.8}$$

where, F = wire tension (N), y = y(z) = wire deflection (m), q_0 = specific external load or body force per unit length of the wire (N/m),

The solution of the Eq. (4.8) will appear as follows:

$$y = -\frac{q_0}{2F}z(l - z)$$

This is an equation of a parabola which indicates that the shape of the wire within the work piece will be parabolic during machining operation.

Also, $y_{\max} = -\frac{q_0}{2F}H(h+H) - \frac{q_0 h^2}{8F} = y_1 + y_2$

where, $y_1 = -\frac{q_0}{2F}H(h+H)$, and $y_2 = -\frac{q_0 h^2}{8F}$

On the other hand, the solution of equation $F\frac{\partial^2 y}{\partial z^2} = q_0$ outside the work piece and within the wire span between the guides (q_0 = 0) will appear as,

$$y = -\frac{q_0}{2F}(h+H)\cdot z$$

This equation describes a straight line, i.e., the wire outside the work piece will be straight up to the guide during machining. However, as the ‘y’ appears to be negative, it ascertains that the static deflection or wire lagging occurs in the opposite

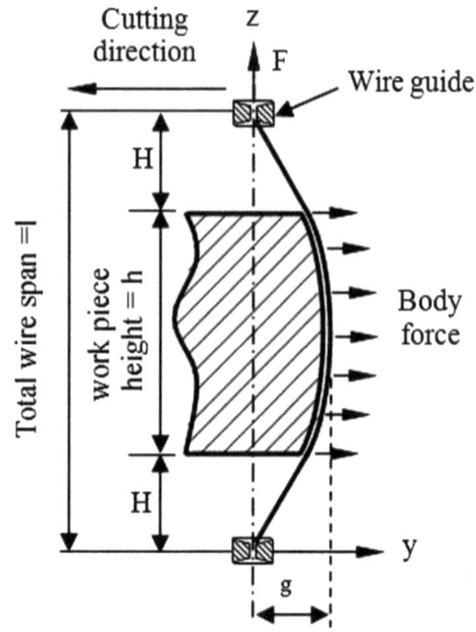

Fig. 4.12 Illustration of wire lag phenomenon during WEDM operation [12]

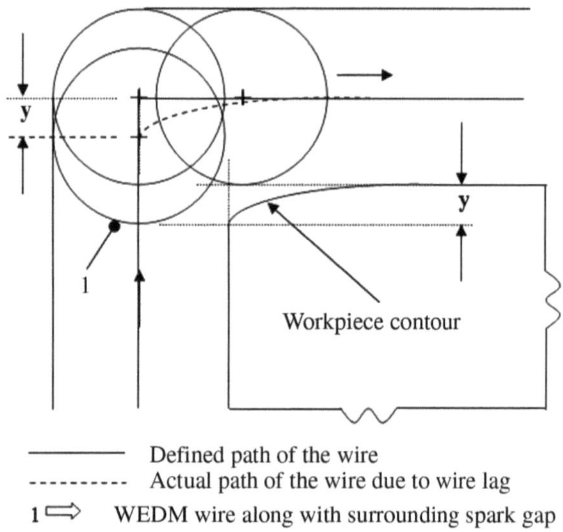

Fig. 4.13 Illustration of the effect of wire lag on the work-piece contour in WEDM

direction of cutting. The geometrical error resulting out of this wire lag should affect any contour cutting as depicted in Fig. 4.13, which clarifies that the magnitude of wire lag (y) is equal to the maximum deviation of the traced path from the defined path of the wire-tool in WEDM. However, as the pulse energy is very less in MWEDM, the lag effect is less pronounced with appropriate machining conditions and magnitude of wire lag may be made significantly less as would be seen later in Sect. 4.8.

4.7.1 Equation of the Traced Path of the Wire for Right Angle Corner Cutting

The lag effect in a WEDM rough cutting process affects the generation of contour of a product or part directly. This has been further illustrated in Fig. 4.14 where generation of a 90° corner is exhibited. The programmed path is A-o-A_2. The guide position would be at 'o' at the extreme corner and the ideal position of the wire along with the surrounding spark gap would be the dotted circle with its center at 'o'. But, due to the wire lag, the actual position of the wire will be the circle C with center at c where $\overline{oc}$ is the wire lag (=y). Thus, while the guides trace the path o-A_2, the center of the wire traces the path c-c_1-c_2-c_n-A_2. This phenomenon directly affects the generation of the job profile. In Fig. 4.14, the dotted circles (O and O_n, O_1 and O_2 are not shown in the figure to avoid clumsiness) are the ideal positions of the wire in the programmed path and circles C, C_1, C_2 and C_n (with their centres at c, c_1, c_2 and c_n respectively) are actual positions of the wire on the traced path. It is assumed that the constant body force acts on the wire always in the reverse direction of cutting, then the distances oc, o_1c_1, o_2c_2, and o_nc_n are equal and constant. Further, if a circle is imagined with radius $\overline{oc}$ with its center at 'o', which moves along a path oA_2 and the circle rolls over the line cA_3 (Fig. 4.15), befitting to the exact situation described above, then the path c-c_1-c_2-c_n will be a cycloid. The situation may be compared with a heavy ball tied up with a rope. The free end of the rope is at 'o' and is moving on the line oA_2 while the center of the ball is at 'c'. Let, the point C is the origin (0, 0) and the straight line $\overline{CFD}$ is X-axis as shown in Fig. 4.16. While rolling at an instant, say, $\overline{CD} = \widehat{C_1D} = r\theta$, where, r = radius of the circle, and θ = $\angle C_1O_1D$.

Let, the coordinate of C_1 is (x, y). C_1E is perpendicular on O_1D and C_1F is perpendicular on CD. Hence,

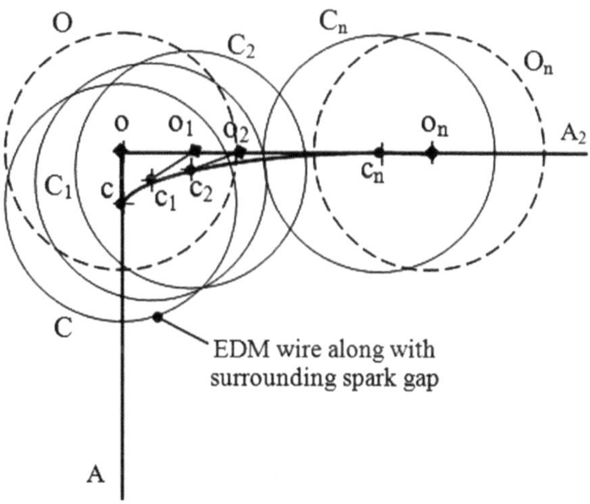

Fig. 4.14 Illustration of the movements of centres of guides and wire during machining

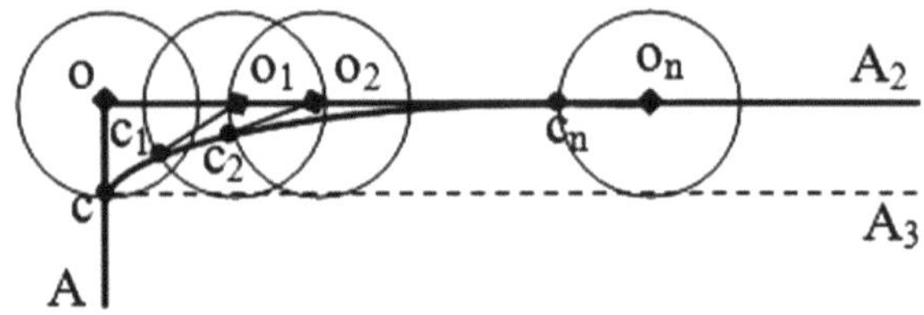

Fig. 4.15 Illustration for the locus of the center of the wire as cycloid

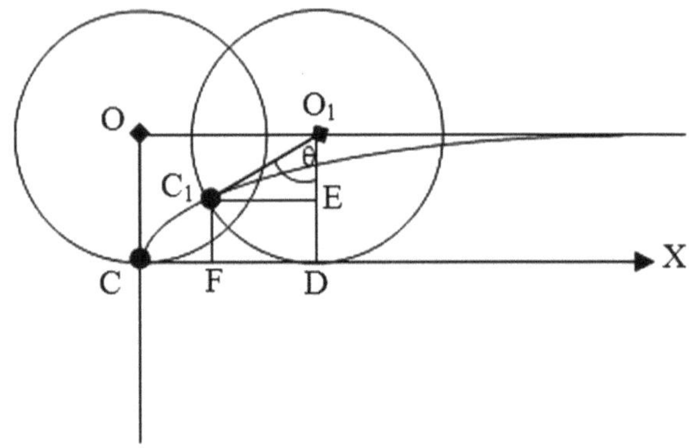

Fig. 4.16 Determining the equation from analytical geometry

$$x = CF = CD - FD = r(\theta - \sin\theta) \tag{4.9}$$

$$\text{And,}\quad y = C_1F = O_1D - O_1E = r\,(1 - \cos\theta) \tag{4.10}$$

Thus, if the circle starts to roll from the origin on X-axis (base line), then the parametric equation of the locus of its center may be described as above. The greatest distance from the base (vertex) will be obtained for $\theta = \pi/2$. Thus, for the vertex;

$$x = r\,(\theta - \sin\theta) = r(\pi/2 - 1) \quad \text{and} \quad y = r\,(1 - \cos\theta) = r.$$

It is worth to mention here that when the corner angle will be lesser or higher than 90°, the wire will trace only a part of a full cycloid whose vertex will depend on the corner angle.

4.8 The Kerf Produced in Micro WEDM

A micro part can be machined with good dimensional accuracy and surface finish only when the kerf (slot) width and corner radius become considerably smaller compared to that obtainable in conventional WEDM. Truly speaking, the cutting tool in WEDM/MWEDM is not the wire, but it is the pulse (electrical discharge). In MWEDM, both the workpiece and the wire are eroded although the erosion of

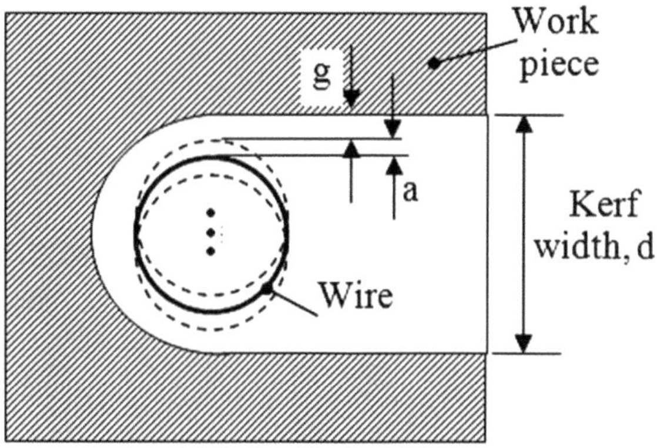

Fig. 4.17 Kerf produced by the wire in MWEDM

the wire does not affect the accuracy as the wire passes through the work only once. The size of the kerf is always more than the diameter of the wire used. This is due to the presence of side sparks as shown in Fig. 4.17. The width of the cut (b) depends on the wire diameter (d), the spark gap (g) and the amplitude of vibration (a). Thus,

$$\text{width of cut,} \quad b = d + 2g + 2a \tag{4.11}$$

The spark gap is controlled by setting servo reference voltage while the amplitude of vibration mainly depends on pulse energy, pulse on time, pulse frequency, type of pulse, wire tension and distance between the wire guides. Naturally, a low pulse energy, normal pulse with suitable ignition delay time, minimum distance between wire guides and a very low flushing pressure are absolutely necessary for obtaining a minimum kerf width in a particular micromachining setup. Di et al. [11] have obtained a kerf width as low as 30.8 μm and a corner radius 15.8 μm using a 30 μm wire tool diameter in their self-developed MWEDM system. This is shown in Fig. 4.18. Figure 4.19 shows the variation of kerf width with discharge energy obtained by varying the capacitances as obtained with experimental condition (II) shown in Table 4.3.

4.9 Wire Transportation System in MWEDM

Wire tension plays the major role in minimising process inaccuracy. The minimum kerf width and corner radius are governed by two process phenomena, namely wire-tool vibration and the wire lag. Both the amplitude of vibration and the (magnitude of) wire lag vary inversely with the wire tension. Thus, the tension in the wire should be as maximum as it could bear. On the other hand, the wire tension is set during machining depending on the feed velocity of the wire and the thermal

Fig. 4.18 A minimum kerf width and corner radius with 30 μm diameter tungsten wire [11]

Fig. 4.19 Kerf width versus pulse energy (capacitance) in MWEDM [11]

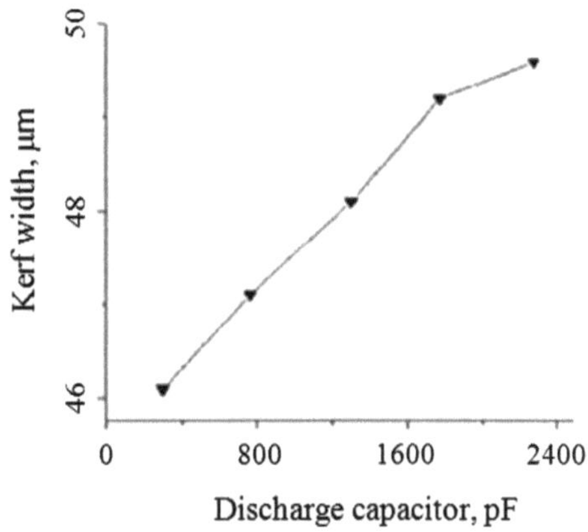

load on the wire which is basically governed by pulse on time, pulse types and pulse frequency. Thus, the tension of the micro wire may require to be changed during machining as and when necessary. Hence, the wire transportation system in WEDM/MWEDM should be a precise closed loop controlled system necessarily.

The wire transportation system in micro WEDM has the general layout as shown in Fig. 4.20, which is widely used for keeping the wire tension at a constant level [13, 14]. It consists of a wire electrode supply spool, a wire feed control apparatus and a wire tension control apparatus. The feeding wire is directly coupled to a pair of feeding rollers, which determine the feeding velocity of the wire electrode. While the wire electrode is pulled out from the wire spool, it passes through another pair of tension control rollers driven by an electromagnetic brake gear causing a braking force on the wire.

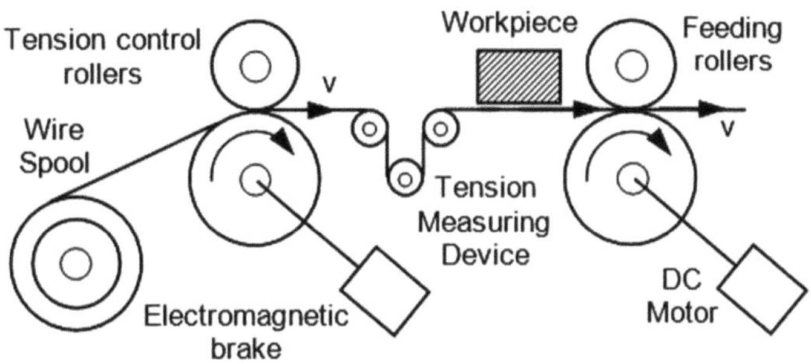

Fig. 4.20 The wire transportation system in MWEDM [14]

Closed-loop wire tension control system has been developed by many researchers [13–15] for the purpose of reducing wire vibration and wire lag. Han et al. [14] devised an optimal tension control of micro wire electrode based on thermo-mechanical analysis and strategically with the aid of a swinging roll between the feed rollers and the tension control rollers. In their experiment, a rectangular slot was cut in the workpiece and the finished workpiece was split into two parts by the middle section. The corner shapes on both the end surface and the middle surface were analysed. The wire lag reflected as the machining error in corner cutting has been represented by the corner radius on the end surface as 'r_e' and on the middle surface as 'r_m', respectively. The experimental condition was: pulse on time—3 μs, pulse off time—15 μs, wire feed speed—2 mm/min, workpiece material—2A12 aluminium alloy, wire material—tungsten (φ 100 μm), and wire tension—0.4 N. In rough cutting r_e and r_m were found to be 0.103 and 0.120 mm respectively while in the finish cutting they were reduced to 0.058 and 0.071 mm. This is shown in Fig. 4.21.

Yan et al. [15] developed a wire transport system by installing a fuzzy logic controller and used it in their developed prototype MWEDM setup. The overall block diagram of the fuzzy logic controller (FLC) for the wire transport system is shown in Fig. 4.22. In this controller, the measured wire tension (*y*) is compared with the reference wire tension (*r*) from which the wire tension error (*e*) and the change in wire tension error (*ce*) are obtained. These two signals, *e* and *ce* are then sent to the fuzzy logic controller for generating the control voltage of the brake motor. The fuzzy logic controller provided smooth wire transport with a constant wire tension value.

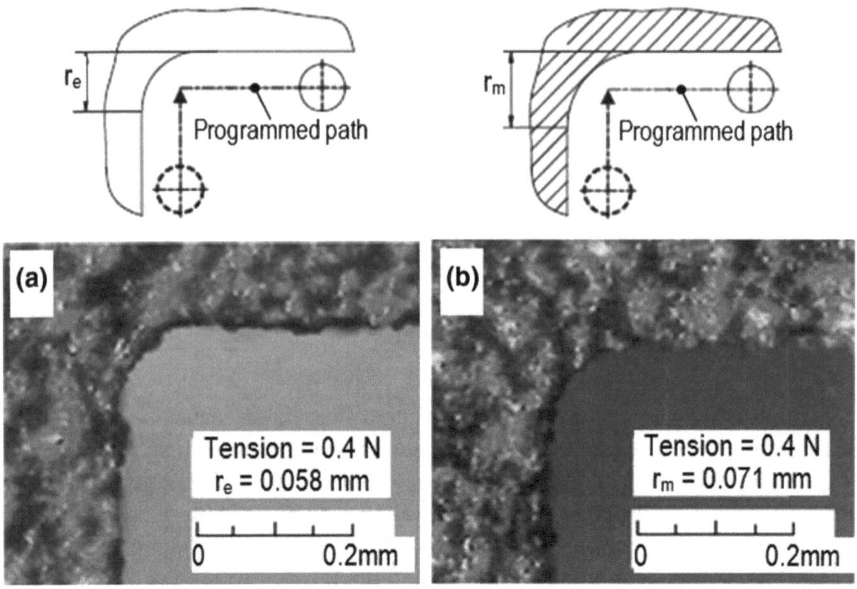

Fig. 4.21 Corner cutting error **a** corner shape at end surface (r_e) and **b** corner shape at middle surface (r_m) [14]

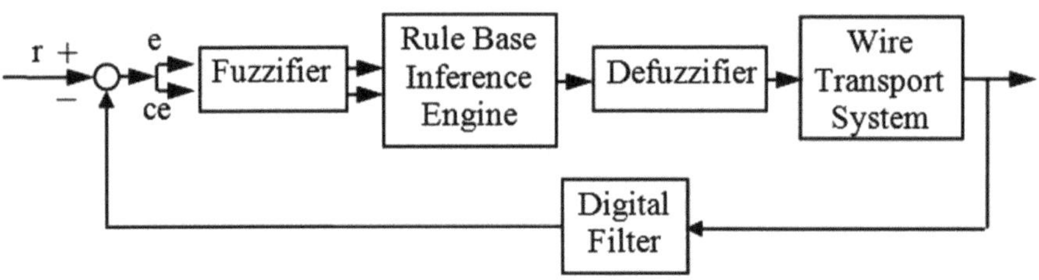

Fig. 4.22 A fuzzy logic controller for the wire transport system [15]

4.10 Process Performances in MWEDM

4.10.1 White Layer

A multi-layered heat affected zone is formed at the machined surface of the workpiece while machining it with WEDM. The upper recast layer of this zone is called 'white layer'. The workpiece material which gets melted is not completely expelled during the process. The residual material resolidifies on the machined surface to form a hard skin on the workpiece. In MWEDM, the formation of white layer can not be eliminated but can be minimised. The sub-surface characteristics occur in various layers or zones, which are usually termed as altered material zones (AMZ).

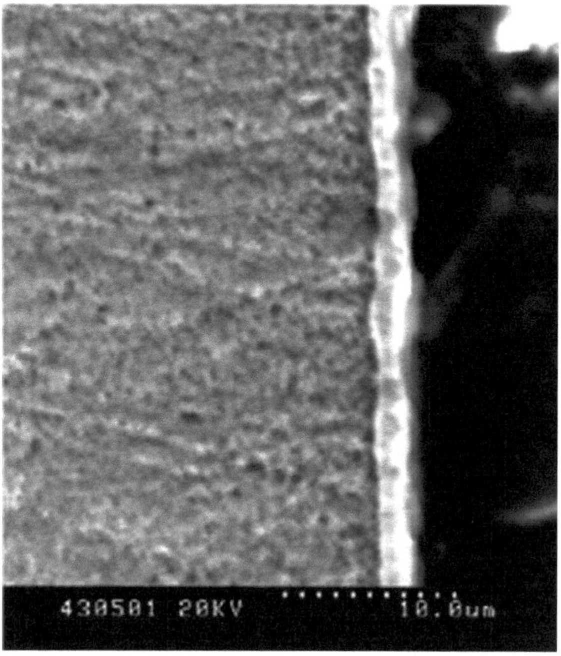

Fig. 4.23 White layer in Micro WEDM [16]

These alterations caused by the thermal energy of spark discharges are generally in the forms of micro-cracks, spalling, increase in hardness, residual stresses, metallurgical transformations and, of course, heat affected zone (HAZ). Thermal spalling is a mechanical failure of the material by overcoming the bond strength of the material due to the generated internal stresses without being melted. The thickness of white layer depends on pulse on time, frequency, pulse energy and the properties of work material. The scanning electron microscope (SEM) photograph in Fig. 4.23 shows the white layer has been reduced to as less as 2 μm for a 1 mm thick stainless steel sheet [16]. The micro hardness of the surface white layer (recast layer) is three to four times higher than the base material in Micro-WEDM. This is basically due to an increase in amount of cementite dendrites that result from absorption of carbon originating from the pyrolysis of the hydrocarbon oil dielectric. This energy dispersive spectroscopy (EDS) study of the white layer suggests that some amount ($\sim$2.28%) of wire material (i.e., tungsten) get deposited on the workpiece surface. However, instead of having detrimental effect, the existence of W element in micro die surface improves quality of the micro part by increasing wear and corrosion resistance.

Fig. 4.24 AFM image of a surface topography in MWEDM [16]

4.10.2 *Surface Roughness and Machining Accuracy*

It is a well known fact that the surface finish obtainable is not good for a spark eroded surface in EDM. For EDM process the CLA values (R_a) of surface roughness varies between 1 and 3.5 μm whereas in conventinal WEDM it varies between 0.8 and 2 μm. However, in a carefully cut specimen, the same may be obtained as minimum as 0.1 μm in MWEDM [5]. Hence, micro parts with such an excellent finish may quite suit for fabrication of miniature products and devices. The machining accuracy (tolerance) of micro WEDM can be limited in a range ±0.2 μm with a worktable's position accuracy ±0.1 μm. Figure 4.24 shows an AFM image (atomic force microscopy) of topography of a surface obtained in micro WEDM [16]. The surface roughness (R_a) obtained by AFM is ~0.1 μm for a 1 mm thick stainless steel sheet machined appropriately by Micro WEDM.

4.10.3 *Assisted Vibration in MWEDM*

One effective way of improving the flushing condition in EDM has been using vibration-assisted tool, which is called vibration-assisted EDM. Applications of ultrasonic vibration to WEDM had been studied by many researchers [17–19]. With ultrasonic vibration applied to the wire, the MRR is found to increase by 30%. The surface roughness and the surface residual stress are also improved. By studying the mechanism of the system, it is proved that the improvement of surface finish and material removal rate attribute to the formation of multi-channel discharges and the increased energy utilization ratio [17, 18]. Hoang and Yang [20] developed a vibration assisted method for micro-WEDM systems wherein vibration of both the wire and workpiece were considered. Effects of imparted vibration on machining

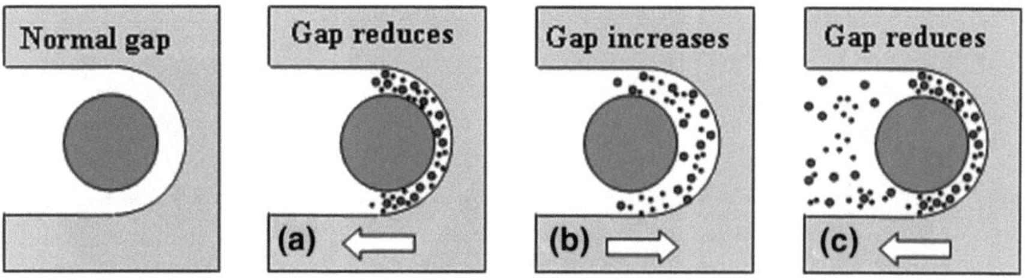

Fig. 4.25 Mechanism of improved flushing with the vibration of workpiece

characteristics are analyzed with various vibration parameters and workpiece thickness.

In WEDM/MWEDM, when the workpiece is vibrated, it is displaced in a periodic motion. Thus, in one half of the cycle it is very close to the wire as indicated in Fig. 4.25a, and in another half of the cycle it is far away from the wire as indicated in Fig. 4.25b. In position (a), the discharge process will take place and the gap will be filled with molten debris particles. The very next moment, i.e., in the second half of the cycle, the workpiece is pulled back causing suction of fresh dielectric in the gap due to high pressure variation as in position (b). In the first half of the next cycle, when the workpiece again come closer to the wire, the debris (melted materials) is flushed away from the machining zone under high pressure. This is shown in Fig. 4.25c.

It has been observed that MRR increases when vibration frequency increases in MWEDM. Moreover, vibration on the workpiece provides more improvement than vibration on the wire. When vibration is applied on the workpiece, the maximum recorded cutting speed is about 2.5 times faster than without vibration. When vibration is applied on the wire electrode the improvement is approximately 1.7 times [20]. This is because the pressure variation created by the wire is much smaller than created by the workpiece. This implies that, the dielectric circulation and the flushing of debris are less effective when vibration is applied to the wire than when vibration is applied to the workpiece. The workpiece thickness is also found to play a role in MRR improvement in MWEDM. In case of thicker workpiece, the flushing of the debris is more difficult, hence a higher improvement of MRR can be observed.

Surface roughness is also significantly improved with both vibration methods [20]. It is understood that when the vibration is applied to the wire electrode, the wire will vibrate in forced vibration mode and will generate nodes and antinodes. Since the displacements of the wire at the position of the nodes and antinodes are different along the length of the wire, surface roughness is varied greatly over the machined surface. Therefore it is difficult to obtain the desired surface quality for the case in which the vibration is applied to the wire. On the other hand, the desired surface finish can be obtained more easily with vibration applied to the workpiece.

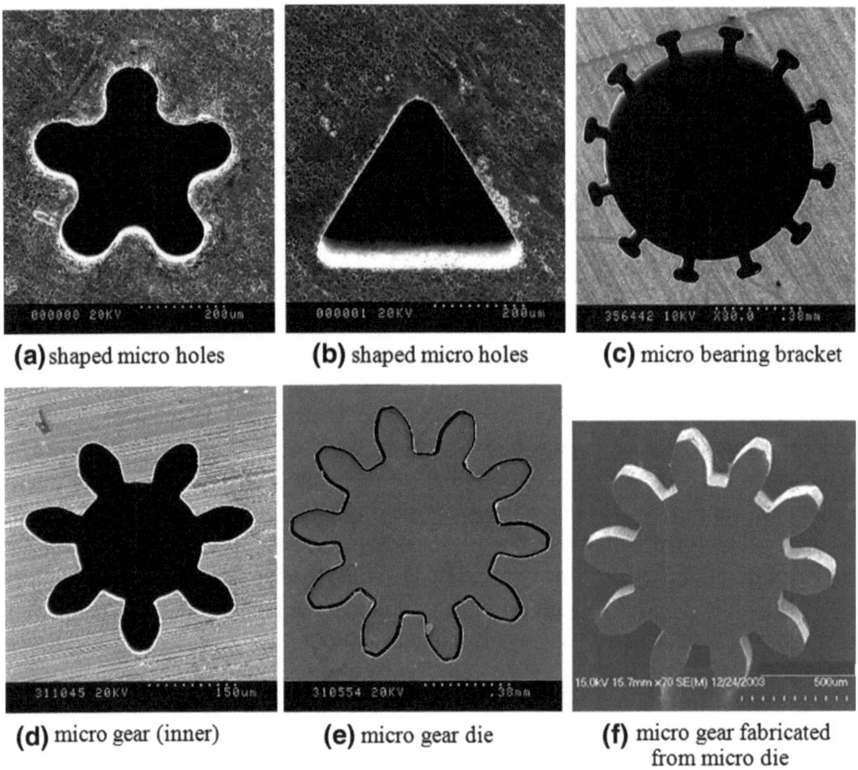

Fig. 4.26 SEM photographs of various micro parts machined by MWEDM and used in micro fabrication [16]

4.11 Applicability of Micro WEDM

Apart from making micro grooves, micro channels and micro beams, micro WEDM is successfully used to manufacture shaped micro holes, micro bearing brackets, micro gears and micro-gear dies with improved software technology supplemented with suitable NC interpolation functions and programming. The SEM photographs of a few machine components are shown in Fig. 4.26 [16]. The shaped holes are made on stainless steel sheet with thickness 1 mm. These holes are widely used for spinneret orifice in textile industries and in fabrication of micro dies. The micro bearing bracket is made on SS sheet of thickness 2.5 mm. These brackets are used in MEMS (micro electro mechanical system) for housing bearings. The micro gears are machined in stainless SS sheet of thickness 1 mm

Table 4.4 Machining conditions in electrostatic induction feeding method [1]

1	Pulse voltage [V]	150
2	Pulse on time [μs]	200
3	Duty factor	50%
4	Feeding capacitance C [pF]	50
5	Workpiece	Steel feeler gauge of 90 μm thickness
6	Wire electrode	Tungsten Φ30 μm
7	Wire running speed [m/min]	0.62
8	Wire tension [N]	1.4
9	Dielectric	air (slit), EDM oil (beam)

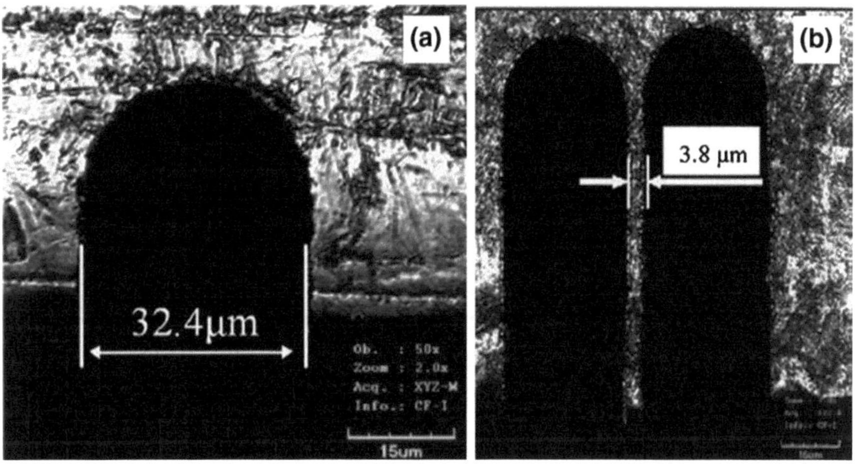

Fig. 4.27 **a** Micro slit (groove) and **b** micro beam machined by MWEDM with electrostatic induction feeding method [1]

with a module as less as 40 μm. Figure 4.26f shows an aluminum alloy micro gear fabricated from micro-gear die as shown in Fig. 4.26e, by micro-forming process.

Yang et al. [1] used electrostatic induction feeding method to develop an MWEDM setup. The construction and working principle of the pulse generator using electrostatic induction feeding method is completely different in regard to the conventional ones. The electrostatic induction feeding method essentially requires the use of oil dielectric (sometimes air), because the open voltage of the working gap decreases the discharge delay time, since the leak current flows into the water dielectric, if used. However, using this method and experimental condition as stated in Table 4.4, micro groove of a minimum width 32.4 μm and straight micro-beam of a minimum thickness 3.8 μm with 100 μm length were obtained (Fig. 4.27). However, further research is required to apply electrostatic induction feeding method to develop a micro pulse generator for MWEDM setup.

References

1. Xiaodong Yang, Chunwei Xu, Masanori Kunieda, Miniaturization of WEDM using electrostatic induction feeding method, Precision Engineering 34 (2010) 279–285
2. Website: https://en.wikipedia.org
3. Book: Nonconventional Machining, by P K Mishra, 3rd edition, Narosa Publishing House, India
4. Mu-Tian Yan, Hsing-Tsung Chien, Monitoring and control of the micro wire-EDM process, International Journal of Machine Tools & Manufacture 47 (2007) 148–157
5. Bruyn,H.E., Has the Delay time Influence on the EDM-Process?, Annals of the CIRP, Vol. 31/1/1982, pp. 103–106
6. S. M. Pandit, K. P. Rajurkar, Data Dependent Systems Approach to EDM Process Modeling from Surface Roughness Profiles, Annals of the CIRP, vol.29/1, 1980, pp. 107–112
7. D. F. Dauw, H. Sthioul, R. Dipretti, C. Tricarico, Wire Analysis and Control for Precision EDM Cutting, Annals of the CIRP, 38/1, 1989, pp 191–194
8. N. Mohri, H. Yamada, K. Furutani, T. Narikiyo, T. Magara, System Identification of Wire Electrical Discharge Machining, Annals of the CIRP, 47/1, 1998, pp 173–176
9. A. B. Puri and Prof. B. Bhattacharya, Modelling and Analysis of the Wire Tool Vibration in Wire-cut EDM, Journal of Materials Processing Technology, Vol 141, Issue 3 (2003), pp 295–301
10. D. F. Dauw, I. Beltrami, High-Precision Wire-EDM by Online Wire Positioning Control, Annals of the CIRP, 43/1, 1994, pp 193–197
11. Di Shichun, Chu Xuyang, Wei Dongbo,Wang Zhenlong, Chi Guanxin, Liu Yuan, Analysis of kerf width in micro-WEDM, International Journal of Machine Tools & Manufacture 49 (2009) 788–792
12. I. Betrami, A. Bertholds, and D. Dauw, A Simplified Post Process for Wire cut EDM, *Jl. of Mat. Proc. Tech*, 1996, vol.58, pp. 385–389
13. Mu-Tian Yan, Pin-Hsum Huang, Accuracy improvement of wire-EDM by real-time wire tension control, International Journal of Machine Tools & Manufacture 44 (2004) 807–81
14. Fuzhu Han, Gang Chenga, Zhijing Fenga, Soichiro Isagoc, Thermo-mechanical analysis and optimal tension control of micro wire electrode, International Journal of Machine Tools & Manufacture 48 (2008) 922–931
15. Mu-Tian Yan, Chen-Wei Huang, Chi-Cheng Fang, Chia-Xuan Chang, Development of a prototype Micro-Wire-EDM machine, Journal of Materials Processing Technology 149 (2004) 99–105
16. Shichun Di, Ruining Huang, Guanxin Chi, Study on Micro-machining by Micro-WEDM, Proceedings of the 1st IEEE International Conference on Nano/Micro Engineered and Molecular Systems January 18–21, 2006, Zhuhai, China
17. Z.N. Guo, T.C. Lee, T.M. Yue and W. S. Lau, A Study of Ultrasonic-aided Wire Electrical Discharge Machining, Journal ofMaterials Processing Technology 63 (1997) 823–828
18. Guo, Z.N., Lee, T.C., Yue, T.M., Lau, W.S., Study on the machining mechanism of WEDM with ultrasonic vibration of the wire 1997.. Journal of Materials Processing Technology 69, 212–221
19. Exploration of shaping conductive ceramic by ultrasonic aided wire electrical discharge machining, Lee, T.C., Guo, Z.N., Yue, T.M., Lau, W.S., 1997, Materials and Manufacturing Processes 12, 1133–1148
20. K.T. Hoang, S.H. Yang, A study on the effect of different vibration-assisted methods in micro-WEDM, Journal of Materials Processing Technology 213 (2013) 1616–1622

Chapter 5
Laser Micro-turning Process of Aluminium Oxide Ceramic Using Pulsed Nd:YAG Laser

Golam Kibria, B. Doloi and B. Bhattacharyya

Abstract One of the newly developed laser micromachining processes for generating micro-turn surface on cylindrical work sample is laser micro-turning process. To explore the capability of this laser micromachining process for achieving particular surface profile and dimensional accuracy of machined parts, authors considered a number of experimental investigation to find the effect of process parameters. During investigation and analysis, a number of experimental designs are applied to in-depth analyse the effect of process parameters on surface roughness (Ra and Rz) and depth deviation. The governing equations of spot and circumferential overlap were developed for investigating the effect of these overlaps on surface criteria. By adopting statistical design of experiments approaches such as Response Surface Methodology, the influence of process parameters on process performance were studied. Moreover, novel machining strategy of laser defocusing condition was also implemented for improving the micro-turning surface features. For qualitative assessment of important process parameters, scanning electron microscopic images of machined surface were analysed for better understanding the process.

G. Kibria (✉)
Mechanical Engineering Department, Aliah University, Kolkata 700156, India
e-mail: prince_me16@rediffmail.com

B. Doloi · B. Bhattacharyya
Production Engineering Department, Jadavpur University, Kolkata 700032, India
e-mail: bdoloionline@rediffmail.com

B. Bhattacharyya
e-mail: bb13@rediffmail.com

G. Kibria et al. (eds.), *Non-traditional Micromachining Processes*,
Materials Forming, Machining and Tribology,
DOI 10.1007/978-3-319-52009-4_5

5.1 Introduction

To fulfil the demand of high-end products with specific features in terms of high precision, accuracy and surface topography, researchers and technocrats are always engaged with micromanufacturing of macro as well as micro-components with which continuous miniaturization is developing day-by-day. It is also a challenging job for the manufacturing engineers and researchers to develop suitable micromachining process which can be applied to machine hard-to-cut materials such as alloys, composites etc. effectively with unique accuracy and productivity. For achieving effective machining accuracies and desired surface finish, a number of traditional and advanced micro-manufacturing processes have been developed for applying these machining techniques for product manufacturing of wide varieties. Moreover, various micro-devices and miniature MEMS need specific functional characterization which necessitate the incorporation of micro-features in diverse range of products. Therefore, at this stage, it is urgently needed to implement novel machining techniques, besides traditional processes, which can fulfil these demands. As most of the non-traditional micromachining techniques utilize different kind of energy for material removal, the material removal processes are irrespective of material's mechanical and physical properties. Amongst various non-traditional processes, laser beam machining (LBM) is a widely applied and successful material removal process for fabrication of different geometries with close tolerances at high production rates.

Based on laser types, type of operations and different wavelength, LBM is classified at various types. Pulsed and continuous wave Nd:YAG laser, Fiber laser and Excimer laser are widely applied laser sources for manufacturing of different features in components. In the pulsed mode laser operation, the material instantly melts as well as vaporises with very short intervals. Therefore, the conduction of heat into the bulk material does not occur. The compilation of Nd:YAG laser with Q-switch mode offers exceptional machining geometries during micro-machining processes such as micro-milling, micro-turning, micro-drilling and micro-cutting. One of the newly developed Nd:YAG laser micro-machining processes is laser micro-turning operation in which the cylindrical samples can be machined successfully to micro-turn the material at small depth with specific surface quality and accuracy. As pulsed mode of Nd:YAG laser has several advantages during processing of materials that include instant material melting and vaporization, small resolidified layer and less damage of machined surface due to micro-cracks formation. In laser micro-turning process, very thin depth from material's upper surface is removed due to laser irradiation while rotating the cylindrical workpiece. Thus, the process can efficiently be applied for micro-turning of several exotic materials such as ceramics and composites. The length of micro-turn surface is controlled by axial feed of workpiece. The micro-turning surface with specific surface properties has huge high precision applications such as self assemblies, ceramic spikes, ceramic bearing ring etc. Furthermore, due to long time operation of grinding wheel, the cutting ability of micro-abrasives reduces. The generation of cutting surface by laser irradiation on blunt grinding wheel surface can be carried out by laser micro-turning process.

5.2 Need for Micro-machining of Advanced Ceramic Materials

Chemically, with the exception of carbon, ceramics are nonmetallic, inorganic compounds. The high structural advanced ceramics were commercially came into market in 1980s for implementing these in place of metals and alloys to sustain in elevated temperature environments. Structural ceramics is an emerging class of engineering materials with variety of current applications and with the potential for much wider applications, especially at high temperatures. High performance structural ceramics are uniquely combined strength stable at high temperatures, hardness, dimensional constancy, high degree of erosion and corrosion behavior, elevated elastic modulus and small mass density. Structural ceramics are used as monolithic parts as composites and protective coatings. Monolithic structural ceramics are currently based primarily on silicon carbide, silicon nitride, and partially stabilized zirconium oxide, aluminum oxide and aluminum titanate. For enhanced performance, ceramic materials combined with particulate, whiskers or fibers of a different ceramic compound or metal produce composites those have several times toughness than that of monolithic ceramics [1–3].

5.3 Suitability of Nd:YAG Laser for Micromachining

Today, engineering ceramics are widely used in a number of applications due to these ceramics show additional mechanical properties, withstand for oxidation, or opto-electro-mechanical as well as magnetic properties. The various applications of ceramics are as follows [1, 2]:

(i) Heat engines,
(ii) Bearings,
(iii) I.C engines components like nozzles, cylinder etc.,
(iv) Turbines components like blade of fan,
(v) Missile and military applications,
(vi) Gas and steam turbines,
(vii) Metal cutting tools,
(viii) Industrial wear components,
(ix) Bio-ceramics for biomedical applications,
(x) Military applications,
(xi) Fabrication of integrated chips and resistors,
(xii) Microelectronics industry,
(xiii) As decorative tiles and pottery,
(xiv) Artificial human joints and many more.

Due to unmatched thermo-physical and chemical properties, now-a-days, engineering ceramics are widely used in micro domain of applications like automotive parts of IC engines, various parts for aerospace vehicles, computer technology, sensors, electrical and electronic parts and medical equipments etc. Moreover, these ceramics are used to manufacture artificial tooth root, artificial hip joints, implant and other medical purposes. Therefore, there are urgent needs of micro machining of advanced engineering ceramics to manufacture micro-features on macro as well as micro-products with high degree of dimensional accuracy and surface quality characteristics. Machining and fabrication of complicated shaped parts made of ceramic materials are difficult especially when it is in the microscopic domain. Lot of research is continuing on in the area of micro machining of advanced ceramic for the successful utilization of these types of material in various applications.

5.4 Laser Beam Micro-machining Processes

Lasers are finding continuously increasing utilization in the field of micromachining processes. Figure 5.1 shows the schematic views of various types of laser beam machining methods. The processes can be classified depending upon the shape of the erosion front of the material removed from workpiece. In one type of laser micromachining process, the laser beam is stationary relative to the workpiece to be machined. There is no relative movement of laser beam and the workpiece surface. A typical example of this type of machining is laser micro-drilling process. The erosion front is located at the bottom of micro-drilled hole, propagated in the direction of line source in order to remove material. In another type of laser micromachining process, laser beam has relative movement in one direction with respect to workpiece. The erosion front is located at the leading edge of laser beam. Therefore, material removal occurs by repetitive movement of the erosion front in the direction of depth of workpiece. A typical example of this laser beam micro-machining is laser micro-cutting. In the third type of laser micromachining, either one or two laser beams are used. Each of the beams produces a cutting front by providing relative movement of workpiece in respect of laser beams. The erosion front for each of the cutting processes is found at the leading edge of respective laser beam. A typical example of this type of laser beam micromachining is laser

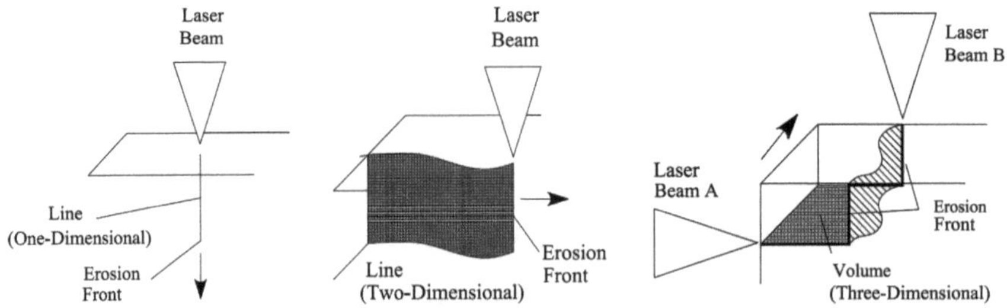

Fig. 5.1 Schematic views of various categories of laser micromachining processes

micro-turning. Some of the major laser micro machining techniques is highlighted in the following sections.

5.4.1 Laser Micro-drilling

Laser micro-drilling is oldest laser materials processing technique. Laser micro-drilling is widely used in aerospace, medicals, and automotive sectors [4–6]. The generation of high aspect ratio micro cooling holes in several components such as turbine blades, flow orifice, electron beam system, optical instruments is carried out by laser micro-drilling process [7, 8]. In this process, high accurate micro-sized holes with high degree of aspect ratio can be generated in variety of hard-to-machine materials such as ceramics, composites, alloys etc. without tool wear as in case of EDM or USM [9]. Laser micro-drilling process has huge application in drilling of carbon fiber composites for manufacturing of aircraft components due to significant advantages of laser beam processing of materials [10].

5.4.2 Laser Micro-cutting

Laser micro-cutting can be called as two-dimensional micromachining technique. In this process, material removal takes place by controlling the intense laser beam as well as by moving the workpiece at particular direction. Laser cutting is most popular cutting process in various industries. Due to heating and melting as well as vaporization of material due to high intense laser beam, material is removed from the cutting front. The melted materials and vaporized gases are expelled from irradiated zone using jet of assist gas (air). The assist gas also helps for enhancing the chemical reactions in the cutting zone. During laser micro-cutting operation, the workpiece is moved in particular direction as while feed motion of laser focusing point vertically downward for facilitating the cutting of thick materials.

5.4.3 Laser Micro-grooving

Laser grooving process is carried out by scanning the irradiated laser beam over path of surface where the micro-groove is to be generated. There are two methods of laser micro-grooving process, (i) using a single laser beam and (ii) using two laser beams. In case of single laser beam, the desired width of groove is machined by producing multiple micro-grooves side-by-side causing overlap between two adjacent groove widths. This process produces trapezoidal or rectangular shaped micro-groove on desired workpiece surface. However, when two intersecting laser beams are used simultaneously, each of the laser generate cutting surface and the desired shaped micro-groove is formed by removing a certain volume of material in bulk form. The mechanism of laser grooving is similar to laser cutting process; only the difference is that laser grooving process does not produce a through cut up to the thickness of work material. The groove depth defines the boundary of the volume of material removed. The quality of the machined surface is related to heat affected zone and the formation of recast layer at the groove walls or surfaces. Dimensional accuracy is one of the important criteria for laser grooving process. Dimensional accuracy is related to deviation of groove width and depth from desired quantity of dimensions. The groove depth is fluctuated due to several phenomena such as change in laser beam characteristics, mechanical vibration, material impurity, gas jet fluctuations, plasma formation and ejection of metal vapour, etc.

5.4.4 Laser Micro-turning

One of the latest and emerging techniques of laser materials processing in micro-machining domain is laser micro-turning process. This particular micro-machining process is applied to machine cylindrical shaped engineering material for removing micron level material from workpiece surface. For removal of bulk form (like a ring shape or tapered groove) of material from workpiece, laser micro-turning process is applied by using two intersecting laser beams [11]. Figure 5.2 shows the schematic view of laser turning process by using two intersecting laser beams. However, for carrying out micromachining process such as laser micro-turning, a single laser beam should be used to remove very thin planer layer from work material surface by irradiating and focusing the laser beam onto the machining surface of rotating workpiece through the desired length of turn along the axis of work sample. Figure 5.3 shows schematic view of laser turning process using single beam while rotating cylindrical workpiece and moving it along axial direction.

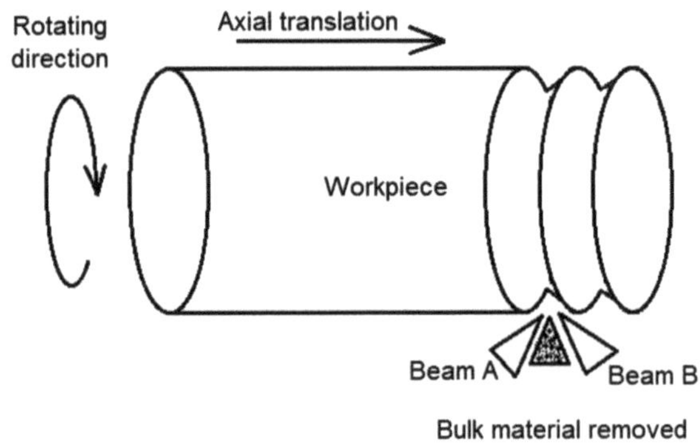

Fig. 5.2 Schematic view of laser turning (thread generation) employing two intersecting laser beams

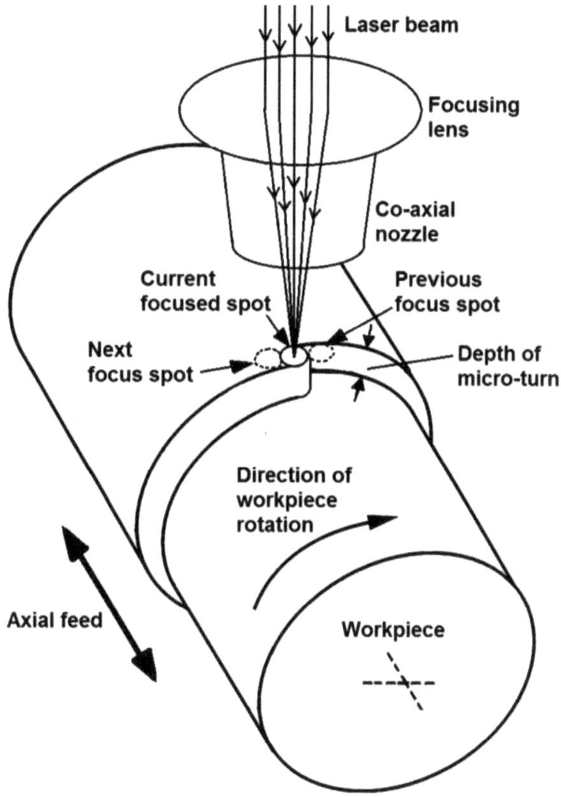

Fig. 5.3 Schematic view of laser micro-turning process using single laser beam

5.5 Difficulties of Micro-turning of Ceramic Materials

Nowadays, engineering micro-components used for fabricating micro as well as macro-systems are benefited by advanced engineering ceramics. This is due to the fact that these ceramics have their diversified mechanical as well as thermo-physical properties such as corrosion and thermal resistance, hardness, fracture toughness, wear resistance, etc. Due to these extraordinary properties of ceramics, the demands of parts made of these advanced ceramics are increasing day-by-day as the lifespan of the parts are high. Micro-turning using diamond turning tool is one of the emerging machining processes for producing micro-cylindrical parts made of difficult-to-cut alloys and ceramics. However, this process has several drawbacks such as irregular wear of highly expensive cutting tools which causes high production cost, defects in machined surfaces and sub-surfaces due to brittleness properties of ceramics, excessive vibration of mechanical parts of machine tools, etc. [12, 13]. Wire electrical discharge turning (WEDT) is another micro-turning process in which cylindrical micro-parts can be manufactured such as micro-electrodes, micro-pins etc. [14, 15].

However, WEDT can only be applied for conductive materials. Thus, it is challenging task for the manufacturer engineers and scientist to find out novel material processing technique which can be utilized for micro-turning of ceramics.

Amongst various ceramics used for manufacturing small parts for micro-systems as well as for biomedical implants, Aluminium Oxide (Al_2O_3) is most important and widely used ceramics materials due to its several extraordinary thermal, physical and mechanical properties such as stability at high temperature, high degree of corrosion and wear resistance, extreme hardness, high stiffness, etc. [16]. Micromachining of ceramic materials using laser beam is complex and dynamic in nature and the large number of process parameters such as beam power, pulse repetition rate, pulse duration, beam scanning speed etc. have non-linear behaviours during fabrication of complex geometries and surface features [17]. Moreover, the optical and physical properties of the workpiece materials to be machined provide significant variations of the output characteristics during laser micro-machining of engineering ceramic materials [17, 18]. Therefore, at this stage, it is urgently need to develop a novel non contact type micro manufacturing process which can efficiently and effectively micro-turn ceramics as well as difficult-to-machine superalloy. Past research revealed that research in the direction of numerical model development and experimentation have been carried out by number of researchers across the globe in several laser micro machining processes such as micro-drilling, micro-cutting, micro-grooving, surface modification etc. [19, 20]. Thus, to machine or micro-turn cylindrical shaped aluminium oxide ceramic utilizing laser beam, laser micro-turning process has been developed.

5.6 Process Mechanism of Laser Micro-turning

In 1988, Chrysolouris et al. [21] first demonstrated the capability of laser beam for three dimensional machining employing two intersecting beams. However, two such intersecting laser beams can remove material from work sample in bulk form. To simplify the laser micro-turning process, single laser can effectively be utilized where material is removed from the rotating cylindrical work sample in layer-by-layer form. Basically, laser micro-turning process is generation of spiral micro-grooves over a defined length of sample where the each groove has some overlap onto the previous groove [22]. Thus, planar surface of thickness in micron range is removed from the top cylindrical surface. The quantity of overlap between two consecutive micro grooves are controlled by mainly two factors, axial feed rate and rotating speed of workpiece. For obtaining desired depth of micro-turning, continuous scanning of laser beam is carried out while rotating of work sample. The surface characteristics i.e. quality of surface profile of laser micro-turning is governed by two overlap factors, spot overlap (SOp) and circumferential overlap (COp). To achieve quality surface features during laser micro-turning process, the values of these overlap factors must be high as possible. Thawari et al. [23]

formulated the relationship between spot overlap, pulse frequency, spot diameter and scanning speed of laser beam as depicted in Eq. (5.1).

$$(\text{Spot diameter}) \times (1 - \text{spot overlap}) \times (\text{pulse frequency}) = \text{linear speed} \quad (5.1)$$

Thus

$$\text{Spot overlap } (SO_p) = (1 - \frac{v}{D \times F_p}) \times 100\% \quad (5.2)$$

In the above equation, D is the spot diameter (in mm), F_p is the pulse frequency (in Hz) and v is the linear scanning speed of laser beam (in mm/s). The relationship of linear scanning speed with workpiece rotating speed is

$$v = \frac{\pi \times D_s \times N}{60} \quad (5.3)$$

In this equation, Ds is the diameter of workpiece (in mm) and N is the workpiece rotating speed (in rpm).

$$\text{Spot overlap } (SO_p) = (1 - \frac{\pi \times D_s \times N}{60 \times D \times F_p}) \times 100\% \quad (5.4)$$

The schematic representation of overlap criteria between two consecutive spots during laser beam scanning onto the workpiece surface is shown in Fig. 5.4. The overlap distance between two successive laser spots i.e. spot 1 and spot 2 is x_s shown by hatched area. Due to the Gaussian energy distribution of laser beam, the crater profiles are generated on the workpiece surface making overlap between crater areas. Thus, this amount of overlap area can be increased by either increasing number of pulses per unit time or reducing the workpiece rotating speed as depicted in Eq. (5.4).

Another crucial criterion of laser micro-turning process for obtaining quality machined surface is circumferential overlap (COp). It is the amount of overlap between two consecutive laser scan widths generated during two consecutive rotation of cylindrical workpiece. The length of circumferential of a cylindrical workpiece of diameter D_s is $L = \pi \times D_s$. If, t_s is the time (in seconds) to scan this length L over the one complete periphery of cylindrical work surface, then $t_s = L/v = 60/N$. Now, if the axial feed rate (in mm/s) of work sample is f, then axial travel of workpiece in t_s seconds is $l_c = f \times t_s = (60 \times f)/N$. Therefore, the circumferential overlap is

$$CO_p = \frac{\text{overlap length}}{\text{laser spot diameter}} = \frac{x_c}{D} = (1 - \frac{l_c}{D}) = (1 - \frac{60 \times f}{D \times N}) \times 100\% \quad (5.5)$$

In this equation, x_c is the overlap length (in mm) between two laser spots produced in two consecutive rotations of workpiece. Figure 5.5 shows the

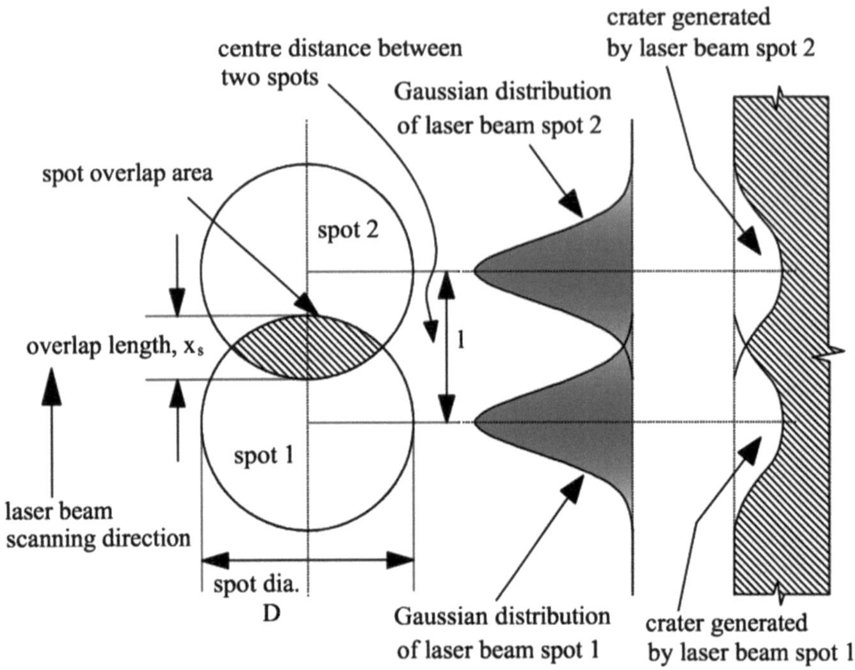

Fig. 5.4 Schematic view of overlap between two consecutive spots produced by laser scanning onto the workpiece showing Gaussian energy distribution and crater depth

schematic representation of circumferential overlap criteria between two laser spots produced side-by-side onto the workpiece surface during successive rotation of work sample. In this figure, the overlap length between two spots is x_c and the overlap area is shown by hatched region and the Gaussian energy distribution for these laser spots are shown. The crater depth profiles produced due to melting and/or evaporation from the laser spot regions have an amount of circumferential overlap. This amount of overlap can be increased by either reducing the axial feed rate of workpiece or increasing number of rotation per minute as also depicted in Eq. (5.5).

5.7 Development of Laser Micro-turning Set Up

Maintaining the standard and performing high quality micro machining operations on advanced engineering material, a high power pulsed Nd:YAG laser set-up (Make: M/s Sahajanand Laser technology, Model no: SLT-SP-2000) with some modifications has been utilized for carrying out various experimentation in laser micro-turning operation. The detailed specifications of this pulsed Nd:YAG laser

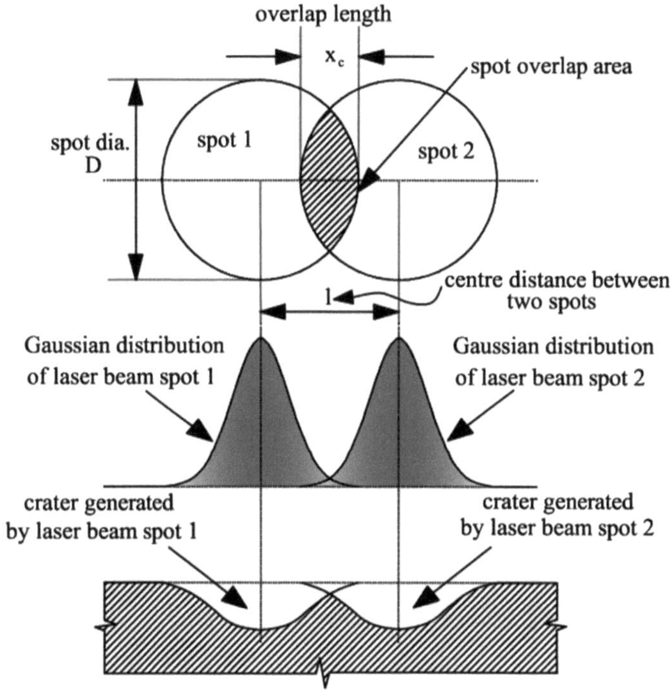

Fig. 5.5 Schematic view of overlap between two spots produced by two consecutive rotations of workpiece showing Gaussian energy distribution and crater depth

set-up are listed in the Table 5.1. The beam operates in Gaussian mode (TEM_{00}). The wavelength of Nd:YAG laser beam is 1064 nm. The focusing lens has focal length of 50 mm. To carryout different set of experimentations utilizing various laser machining strategies on flat and cylindrical shaped work samples using the existing laser machining set-up, various fixtures have been developed indigenously. The pulsed Nd:YAG laser beam machining system includes the various subsystems that include laser beam generation unit, cooling system (internal and external), power supply system, CCD camera, CCTV, X-Y-Z controller, assist air/gas supply system and developed fixtures etc. The photographic view of these subsystems of laser beam machining set-up is shown in Fig. 5.6.

To conduct experimentation on laser micro-turning process on different workpiece materials utilizing pulsed Nd:YAG laser micromachining system, workpiece rotating system is developed separately. The rotating system is made of three sub components, (i) servomotor with workpiece holding collet system, (ii) servo amplifier, and (iii) power supply unit. The schematic representation of the workpiece rotating system is shown in Fig. 5.7. In Table 5.2, the details of the servomotor and servo controller, which are utilized for developing the rotating system, are enlisted. The workpiece holding collet can hold workpiece in the range of 2–16 mm diameter sample. The collet was fixed to the axis of motor precisely to

Table 5.1 Detailed specifications of pulsed Nd: YAG laser setup

Specification	Description
Laser type	Nd:YAG laser
Wavelength	1064 nm
Mode of operation	Q-switched (pulsed)
Type of Q-switch	Acousto optic
Mode of laser beam	Fundamental mode (TEM_{00})
Mirror reflectivity	Rear mirror—100% Front mirror—80%
Beam diameter $1/e^2$	1 mm
Laser beam spot diameter	100 μm
Pulse width	120–150 ns

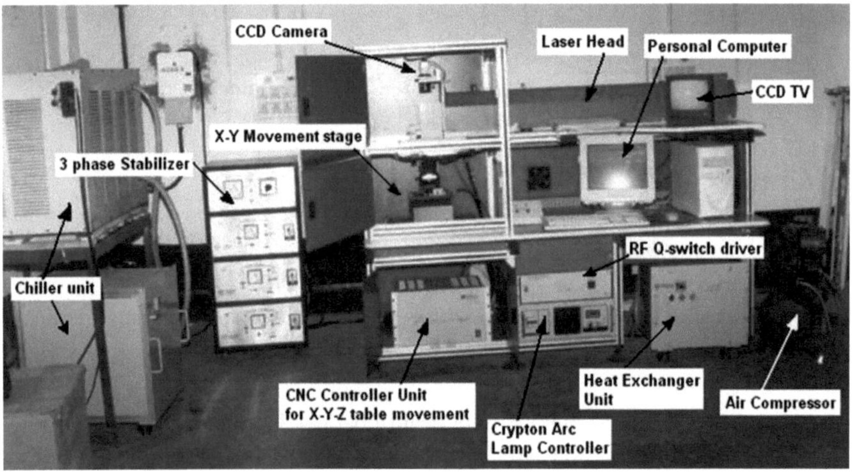

Fig. 5.6 Photographic view of the pulsed Nd:YAG laser machining system indicating various sub-systems [24]

reduce the eccentricity error. The eccentricity of rotation was found as 3 μm, which is acceptable. The whole rotating system is placed on X-Y worktable of the laser machine as shown in Fig. 5.7.

To rotate the workpiece at desired rotational speed, all the sub systems are inter connected. The power supply system has several switches and knobs to change the rotating speed of servomotor as well as to change the rotational direction of servomotor. After placing the rotating system underneath the laser beam, the axial feed of workpiece is given using the software installed in the personal computer attached to the laser beam machining system. Multi pass micro-turning is carried out by scanning the desired length of rotating workpiece surface and also by focusing the laser beam onto the new surface generated by previous laser beam scan. Therefore, laser micro-turning process of desired axial length and micro-turn depth is achieved on cylindrical workpiece.

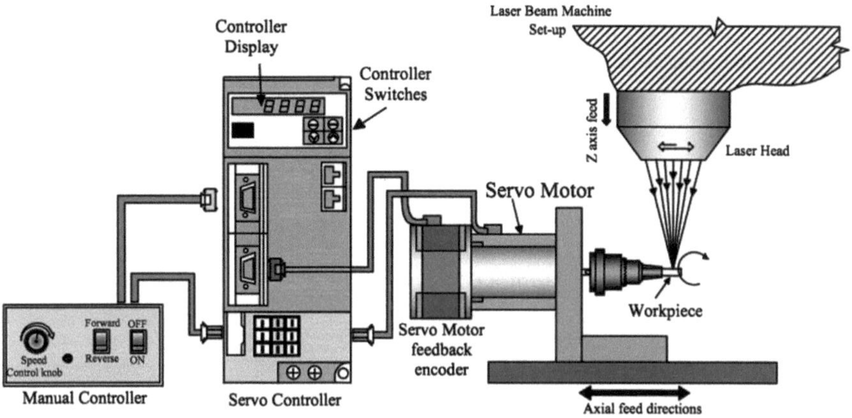

Fig. 5.7 Schematic diagram of the developed workpiece rotating system with connections between sub-components [25]

Table 5.2 Detailed specifications of servomotor and servo controller

Details of servomotor		Details of servo controller	
Specifications	Description	Amplifier type	RYC101D3-VVT2
Servomotor type	GYS101DC2-T2A	Phase	Single
Input voltage	AC 200 V	Input voltage	200–230 V
Rated output	100 W	Rated output	100 W
Rated speed	3000 rpm	Circuit breaker	EA33AC/5

5.8 Experimentation and Measurement Schemes

For conducting research investigation of the novel laser micromachining technique i.e. laser micro-turning process, a well planned research scheme is organized to study the influence of various controllable process parameters using various statistical techniques such as Response Surface Methodology etc., the effect of spot and circumferential overlaps, the influence of various laser defocusing conditions on machining criteria such as surface roughness (Ra, Rt) and micro-turning depth. The axial feed and Z feed rate is provided using Multisawing software, which is installed in personal computer connected with the laser micromachining system. All the experiments were conducted using cylindrical shaped 99% pure aluminium oxide ceramic material (length 40 mm and diameter 10 mm). The detailed thermo-physical properties of 99% pure aluminium oxide ceramic are enlisted in Table 5.3. The process parameters which are varied during laser micro-turning are laser average power, pulse frequency, workpiece rotational speed, assist air pressure, Y feed rate, defocusing positions and number of passes.

Using a laser power measuring device (make: Ophir Optronics Solutions Ltd, Israel), the laser beam average power was controlled. Several experimental

Table 5.3 Detailed thermo-physical properties of 99% pure aluminium oxide ceramics

Properties	Values and units
Melting temperature	2050 °C
Creep rate	10^{-9} s^{-1} at 150 MPa
Density	3.984 g/cm^3
Specific heat	755 J/kg K
Activation energy	325(25) kJ/mol
Thermal conductivity	33 W/mK
Tensile strength	267 MPa
Thermal diffusivity	0.111 cm^2/s
Bulk modulus	257 GPa
Shear modulus	167 GPa
Grain size	5 μm

methodologies were applied to conduct experiments varying the above-mentioned process parameters. Before selecting the actual range of process parameters, a number of pilot experiments were conducted. According to experimental schemes, laser micro-turning process is carried out at various parametric combinations and performance criteria such as machining depth and surface roughness (Ra and Rt) were measured using precision measuring instruments. The micro-turn depth was measured by differentiating the Z axis movements of machined (A) and unmachined surface (B) of workpiece using high precision optical measuring microscope (resolutions of X and Y are 0.5 μm each) with 10× magnification lens by a high precision dial gauge attached as indicated in Fig. 5.8. The target machining depth is 100 μm. The micro-turning depth deviation is calculated by differentiating the machined depth and target depth. The roughness values (Ra and Rt) of laser micro-turning surface is measured using surface roughness measuring instrument (Model: SURFCOM 120A-TSK, Make: Zeiss, Germany). The methodology of measuring the surface roughness values is depicted in Fig. 5.9. Roughness of machined surface was measured at several locations of cylindrical job by rotating at 60° angle each. As the micro-turning length along the workpiece is selected as 5 mm, therefore, surface roughness was measured for a length of 2.5 mm exactly at middle segment of machined surface. During roughness measurement, the cutoff length (λ_c) is selected as 0.25 mm because form errors can be eliminated and micro dimensional irregularities can be taken into account during surface roughness assessment.

5.9 Results and Discussion

Based on the various schemes, experimentation have been conducted at various parametric combinations considering different process parameters such as laser average power, pulse frequency, workpiece rotational speed, assist air pressure,

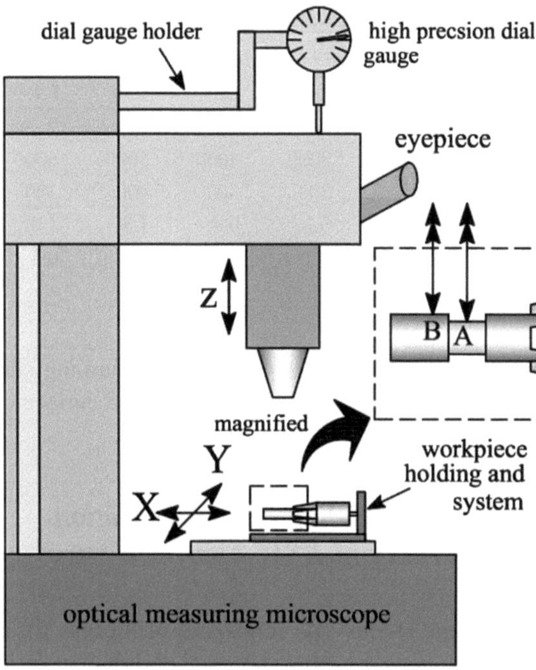

Fig. 5.8 Methodology of micro-turn depth measurement utilizing optical measuring microscope

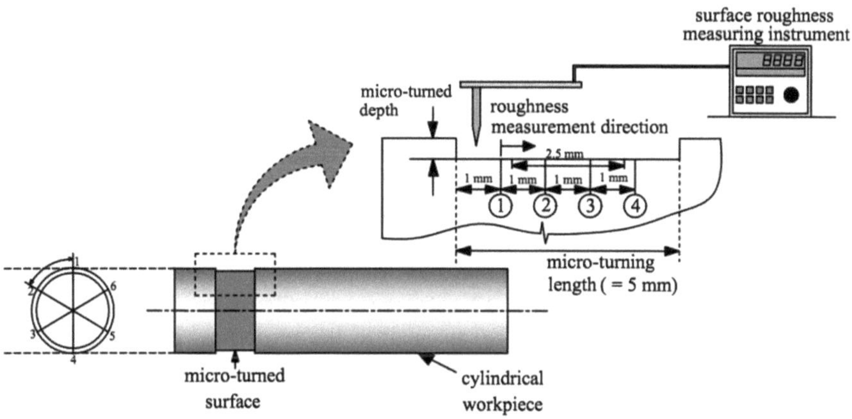

Fig. 5.9 Measurement scheme of surface roughness criteria

Y feed rate, defocusing positions and number of passes. Performance criteria were measured at all machining parametric settings and the test results were analysed through various plots and analytical tools. The results of validation experiments show good sign of validity of empirical models and optimizations. Optical and

Table 5.4 Process parameters and various levels for RSM based experimentation [25]

Parameters	Unit	Symbol	Levels				
			L1	L2	L3	L4	L5
Average power	W	X_1	6	7	8	9	10
Pulse frequency	Hz	X_2	3000	4000	5000	6000	7000
Rotational speed	rpm	X_3	200	300	400	500	600
Air pressure	kgf/cm^2	X_4	0.3	0.8	1.3	1.8	2.3
Y feed rate	mm/s	X_5	0.1	0.2	0.3	0.4	0.5
Coded value			−2	−1	0	+1	+2

scanning electron microscopic (SEM) images of laser micro-turning surface were also analysed to acquire clear depiction of quality of machined surface conditions.

5.9.1 Parametric Studies, Modeling and Optimization of Laser Micro-turning in RSM Approach

In this section, basic parametric studies and influences of process parameters on laser micro-turning process are analysed using statistical design of experiments tool known as response surface methodology (RSM). Rotatable Central Composite Design (RCCD) based on RSM is a well-designed and established experimental plan to considerably trim down the total quantity of experiments required to perform for making design of experiment cost effective, time saving and also for achieving high accuracy. In the present experimental investigation, five number of parameters such as average power, pulse frequency, rotational speed of workpiece, air pressure and Y feed rate have been considered. Based upon the various input factors and their levels as mentioned in Table 5.4, a uniform rotatable CC Design for developing response models of laser micro-turning process has been designed. After conducting each of the experiments, responses i.e. micro-turning depth deviation and surface roughness (Ra) have been measured and the results of responses for each parametric combination of process parameters are enlisted in Table 5.5. MINITAB of version 13 software was utilized to plan the design of matrix of experimentation and also to analyze the responses. Based on the test results, mathematical models have been developed for correlating the outputs with the process parameters. Further, response surface plots were analysed for investigating the influence of process parameters on responses.

Considering the proposed second order polynomial regression model, the empirical models on micro-turning depth deviation and surface roughness (Ra) have been established by computing the values of various coefficients of second order polynomial equation. The developed mathematical relationship correlating the response and process variables are shown in Eqs. (5.6) and (5.7).

Table 5.5 Experimental results based on RSM experimentation design approach

Expt. no.	Coded values					Responses	
	X_1 (W)	X_2 (Hz)	X_3 (rpm)	X_4 (kgf/cm^2)	X_5 (mm/s)	Depth deviation, mm	Surface roughness (Ra), μm
1	8	5000	400	1.3	0.3	0.01192	6.59
2	8	5000	400	0.3	0.3	0.00002	5.51
3	8	5000	400	1.3	0.3	0.02922	6.31
4	7	6000	500	0.8	0.4	0.01271	6.61
5	8	5000	400	1.3	0.5	0.00030	5.77
6	7	4000	300	0.8	0.4	−0.03439	5.86
7	9	6000	500	0.8	0.2	0.07586	7.54
8	9	4000	300	1.8	0.4	0.01579	5.45
9	7	6000	300	1.8	0.4	−0.02178	6.43
10	8	7000	400	1.3	0.3	0.02543	7.12
11	9	4000	500	1.8	0.2	0.04314	7.21
12	7	4000	500	0.8	0.2	0.03639	7.00
13	8	5000	400	2.3	0.3	0.00710	6.13
14	8	5000	400	1.3	0.1	0.09000	8.39
15	8	5000	600	1.3	0.3	0.01678	6.33
16	7	6000	500	1.8	0.2	0.03755	7.49
17	10	5000	400	1.3	0.3	0.03883	5.46
18	6	5000	400	1.3	0.3	0.00631	6.19
19	8	5000	400	1.3	0.3	0.01333	6.68
20	9	4000	500	0.8	0.4	−0.01648	5.16
21	7	4000	500	1.8	0.4	0.00273	5.88
22	9	6000	300	0.8	0.4	0.01954	5.73
23	8	5000	200	1.3	0.3	0.02786	6.97
24	7	4000	300	1.8	0.2	0.04686	6.74
25	9	6000	500	1.8	0.4	0.00407	5.54
26	9	6000	300	1.8	0.2	0.08448	7.94
27	8	5000	400	1.3	0.3	0.00958	6.15
28	8	3000	400	1.3	0.3	−0.04500	5.67
29	8	5000	400	1.3	0.3	0.00963	5.99
30	8	5000	400	1.3	0.3	0.01128	6.13
31	7	6000	300	0.8	0.2	0.05218	7.35
32	9	4000	300	0.8	0.2	0.05324	7.23

$$Y_{depth\,dev.} = 0.053142 - 0.038222 \times X_1 + 0.000048 \times X_2 - 0.000191 \times X_3 + 0.094368 \times X_4 - 0.852803 \times X_5 + 0.003616 \times X_1^2 - 0.000000 \times X_2^2 + 0.000000 \times X_3^2 - 0.004544 \times X_4^2 + 0.926106 \times X_5^2 + 0.000004 \times X_1X_2 - 0.000071 \times X_1X_3 + 0.002107 \times X_1X_4 - 0.012561 \times X_1X_5 + 0.000000 \times X_2X_3 - 0.000016 \times X_2X_4 - 0.000015 \times X_2X_5 - 0.000070 \times X_3X_4 + 0.000423 \times X_3X_5 + 0.031346 \times X_4X_5 \quad (5.6)$$

$$Y_{Ra} = -3.3853 + 2.4496 \times X_1 + 0.0001 \times X_2 - 0.0008 \times X_3 + 0.5522 \times X_4 + 1.9032 \times X_5 - 0.0825 \times X_1^2 + 0.0000 \times X_2^2 + 0.0000 \times X_3^2 - 0.3400 \times X_4^2 + 23.0837 \times X_5^2 - 0.0000 \times X_1X_2 - 0.0009 \times X_1X_3 + 0.0975 \times X_1X_4 - 2.6381 \times X_1X_5 - 0.0000 \times X_2X_3 + 0.0000 \times X_2X_4 - 0.0001 \times X_2X_5 - 0.0007 \times X_3X_4 - 0.0016 \times X_3X_5 - 0.4104 \times X_4X_5 \quad (5.7)$$

Here, X_1, X_2, X_3, X_4 and X_5 are the uncoded values of input process parameters i.e. average power, pulse frequency, workpiece rotational speed, assist air pressure and Y feed rate, respectively. The ranges of these process parameters are as $6 \leq X_1 \leq 10$ (W), $3000 \leq X_2 \leq 7000$ (Hz), $200 \leq X_3 \leq 600$ (rpm), $0.3 \leq X_4 \leq 2.3$ (kgf/cm^2) and $0.1 \leq X_5 \leq 0.5$ (mm/s). The analysis of variance (ANOVA) results for micro-turning depth deviation and surface roughness (Ra) models are shown in Tables 5.6 and 5.7. The standard F-value for lack-of-fit is 4.06 for 95% confidence level. However, the calculated F-values for depth deviation and surface roughness (Ra) are 3.11 and 1.18, which are lower than the standard F-value. These imply that the developed mathematical models are adequate at 95% confidence level. The values of adjusted R^2 are 87.07 and 86.90% and these values indicate the degree of accuracy of the developed models. The p-values for lack-of-fit are 0.117 and 0.436, indicating that the models adequately fit the data and ensuring that these models can well correlate the responses.

In Table 5.8, the results depth deviation and surface roughness of eight verification experiments obtained from experimentation are enlisted. The parametric combinations were chosen in the considered range of each process parameters. In

Table 5.6 Analysis of variance test for depth deviation [25]

Source	DF	Seq SS	Adj SS	Adj MS	F	*p*
Regression	20	0.027766	0.029966	0.001388	11.44	0.000
Linear	5	0.021140	0.001785	0.000357	2.94	0.063
Square	5	0.004014	0.004014	0.000803	6.62	0.004
Interaction	10	0.002611	0.002611	0.000261	2.15	0.112
Residual error	11	0.001335	0.001335	0.000121		
Lack-of-fit	6	0.001053	0.001053	0.000175	3.11	0.117
Pure error	5	0.000282	0.000282	0.000056		
Total	31	0.029101				

Table 5.7 Analysis of variance test for surface roughness (Ra) [25]

Source	DF	Seq SS	Adj SS	Adj MS	F	*p*
Regression	20	18.7631	18.7631	0.93816	11.28	0.000
Linear	5	14.7690	0.5454	0.10908	1.31	0.328
Square	5	2.6320	2.6320	0.52641	6.33	0.005
Interaction	10	1.3620	1.3620	0.13620	1.64	0.215
Residual error	11	0.9148	0.9148	0.08316		
Lack-of-fit	6	0.5367	0.5367	0.08945	1.18	0.436
Pure error	5	0.3781	0.3781	0.07561		
Total	31	19.6779				

Table 5.8 Verification experimentation showing the test results of responses

Expt. no.	Process parameters					Responses obtained from experiments	
	Average power (W)	Pulse frequency (Hz)	Rotating speed (rpm)	Air pressure (kgf/cm^2)	Y feed rate (mm/s)	Micro-turning depth deviation (mm)	Surface roughness (μm)
1	7	7000	500	0.8	0.3	0.0314	7.51
2	8	4000	500	1.8	0.2	0.0352	6.16
3	8	5000	500	1.8	0.4	0.0094	7.00
4	9	5000	400	0.8	0.2	0.0480	6.26
5	9	6000	200	0.8	0.3	0.0517	7.23
6	8	6000	500	1.8	0.2	0.0253	6.74
7	9	4000	500	1.3	0.3	0.0002	6.09
8	7	6000	500	1.3	0.1	0.0017	6.89

Table 5.9, the comparative results of prediction errors of responses for experimental and RSM predicted results are shown for eight verification experiments. The prediction errors for depth deviation and surface roughness are 3.12 and 3.19%, respectively. The overall prediction error is 3.16%. These percentage errors of prediction are within acceptable limit and it is concluded that the developed models are valid within the parametric ranges.

$$Prediction\ error\ (\%) = \frac{(Experimental\ result - Predicted\ result)}{Experimental\ result} \times 100 \quad (5.8)$$

Utilizing various surface plots which represent the variation of response during simultaneous increasing of two process parameters, the responses were analysed. In Fig. 5.10, the influence of variation of laser average power and laser pulse frequency on depth deviation is shown. The figure shows that the depth deviation is increasing with laser beam average power. Due to high intense energy in the laser

Table 5.9 Comparison between experimental and RSM predicted results of test data

Expt. no.	Experimental results		RSM predicted results		% of prediction error	
	Micro-turning depth deviation mm)	Surface roughness, Ra (μm)	Micro-turning depth deviation (mm)	Surface roughness, Ra (μm)	Micro-turning depth deviation	Surface roughness (Ra)
1	0.0314	7.51	0.0303	7.28	3.66	2.95
2	0.0352	6.16	0.0341	5.96	3.25	3.19
3	0.0094	7.00	0.0091	6.74	3.14	3.64
4	0.0480	6.26	0.0468	6.09	2.48	2.70
5	0.0517	7.23	0.0504	6.96	2.39	3.83
6	0.0253	6.74	0.0243	6.56	4.17	2.66
7	0.0002	6.09	0.0002	5.84	3.89	4.03
8	0.0017	6.89	0.0016	6.72	2.01	2.53
Average percentage of prediction error					3.12	3.19
Overall percentage of prediction error					3.16	

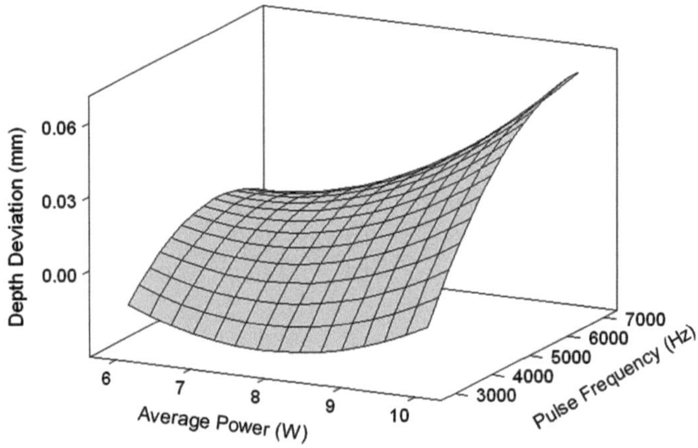

Fig. 5.10 Response surface plot of depth deviation for varying average power and pulse frequency [25]

beam at high average power, the amount of material evaporation is increasing and this result in more depth deviation. It is also observed that at low value of average power, the depth deviation is not so high and it is due to the fact that pulse off duration decreases with increase of pulse frequency and due to this, number of pulses reaching the workpiece surface increases. These huge numbers of pulses cause more material to remove from workpiece surface. However, at higher setting of laser average power, the increase of pulse frequency results in more depth deviation. With increase of pulse frequency, according to Eq. (5.9), peak power of

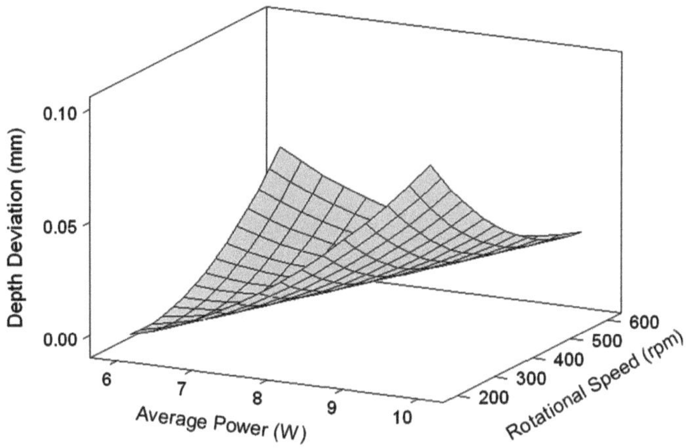

Fig. 5.11 Response surface plot of depth deviation for varying average power and workpiece rotational speed [25]

laser beam decreases. But, at the same time, the duration of successive pulses decreases and large amount of material is removed from workpiece surface.

$$Peak\ power\ (P_P) = \frac{Average\,power\ (P_A)}{Pulse\,frequency\ (F_P) \times Pulse\ duration\ (\mu)} \tag{5.9}$$

In Fig. 5.11, the effect of average power and rotational speed of workpiece on depth deviation is shown. The surface plot shows that with the increase of laser average power, the depth deviation increases. As per Eq. (5.9), the peak power of laser beam increases with the increase in average power and thus, the amount of material melted and evaporated is more. It is observed that increase in workpiece rotating speed results in increase in depth deviation. At high average power, increase in workpiece rotational speed results in reduction in depth deviation. Due to increase in rotating speed, the amount of absorbed energy by cylindrical material is low and consequently, the amount of material removal is high enough and more depth deviation.

In Fig. 5.12, the influence of air pressure and Y feed rate on depth deviation is shown. By closely observing the surface plot, it is obvious that the parameter assist air pressure has less effect on variation of depth deviation. Moreover, it is also observed that with the increase in Y feed rate, the depth deviation decreases. It is due to the fact that due to higher feed rate value, the time duration of material melting and evaporation is less per laser scan area and thus, material removal is moderate.

In Fig. 5.13, the surface plot shows the effect of variation of laser average power and pulse frequency on roughness of machined surface. It is seen that surface roughness first increases and after that, it decreases due to varying the laser average power. As per Eq. (5.9), with increase in laser average power, the peak power also

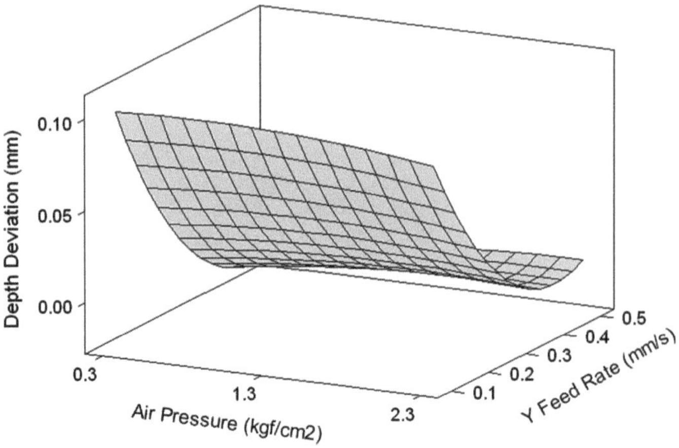

Fig. 5.12 Response surface plot of depth deviation for varying air pressure and Y feed rate [25]

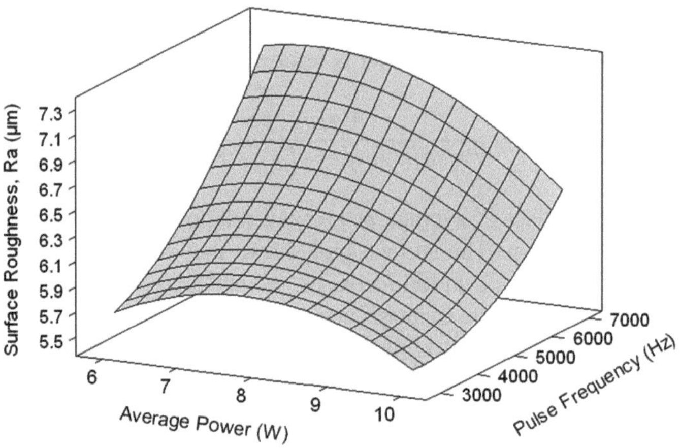

Fig. 5.13 Response surface plot of surface roughness (Ra) for varying average power and pulse frequency [25]

increases and therefore, the amount of material melted and vaporized is more due to more incident energy and ultimately, the roughness of machined surface increases. At high setting of average power, due to removal of more material and wider crater formation on the workpiece surface, the surface roughness decreases. From the same surface plot, it is also seen that with increase in pulse frequency, the roughness of machined surface increases. Although, the value of peak power of laser beam decreases with pulse frequency according to Eq. (5.9), the duration of consecutive laser pulses is small and the upper surface of material melts and

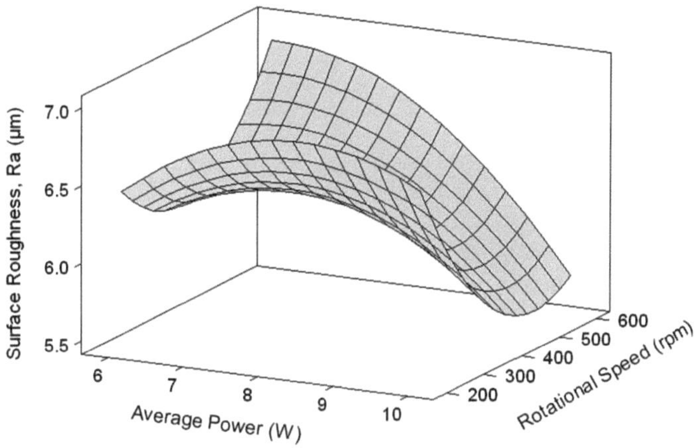

Fig. 5.14 Response surface plot of surface roughness (Ra) for varying average power and workpiece rotational speed [25]

vaporizes instantly. Due to this, more peaks and valleys are generated on the machined surface and cause more roughness.

The surface plot of variation of laser average power and rotating speed of workpiece on surface roughness is shown in Fig. 5.14. The minimum value of surface roughness is obtained at high setting of average power and moderate value of rotational speed of workpiece. It is obvious from this surface plot that the increase in workpiece speed causes the absorbed amount of laser energy per unit scan area is less and it further reduces the amount of material removal from workpiece. This causes fewer irregularities on the machined surface. The surface plot also depicts that when the workpiece rotating speed is high, the surface roughness is less for varying laser average power. According to Eq. (5.4), it is obvious that spot overlap decreases with increasing value of workpiece rotational speed. This causes more waviness in circumferential direction of work sample. Moreover, according to Eq. (5.5), the circumferential overlap increases with workpiece rotational speed and therefore, the value of surface roughness decreases.

In Fig. 5.15, the surface plot shows the effect of assist air pressure and Y feed rate on roughness criterion. The minimum value of surface roughness is achieved at low value of assist air pressure and large setting of Y feed rate. At low value of Y feed rate, as the time duration of laser beam material interaction is high, therefore, the absorbed energy per laser scan area is high. Due to this, more material is removed from the laser irradiated zone creating large micro-peaks, which ultimately causes high value of roughness. From the same plot, it is also observed that the variation in assist air pressure has very less effect on surface roughness criterion. However, it is seen that with increase in air pressure, surface roughness value at first increases; however, at high air pressure, it decreases the roughness value of machined surface. At high air pressure, the melted and vaporized material can be

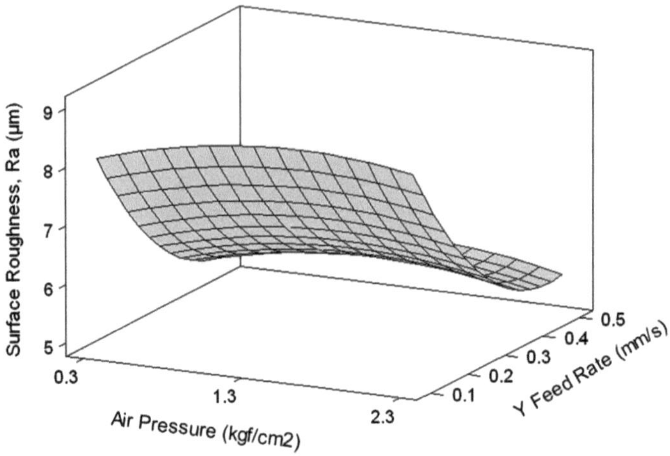

Fig. 5.15 Response surface plot of surface roughness (Ra) for varying air pressure and Y feed rate [25]

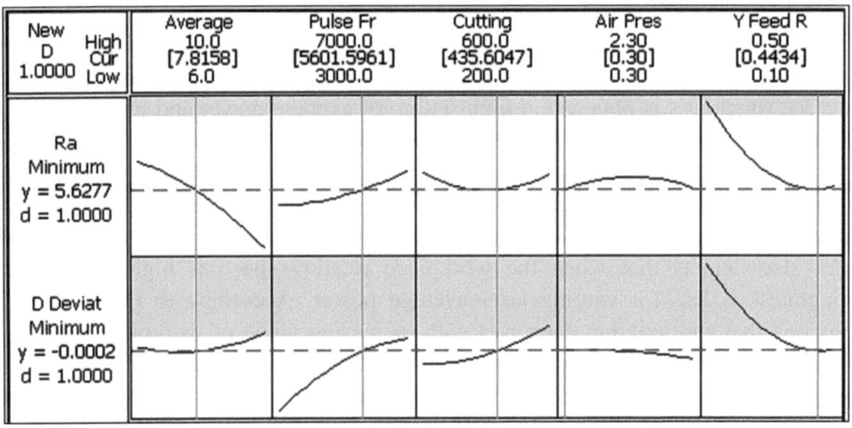

Fig. 5.16 Results of multi objective optimization for minimum depth deviation and surface roughness (Ra) [25]

removed uniformly and completely from machining zone; therefore, the irregularities on the surface are quite less.

For obtaining the accurate geometrical dimensions of machined parts as well as for achieving quality machined surface, multi performance optimization was carried out using MINITAB software. The results of multi objective optimization for achieving the minimum value of depth deviation and surface roughness are shown in Fig. 5.16. The parametric setting at which the target machined depth and minimum surface roughness was achieved as average power at 7.81 W, pulse frequency at 5601.59 Hz, rotating speed at 435.60 rpm, assist air pressure at 0.30 kgf/cm^2 and

Table 5.10 Results of confirmation experiment at optimal parametric combination

Optimization type	Experimental parameters combination	Experimental results: Surface roughness, Ra = 5.91 μm Depth deviation = −0.00021 mm RSM predicted results: Surface roughness, Ra = 5.63 μm Depth deviation = −0.00020 mm Percentage of errors: For surface roughness = 4.76% For depth deviation = 4.78%
Multi-objective optimization	Average power = 7.81 W Pulse frequency = 5600 Hz Rotating speed = 436 rpm Air pressure = 0.30 kgf/cm^2 Y feed rate = 0.443 mm/s	

Y feed rate at 0.443 mm/s. Verification experiments were performed to confirm the multi-objective optimized results at parametric combination that can be set to the nearer feasible setting achieved in RSM based optimization. Table 5.10 shows the results of verification experiment and calculated error percentage for both responses. It is observed from the results of percentage of prediction error calculation that the prediction errors were 4.76 and 4.78% for surface roughness and depth deviation. These errors are within acceptable limit and it can be concluded that the multi objective optimization is valid within the parametric range of consideration.

5.9.2 Influence of Overlap Factors on Machining Criteria

It has been found in the previous section that there are huge influences of various process parameters on surface roughness (Ra) and depth deviation for obtaining desired surface features and dimensional geometries during laser micro-turning process of alumina ceramic. However, for achieving better surface characteristics, it is urgently to know the basic understanding of surface roughness variation as the surface quality and irregularity is governed by two overlap factors i.e. spot overlap and circumferential overlap. As the values of these overlap factors depend on range and value of various process parameters, therefore, in this section, attempts have been made to carry out in-depth studies of overlap factors on surface features during machining.

For experimentation, same pulsed Nd:YAG laser micromachining system and workpiece rotating fixture is utilized. The ranges of various process parameters considered are enlisted in Table 5.11. For each of the experiments, the number of laser scan passes was 100. In this experimental investigation, the values of workpiece rotating speed and pulse frequency were selected cautiously so that the percentage values of spot overlap (SOp) lay as some distinct and positive value. In

Table 5.11 Experimental details and other conditions

Condition	Description
Average power, W	7, 8
Pulse frequency, Hz	3000, 4000, 5000, 6000, 7000
Workpiece rotating speed, rpm	200, 300, 400, 500, 600
Y feed rate, mm/s	0.1, 0.2, 0.3, 0.4, 0.5
Pulse width, % of duty cycle	3%
Air pressure, kgf/cm^2	1.3
Z feed rate, mm/s	0.01

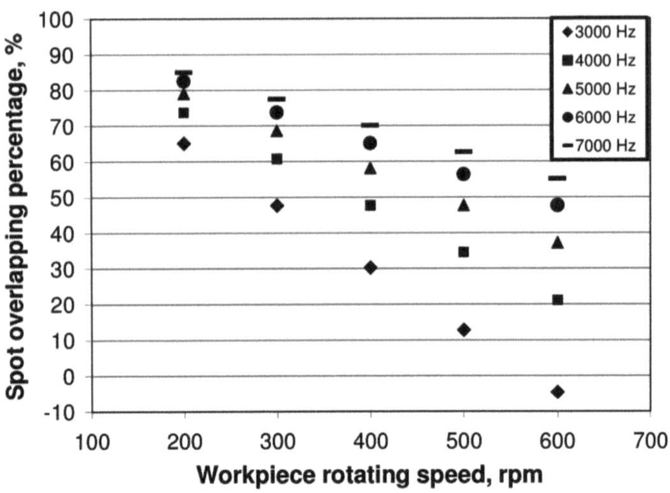

Fig. 5.17 Influence of rotational speed of workpiece and pulse frequency on spot overlap

Fig. 5.17, the graph represents the variation of spot overlap (%) at various pulse frequency and rotational speed of workpiece. During trial runs, the parametric range of Y feed rate was chosen very cautiously so as to keep the circumferential overlap within acceptable limit. In Fig. 5.18, the influence of rotational speed of workpiece and Y feed rate on circumferential overlap (%) is depicted. By observing these two plots, it is concluded that the variation of significant process parameters have huge influence on these overlaps. In these graphs, some data points were taken in which there are no overlaps to analyse its effect on response criteria.

The graphical plots shown in Fig. 5.19 and 5.20 represent the effect of spot overlap on surface roughness criterion due to variation in pulse frequency and rotational speed of workpiece, correspondingly. As per Eq. (5.4), the spot overlap can be varied using two schemes, one by varying rotational speed of workpiece while pulse frequency kept at particular value, and second by varying pulse frequency while rotational speed kept at particular value. For the Fig. 5.19, the constant parameters were average power at 7 W and Y feed rate at 0.3 mm/s. From this

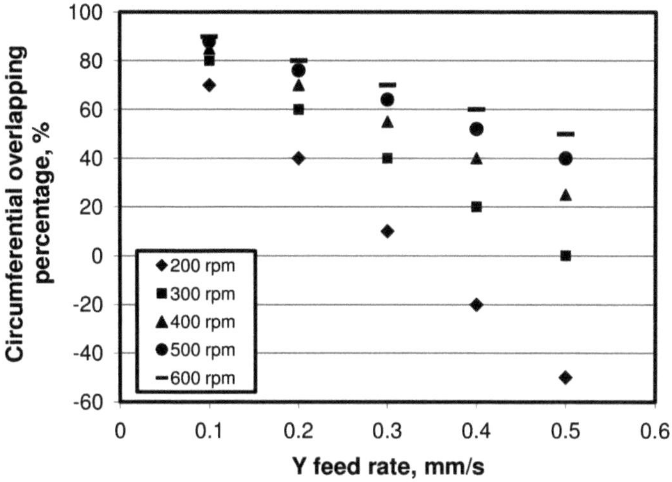

Fig. 5.18 Influence of rotational speed of workpiece and Y feed rate on circumferential overlap

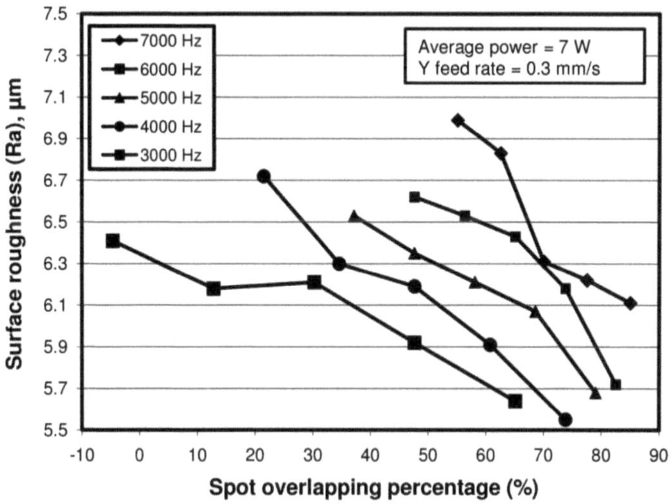

Fig. 5.19 Effect of spot overlap on surface roughness (Ra) at various pulse frequencies [26]

plot, it is obvious that increase in spot overlap results in decrease in surface roughness value at all pulse frequency values. Due to high amount of spot overlap, the machined surface become even with least number of irregularities. From the same plot, it is also observed that for any value of spot overlap, increase in pulse frequency results in increase in surface roughness values. According to Eq. (5.4), the spot overlap increases with pulse frequency. Yet, at higher value of pulse

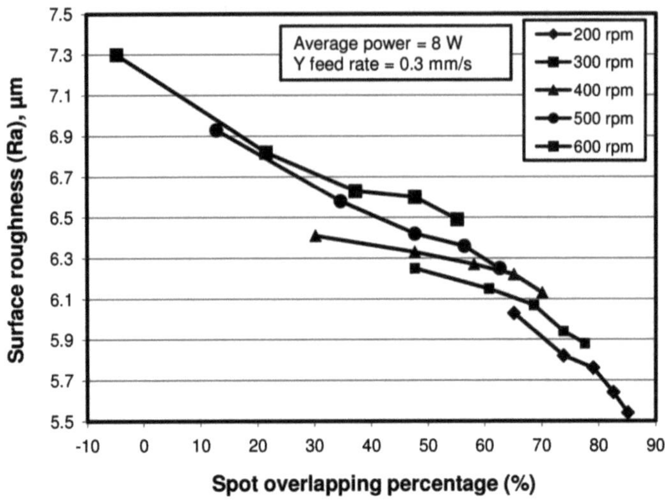

Fig. 5.20 Effect of spot overlap on surface roughness (Ra) at various rotational speed of workpiece [26]

frequency setting, the time interval between two consecutive pulses is low. The laser irradiated material get evaporation energy instantly and ultimately, the total energy absorbed by the material surface is high. Consequently, depth of craters obtained is high and ultimately increases surface roughness values. The graph shown in Fig. 5.20 was drawn by keeping the average power at 8 W and Y feed rate at 0.3 mm/s. From this figure, it is clear that the value of surface roughness is decreasing with increase in spot overlap at all values of rotational speed of workpiece. As more material is removed from the laser irradiated zone due to high amount of spot overlap, therefore, uniform machined surface is achieved. For any particular spot overlap value, the roughness of machined surface is high at higher value setting of workpiece rotational speed. The value of spot overlap is less at high rotating speed and thus, the uneven machined surface is obtained. Figures 5.21 and 5.22 represent the variation of surface roughness values due to varying the circumferential overlap (%) drawn at different values of Y feed rate and rotational speed of workpiece, consecutively. Figure 5.21 is drawn by keeping parameters at constant as average power at 8 W and pulse frequency at 5000 Hz. It is clear from Eq. (5.5) that circumferential overlap is governed by two significant process parameters i.e. workpiece rotational speed and Y feed rate. The increasing value of rotating speed of workpiece at constant Y feed rate setting results in increase in circumferential overlap and surface roughness value decreases. As per Eq. (5.5), with decreasing value of workpiece rotational speed, the value of circumferential overlap increases and this result in reduce in roughness value of machined surface due to uniform material removal. It is also obvious from Fig. 5.21 that for any particular setting of circumferential overlap, higher value of surface roughness is

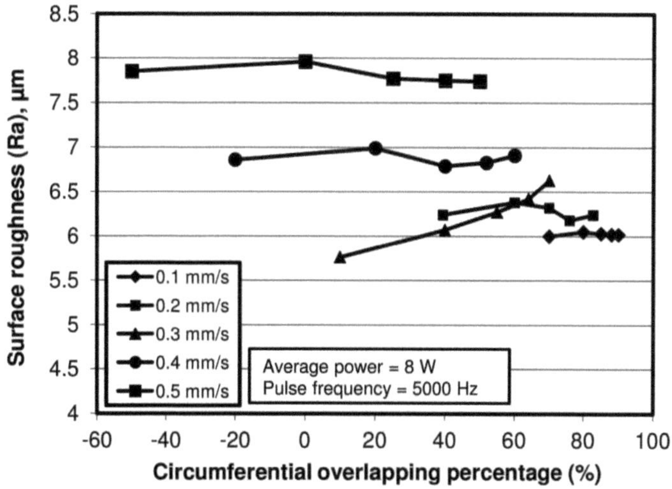

Fig. 5.21 Effect of circumferential overlap on surface roughness (Ra) at various Y feed rate [26]

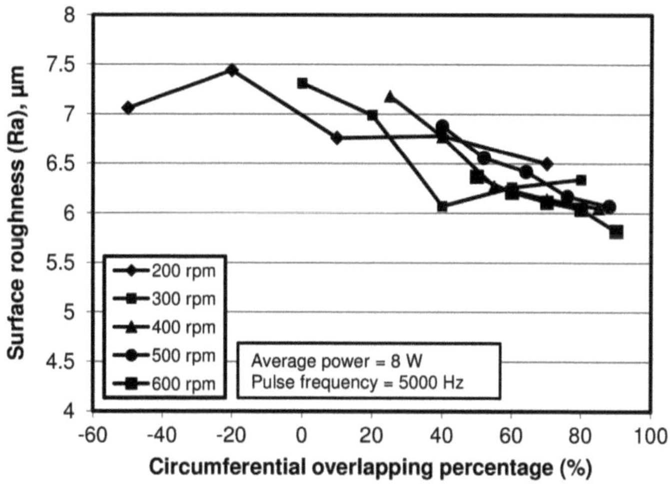

Fig. 5.22 Effect of circumferential overlap on surface roughness (Ra) at various rotational speed of workpiece [26]

obtained for increasing the Y feed rate. It is due to the fact that lot of null spacing is generated for successive laser scanning and uneven machined surface is generated.

The graphical plot shown in Fig. 5.22 depicts the variation in surface roughness due to increase in circumferential overlap. The constant process parameters were average power at 8 W and pulse frequency at 5000 Hz, correspondingly. From this plot, it is obvious that roughness of micro-turn surface reduces due to increase in circumferential overlap for all values of rotational speed of workpiece. At higher setting of circumferential overlap, the consecutive laser scanned widths comes nearer and due to this, the amount of irregularities reduce. It is also clear from the plot that at higher setting of workpiece rotational speed, the values of circumferential overlap increases and in turn, roughness of machined surface reduces by reducing the surface unevenness.

5.9.3 Influence of Laser Defocusing Conditions on Machining Criteria

In previous sections, laser micro-turning process has been carried out based on response surface methodology (RSM) approach to explore the effect of process parameters on machining criteria and also to achieve the optimum machining parametric combination to obtain minimum value of micro-turning depth deviation and surface roughness simultaneously. Moreover, in-depth investigation and studies of spot and circumferential overlaps have also been done. The values of surface roughness (Ra) show that the laser micro-turning surface has lot of irregular micro-peaks, which have been generated due to variation in spot as well as circumferential overlaps and also due to sudden solidification of melted material. However, manufactured products must be geometrically and dimensionally accurate with high quality machined surface. Therefore, at this stage, further experiments must be conducted taking novel machining strategies considering defocusing conditions to remove the irregular micro-peaks from laser micro-turn surface without much impeding the geometrical dimensions of machined parts. At defocusing condition, due to enlargement of laser beam spot, the energy density of the laser beam is reduced and the micro-peaks get sufficient energy to melt and evaporate from the upper surface without affecting the base material. As it is not possible to completely remove the micron-sized peaks from the surface in one pass, therefore, multi pass laser micro-turning operation at various defocusing conditions has been carried out and the test results were analyzed through graphical plots. Figure 5.23 shows the schematic representation of scheme laser micro-turning process for removing micro-peaks from laser micro-turn rough surface. The schematic view of defocusing conditions of laser beam for machining micro-peaks from rough surface is shown in Fig. 5.24. The defocus position is termed as upward when the laser focus plane is above the workpiece surface, whereas defocus position is termed as downward when the laser focus plane is inside the workpiece or below the workpiece surface. As the present experimentation is dealing with surface features, therefore, surface roughness (Ra and Rt) have been taken into consideration as responses for the analysis. All the experiments have been carried out at nearer feasible process parametric setting of RSM based multi-objective optimization. Eight number of job surfaces were prepared at parametric combination of average power, pulse frequency, workpiece rotating speed, assist air pressure and Y feed rate at 7.81 W/5600 Hz/436 rpm/0.30 kgf-cm^{-2}/0.443 mm/s. The

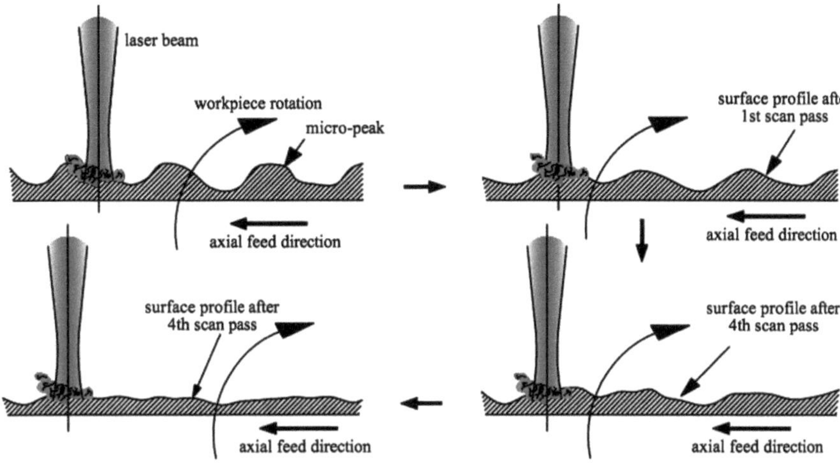

Fig. 5.23 Mechanism of elimination of surface irregularities during laser defocusing

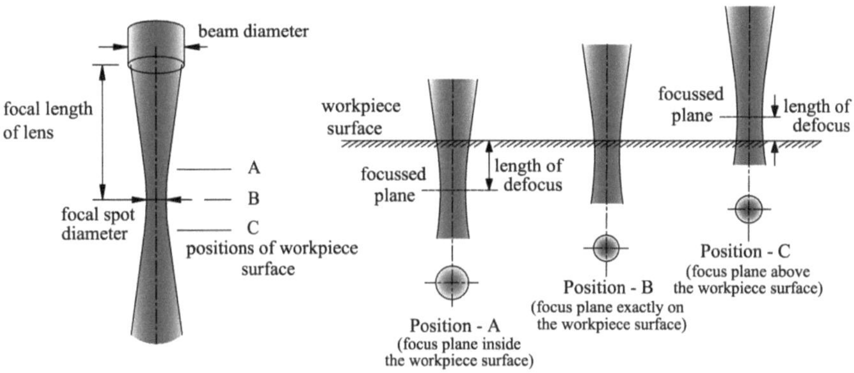

Fig. 5.24 Representation of various defocusing conditions [24]

values of surface roughness (Ra and Rt) measured for these eight experiments are listed in Table 5.12. These eight machined work samples were used for carrying out experimentation by varying the two process parameters i.e. defocus positions and number of passes. The values of these two process parameters are enlisted in Table 5.13. For each focusing position, each experiment is carried out by increasing the number of laser scan passes (up to 10). After conducting each experiment by increasing the number of laser scan pass, the values of surface roughness (Ra and Rt) were measured by the same surface roughness measuring instrument. The variation of surface roughness values was graphically plotted and analysed.

The variation of surface roughness (Ra) with number of laser passes for upward laser defocusing positions is shown in Fig. 5.25. From this figure, it is obvious that

Table 5.12 Results of surface roughness (Ra and Rt) machined at RSM based optimal parametric combination during preparation of surface

Expt. No	Process parametric combination	Experimental results	
		Surface roughness	
		Ra (μm)	Rt (μm)
1	Average power = 7.81 W Pulse frequency = 5600 Hz Workpiece rotating speed = 436 rpm Air pressure = 0.30 kgf/cm^2 Y feed rate = 0.443 mm/s	5.94	45.38
2		5.87	46.21
3		5.84	46.02
4		5.89	45.68
5		5.98	45.69
6		6.00	45.88
7		5.88	46.39
8		5.81	46.01

Table 5.13 List of parameters and different levels for defocusing experiments [24]

Process parameters	Unit	Values
Defocus conditions	mm	−0.2, −0.4, −0.6, −0.8, +0.2, +0.4, +0.6, +0.8
Number of passes	–	1, 2, 3, 4, 5, 6, 7, 8, 9, 10

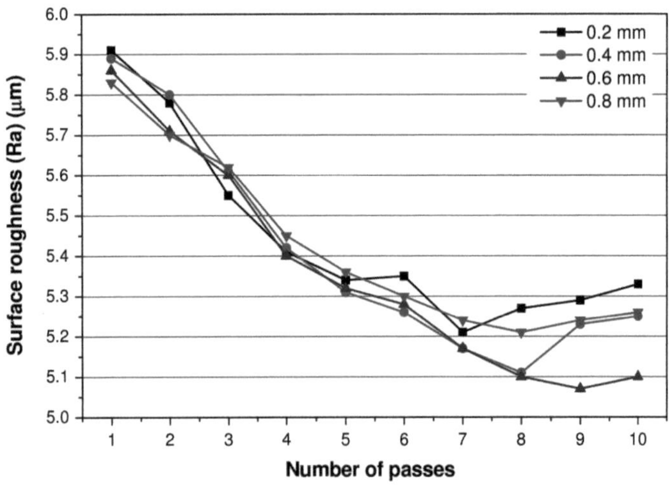

Fig. 5.25 Influence of number of laser scan passes on surface roughness (Ra) at various upward defocus conditions [24]

surface roughness is decreasing with scan passes for all downward defocusing conditions. With number of laser scan passes, the height of micron sized peaks decreases due to removal of materials from the peaks. From this figure, it is also

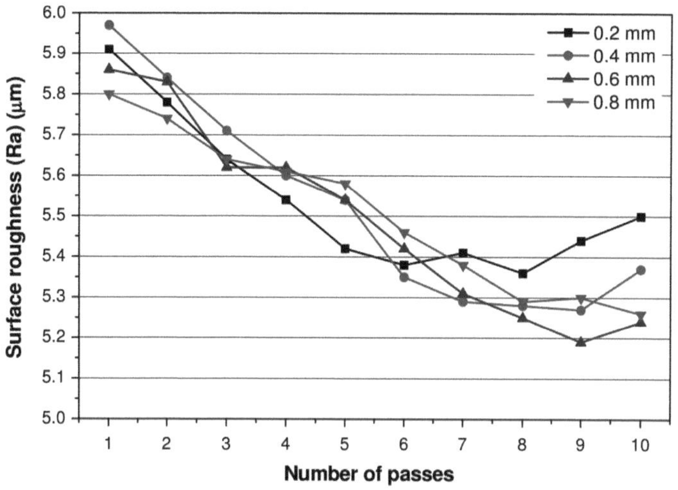

Fig. 5.26 Influence of number of laser scan passes on surface roughness (Ra) at various downward defocus conditions [24]

revealed that for any value of laser scan pass, the surface roughness obtained are different for various defocusing conditions. With higher value of defocusing, roughness of machined surface is low due to finishing of irregular micro-peaks for delivering adequate beam energy of laser. It is also obvious from the same figure that at higher setting defocus, the surface roughness is high due to inadequate energy density of laser beam that cannot reduce the height of micro-peaks.

In Fig. 5.26, the variation of surface roughness (Ra) with laser scan passes with downward defocus conditions is shown in Fig. 5.26. From this graph, it is obvious that with increase in scan passes, the value of surface roughness is decreasing due to removal of micro-peaks from the surface. It is seen from the plot that surface roughness again increases after certain number of laser scan passes. It is due to the fact that the removal of irregular peaks from machined surface is reducing. At high value of defocusing, the spot diameter of laser beam on the material is high and due to this, the laser energy density is less. Due to this fact, the laser beam has incapable for removal of micro-peaks from machined surface and consequently, the surface roughness is high.

In Fig. 5.27, the variation in surface roughness (Rt) with laser scan passes is shown for different values of upward defocus conditions. From this plot, it is obvious that surface roughness (Rt) is decreasing for increasing value of scan passes. As the laser scan pass is increasing, the total height of micro-peaks are decreasing and this phenomena has direct effect on surface roughness (Rt) values. For any particular value of laser scan pass, low value of defocus results in rough machined surface and this is due to variation in energy densities of laser beam at different defocus conditions. At high value of defocus i.e. at 0.8 mm, Rt is high compared to other defocus conditions and this is due to less energy density at high setting of defocus.

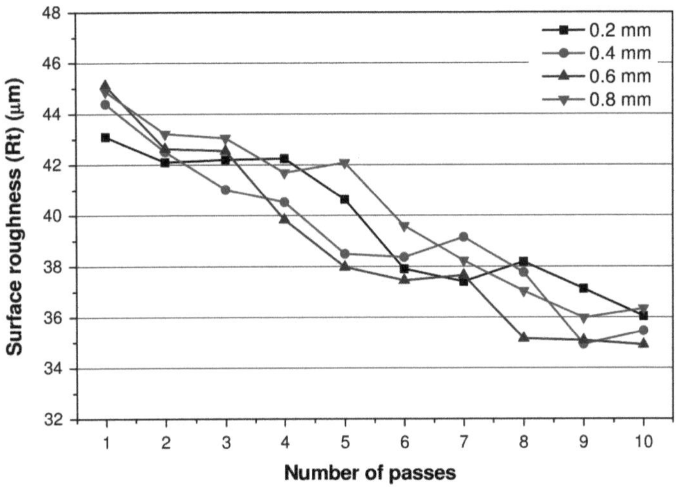

Fig. 5.27 Influence of number of laser scan passes on surface roughness (Rt) at various upward defocus conditions [24]

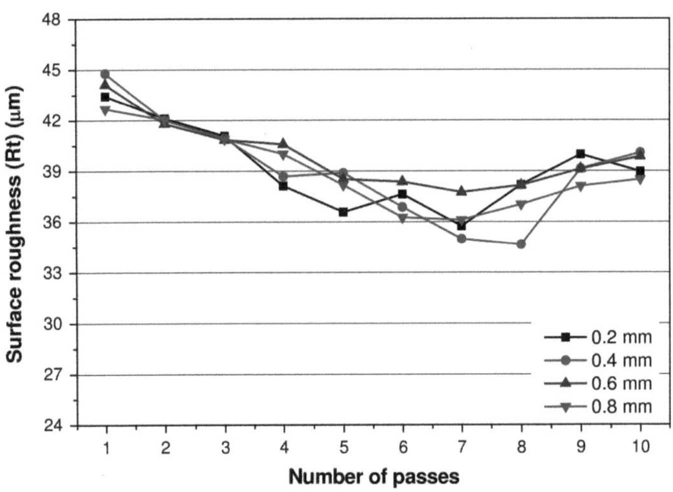

Fig. 5.28 Influence of number of laser scan passes on surface roughness (Rt) at various downward defocus conditions [24]

In Fig. 5.28, the variation in surface roughness (Rt) is shown with number of laser scan pass at different downward defocus conditions. The graph shows that the roughness of machined surface decreases up to definite value of scan passes. This is due to decrease in overall height of micro-sized peaks on the machined surface. It is seen from the plot that after few number of laser scan passes, there is no such

Table 5.14 Results of parameter settings to obtain least surface roughness (Ra and Rt) values

Responses	Parameter settings of average power, pulse frequency, workpiece rotating speed, air pressure and Y feed rate	Parametric setting of additional parameters		Minimum value of responses	
		Number of passes	Defocus position	Surface roughness, Ra	Surface roughness, Rt
Surface roughness, Ra	7.81 W/5600 Hz/436 rpm/0.30 kgf-cm-2/0.443 mm/s	9	0.6 mm (upward)	5.07 μm	35.08 μm
Surface roughness, Rt		8	0.4 mm (downward)	5.28 μm	34.65 μm

improvement in surface roughness. At defocus positions of 0.4 and 0.8 mm, the value of surface roughness slightly increases afterward 7th laser scan pass. It is also obvious from the plot that the values of surface roughness obtained are random and least value of Rt is achieved at 8th number of laser scan pass and at defocus condition of 0.4 mm.

By analyzing various plots of Ra and Rt, it can be concluded that the values of surface roughness (Ra and Rt) are decreasing with number of laser scan pass for all values of upward defocus conditions. Moreover, it is also observed that with same value of defocus condition, Ra values are decreasing, while the values of Rt are moderately changing. In Table 5.14, the least values of surface roughness (Ra and Rt) obtained at different parametric settings are enlisted. The minimum value of Ra is obtained as 5.07 μm (corresponding Rt is 35.08 μm) for upward defocus whereas the least value of Rt is obtained as 34.65 μm (corresponding Ra is 5.28 μm) for downward defocus.

5.9.4 Comparative Study of Focused and Various Defocus Conditions Machining

In the previous section, it has been found that the laser micro-turning process at defocused positions have great influence for removing the micro-peaks from the rough machined surface. In this section, an attempt has been made to comparative study and analysis of laser micro-turning process at focused and defocused conditions considering four process parameters such as laser beam average power, pulse frequency, workpiece rotating speed and Y feed rate. In Table 5.15, the details of constant and varying process parameters are shown. The target micro-turning depth is same as 100 μm. All the experiments were carried out using one factor at a time methodology. It is revealed from the previous section that minimum surface roughness (Ra) is achieved at upward defocusing position. Therefore, the present experimentation is carried out at focused as well as upward

Table 5.15 Details of machining conditions

Process parameters	Values
Variable parameters	
Average power	5, 6, 7, 8, 9, 10 W
Pulse frequency	4000, 4500, 5000, 5500, 6000 Hz
Rotational speed	200, 300, 400, 500, 600 rpm
Y feed rate	0.2, 0.3, 0.4, 0.5, 0.6 mm/s
Condition of lens	Focused, defocused
Defocused	0.2, 0.4, 0.6 mm upward
Constant parameters	
Air pressure	0.3 kgf/cm^2

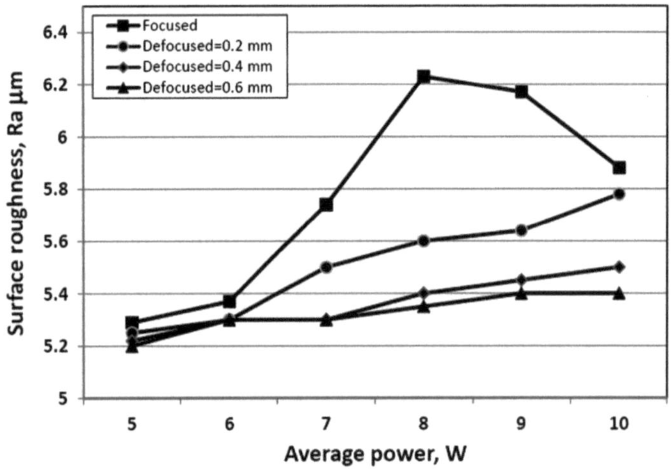

Fig. 5.29 Effect of variation of average power on surface roughness (Ra) at focused and different upward defocus positions

defocusing conditions. Surface roughness (Ra and Rt) were measured after conducting each experiment and the test results were analyzed through various graphical plots.

The variation of surface roughness (Ra) with laser beam average power for focused and defocused (upward) positions is shown in Fig. 5.29. The constant process parameters were pulse frequency at 5000 Hz, rotational speed of workpiece at 400 rpm and Y feed rate at 0.4 mm/s. It is seen from this plot that the quality surface is achieved at lower setting of average power. This is due to less amount of material ablated from workpiece surface. However, with laser beam average power, the roughness of machined surface increases. From this comparative graph, it is also observed that the values of surface roughness are low for the machined surface generated during laser defocused positions than focused condition. It is due to the laser beam energy density to be low and consequently, adequate laser energy for melting and evaporation is received by material surface which efficiently remove

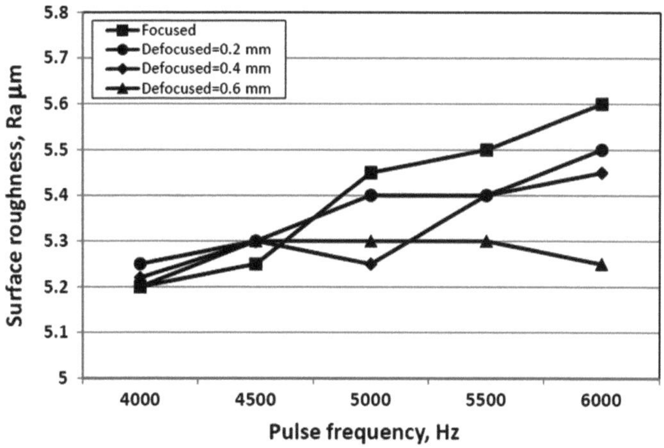

Fig. 5.30 Effect of variation of pulse frequency on surface roughness (Ra) at focused and various upward defocus positions

micro-peaks generated on the surface. For a particular value of average power, the values of surface roughness decreases with increasing value of defocusing and it is only due to efficient removal of micro-peaks from the machined surface.

The variation in surface roughness (Ra) with laser pulse frequency at focusing and defocusing positions is shown in Fig. 5.30. The average power, rotational speed of worpiece and Y feed rate were kept at 8 W, 400 rpm and 0.4 mm/s, correspondingly. The values of surface roughness (Ra) are found to increase slightly at focused and defocused machining conditions. With increase in pulse frequency, peak power is decreasing. However, the time duration between two pulses is very short and the material from top surface gets adequate thermal energy to remove the material rapidly creating higher peaks and valleys. From the same plot, it is observed that the roughness values are less at same parameters setting when machining is done at defocus positions when compared to focus position and this is due to adequate thermal energy which is received on the material surface and consequently, the height differences of peaks and valleys are less. When applying higher setting of defocus during machining keeping other factors constant, it is observed to obtain less value for surface roughness.

The variation of surface roughness (Ra) with increasing value of rotational speed of workpiece at different upward defocus positions and focused condition is observed in Fig. 5.31. The values of other process parameters were constant as average power of laser beam at 8 W, laser pulse frequency at 5000 Hz and Y feed rate at 0.4 mm/s. The plot shows that the roughness of micro-turning surface is less at defocus conditions when compared with roughness value for focus condition machining. As it is a known phenomenon that at defocus condition, the laser energy density is reduced and therefore, the height differences of generated peaks and valleys are less. With increase in rotating speed, it is seen that the roughness of

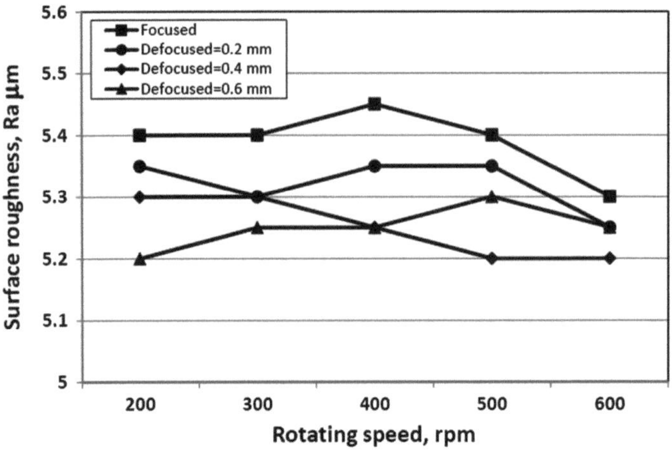

Fig. 5.31 Effect of variation of rotational speed of workpiece on surface roughness (Ra) at focused and various upward defocus positions

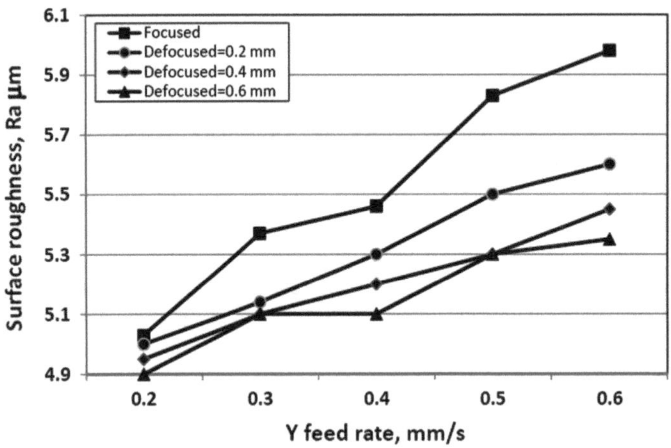

Fig. 5.32 Effect of variation of Y feed rate on surface roughness (Ra) at focused and various upward defocus positions

machined surface achieved is low and it is due to increase of circumferential overlap percentages as per Eq. (5.5). Moreover, surface roughness values are less at higher setting of defocused condition and this is due to instantaneous melting and afterward solidification of material on material surface.

In Fig. 5.32, the effect of variation of Y feed rate on roughness (Ra) is observed at different defocus and focused conditions. The constant parameters were average power at 8 W, laser pulse frequency at 5000 Hz and rotational speed of workpiece

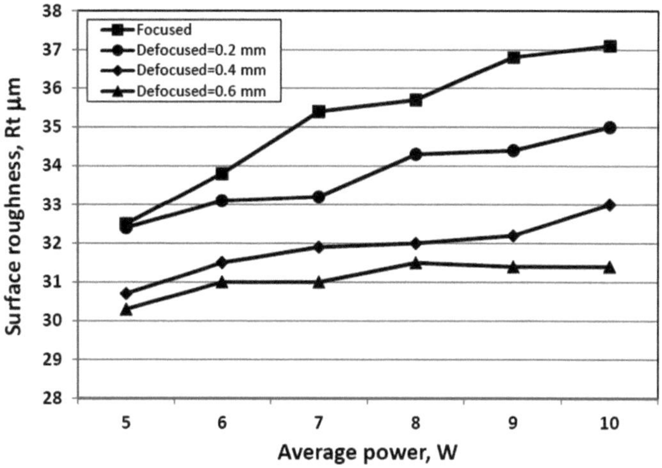

Fig. 5.33 Effect of variation of average power on surface roughness (Rt) at focused and various upward defocus positions

at 400 rpm. The plot shows that the roughness of machined surface increases with Y feed rate and this is due to decrease in circumferential overlap values for higher values of Y feed rate. Due to this, the machined surface obtained as uneven and irregular. Here also, it is seen that at focused condition machining, the surface roughness value is high when compared to defocused conditions.

The variation of surface roughness (Rt) with laser beam average power at defocus conditions and focused machining condition is depicted in Fig. 5.33. The pulse frequency, rotational speed of workpiece and Y feed rate were kept constant at 5000 Hz, 400 rpm and 0.4 mm/s, respectively. The plot shows that the roughness of machined surface increases with average power due to generation of micro-peaks on the material surface. At defocused position of laser beam, with constant value of average power, surface roughness achieved as low compared to focus machining condition. Moreover, for higher defocus values, there is no such effect of average power on Rt values.

In Fig. 5.34, the graph shows the variation in surface roughness (Rt) with pulse frequency at focused and defocused conditions while keeping the other process parameters as constant at average power at 8 W, rotational speed of workpiece at 400 rpm and Y feed rate at 0.4 mm/s. From the plot, it is obvious that Rt has no such variation with pulse frequency at defocused positions. However, at focused condition, Rt is slightly increasing. The values of Rt is higher at all defocus conditions compared to focused positional micro-turning.

The variation in Rt with respect to rotational speed of workpiece is depicted in Fig. 5.35 at focused and different defocus positions. The constant parameters were average power at 8 W, pulse frequency at 5000 Hz and Y feed rate at 0.4 mm/s. The figure shows that the roughness of machined surface decreases with rotating speed of work sample and it is due to increase in circumferential overlap with

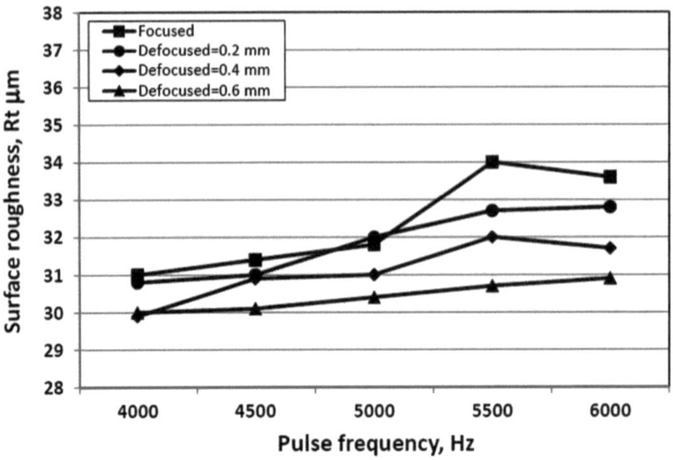

Fig. 5.34 Effect of variation of pulse frequency on surface roughness (Rt) at focused and various upward defocus positions

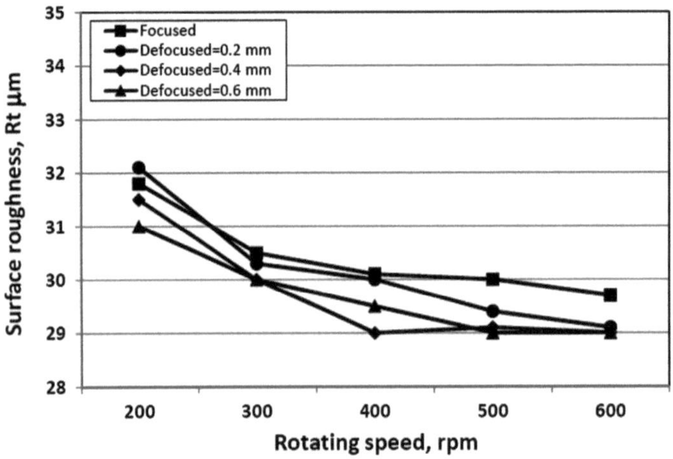

Fig. 5.35 Effect of variation of rotational speed of workpiece on surface roughness (Rt) at focused and various upward defocus positions

rotating speed. The plot also shows that there is no such variation in Rt at defocus conditions. The Rt values obtained is higher in focus positional micro-turning than roughness at defocus positions.

The variation in surface roughness (Rt) with Y feed rate at various defocus conditions and focused position is graphically shown in Fig. 5.36. The pulse frequency was 5000 Hz, average power was 8 W and rotational speed of work sample was 400 rpm. The graph shows that the Rt values are increasing with Y feed rate

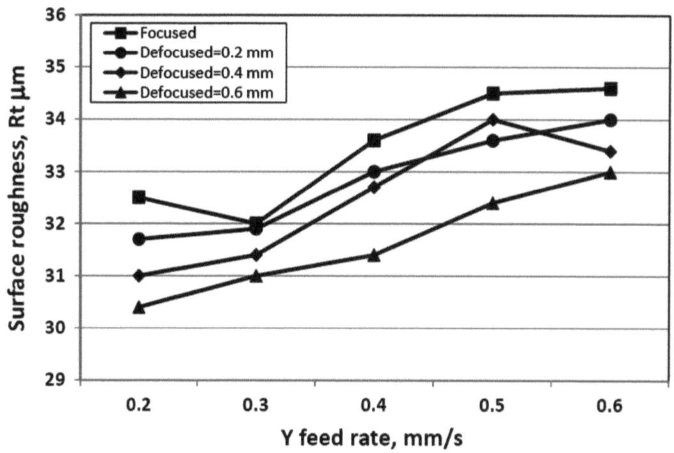

Fig. 5.36 Effect of variation of Y feed rate on surface roughness (Rt) at focused and various upward defocus positions

and it is due to lower value of circumferential overlap at higher setting of Y feed rate as per Eq. (5.5). It is also observed from the plot that the values of Rt are low in defocused positions when compared to focused condition micro-turning.

5.9.5 Microscopic Observations of Laser Micro-turning Surfaces

In this section, detailed surface features and conditions of machined micro-turning surface of alumina work material have been observed using Scanning Electron Microscope (SEM) during machining at various experimental schemes as mentioned in previous sections. SEM micrographs shown in Fig. 5.37a–f demonstrate surface morphologies of the machined surface taken at 27× and 200× magnifications and machined at various parametric combinations utilizing Response Surface Methodology (RSM) approach. The SEM micrographs shown in Fig. 5.37a has been taken for the micro-turned surface machined at 9/4000/500/1.8/0.2 of average power (W), pulse frequency (Hz), workpiece rotating speed (rpm), assist air pressure (kgf/cm^2) and Y feed rate (mm/s) parameters and the surface roughness (Ra) is achieved as 7.20 µm. Similarly, Fig. 5.37c, e show the SEM micrographs of machined surface machined at 9/4000/500/0.8/0.4 and 7.81/5600/436/0.3/0.443 (at nearer feasible parametric setting which can be set in the Nd:YAG laser set-up) of the process parameters combinations, respectively. The surface roughness (Ra) achieved for these parametric combinations are 5.15 and 5.91 µm, respectively. It is observed from Fig. 5.37b, d that for lower setting of assist air pressure and higher setting of Y feed rate the roughness of micro-turned surface achieved is

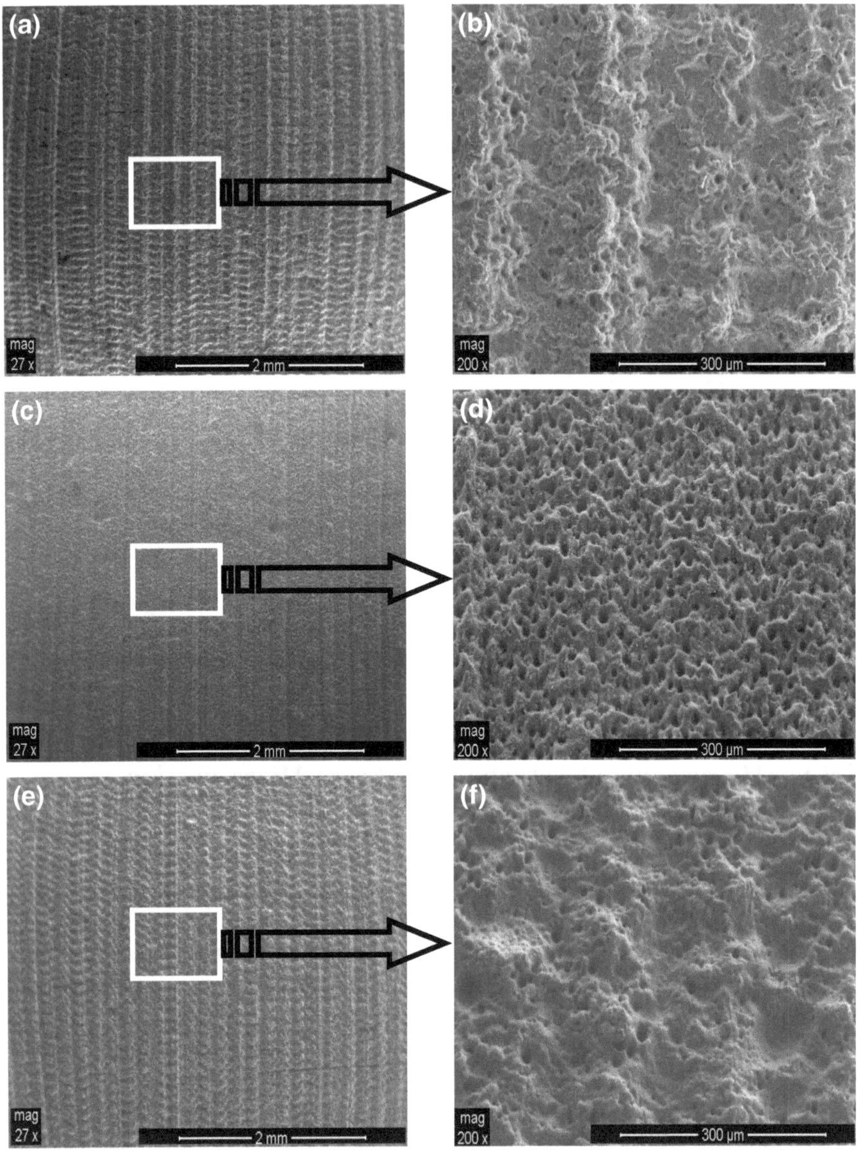

Fig. 5.37 SEM graphs of the surfaces machined at **a**, **b** 9/4000/500/1.8/0.2; **c**, **d** 9/4000/500/0.8/0.4 and **e**, **f** 7.81/5600/436/0.3/0.443 of average power (W), pulse frequency (Hz), workpiece rotating speed (rpm), assist air pressure (kgf/cm^2) and Y feed rate (mm/s) [25]

comparatively low. Moreover, the SEM micrograph shown in Fig. 5.37d shows low degree of peaks and valleys on the machined surface compared to micrograph shown in Fig. 5.37b. Comparing the SEM micrograph shown in Fig. 5.37b, d, f, it is observed that the there are small amount of irregularities with some peaks and

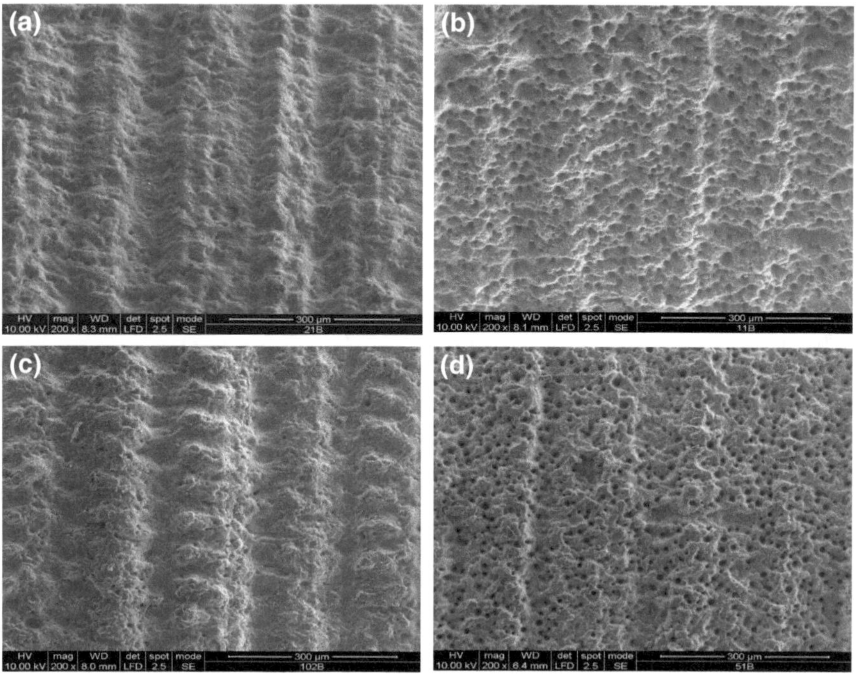

Fig. 5.38 Micrographs of the surfaces machined at average power, pulse frequency, workpiece rotating speed and Y feed rate of **a** 7 W/4000 Hz/600 rpm/0.3 mm/s (Ra is 6.72 μm), **b** 7 W/6000 Hz/300 rpm/0.3 mm/s (Ra is 6.18 μm), **c** 8 W/5000 Hz/400 rpm/0.5 mm/s (Ra is 7.77 μm) and **d** 8 W/5000 Hz/600 rpm/0.1 mm/s (Ra is 6.02 μm) [26]

valleys on the machined surface machined at optimal parametric combination. Although, at optimal parametric setting, the micro-turning depth deviation is comparatively low as −0.00021 mm than the values achieved in other two parametric combinations i.e. 0.04314 mm (Fig. 5.37b) and −0.01648 mm (Fig. 5.37d).

In Fig. 5.38a–d, the SEM images of micro-turning surfaces machined at different parametric combinations are shown. By comparing images shown in Fig. 5.38a, b, it is obvious that quality of machined surface is better in Fig. 5.38b compared to shown in Fig. 5.38a. The corresponding surface roughness (Ra) values are 6.72 μm (Fig. 5.38a) and 6.18 μm (Fig. 5.38b). This variation in surface roughness value is resulted due to combining effects of increasing pulse frequency from 4000 to 6000 Hz and decreasing rotational speed of workpiece from 600 to 300 rpm. Due to this variation in process parameters, the spot overlap has increased from 21.43 to 73.81%. In Fig. 5.38c, d, the SEM images shows the machined surfaces obtained at different parametric combinations in which the value of circumferential overlap has increased to 90% from 25% due to combined variation in Y feed rate to 0.1 mm/s from 0.5 mm/s and rotational speed of work sample to 600 rpm from 400 rpm. The corresponding values of surface roughness (Ra) were obtained as 7.77 μm (Fig. 5.38c) and 6.02 μm (Fig. 5.38d).

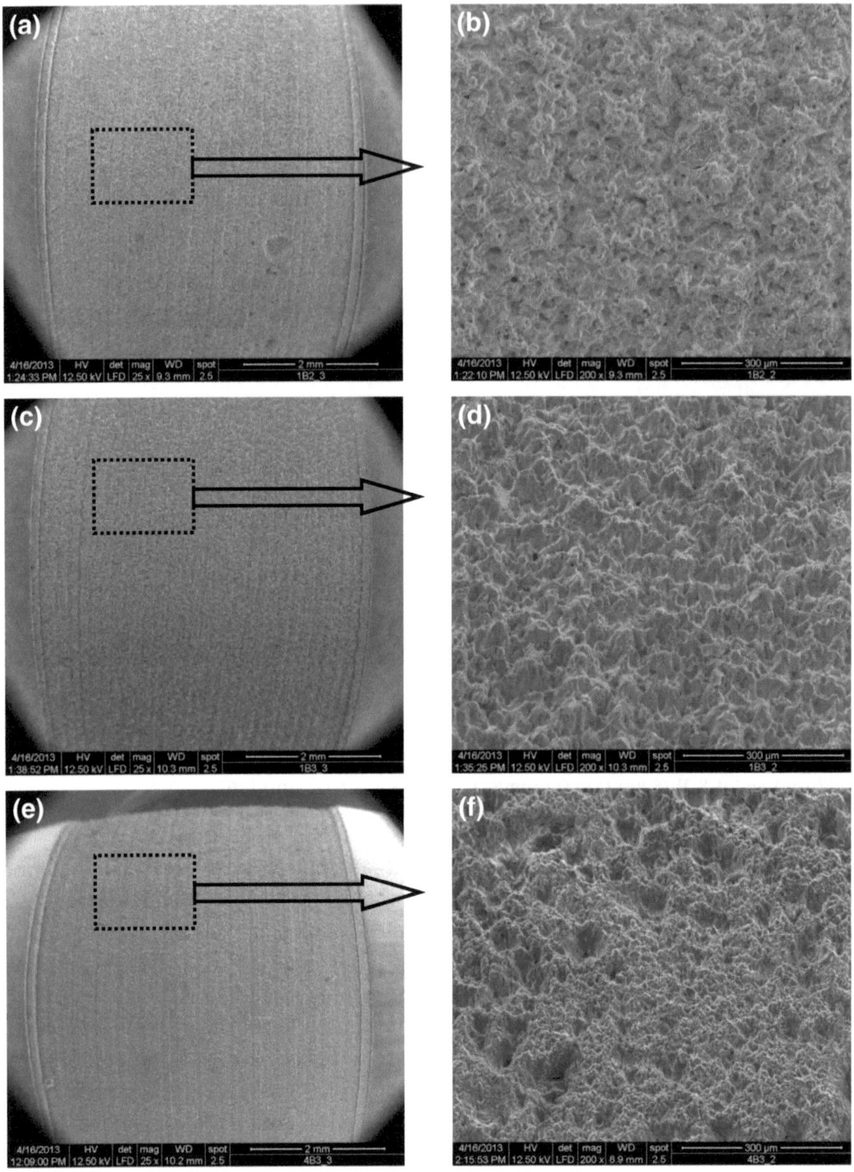

Fig. 5.39 Surface conditions of alumina machined at **a**, **b** defocus of 0.2 mm (*downward*) and 10th laser scan pass (Ra = 5.50 μm and Rt = 38.96 μm); **c**, **d** defocus of 0.8 mm (*upward*) and 10th laser scan pass (Ra = 5.26 μm and Rt = 36.34 μm) and **e**, **f** defocus of 0.6 mm (*upward*) and 10th laser scan pass (Ra = 5.10 μm and Rt = 34.91 μm)

The SEM images at various magnifications of laser micro-turning surface machined at different parametric combinations are shown in Fig. 5.39a–f. The values of surface roughness (Ra and Rt) obtained at these three parametric

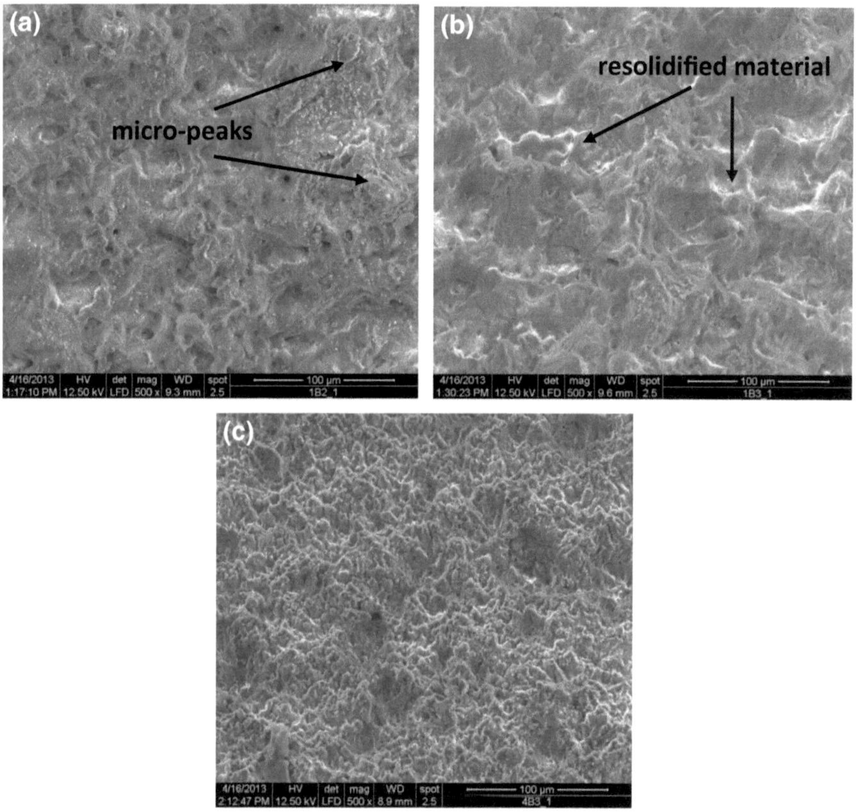

Fig. 5.40 Magnified SEM graphs of surface machined at **a** defocus of 0.2 mm (*downward*) and 10th laser scan pass, **b** defocus of 0.8 mm (*upward*) and 10th laser scan pass and **c** defocus of 0.6 mm (*upward*) and 10th laser scan pass

combinations are Ra as 5.50 μm and Rt as 38.96 μm (Fig. 5.39a, b), Ra as 5.26 μm and Rt as 36.34 μm (Fig. 5.39c, d) and Ra as 5.10 μm and Rt as 34.91 μm (Fig. 5.39e, f). It is concluded from these SEM images that quality machined surface is obtained at defocus condition of 0.6 mm and at 10th number of laser scan passes compared to other two parametric combinations. In Fig. 5.40a–c, the SEM images (500× magnification) of machined surface micro-turned at same above-mentioned three parametric combinations were shown to find out the resolidified region or surface cracks on machined surface. Some micro-peaks can be viewed in the surface shown in Fig. 5.40a. However, these peaks are absent on the machined surface shown in Fig. 5.40b. This is due to the fact that some melted material again resolidified on the surface itself and reduce the chances of generation of micro-peaks on micro-turning surface. However, the machined surface shown in Fig. 5.40c has no such micro-peaks and also resolidified material due to adequate laser energy received by the material during machining.

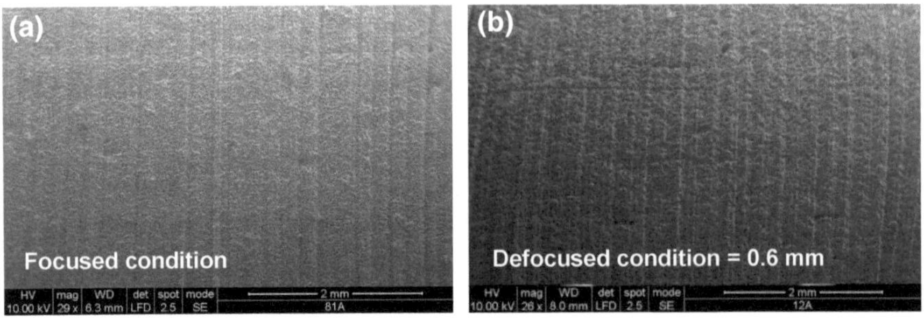

Fig. 5.41 Quality of machined surfaces at average power at 8 W, rotational speed of workpiece at 400 rpm, pulse frequency at 5000 Hz and Y feed rate at 0.4 mm/s at **a** focused condition (Ra = 5.46 μm and Rt = 33.6 μm) and **b** defocused position (Ra = 5.1 μm and Rt = 31.4 μm)

The exhibited SEM images shown in Fig. 5.41a, b represent the machined surface obtained at focused condition (Ra and Rt were obtained as 5.46 and 33.6 μm) and 0.6 mm defocused position (Ra and Rt were obtained as 5.1 and 31.4 μm) at parametric combination of laser average power at 8 W, pulse frequency at 5000 Hz, rotational speed of work sample at 400 rpm and Y feed rate at 0.4 mm/s. By comparing these values of surface roughness as well as observing the SEM images, it is concluded that the machined surface obtained at defocus condition is much better in quality than machined at focused conditional machining. Furthermore, there are lot of micro-peaks found on the machined surface generated at focused condition than obtained in defocused positions.

5.10 Conclusions

In the present chapter, the capability of pulsed Nd:YAG laser for carrying out laser micro-turning operation on difficult-to-machine 99% pure aluminum oxide (Al_2O_3) ceramic has been investigated in-depth through various machining strategies and experimental schemes. Through various experimental schemes and novel machining strategies, it is found that pulsed Nd:YAG laser has the potential to achieve desired micro-turn depth and surface finish of the cylindrical aluminum oxide (Al_2O_3) workpiece by proper controlling the process parameters. Moreover, formulations of spot and circumferential overlaps have been done and effect of overlap factors on machined surface criteria were studied to achieve accurate geometrical dimensional parts with desired surface features. The outcomes of the present research based on various experimental schemes in the area of the pulsed Nd:YAG laser micro-machining of advanced ceramic material will be effectively utilized to predict the surface conditions and dimensional features to fulfill the urgent requirement for high precision micro-machining and micro-fabrication of ceramic materials. Furthermore, author trusts that the research findings will definitely provide important information to the precision manufacturing engineers in clear understanding the influence of various process parameters of pulsed Nd:YAG laser for successful application of this novel technique of laser turning in micromachining domains.

Acknowledgements The authors acknowledge the financial support and assistance provided by CAS Ph-IV programme of Production Engineering Department of Jadavpur University under University Grants Commission (UGC), New Delhi, India.

References

1. Tilloca G (1991) Thermal stabilization of aluminum titanate and properties of aluminum titanate solid solutions. Journal of Materials Science 26(10):2809–2814
2. Kingery WD, Browen HK, Uhlmann DR (1976) Introduction to Ceramics. John Wiley & Sons Inc, New York
3. Callister WD (1975) Materials Science and Engineering: An Introduction. John Wiley and Sons, Singapore
4. Yeo CY, Tam SC, Jana S, Lau MWS (1994) A technical review of the laser drilling of aerospace materials. Journal of Materials Processing Technology 42(1):15–49
5. Benes JJ (1996) Technology adds a new twist to difficult drilling. American Machinist 140:78–79
6. Giering A, Beck M, Bahnmuller J (1999) Laser drilling of aerospace and automotive components. Proceedings of the International Congress on Laser Applications and Electro-optics, In: ICALEO 1999 proceedings, Laser Institute of America, Orlando, FL, 87:80–87
7. Knowles M (2000) Micro-ablation with high power pulsed copper vapor lasers. Optics Express 7(2):50–55
8. Wu CY, Shu CW, Yeh ZC (2006) Effects of excimer laser illumination on microdrilling into an oblique polymer surface. Optics and Lasers in Engineering 44(8):842–857
9. Voisey KT, Clyne TW (2004) Laser drilling of cooling holes through plasma sprayed thermal barrier coatings. Surface and Coatings Technology 176:296–306
10. Rodden WS, Kudesia SS, Hand DP, Jones JDC (2002) A comprehensive study of the long pulse Nd:YAG laser drilling of multi-layer carbon fibre composites. Optic Communications 210(3–6):319–328
11. Chryssolouris G, Schonbeck J, Choi W, Sheng P (1989) Advances in three-dimensional laser machining Proc. ASME Winter Annual Meeting,1–7
12. Buttery TC, Statham A, Percival JB, Hamed MS (1979) Some effects of dressing on grinding performance. Wear 55:195–219
13. Shih AJ (2000) An experimental investigation of rotary diamond truing and dressing of vitreous bond wheels for ceramic grinding. International Journal of Machine Tools and Manufacture 40:1755–1774
14. Haddad MJ, Alihoseini F, Hadi M, Hadad M, Tehrani AF, Mohammadi A (2010) An experimental investigation of cylindrical wire electrical discharge turning process. International Journal of Advanced Manufacturing Technology 46:1119–1132
15. Haddad MJ, Tehrani AF (2008) Investigation of cylindrical wire electrical discharge turning (CWEDT) of AISI D3 tool steel based on statistical analysis. Journal of Materials Processing Technology 198:77–85
16. Jahanmir (1999) Machining of Ceramics and Composites, CRC Press, New York
17. Lallemand G, Jacrot G, Cicala E, Grevey DF (2000) Grooving by Nd:YAG laser treatment. J Mater Process Technol 99(1–3):32–37

18. Chryssolouris G (1991) Laser machining: theory and practice. Springer-Verlag, New York
19. Dubey AK, Yadava V (2008) Experimental study of Nd:YAG laser beam machining—an overview. Journal of Materials Processing Technology 195:15–26
20. Dahotre NB, Harimkar SP (2008) Laser machining and fabrication of materials. Springer Science, New York
21. Chrysolouris G, Bredt J, Kordas S, Wilson E (1988) Theoretical aspects of a laser machine tool. Journal of Engineering for Industry 110(1):65–70
22. Kruth JP, Yasa E, Vanparys M, Vaerenbergh JV (2007) The enhancement of micro-machining ability of selective laser melting by selective laser erosion. International Conference on Polymers and Moulds Innovations, PMI Gent, Belgium
23. Thawari G, Sundar JKS, Sundararajan G, Joshi SV (2005) Influences of process parameters during pulsed Nd:YAG laser cutting of nickel-base superalloys. Journal of Materials Processing Technology 170(1–2):229–239
24. Kibria G, Doloi B, Bhattacharyya B (2015) Investigation and analysis on pulsed Nd:YAG laser micro-turning process of aluminium oxide (Al_2O_3) ceramic at various laser defocusing conditions. The International Journal of Advanced Manufacturing Technology 76(1–4):17–27
25. Kibria G, Doloi B, Bhattacharyya B (2013) Predictive Model and Process Parameters Optimization of Nd:YAG Laser Micro-turning of Ceramics. International Journal of Advanced Manufacturing Technology 65(1–4):213–229
26. Kibria G, Doloi B, Bhattacharyya B (2014) Investigation into the effect of overlap factors and process parameters on surface roughness and machined depth during micro-turning process with Nd:YAG laser. Journal of Optics and Laser Technology 60:90–98

Chapter 6
Fiber Laser Micro-machining of Engineering Materials

A. Sen, B. Doloi and B. Bhattacharyya

Abstract In the present era, fiber lasers have been successfully replacing Nd-YAG and CO_2 lasers along with other conventional laser systems for various micro-machining applications such as micro-cutting of stents, thin sheet of ferrous and non-metals in terms of cutting speed, cut edge quality and the length of micro cracks. The usage of fiber laser can also be observed in field of micro-machining of various engineering materials owing to the characteristics of short pulse lengths which range from millisecond to picosecond or even femtosecond. The present chapter aims to carry out an in depth study of the fiber laser micro-machining, i.e., micro-cutting, micro-drilling, engraving, marking, etc., of engineering materials ranging from polymer to ceramics. The aim of the chapter is also to include an overall concept of fiber laser micro-machining system in the present scenario and its applications along with the occurring physical phenomena and influence of various process parameters on the fiber laser generated micro-features.

6.1 Introduction

In the present era of commercialization and advancements in technology, smaller but effective products are in utmost demand in places starting from the household to heavy industrial areas. The booming interests towards new products with new functionality in each passing day, not only demand stringent research, but also

A. Sen (✉) · B. Doloi · B. Bhattacharyya
Production Engineering Department, Jadavpur University, Kolkata 700032, India
e-mail: abhishek.sen1986@gmail.com

B. Doloi
e-mail: bdoloionline@rediffmail.com

B. Bhattacharyya
e-mail: bb13@rediffmail.com

G. Kibria et al. (eds.), *Non-traditional Micromachining Processes*,
Materials Forming, Machining and Tribology,
DOI 10.1007/978-3-319-52009-4_6

enhancement of the present machining facility available in the world. Conventional machining approaches towards various engineering materials with high ductility and brittleness are not only cumbersome, but also suffer from the prolonged machining time along with cost ineffectiveness. These result into poor machining surfaces with high tolerances which fail to match up with the present need of stringent quality control. On the contrary, non-traditional machining (NTM) approaches with different form of energy as an input energy, i.e., thermal, mechanical, etc., are more suitable to meet up the challenges towards achieving a high surface finish and astoundingly good machining quality.

Laser has emerged as one of the best machining solution from the NTM family for achieving thick material cutting to solar cell scribing with the least amount of time and flexibility of the process along with cost effectiveness. Laser micro-machining process can be attributed as an application which requires a substantial amount of energy (in pulse mode operation) to induce the change in state along with phase transformation in a small volume for a given material [1]. Among the recent advancements in the laser systems, fiber lasers have observed a rapid and consistent growth from the stent cutting to the thick and thin material cutting, drilling, etc.

The pulsed mode of the fiber laser system, with an output power less than 200 W, is most suitable for micro-machining of a wide range of engineering materials owing to its superior features such as simplicity, ruggedness, cost effectiveness, low maintenance, higher efficiency, higher reliability and smaller spot size [2]. Fiber laser micro-machining approaches towards various difficult to machine materials for generating precise micro-features such as micro-channels, micro-grooves, micro-holes, etc., mainly in the domain of bio-medical, automotive and aerospace engineering, have aided to become the most versatile and rapid growing laser systems in the last decade.

6.1.1 Fundamentals of Fiber Laser Generation

Fiber laser, since its inception in 1961, moved along the path of technological advancement from the developments of high power diode lasers, low loss doped glass fiber, diode-fiber coupling devices, Bragg gratings, etc. [3]. A simple fiber laser can be constructed by placing a fiber amplifier inside a cavity designed to provide optical feedback. Laser is generated by pumping the diodes to excite lower energy molecules to the excited states. Fibers, having one core and one cladding along with the protective coating, are usually made of silica glasses for the purity. The propagation of the lasing mode is in the core, whereas, the propagation of the pump beam is by the outer layer cladding. The primary function of the outer layer cladding is to confine the pump light, which allows the pumping of the core with a high power beam. Fiber glass usually has a higher refractive index core which is followed by a low refractive index of the cladding. Further, this particular property of the outer layer cladding converts pump light of relatively low brightness to a

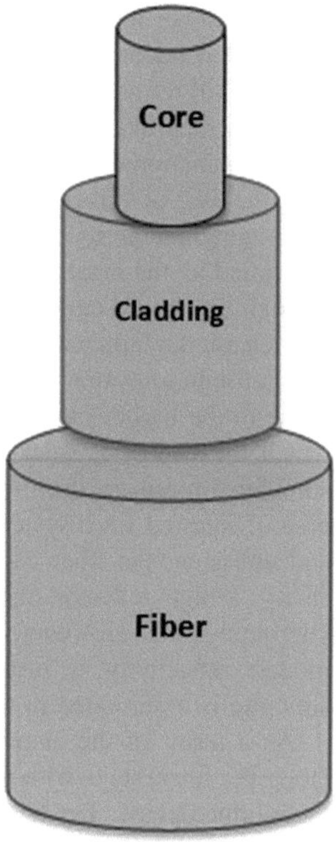

Fig. 6.1 Schematic representation of a simple fiber structure

much brighter laser beam. A schematic representation of a fiber is shown in the Fig. 6.1.

The most popular rare-earth elements that are doped with the optical fiber are neodymium (Nd^{3+}), ytterbium (Yb^{3+}) and erbium (Eb^{3+}), which are further pumped by diode lasers operating at a wavelength range of 950–980 nm. Out of the above mentioned rare earth doped elements, ytterbium (Yb^{3+}) is considered to be the most widely accepted element owing to its low quantum defects, and high quantum efficiency (close to 95%) [3]. In addition to this, Yb^{3+} doped optical fibers are useful to control the refractive index along with the photon absorptions. Phosphate glasses have provided a good alternative solution than silica glasses by providing a higher Yb^{3+} emission cross-section and laser efficiency as compared to silica glasses [4]. Stępień et al. [4] demonstrated a development of the thermally stable Yb_2O_3-high-doped phosphate glass synthesized in a P_2O_5–Al_2O_3–Yb_2O_3–BaO–ZnO–MgO–Na_2O oxide system. The characteristics of high transmittance along with a low absorption coefficient in a wide spectral range of 320–2700 nm is associated with Yb^{3+} doped developed glasses.

The numerical aperture by virtue of which the effective angle of light can be altered, is associated with the acceptance of light, i.e., higher the numerical aperture-higher the broad light of acceptance. The light propagation into the fiber itself is dependent on five characteristics; (a) fiber type, (b) fiber size, (c) numerical aperture, (d) refractive index and (e) doping of the fiber. Fiber Bragg gratings act as mirrors which also control the amount of reflections for a particular wavelength. The combination of fiber with Nd^{3+}, Yb^{3+} etc., along with the fiber Bragg gratings form the base of a fiber laser. The wavelength of the reflected light is determined by the longitudinal period of the grating and the percentage of the reflected radiation is controlled by the magnitude of the refractive index [3]. Fiber lasers reflect the light through the optical cavity so that the string of photons stimulates the atoms to store and release the light energy by means of useful wavelengths. The photons absorbed by ytterbium ions, may disappear. The electrons moving around the atomic nucleus move to the higher orbit by absorbing energy. Pumping is a process by virtue of which the energy is injected and stored by the fiber. Within a millisecond, the stimulated electrons drop to ground states (for example, drops to 1064 nm in the case of infrared laser system). In order to maximize the coordination of pumping and emission, the fiber cavity resonator is formed [3]. At the pump end of the cavity, a high reflector fiber Bragg grating is sliced to the fiber doped with the ytterbium atoms. However, at the output end, a similar kind of Bragg grating with a modest reflectivity is installed. A schematic representation of the fiber laser structure is represented in the Fig. 6.2.

As a result of the above mentioned phenomenon, the light which propagates along the fiber, stays with the fiber. Such a design of the fiber laser is termed as a monolithic device. The length of the doped fiber is important as it determines the amount of light that can be absorbed within the fiber. The long length of the fiber results in a high surface to volume ratio, which aids in removing the excessive amount of heat from the fiber. Usually, fiber laser has an absorption length, which is 95% of the pump energy [5].

The number of diodes for pumping is dependent on the output power of the fiber laser. Large numerical aperture diodes with high power along with the larger diameter output fiber are the most preferable for the construction of the high power fiber laser. In the recent years, invention of double clad fiber lasers leads to fiber lasers with high power, greater electrical efficiency and astoundingly high brightness. Double clad fiber lasers have higher numerical aperture and larger in diameter as compared to the single mode fiber lasers. When the laser is assembled and pumped power to the diode, much of pump light passes down the cladding without intersecting the dopant ions in the core of the fiber. Without the intersection and

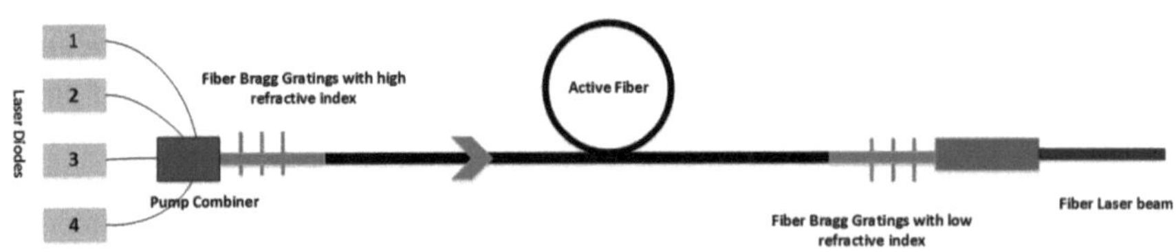

Fig. 6.2 Schematic representation of fiber laser structure inside the laser head

transformation of the power, the laser cannot be generated. To overcome this problem, extra reflective facets on the outside of the first cladding can be used. During the laser generation and propagation of laser along the length of the fiber, two kinds of non-linear effects, i.e., stimulated Brillouin scattering and stimulated Raman scattering [6], can be observed. To overcome these phenomena, the length of the fiber along with the diameter of the core can be increased. However, it may lead to the reduction of the laser light intensity in a considerable amount compared to the single mode beam quality. Bending of fiber to some extent may change the acceptance of light by the fiber and improve the beam quality. In addition to this, a residual pump stripper with high refractive index can be utilized at the end of the fiber to overcome the loss of thermal energy.

6.1.2 Different Types of Fiber Laser Systems

Fiber laser power output up to 1 kW range are most desirable for material processing applications, such as metal cutting, welding, brazing and marking [7]. In the present era, ultrafast fiber lasers, based on passive mode-locked fiber oscillators, have been dominating both the industrial and scientific realms. The amplification during the several stages of ultrafast fiber laser generation and beam deliver system have allowed them to reach at the desired output pulse energies. A large variety of fiber components rely on all-fiber systems to be constructed rather being utilized only for the beam delivery system. Core diameters of a conventional rare earth doped fiber amplifiers are typically in the range of 5–10 μm. Chirped pulse amplification (CPA) technique for ultrafast fiber laser systems, is a new design concept designed for preventing nonlinear processes in the fiber amplifiers for better utilization in various micro-machining applications [8, 9]. Over the years, the fiber mode area has undergone several modifications to achieve a continuous increase in achievable pulse energies.

In general, fiber laser can be broadly classified into two types: (a) continuous fiber laser and (b) pulse fiber laser. Fiber laser can be further classified into single frequency fiber laser, Q-switched fiber laser and mode locked fiber laser. A continuous fiber laser consists of high reflector (HR) and output coupler (OC) along with fiber Bragg gratings [10].

Usually, a single-frequency fiber laser is pumped by a single-mode pump diode in order to obtain an output power, ranging from a few tens up to hundreds of milliwatts. In Q-switched fiber lasers, longer pulses are obtained in longer fibers. However, if the pulse duration becomes longer than round-trip cavity time, pulse breakup may occur. On the flip side, optimal utilization of the intrinsic cavity mode structure in an optical fiber oscillator is utilized by the mode locked fiber laser, in order to produce short pulse, as depicted in the Fig. 6.3. In the recent years,

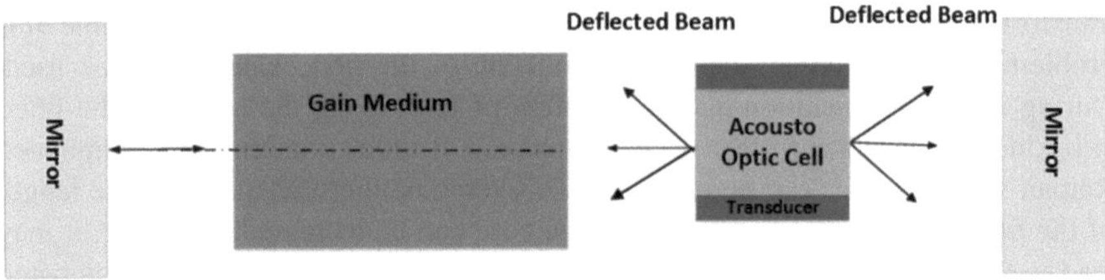

Fig. 6.3 Schematic representation of a mode locked fiber laser with acousto optic modulator

end-to-end all-fusion-spliced-fiber master oscillator power amplifier (MOPA) designs (pulsed operation is obtained by means of an external electronically controlled modulator) are also utilized in fiber-coupled semiconductor lasers [10]. However, for the thick material cutting, continuous fiber lasers are more suitable, whereas pulse fiber lasers are commonly employed to produce high peak intensity laser beam for achieving high aspect ratio micro-channels, micro-holes, etc. Depending on the regime of the pulse width of the fiber laser, it can be also termed as nanosecond, picosecond or femtosecond pulsed fiber laser systems.

6.1.3 Importance and Advantages of Fiber Laser in Micro-machining Domain

The importance of fiber laser system for the micro-machining domain is inevitable. Fiber lasers are ideally suited for the generation of micro-features required for the bio-medical applications, electronics industries, automotive and aerospace industries. The importance of fiber laser for the fabrication of micro-features on a wide range of engineering materials lies in its characteristic features. The astounding features of fiber laser such as compactness, elimination of complex cooling systems, excellent constant beam properties, long focal lengths, and rapid warm-up make it ideal for the applications in the aforesaid manufacturing industries. Fabrication of stents and seam welding pacemakers and implantable batteries for medical applications are fewer possibilities of the fiber laser [11].

6.1.3.1 Superior Beam Quality

The laser beam quality is determined by the wavelength, beam waist radius along with far-field divergence and further, it is defined by M^2 value. Further, the highest beam quality lasers will be defined as M^2 close to 1. The high quality laser beams are formed due to the long and thin gain medium. Fiber lasers which are commercially available are high quality beams ranging up to a few kW.

The primary characteristics of the high quality beams are that, it can be focused on a small spot on the workpiece, providing high power density thereby leading to faster processing time. Compared to the other laser technologies, fiber laser beams have a longer focal length (working distance) along with greater depth of focus (workpiece positioning tolerance) [11]. Further, the beam diameter can be smaller if there is a fixed focal length and focus spot size. Therefore, a larger area of parameter space is achieved because of the high beam quality and hence, the optical parameters for a process are easily optimized.

6.1.3.2 High Wall Plug Efficiency and Reduction of Operating Cost

The wall plug efficiency of a laser is defined by the ratio of the electrical power efficiency to optical power efficiency. High wall plug efficiency around 25–30% can be achieved for the fiber lasers due to the high efficiency of the pump source and the extraction from the gain medium. Therefore, this not only involves lower operating cost but also requires reduced electrical supply along with reduced cooling requirements which is up to 300 W, compact laser head design.

Further, fewer mechanical components are required in the construction of the laser as compared to the other solid state lasers. This invariably helps the fiber laser to be cost competitive compared to the other kW class, such as CO_2 lasers, as used in several flat-bed cutting applications.

6.1.3.3 High Reliability and Low Maintenance

Fiber Laser products in the present market are offering 100,000 h MTBF (mean time before failures) for their pump sources which is equivalent to over 11 years of operation in total [11]. At present, the fiber laser products have demonstrated over 40,000 h of operation already and they are still being used today, owing to its high reliability. Therefore, it will be difficult for any industrial technology to currently replace any laser diodes primarily due to the cost and time effectiveness of the system. Another interesting feature that adds a feather in the cap of the diode pumped fiber laser system is the low routine maintenance. This feature is primarily because of its all fiber construction, which means that no beam delivery optics or alignment of the resonator is required. In addition to this, minimum or almost no cooling system is required for the running of the fiber laser system as compared to Nd-YAG laser system.

6.1.3.4 Ease of Beam Delivery

Apart from being highly reliable and cost effective, fiber lasers possess easy operating features that are the light that generates inside the fiber, delivers to the work material within the fiber. This particular phenomenon enables a stable setup and eradicates any issue relating to the number of beam delivery mirrors or problems arising out of coupling laser beam into a beam delivery fiber.

6.1.4 Basic Principle of Fiber Laser Micro-machining Process

The working range of the fiber laser is either in the near infrared (NIR) or infrared region (IR). Fiber laser micro-machining process is based on the interaction of the laser light with the top surface of the working material in order to remove a small amount of material from the surface. It is a well known fact that the photon energy is inversely proportional to the laser working wavelength, i.e., shorter the wavelength-higher the photon energy. When the working wavelength (1064 nm) of the fiber laser is shortened either to the first order harmonic wavelength (532 nm) or second order harmonic wavelength (264 nm), the thermal process changes into a chemical process, which eventually break the bonds of the workpiece material [12]. As a result of this, fiber laser micro-machining process can be classified into two distinct features, i.e., photo-thermal process and photo-chemical process. Photo-thermal processes are identified mostly in metals, ceramics, alloys, etc. Photo-thermal processes are characterized by a rapid thermal cycle, heating, melting and partly evaporation of the heated volume [13]. Whereas in photo-chemical process, the photon energy is sufficient enough to break the chemical bonds, which is mostly suitable for polymers and plastics.

Fiber laser micro-machining process for various engineering materials ranges from nanosecond regime to femtosecond regime. Depending on the laser-material interaction time, the micro-machining quality is varied. Based on the laser-material interaction time regime, three processes can be distinguished. In all the three processes, first, absorption of the photon by an electron occurs and then the rest of the set of other physical steps are varied depending on the time regime. In nanosecond ablation of the materials (for metals), the absorbed laser energy first, heats the workpiece to its melting point and further to the vaporization temperature. During the interaction time, the loss of energy is associated with the heat conduction into the solid material. Nanosecond pulse ablation is associated with the detrimental heat affected zone (HAZ) which can be partially overcome by picosecond ablation. In picosecond ablation, the pulse length is in the same order during the time of transferring the energy from the electrons to the lattice. However, the picosecond ablation also suffers from a considerable amount of evaporation and the formation of the melted zone inside the material [13]. This pertinent problem can be completely eliminated by femtosecond ablation as no time is available for the heat transfer to the lattice during this series of processes in femtosecond ablation. This results in no heat affected zone and precise micro-machining characteristics.

6.1.5 Applications of Fiber Laser in the Micro-machining Domain

In the field of medical sciences, fiber laser micro-cutting process has been widely accepted process instead of lamp pumped pulsed Nd-YAG lasers. Manufacturing of medical devices is a daunting task because of their relatively small size, use of expensive materials and ability to fit into small areas. In this scenario, fiber lasers can operate in the areas which can tolerate the power of only a few microns for being an effective solution for costly and smaller devices.

Fiber laser micro-machining has emerged through the micro-cutting of cardio-vascular stents over the years. Fiber lasers have obtained a decent market share in the present era, in the field of fine and precise micro-cutting which combines both continuous mode as well as pulse mode, aided by fusion cutting (inert gas to sublimation cutting) [14]. Fiber lasers are commonly employed in thin material cutting, whereas the other solid state lasers and gas lasers such as Nd-YAG and CO_2 lasers are mostly utilized for the thick material cutting. Although, experimental studies carried out by Olsen et al. [15] demonstrated that the fiber laser cutting efficiency for thick materials is better than that of a CO_2 laser. Comparative experiments were conducted by Kleine et al. [16] and Meng et al. [17] in order to analyze the quality of cutting between the fiber lasers and the Nd-YAG lasers. The result of the analysis was that, with the fiber laser cutting, cut quality was comparatively better than the Nd-YAG lasers, i.e., in terms of roughness, heat affected zone (HAZ) and kerf width. The number of research works employing fiber laser in order to micro-machining of difficult to machine materials with the aid of fiber laser systems, is increasing rapidly in the recent years. Fiber laser ablated micro-holes in various super alloys such as Ti-6Al-4V, nickel super alloys are utilized for various automotive and aerospace applications.

Fiber laser marking process is one of the key areas that have emerged through the marking of electronic goods to the daily household products. Hard coated plastic cases have been adopted by several manufacturers in order to provide scratch resistance. However, laser marking over such surfaces without damaging it is typically daunting task because in this case, a précised balance between the peak power, average power as well as the pulse energy has to be maintained. As a result of this, fiber lasers are replacing green [18] and other ultrashort lasers which are not cost effective compared to the fiber lasers.

An increase in the touch screen usage has created a more functional thin-film processing application for fiber laser systems. Multi-pass processing enables this application, where the initial pass removes the bulk of the film and the secondary passes cleanses the residues by using a combination of different process parameters, short pulses, high pulse frequency and high speed. Fiber lasers have the ability to change the parameters such as pulse energy, pulse duration within few microseconds. Controlled heating characteristics that provide better and faster marking can be obtained by the use of long duration pulses and high repetition rates. These long duration pulses and high repetition rates strongly resemble gated continuous wave pulse fiber laser than the high peak power nanosecond pulses. Fiber laser engraving is another emerging area of fiber laser in the field of micro-machining. Other key features of fiber lasers are hallmarking, texturing, and personalized marking on a wide range of precious materials.

6.2 Different Units of Diode Pumped Fiber Laser Micro-machining System

Details of a 50 W ytterbium doped fiber laser system with a wavelength of 1064 nm working in TEM_{00} mode of operation, made by M/S Sahajanand Laser Technology Limited (Model: AKSHAR Fiber Pro-309), have been discussed. A diode pumped fiber laser micro-machining system is comprised of (a) power supply unit, (b) laser head, (c) collimator, (d) beam bender, (e) beam delivery unit and focusing lens, (f) assist gas supply unit and (g) computer numerical control (CNC) controller for X–Y–Z movement. Fiber laser is generated within the laser head with the aid of power supply unit. The rest of the aforesaid sub-systems of a fiber laser are utilized in the beam delivery system. The schematic diagram of the fiber laser micro-machining setup is shown in the Fig. 6.4. In the aforesaid fiber laser system, the working range of pulse frequency is 50–120 kHz, whereas pulse width is in the regime of 120 ns.

6.2.1 *Power Supply Unit*

The power supply unit consists of a 1.5 kVa of isolation transformer and 1 kVa of uninterruptible power source (UPS). The isolation transformer is connected with the main power supply unit and the UPS is connected to the isolation transformer. The pumping of laser diodes inside the laser head is achieved via the power supply unit.

6.2.2 *Laser Head*

The diode pumped fiber laser system is comprised of a fiber itself doped with Yb^{3+}, fiber Bragg grating and number of diodes. The number of laser diodes is dependent on the total output power of the system. A total number of 8 diodes are utilized in the present system for pumping. Fiber Bragg gratings are sliced into two ends of the fiber optics so that the laser can be generated. The total length of the fiber can be up to 3 mm depending upon the output power generated by the fiber laser.

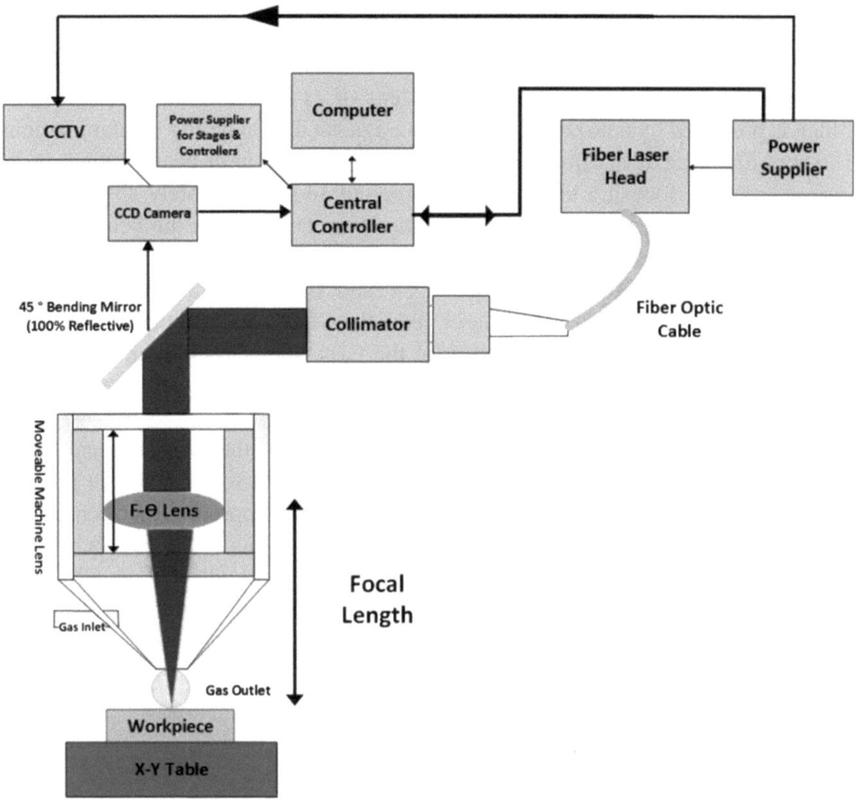

Fig. 6.4 Schematic diagram of a fiber laser micro-machining set up

6.2.3 Collimator

After the generation of the laser within the optical fiber, the laser is transferred into a collimator. In the collimator, two prisms are fitted at the two ends of it. This collimator acts as a beam expander, i.e., the laser can propagate to the required distance. The diameter of the laser beam is 9 mm at the end of the collimator.

6.2.4 Beam Bender

After the collimator, a beam bender with 100% reflectivity is placed at an angle of 45° with the horizontal plane so that the laser can be perpendicular to the focus lens. At the top of the beam bender, a charge couple device camera (CCD) is placed which is further connected to a CCTV.

6.2.5 Beam Delivery Unit and Focusing Lens

The laser finally propagates through a f-θ lens of 71 mm of diameter (focus lens) which is protected by a nozzle for preventing against dust and other contaminations. The spot diameter of the fiber laser beam is 21 μm. The laser beam falling on the surface from the focusing lens has to be précised and appropriately aligned. If the center of the lens does not coincide with the beam center, then the beam that will pass through the lens will not be linear, resulting in low energy of the laser beam and less micro-machining efficiency. CCTV and CCD camera are the two units used for focusing. At the top of the laser head, the CCD camera is placed in order to capture the image during the operation of the laser micro-machining. Further, the beam should also be focused on the surface so as to receive optimum micro-machining operations.

The function of the CCTV is to adjust the position of the focused beam on the workpiece. This device is connected to the CCD camera so as to obtain précised laser beam focusing condition. After the laser beam is appropriately focused, the material melts at a faster rate and thereby evaporates. This is the reason why the focal point position of the laser beam should be properly adjusted. A CNC controlled arrangement is provided in the experimental setup, in order to effectively adjust the focal point of controlling the lens movement position along the Z axis.

6.2.6 Assist Gas Supply Unit

Inert gases such as nitrogen, argon, helium, etc., along with compressed air can be supplied through a co-axial nozzle attached to the beam delivery system depending upon the selection of micro-machining process and materials. The jet flow of assisting gas assists in removing the molten material from the ablated surface to partially overcome re-solidification of the molten material from the micro-machining zone. If any compressed air flow is required, then the supply line must pass through a moisture separator and should be connected further to a pressure regulating valve. It results in the jet flow of dry pressurized air to the laser micro-machining zone. The laboratory setup has a vacuum job fixture which can hold a workpiece up to 1 mm of thickness when the compressed gas pressure is 4 kgf/cm^2.

6.2.7 CNC Controller for X–Y–Z Movement

A CNC controller unit not only controls the movement of the worktable along the X-Y axis, but also controls the movement of the laser nozzle along the z axis. Servo motors are attached to each of the axis and are also connected to the servo interfacing unit. This servo controller is connected to the computer system (interface software-I mark plus) from where the axis movements of the X–Y worktable are controlled. If any movement of the X, Y or Z axis is required, then the computer system sends a command to the servo interfacing unit which is followed by receiving a command for the respective servo motor movement in the specific directions.

The axis of the fiber laser system can be governed by either CNC based motor system or galvanometer type motor system, depending upon the applications of the fiber laser systems. For marking and scribing applications, a galvanometer type system is most preferable. In contrast to the aforesaid system, CNC based system is well recognized for cutting, drilling and various micro-machining applications. Generally, in CNC based fiber laser systems, the table can only move in the X–Y direction, whereas the laser head can only travel in Z direction. In a sharp contrast, galvanometer type laser systems rely only on the movement of the laser head in X-Y-Z direction. As a result of this, scanning speed can be very high, sometimes even up to 3000 mm/s. In the aforesaid fiber laser micro-machining system, a CNC type system is utilized where the maximum cutting speed ranges up to 70 mm/s. Although, for the optimum results of the system, maximum cutting speed of 40 mm/s can only be utilized.

6.3 Various Fiber Laser Micro-machining Operations on Engineering Materials

Commercial acceptance and application of fiber laser micro-machining has increased in the recent years due to the astounding mode quality and high focusing ability, which are necessary for achieving small size features. The growth of more efficient, compact, high beam quality fiber lasers in the recent years is due to the increased demand for extremely fine machining. Over the years, the widely used commercial laser was Nd-YAG lasers because of its ability to produce a fine cut quality of a variety of engineering materials along with their drilling and welding ability. Fiber lasers as compared to the CO_2 lasers, at 1 μm wavelength, have better and smaller focusing ability in order to achieve smaller spot sizes. In the present era, several optimizing research works have been carried out in order to improve the micro-cutting quality.

Compared to the conventional lasers, fiber lasers function at near infrared (IR) spectral region and provide a range of advantages along with the promise to create avenues for new micro-machining applications. Fiber laser micro-machining of metals, alloys are governed mainly by the photo-thermal process. Various process input parameters such as laser power, laser beam focal position, composition and thickness of the workpiece material, scanning speed/cutting speed, pulse width, pulse repetition frequency, number of passes, type of assisting gasses along with assisting gas pressure, etc. are the major governing variables considered for the fiber laser micro-machining of a wide range of engineering materials. The quality characteristics (or process performance) of interest in fiber laser micro-machining are material removal rate (MRR), machined geometry (kerf width, hole diameter, taper), surface quality (surface roughness, surface morphology), metallurgical characteristics (recast layer, heat affected zone, dross inclusion) and mechanical properties (hardness, strength, etc.) However, fiber laser micro-machining of metals and alloys are commonly associated with the formation of striation, dross, recast layer, HAZ, micro-cracks along with the micro-machining zone [19].

6.3.1 Micro-cutting

Okamoto et al. [20] partially tried to overcome dross free fiber laser micro-cutting of a copper plate of 100 μm of thickness in the presence of nitrogen assists gas with the aid different set of nozzles. It can be summarized from the experimental results that not only the maximum cutting speed varied with the different set of nozzles, but also straight cut profiles were achieved with minimum HAZ. Another investigation on laser cutting of various micro-patterns on carbon fiber reinforced polymer (CFRP) with a continuous wave fiber near-IR laser (λ = 1090 nm) to achieve précised cutting of CFRP, devoid of debris and thermal damages around the grooves was conducted by Niino et al. [21]. A fast beam galvanometer scanning process with the multiple-scan-pass method was conducted along with the fiber laser irradiation on CFRP. The observations that were made out of the experiment was that, fine cutting of 3 mm CFRP sample at fiber laser irradiation with a 15 multiple-scan-pass (scanning speed: 0.8 m/s) was free of debris and other thermal damages around the periphery of the micro-machining zone. Further, around 600 μm of the kerf width of the groove on the laser beam incident surface was observed. Thin film solar cell micro-machining with the aid of high power ultra-short fiber laser was carried out by Lecourt et al. [22], where the pulse repetition rate was varied from 100 to 1 MHz. Another set of comparative study [23] showcased the advantages of using bursts of picosecond pulse for micro-machining of materials compared to the nanosecond source, having the same pulse energy along with pulse width and pulse shape. The result of the experiment was satisfactory when bursts of picosecond pulses were used, as the material removal along with the surface quality was as high as 25% and more in comparison to the nanosecond pulses having the same pulse energy along with the combination of pulse width and pulse shape.

In a set of comparative study, femtosecond pulsed micro-machining with 1.56 μm erbium-doped fiber-laser oscillator-amplifier system along with direct amplification and pulse extraction was carried out by Mizunami et al. [24]. For the pulse extraction and amplification, a peak power, as high as 25 kW was obtained. Further, with the process of pulse extraction and amplification, a 6 μm trace of ablation was observed on an amorphous carbon film. The experimental result was narrower than that of 16 μm of direct amplification. On the flip side, when polyvinyl chloride film underwent micro-machining with the aid of a fiber laser, it was observed that direct amplification was more suitable and a cutting width as low as of 4 μm was obtained from the experiment. Further, for an increased repetition rate of 100 kHz, optical damage was observed to be less, owing to the lower peak power of the fiber laser. However, the same phenomenon was not observed in the case of direct amplification. Another study of precise laser micro-machining on multiple layers of ceramic/metal (thermal barrier coated super alloy) was conducted by Qi et al. [25]. In this experiment, an industrial-grade economic nanosecond pulsed fiber laser was utilized in order to study the relationship between the process parameters and machining geometric features of stainless steel (SS) 304, copper, Inconel 718 and TBC coating samples. It was found that the depth of the trench was dependent on the laser peak fluence in a logarithmic relation. When the beam waist was 50 μm for SS304 and 30 μm for copper, the volume removal rate was high. It was also observed that a similar trend was followed by the finite element model (FEM) model predicted volume removal rate along with the beam waist variation. Finally, for TBC coating, fiber laser micro-machined surface roughness using laser scanning was measured 5.96 μm (Ra), whereas for Inconel 718, surface roughness turned out to be 7.32 μm (Ra).

6.3.1.1 Stent Cutting

The most promising possibility of fiber laser in the field of bio-medical lies in profile cutting for cardiovascular stent, which is a lattice-shaped metal tube which is utilized for the implantation into arteries [26]. Cardiovascular stents can be made of various materials such as nickel titanium (NiTi) alloy, stainless steel. Liu et al. [27] analyzed the effect of fiber laser process parameters such as surface roughness, kerf width, heat-affected zone (HAZ) and formation of dross for the fabrication of vascular stent on NiTi alloy tube of outer diameter of 1.8 mm and wall thickness of 240 μm. The experiments were performed with the continuous water flowing through the inside walls of the tube in the presence of argon as an inert gas (constant pressure of 0.45 bar). It was observed that due to the wet cutting conditions, pure striation topography was not formed. The optimum choice was the mixed topography striation ratio for the micro-cuts having lower surface roughness, low kerf width and corresponding small recast layer than the micro-cuts bearing other topographies. Kerf width geometry was influenced by the power density. An increment in the power density causes an increase in pulse energy. This in turn resulted in the increment of the kerf width dimensions. The entry cut width was found to be wider than 30 μm or even 35 μm, when the power density varied from 2900 to 8000 J/nm. Further, at an intermediate power density along with the low cutting speed, smallest entry width of 25.53 μm could be achieved. On the other hand, at high power density followed by fast cutting speed, entry kerf width was obtained which was 38.59 μm (maximum dimensions for the kerf width). With the increase in the power density, it was also observed that the surface roughness along with the kerf width had the tendency to increase. However, with the settings of different combinations of the aforesaid parameters, the results varied. Further, it was observed that the cutting speed was the most critical parameter for obtaining good surface finish compared to the other process parameters. In addition to this, surface topographies also varied with the cutting speed, while keeping the power density as constant. Spot overlapping factor determines the uniformity and smoothness of the micro-machined profiles, i.e., higher the spot overlapping factor-higher the uniformity and smoothness of the profiles. Cutting speed is also associated with the spot overlapping factor [14]. When the cutting speed is high, the spot overlapping factor is low. The phenomenon reverses back with the increment in the cutting speed. Meng et al. [17] utilized a 50 W fiber laser system for the micro-cutting of 316L stainless steel tube (thickness 110 mm, diameter 2 mm) cardiovascular stents in order to observe the effects of various micro-cutting parameters including laser output power, pulse length, repeat frequency, cutting speed and assisting gas pressure on the kerf width profiles. The kerf width was the primary function of the cutting speed in order to obtain different laser power outputs. The repeat frequency was observed to be 1500 Hz, followed by 0.15 ms of pulse lengths along with 0.3 MPa of assisting oxygen gas pressure. The kerf width dimensions increased with the increment of laser output powers owing to high power density. The phenomenon reversed back when the cutting speed increased, because of the low power density. The kerf width size also increased when both the pulse length and repeat frequency were increased. It was observed that the kerf width was less than 20 μm along with low roughness and HAZ when the fiber laser process parameters were, laser output power of 7 W, pulse length of 0.15 ms, the repeat frequency of 1500 Hz, cutting speed of 8 mm/s and assisting oxygen pressure 0.3 MPa. Another experiment carried out by Muhammed et al. [28], showed the effect of both wet and dry 316L stainless steel with the aid of fiber laser system. It was concluded from the experiment that in the case of wet stainless steel cutting, the results were comparatively better in terms of narrower kerf width along with low surface roughness, less dross, no back wall damages and smaller HAZ. All these positive features would lead to lower cost of post-processing of the material. In order to control the micro-cutting quality, low average power with the combination of low pulse width, play a significant role. To achieve high aspect ratio profiles, pulse width should be lower in order to increase the laser beam penetration rate. However, this in turn will lower the pulse energy irradiated on the workpiece. The controlling of all the process parameters is critical and can be achieved by the extensive research works.

6.3.2 Micro-grooving

Laser grooving process can be characterized by the jet flow of gas in order to remove excessive amount of molten material from the micro-grooved zone. The characteristics of three-dimensional heat transfer, moving boundary, a multi-phase and spatially distributed heat source are associated with the fiber laser micro-grooving process [29]. Sen et al. [14] carried out a total of 49 experiments of

Fig. 6.5 Microscopic view of the micro-groove width at different number of passes [14]

fiber laser micro-grooving on Ti-6Al-4V in an open atmosphere to find out the influences of various process parameters such as laser power, pulse frequency, scan speed and number of passes the geometrical aspects of micro-grooves, i.e., width, depth and surface roughness. The experimental results showed that the depth and width dimensions tend to decrease with the higher scan speed with the combination of low pulse frequency, whereas with the increase of pulse frequency combining moderate scan speed, reduction in the dimensions of width and depth were observed. It was also found that depth tend to increase while width decreased with the increment of the number of passes. With the more heat input, width and the depth of the micro-grooves increased simultaneously. Smooth micro-groove profiles were obtained with the combination of low scanning speed and high pulse frequency. A microscopic view of a fiber laser ablated micro-groove on Ti-6Al-4V is shown in the Fig. 6.5. Another research work carried out by Biffi et al. [30] revealed that a nanosecond pulse fiber laser could be utilized in order to fabricate linear grooves of on poly crystalline FSMA$Ni_{45}Mn_{33}Ga_{22}$. The authors observed that the scanning velocity was a more critical parameter on the micro-grooves width dimensions rather than the number of the laser pulses. On the contrary, micro-groove depth dimensions tend to vary with the change in the laser pulses. However, the formation of micro-cracks in the middle of the micro-grooves was a problem associated with the aforesaid material.

6.3.3 Micro-channeling

Laser ablated micro-channels are growing field of interest for the fabrication of micro-fluidic channels not only in the bio-medical applications, but also in micro electrical mechanical systems (MEMS). Fiber laser technology has moved away beyond the growing interest for the fabrication of micro-fluidic channels using nanosecond pulse or picosecond pulse fiber laser systems. Sen et al. [31]

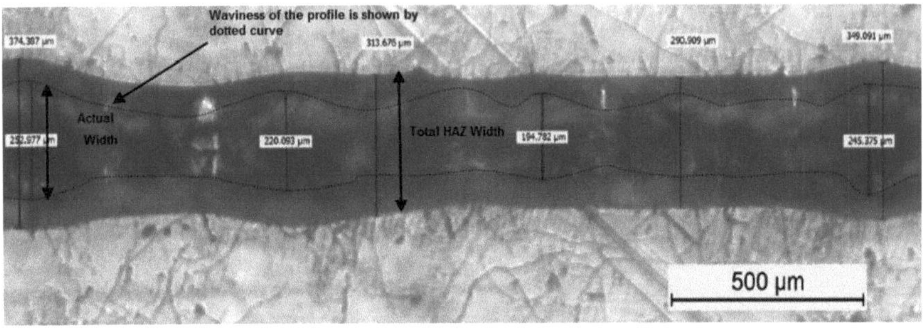

Fig. 6.6 Microscopic view showing the widths of the micro-channel profile at 50 mm/s of scan speed, 15 W of average power, 50 kHz frequency and number of pass of 1 [31]

successfully facilitated a multi diode pumped infrared fiber laser system to fabricate micro-fluidic channels on polymethyl methacrylate (PMMA). Scan speed of 10–300 mm/s, pulse frequency of 50–90 kHz, average power of 5–15 W and number of passes of 1–5; were considered for the utilization of fiber laser in this domain. The author found some of the micro-channel profiles having high aspect ratios such as 2.34, 2.06 and 0.87. In addition to this, the waviness of the micro-channel profiles along with HAZ width was found to be less for some of the micro-channeling conditions. However, some of the micro-channel profiles suffered from the uniformity and re-solidification along the micro-channel profiles which is shown in the Fig. 6.6. Sen et al. [32] also carried out sensitivity analysis of the process variables (scan speed, pulse frequency, number of passes and average power) considered for the fiber laser ablated micro-channel profiles on PMMA. The sensitivity analysis revealed that pulse frequency was the most critical parameter for obtaining desired width and depth dimensions on PMMA. The sensitivity of the other three process parameters, i.e., scan speed, laser average power and number of passes on the depth dimensions were found out to be negative. The author successfully facilitated infrared fiber laser micro-machining system for the fabrication of micro-features on polymer materials.

6.3.4 Micro-drilling

Pulsed mode of fiber laser system is quite suitable for micro-drilling of high ductile and brittle materials also. Many research works are being carried out in this area and the results are being discussed as follows.

From the past research works, it is evident that sputtering along with other metallurgical defects such as cracks, morphological irregularities are always associated with the laser micro-drilling process owing to the rapid solidification of the molten materials [33]. In this regard, Biffi et al. [33] carried out percussion micro-drilling of commercially pure titanium sheets having 0.5 mm of thickness

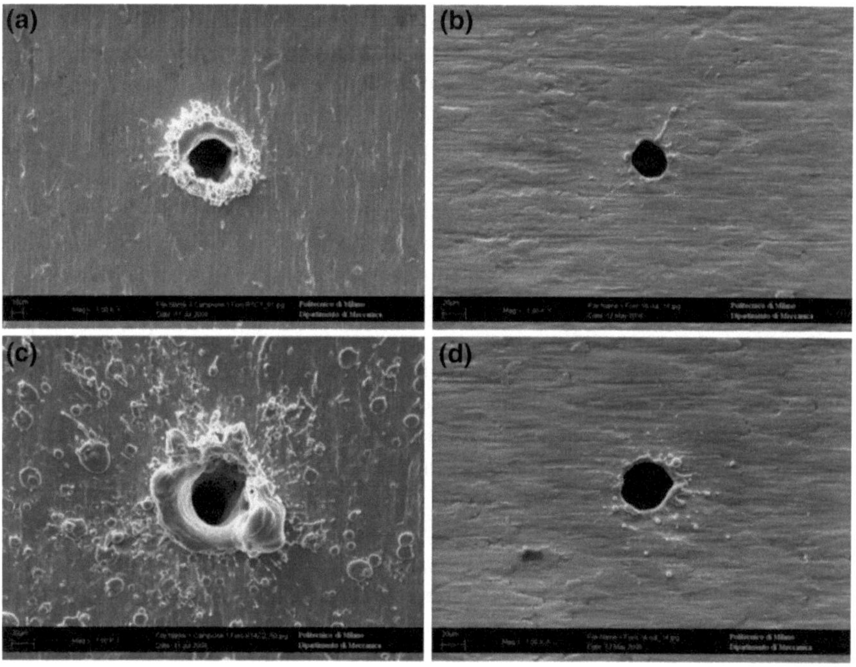

Fig. 6.7 SEM views of drilled holes according to the two boundary process conditions: **a** entrance at pulse frequency of 20 kHz and pulse energy of 0.5 mJ; **b** exit hole at pulse frequency of 20 kHz and pulse energy of 0.5 mJ; **c** entrance at pulse frequency of 50 kHz and pulse energy of 1.2 mJ; **d** exit hole at pulse frequency of 50 kHz, and pulse energy of 1.2 mJ [33]

with the aid of nanosecond pulsed fiber laser system. Figure 6.7 depicts the obtained entrance and the exit holes at the two limit process conditions along with a maximum average power (pulse frequency of 20 kHz, pulse energy of 0.5 mJ) and a minimum average power (pulse frequency of 50 kHz, pulse energy of 1.2 mJ). An annular corona of melted material was observed at both the entrance holes, but on the other hand, the exit holes were almost devoid of drops and sputter (Fig. 6.7b, d). They successfully demonstrated high aspect ratio sputter free holes on the exit side, although the entry side holes suffered from a considerable amount sputter.

When ultrafast fiber laser percussion micro-drilling of high purity copper (99.9%), stainless steel (Fe/Cr18Ni10) and high-carbon steel (C 1095) sheets with 0.25-mm thickness each [34] were utilized, the results were satisfactory compared to Q-switched nanosecond laser. However, the overall accuracy of the ablated structures was not as good as compared to the femtosecond laser micro-drilling of aforesaid materials. Ghosal et al. [35] identified that wait time and modulation frequency were the most important parameters for the material removal rate (MRR) during the micro-drilling of Al/Al_2O_3-MMC by a ytterbium doped fiber laser system. The authors concluded that the material removal rate increased with an increase of N_2 gas pressure and maximum material removal rate observed within the range of 400–475 W of laser power. The minimum taper angle on micro-machined hole was observed with the combination of nitrogen pressure of 18–20 bar, along with 600–680 Hz of modulation frequency. The optimal parametric combination for maximized MRR and minimized taper was identified as 473.12 W of laser power, 604.54 Hz of modulation frequency, 0.18 s of wait time, 19.82 bar of assist gas pressure and 93.47% of duty cycle pulse width.

Huang et al. [36] also fabricated high aspect ratio micro-holes on glass as well as on metal and tissues using a femtosecond fiber laser system. The fabricated micro-holes were free from any kind of debris formation and thermal damages. It was observed by the researchers that in the absence of micro-cracks around the periphery of the micro-holes, high aspect ratio, i.e., 8:1 micro-holes could be generated in both glass and metals with the aid of femtosecond fiber laser system. Fiber laser can also be successfully applied to the fabrication of micro-holes on the ceramic substrates. Adelmann et al. [37] carried out rapid single mode fiber laser micro-drilling of alumina and aluminum nitride in order to analyze the effect of various process parameters such as laser focus position, laser power, gas pressure and laser modulation scheme on the hole diameter of both the aforesaid materials. The experimental observations revealed that single mode fiber laser was capable to generate 50 μm micro-holes with minimum taper angle (as low as 1.57°) and astounding circularity of ≤ 1.07. However, some amount of sputter on the top side of the micro holes was observed, which could be further eliminated by the utilization of pre-pigmenting and ultrasonic cleaning.

6.3.5 Engraving

The number of research works on engraving process with the aid of fiber laser, is an emerging field of interest. An engraving operation was carried out on C45 carbon steel by Genna et al. [38], where, a Q-Switched 20 W Yb: YAG fiber laser with the fundamental wavelength of $\lambda = 1070$ nm was utilized, in order to study the influence of process parameters on surface roughness along with the material removal rate (MRR). The engraved depth was found out to be highly dependent on the scan speed, pulse frequency along with the scan number. It was also observed that the line spacing, pulse frequency along with the scan speed governed the surface roughness of the engraved profiles. However, after around 10 scans, the surface roughness remained unaltered with the combination of aforesaid parameters. It was summarized that the selection of appropriate and accurate process conditions would lead to higher values of MRR and better surface roughness. Finally, when the combination of scan speed of 20 mm/s, pulse frequency ranges of 30–40 kHz along with line spacing of 0.04 mm were utilized, better MRR and surface roughness (R_a) were observed. In comparison with the other process parameters, the effect of the scan speed was found to be negligible. Further, experimental tests were carried out on Ti-6A1-4V alloy sheet having a thickness of 4 mm, using Q-Switched 30 W Yb: YAG fiber laser by Tagliaferri et al. [39].

The aim of the experiment was to identify the effect of the process parameters over the depth of the machined volume along with the quality of the roughness of the machined surface. After the first phase of the experimentation, the effect of the process parameters over the machined volume along with the quality of the roughness of the engraved surface was identified. The results of the experiment revealed that the machined volume was linearly dependent on the total amount of released energy, for each of the pulse frequency.

6.3.6 *Micro-turning*

Laser micro-turning process aids the fabrications of micro-patterns on cylindrical workpiece, ranging from metal to ceramics. However, only a few research works are highlighted on fiber laser micro-turning operation process on the engineering materials. Till date, Nd-YAG laser systems are most preferred systems by the researchers in the micro-turning process instead of fiber laser micro-machining system. Biswas et al. [40] carried out spiral micro-grooving on pure aluminum with the aid of a 50 W ytterbium doped fiber laser system. Laser average power, pulse frequency and axis feed rate were considered as the fiber laser process parameters in order to study the influence of spiral micro-grooves geometries in terms of groove width and groove depth. Around 115.2 μm of maximum micro-groove width was achieved at parametric combination of the average power of 42.5 W along with 55 kHz of pulse frequency and around 0.4 mm of pitch value. The micro-groove depth, as high as 51.95 μm, at a parametric combination of the average power of 45 W was also be achieved along with 70 kHz of pulse frequency and 0.2 mm of pitch value. The outcome of the aforesaid experimental work has the potential to be utilized for various of other engineering materials also.

6.3.7 *Marking*

Fiber laser marking is the key area that has emerged in the recent times. The utilization of fiber laser for marking applications produces clean and high end finishing of a wide variety of materials. Astarita et al. [41], carried out fiber laser marking of titanium cold sprayed coatings on aluminium substrates. The authors observed that the heat input rules the fiber laser marking process in particular three different conditions. Firstly, irregular groove was observed due to a too high energy released for unit length, whereas regular mark with hidden damages along with regular and effective mark was also observed. It was concluded by the authors that regular mark with inadequate groove was observed due to a low heat input during the process. The homogeneity of the cold spray coating played a crucial role in obtaining high end finishing of marking. The discontinuous nature of coating affected the nature of the marking process. Penetration depth more than 0.4 and

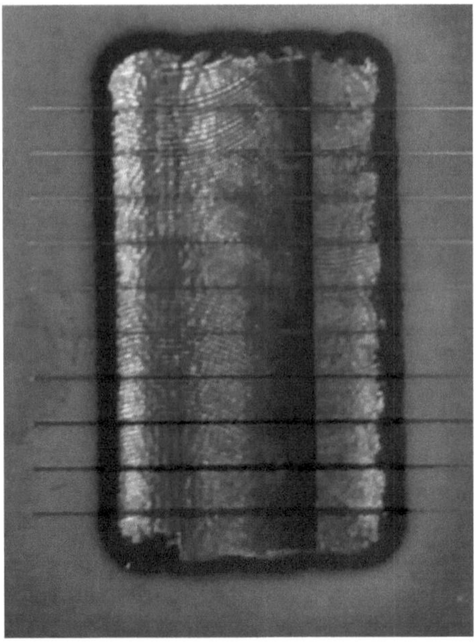

Fig. 6.8 Microscopic view of the markings on the titanium cold sprayed layer at 20 kW of power [41]

0.2 mm were observed on titanium sheet (without coating) and cold sprayed coating, respectively. A microscopic view of markings on the titanium cold sprayed layer performed at output power of 20 kW is shown in the Fig. 6.8.

6.4 Recent Developments and Future Scope of Fiber Laser Micro-machining

In order to exceed the limited peak power ($\sim$10 GW) of fiber systems, fiber-laser pumped parametric amplifiers are being developed by the researchers in the present era. These solutions necessarily engage a range of complexities to the systemic architecture. Further, in order to eradicate any group delay between the individual amplifier sections, these amplifiers need to be appropriately designed. Along with the aforementioned tasks, the optical path delay between the individual amplifiers must be controlled by using proper feedback electronics to reach to a fraction of a wavelength. Pulse energies in the micro-joule to millijoule range are required for the micro-machining of dielectrics and metals. Superior hole quality in micro-drilling is the driving force for the usage of femtosecond pulses in micro-machining applications as compared to the nanosecond or picosecond pulses along with the faster ablation speed, when it is combined with the use of pulse burst. With the introduction of large-mode-area fibers, fiber lasers capable of generating femtosecond pulses with millijoule pulse energies became possible. These fiber lasers have allowed the construction of micro-joule level fiber picoseconds pulse sources that does not utilize the CPA [9]. By the usage of nonlinear spectral compression, high spectral quality can be obtained. In order to expand the mode area of optical fibers, the most suitable method is to increase the core diameter. This phenomenon is effective for the core diameters in the range of 25–50 μm. The fundamental mode can be excited if appropriate mode matching at the input, higher order mode filtering via bending and/or the insertion of satellite cores, which channel out higher-order modes from the central core structure, is carried out; although resulting fibers are multimedia in nature. New applications are invented with the utilization of nonlinear effects in gas-filled fibers. Another interesting area that is still under progress is the combination of coherent beam with the fiber laser system, in order to expand its capabilities in the ultrafast region. Two different forms of multimedia fibers have become widely accepted in the recent years, which are, large-path fibers and leakage channel fibers. The characteristics of both the fibers are, firstly, they are derived from photonic crystal fiber designs, bearing a core which is encompassed by air holes in a cladding to make sure that there is countless single-mode propagation.

In the previous years, research works on the fiber laser systems were primarily based on generating micro-features mainly for biomedical industries and thin material cutting. Although, few researchers have showcased the utility of fiber laser for thick material cutting, but the number of research works in this field are not satisfactory. In the present day scenario, researchers are trying to utilize the fiber laser in various hybrid machining approaches, i.e., laser assisted micro-machining, laser assisted micro-turning, etc. along with the conventional micro-machining approaches for the newly developed materials. Utilization of infrared fiber laser for generating micro-fluidic channels on polymer materials such polymethyl methacrylate, is a new dimension in the domain of fiber laser micro-machining [31]. Research works are also being conducted on fiber laser micro-machining of various engineering materials with the aid of developed set ups for assisting the fiber laser micro-machining process. However, further research works on hybrid fiber laser micro-machining processes need be carried out.

6.5 Summary

Fiber laser has emerged as one of the strong force in the solid state laser systems in the recent years. The utilization of the fiber laser is not only confined to the fabrication of micro-features on a few selected fields, but also successfully managed to diversify its applications in the realm of thick material cutting also. The number of research works is escalating for the developers of fiber laser systems, such as amplifiers, high diode lasers, along with the generation of burr, dross free micro-cuts on a wide spread of materials. The potential of fiber laser system in the micro-machining area is still not completely utilized as a consequence of which the market share of fiber laser is still lagging behind the conventional solid state laser systems such as Nd-YAG laser systems. In the present chapter, a vivid discussion has been conducted with the concept of the fiber laser generation along its diversity of applications in the various micro-machining arenas. Fiber laser diversified from the conventional laser systems only from the fact that the laser is generated within the fiber itself while the other laser systems only utilize fiber as a beam delivery system. The past research studies show that most of the research works with the aid of fiber laser micro-machining are confined only to micro-cutting, micro-drilling and engraving. Only a hand full of work has been conceptualized for hybrid laser machining with the aid of fiber laser systems. Although, the experimental results turn out to be satisfactory compared to CO_2 and Nd-YAG lasers, but even then, heat affected zone, dross formation etc., are inevitable for ultrafast fiber laser systems. The problem can be partially or fully overcome by introducing pico and femtosecond fiber laser systems. However, the aforesaid lasers are not at all cost effective. Efforts are being made by engineers and researchers to not only reduce the overall costs, but also to establish outstanding industrial applications which will be critical factors to converge both the academic and the industrial arenas.

Acknowledgments The authors acknowledge the financial support and assistance provided by CAS Ph-IV program of Production Engineering Department of Jadavpur University under University Grants Commission, New Delhi and TEQIP phase II program of Jadavpur University.

References

1. J. D. Majumdar and I. Manna (2013), Introduction to Laser-Assisted Fabrication of Materials, Laser assisted fabrication of materials, Springer series in material science, Springer-Verlag, 1–68.
2. R. D. Schaeffer (2012), Fundamentals of laser micromachining, CRC Press, Taylor & Francis Group, 5–36.
3. A. K. Nath (2013), High Power Lasers in Material Processing Applications: An Overview of Recent Developments, Laser assisted fabrication of materials, Springer series in material science, Springer-Verlag, 69–112.
4. R. Stępień, M. Franczyk, D. Pysz, I. Kujawa, M. Klimczak and R. Buczyński (2014), Ytterbium-Phosphate Glass for Microstructured Fiber Laser, Materials 2014, 7, 4723–4738.
5. M. Tagliaferri, F. Cannone and D. Nolli (2015), Pulsed fiber laser with double- pass pumping, US 20150010025 A1, Jan 2015.
6. R. G. Smith (1972), Optical power handling capacity of low loss optical fibers as determined by stimulated Raman and Brillouin scattering, Applied Optics, 11, 2489–2494.
7. S. Woods (2009), Understanding materials processing lasers- a comprehensive overview covering the capabilities and applicability of the major systems, Laser Technik Journal, Willey online Library, 6: 5, 23–26.
8. M.L. Stock, A. Galvanauskas, M.E. Fermann, G. Mourou and D. Harter (1993), Generation of high-power femtosecond optical pulses by chirped pulse amplification in erbium doped fibers. Pro. Opt. Soc. Am. Top. Meeting on Nonlinear Guided Wave Phenomena (OSA).
9. M. E. Fermann and I. Hartl (2013), Ultrafast fibre lasers, Review articles focus, Nature photonics, DOI:10.1038/NPHOTON.2013.280.
10. L. Dong and M. E. Fermann (2011), Introduction to optical fiber lasers, High power laser handbook, McGraw-Hill, ISBN: 978-0-07-160902-9, 413-461.

11. M. Naeem, Choice of laser sources for micro-machining applications, SPI Lasers, http://www.spilasers.com/whitepapers/choice-of-laser-sources-for-micromachining applications/.
12. D. Teixidor, F. Orozco, T. Thepsonthi (2013), Effect of process parameters in nanosecond pulsed laser micromachining of PMMA-based microchannels at near-infrared and ultraviolet wavelengths, The International Journal of Advanced Manufacturing Technology, 67 (5), 1651–1664.
13. J. Meijer (2002), Laser micromachining, Micromachining of engineering materials. USA: Marcel Dekker, 203–238.
14. A. Sen, B. Doloi and B. Bhattacharyya (2015), Fiber laser micro-machining of Ti-6Al-4 V, Lasers Based Manufacturing: 5th International and 26th All India Manufacturing Technology, Design and Research Conference, AIMTDR 2014, Springer India, 255–281.
15. F. O. Olsen, K. S. Hansen, and J. S. Nielsen (2009), Multibeam fiber laser cutting. Journal of Laser Applications, 21, 133.
16. K.F. Kleine and K.G. Watkins (2003), Fiber laser for micro cutting of metals. Proc. SPIE 4974, 185.
17. H. Meng, J. Liao and Y. Zhou (2009), Laser micro-processing of cardiovascular stent with fiber laser cutting system. Optics of Laser Technology, 41(3), 300–302.
18. W.P. Risk, T. R. Gosnell and A. V. Nurmikko (2003), Compact blue-green lasers, Cambridge University Press, Cambridge, 1–19.
19. C.H. Fu, J.F. Liu and A. Guo (2015), Statistical characteristics of surface integrity by fiber laser cutting of Nitinol vascular stents. Applied Surface Science. 353, 291–299.
20. Y. Okamoto, N. Kataoka, H. Tahara, K. Shiwayama and Y. Uno (2006), Micro cutting of thin copper plate by fiber laser with laval nozzle, Journal of Laser Micro/Nanoengineering, 1, 3.
21. H. Niino, Y. Kawaguchi, T. Sato, A. Narazaki, R. Kurosaki, M. Muramatsu, Y. Harada, K. Wakabayashi, T. Nagashima, Z. Kase, M. Matsushita, K. Furukawa and M. Nishino (2013), Laser Cutting of Carbon Fiber Reinforced Plastics (CFRP) by Fiber Laser Irradiation, Conference on Lasers and Electro-Optics/Pacific Rim, ISBN: 978-1-4673-6475-1.
22. J.B. Lecourt, C. Duterte, F. Liegeois, D. Lekime and Y. Hernandez (2012), High-power ultrashort fiber laser for solar cells micromachining, Proc. SPIE8237, Fiber Lasers IX: Technology, Systems, and Applications, 823738, DOI:10.1117/12.907951.
23. P. Deladurantaye, A. Cournoyer, M. Drolet, L. Desbiens and D. Lemieux (2011), Material micromachining using bursts of high repetition rate picosecond pulses from a fiber laser source, Proc. SPIE 7914, Fiber Lasers VIII: Technology, Systems, and Applications, 791404, DOI:10.1117/12.875265.
24. Toru Mizunami and and A. Ehara (2011), Femtosecond-pulsed laser micromachining and optical damage by an erbium-doped fiber-laser system, Proceedings of the 36th International Conference on Micro- and Nano-Engineering (MNE), 36th International Conference on Micro- and Nano-Engineering (MNE), Microelectronic Engineering, 88(8), 2334–2337.
25. H. Qi and H. Lai (2012), Micromachining of metals and thermal barrier coatings using a 532 nm nanosecond fiber laser, Physics Procedia, Laser Assisted Net shape Engineering 7, 39, 603–612.
26. YP. Kathuria (2003), Biocompatible metallic stent for medical therapy. SPIE, 5287:52–61.
27. L. Liu, D. B. Li, Y. F.Tong and Y.F.Zh (2016), Applied Physics A, 122:638, DOI 10.1007/s00339-016-0170-0.
28. N. Muhammad, D. Whitehead, A. Boor and L. Li (2010), Comparison of dry and wet fibre laser profile cutting of thin 316L stainless steel tubes for medical device applications, Journal of Materials Processing Technology, 210, 15, 2261–2267.
29. W.C. Choi and G. Chryssolouris (1995), Analysis of the laser grooving and cutting processes, J. Phys., Appl. Phys, 28, 873–878.
30. C.A. Biffi and A. Tuissi (2016), Micro-processing of NiMnGa shape memory alloy by using a nanosecond fiber laser, Optics &LaserTechnology, 78, 42–49.
31. A. Sen, B. Doloi and B. Bhattacharyya (2016), Fibre laser micro-channeling of PMMA, Lasers in Engineering, 35, 1–4, 123–138.

32. A. Sen, B. Doloi and B. Bhattacharyya (2015), Effect of process parameters on IR fiber laser micro-machining of PMMA, International Conference on Precision, Meso, Micro and Nano Engineering, IIT Bombay, India, 10–12 December 2015, 1–6.
33. C.A. Biffi, N. Lecis, B. Previtali, M. Vedani and G. M. Vimercati (2011), Fiber laser microdrilling of titanium and its effect on material microstructure; International journal of Advanced Manufacturing Technology, 54:149–160, DOI 10.1007/s00170-010-2918-6.
34. A. Ancona, D. Nodop, J. Limpert, S. Nolte and A. Tünnermann (2009), Microdrilling of metals with an inexpensive and compact ultra-short-pulse fiber amplified microchip laser, Applied Physics A, 94: 19–24, DOI 10.1007/s00339-008-4906-3.
35. A. Ghosal and A. Manna (2013), Response surface method based optimization of ytterbium fiber laser parameter during machining of Al/Al_2O_3-MMC, Optics and Laser Technology, 46, 67–76.
36. H. Huang, L-M. Yang and J. Liu (2014), Micro-hole drilling and cutting using femtosecond fiber laser, Optical Engineering, 53 (5), 051513.
37. B. Adelmann and R. Hellmann (2015), Rapid micro hole laser drilling in ceramic substrates using singlemode fiber laser, Journal of Materials Processing Technology, 221, 80–86.
38. S. Genna, C. Leone, V. Lopresto, L. Santo and F. Trovalusci (2010), Study of fibre laser machining of C45 steel: influence of process parameters on material removal rate and roughness, International Journal of Material Forming, 1, 1115–1118.
39. F. Tagliaferria, N. Paganob, C. Leoneb, B. Palumbo, Statistical analysis of fiber laser machining of titanium alloy, https://www.researchgate.net/publication/267411159.
40. O. F. Biswas, A. Sen, G. Kibria, B. Doloi and B. Bhattacharyya (2016), Experimental investigation into spiral micro-grooving on aluminium using fiber laser, Proceedings of 6th International & 27th All India Manufacturing Technology, Design and Research Conference (AIMTDR-2016), College of Engineering, India, 16–18 December, 411–415.
41. A. Astarita, S. Genna, C. Leone, F.M.C. Minutolo, A. Squillace and C. Velotti (2016), Study of the laser marking process of cold sprayed titanium coatings on aluminium substrates, Optics & Laser Technology, 83, 168–176.

Chapter 7
Laser Beam Micro-cutting

N. Roy, A.S. Kuar and S. Mitra

Abstract This chapter gives a brief overview of laser beam cutting. An illustration of basic fundamentals of laser material interaction and material removal is given here which considers generation of photon and interaction with electron to generate heat, absorption of heat, phase transformation, plasma generation, ablation and removal mechanism of cut material from the irradiate region at dry condition. An experimental study of laser beam cutting of Inconel 625 superalloy at dry condition is given at the end of this chapter. During this experimental study effect of different process parameters on machining characteristics also discussed in details.

7.1 Introduction

Laser, acronym of, "Light Amplification by the Stimulated Emission of Radiation" is an optical oscillator where amplified radiation is oscillating by the stimulated emission process. Laser is said to be the available purest frequency form of radiation energy to be used as an industrial energy, having a great aftermath on manufacturing. Today laser is omnipresent from our house to machine floor of manufacturing sector to medical industry. In modern manufacturing industries, laser plays a significant contribution in terms of machining of hard to machine materials and development of new processing techniques. Simulated emission produced electromagnetic radiation (laser) is utilized for the processing of a wide range of newly developed engineering materials.

N. Roy (✉) · A.S. Kuar · S. Mitra
Production Engineering Department, Jadavpur University, Kolkata 700032, India
e-mail: nilanjan_83@yahoo.co.in

A.S. Kuar
e-mail: askuar@rediffmail.com

S. Mitra
e-mail: mitra.souren@gmail.com

G. Kibria et al. (eds.), *Non-traditional Micromachining Processes*,
Materials Forming, Machining and Tribology,
DOI 10.1007/978-3-319-52009-4_7

In different kind of unconventional manufacturing processes, laser beam machining can be considered as a machining process with high speed of operation, flexibility, non-contact type in nature and localized processing. A beam of electromagnetic photon is used as a machine tool during laser beam machining operation. High collimation, monochromacity, spatial and temporal coherence along with very less divergence is the reasons behind the production of high intensity laser beam. High focusability with large irradiance (up to 10^{21} W/cm^2) makes it a non conventional machine tool to process wide range of difficult to machine materials i.e. superalloys, nanocomposite, bio-compatible ceramic materials. Absence of wear & tear and chance of contamination of processed materials makes laser beam a sterile tool to process medical and biological components. Laser beam machining operation is independent from physical, mechanical and/or electrical properties of target substrate, whereas it only depends on the opto-thermal properties of the target material. Monochromaticity of the laser beam has a great impact on laser matter interaction as it affects the absorptivity of the single wavelength beam by target surface. Normally Gaussian spatial distribution with TEM_{00} mode is used during laser material processing to get enough energy density at the laser irradiate region, for alteration of irradiate zone's property in a controllable desired manner. Diode laser, CO_2 Laser, Nd:YAG laser, Fibre laser and Excimer lasers are generally used for various material processing applications in industries [1].

7.2 Physics of Laser Material Processing

The phenomenon of electromagnetic radiation Laser-matter interaction is generally a coupled of electronic and thermal events which regardless starts with an electronic excitation followed by local heating due to photon-electron interactions. Reflection, refraction, scattering, transmission and absorption are the physical phenomena that take place whenever the laser beam falls on the surface of work substrate. Figure 7.1 shows a schematic of interaction of laser beam with work material.

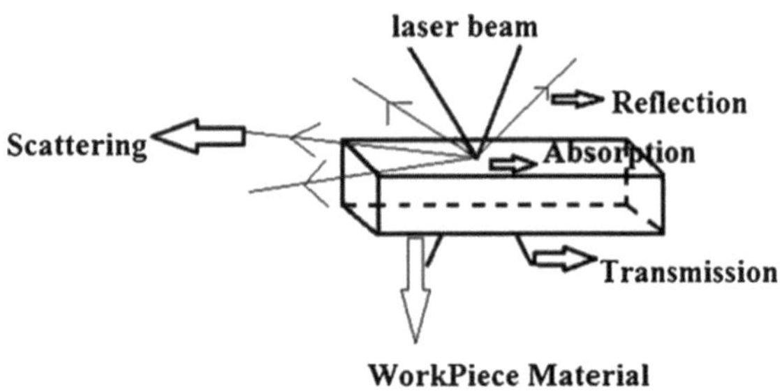

Fig. 7.1 Interaction of laser beam with workpiece material [2]

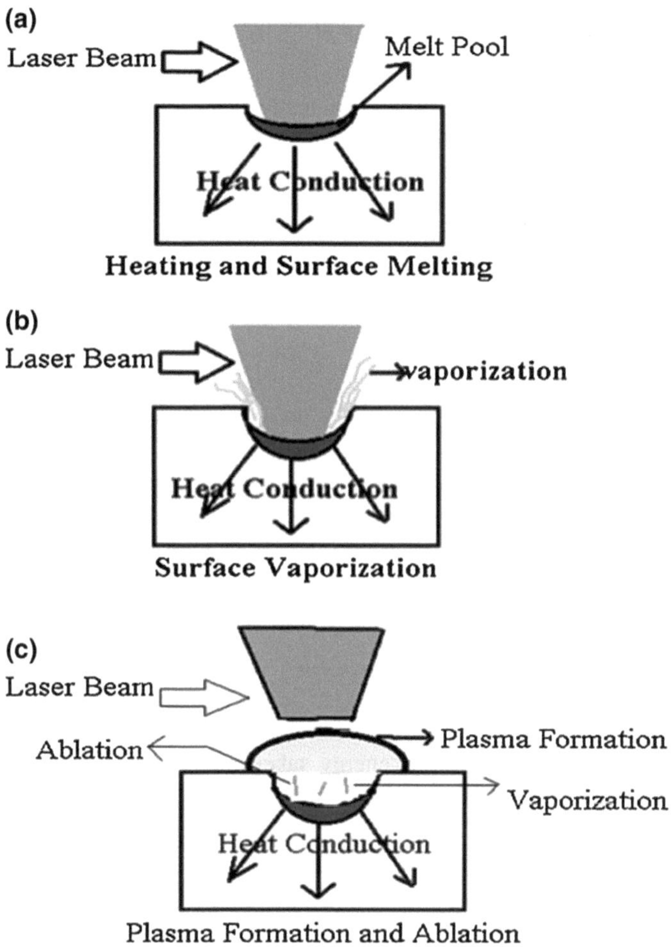

Fig. 7.2 Various governing physical phenomena of material removal mechanism [2]

Reflection, scattering, transmission, and absorption are the physical phenomena occur during solid laser interaction at the workpiece surface. Linear and non-linear absorption are the most important phenomenon for lattice heating. The physical phenomena of material removal mechanism are shown in Fig. 7.2.

Generally high speed electron formed electric cloud is present around positively charged nuclei in atom. This electric cloud is the basic building block to form a molecular structure in atom. When a laser beam is interact with the matter a force is generate by the collision of the electric field of electromagnetic radiation and atoms in the irradiate region which eventually cause the oscillation of nuclei to generate the heat. The force exerted by the electromagnetic radiation of the electron can be expressed,

$$F = eE + e\left(\frac{v}{c} \times H\right)$$

where, E is electric field vector, H is magnetic field vector, v is the velocity of electron and c is the speed of light.

Excited electron–phonon interaction (10^{-11} to 10^{-12} s) is one of the main reasons to transfer laser energy to lattice in metal by energy absorption by conduction band electron. Excitation energy of the bound electrons and kinetic energy of the free electrons are the source of extra energy of charged particles. Absorption of laser energy is governed by Beer-Lambert law [3],

$$I(Z) = I_0 e^{\mu z}$$

where, I_0 is the incident intensity, $I(z)$ is the intensity at depth z, and μ is the absorption coefficient.

Conversion of absorbed energy into heat is happened in very short time period (<pulse width) by degradation of the ordered and localized primary excitation energy. The temperature distribution on laser irradiate zone not only depends on absorbed laser energy but also on thermal diffusivity (D). Thermal diffusivity [4] can be expressed in terms of specific heat (C_p) and thermal conductivity (k),

$$D = k/\rho C_p$$

where, ρ is the density.

Significant attenuation of laser energy takes place during pulse duration in vertical direction which is expressed as,

$$z = (2Dt_p)^{1/2}$$

where, z is vertical distance, t_p is pulse duration.

Depending on the temperature rise at the irradiate region various physical phenomenon like heating, melting and vaporization occurred. To and fro movement of solid liquid interface in a very high velocity(≤ 30 m/s) is the reason behind surface melting and subsequent resolidification at the machining zone [5]. Depth of melting increases with increase in time subsequently reaches at maximum at around pulse time (t_p) when the temperature reaches at boiling point. Further increase in laser power density or pulse on time, results in material removal from irradiate region by vaporization without increase in the depth of melt. The liquid vapour interface moves in vertical direction inside the material during pulse on time after the vaporization is initiated. The highest possible value of surface tension pressure is exceeds by evaporation induced recoil pressure during laser irradiation and material removal starts to take place. Material removal takes place by evaporation above the vapour-liquid interface. Depth of vaporization [3], mass of material removed per unit time and velocity of the liquid–vapour interface can be calculated by

$$\dot{m} = V_s \rho$$

where $\dot{m}$ the mass of material removed per unit time, Vs is the velocity of the liquid–vapour interface and ρ is the density.

$$d = V_s t_p$$

where, t_p is the pulse time and d is the depth of vaporization.

When the incident laser energy on the target surface is sufficiently large to exceed the boiling temperature, rapid vaporization starts. The vapour particle size decreases with decreasing pressure and evaporation rate [6]. Plasma is generated at high laser irradiance ($I \geq 109$ W/cm^2) where the vapour or the ambient gas becomes ionized due to the interactions between the incident laser beam and the produced vapour. Lesser amount of laser energy is available on machining zone due to shielding effect of plasma plume. In photo-thermal ablation process material is removed by thermal stresses and surface vaporization. Whereas in photo-chemical ablation material removal takes place by molecular fragmentation without significant thermal damage by the high energy photon causes the direct bond breaking of the molecular chains in the organic materials [7].

7.3 Laser Beam Cutting

In different types of laser based fabrication techniques, laser cutting is characterized by high speed, reliable and full/semi automated 2-Dimensional material processing method to produce a high aspect ratio and clean cut width. During laser cutting process, a high intensity laser beam is focused on the target region of the workpiece which subsequently heats up the surface of the irradiate zone by absorbing the thermal energy of the laser beam. Cutting front is generally produced by expulsion and vapourization of the melt pool from the machining zone by the assistance of pressurized gas flow coaxially/axially. From the quality and quantitative point of view assist the gas flow has a great impact on the laser cutting process. It helps to get cleaner cut width along with less heat affected zone (HAZ) in terms of quality aspect. Chemical reactions like oxidization at elevated temperature at the machining zone may enhance the material removal which indicates at the quantitative aspect. Now-a days, laser cutting process is applied in the industries to process a wide range of materials such as wood, ferrous and non ferrous metals, polymers, ceramics, composite materials, superalloys to nanocomposte materials also.

7.3.1 Different Types of Laser Beam Cutting

Based on the role of assist gas and interaction between laser beam and workpiece substrate laser beam cutting operation can be classified into four groups, i.e. laser sublimation cutting, controlled fracture technique, fusion cutting and reactive fusion cutting.

7.3.1.1 Laser Evaporative Cutting

During laser evaporative/sublimation cutting, absorbed laser energy heats up the irradiate region to the vaporization point and ablates it to the vapour state up to a subsurface region along the cutting seam (Fig. 7.3).

Materials like polymers and organic substances are suitable to process by this method due to low thermal conductivity and lesser vaporization temperature. In this process, all the thermal energy is used in evaporation in a localized manner. Generally, non-reactive gas jet has been used axially/coaxially to remove the vapour from machining zone [8]. The volume of material removed per unit second per unit area is used to determine the penetration velocity,

$$V = F_0/\rho[L + C_p(T_V - T_0)]$$

where,
F_0 = absorbed power density (W/m^2); ρ = density of solid (kg/m^3); L = latent heat of vaporization (J/kg); Cp = heat capacity of solid (J/kg °C); T_v = vaporization temperature (°C); T_0 = temperature of the material at start (°C).

During laser sublimation cutting, no additional evaporation occurs after cut width reaches its maximum value due to insufficient laser energy at machining zone by conduction and convection loss of thermal energy [9].

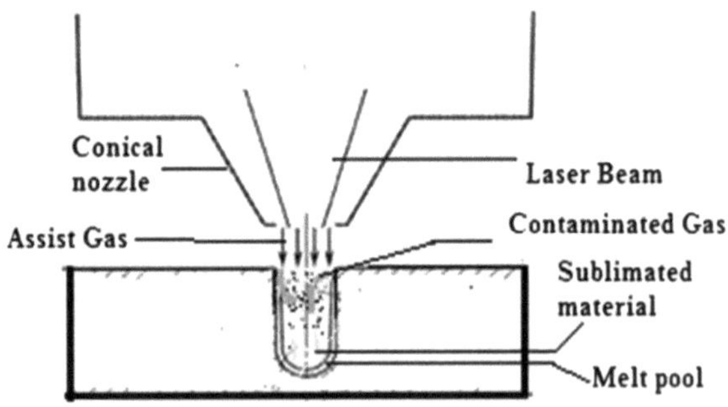

Fig. 7.3 Schematic of evaporation laser Cutting technique

7.3.1.2 Controlled Fracture Technique

In laser induced stress cutting, irradiate laser energy generates mechanical stress on the machining zone, which causes crack propagation with controllable fracture growth, results in material separation on the direction of laser beam movement. In this process, a laser beam moves over the work substrate only to heat up (>melting temperature) the irradiate region. During the cooling period, the thermo-mechanical stress is induced on the irradiate region which produced the crack. A high plastic compressive stress is generated around the laser spot diameter which relaxed with the movement of laser beam, results in generation of residual tensile stress on the upper surface of the irradiate region. This artificial crack then propagates along the path of laser movement for precision separation of materials in a controlled manner. Due to asymmetry of thermo-mechanical stress along with crack generation, the crack trajectory does not follow the laser movement path precisely [10]. Controlled fracture technique requires very less energy due to the absence of melting or vaporization processes. Thus, this technique is very useful to process brittle materials with clean cut with higher cutting speed than other laser cutting techniques.

7.3.1.3 Laser Fusion Cutting

Movement of laser beam relative to workpiece makes a straight or curved cut profile during laser fusion cutting. High intensity laser energy from impinging beam absorbed by the work substrate melts the metal throughout the thickness to create a cutting front. Blind cut, where thickness of work substrate is more than the depth of cut, can also be achieved by fusion cutting. Multiple reflections on the keyhole and sudden increase of absorptivity result in increase in the depth of cut. On the contrary, the combination of reaction forces exerted by the vapour acceleration and change in surface tension on melt pool drags the melt pool towards the direction of laser beam path to generate the kerf. Absorption of thermal energy is mainly governed by re-radiation, Fresnel absorption and plasma absorption. Generally, high pressure, nonreactive assists gasses such as argon, helium etc, transform momentum to the thin melt pool. Drag force of the high pressure assist gas is one of the reasons of melt pool and plasma removal from the machining zone. Melt film is normally ejected from the bottom of the kerf as a droplet due to comparatively thicker melt along with retarding of the film and surface tension at the subsurface region of machining zone. Otherwise dross formation has been occurred [11]. Generation of reverse vortex by assist gas on melt pool, driving the dross upward and leaves the machining zone to form blind cut. The assist gas not only enhances the material removal process, it also acts as a coolant. One part of the pressurised assist gas flows radially outward direction along the work substrate while the other part enters into the kerf. Kerf edges generated by laser fusion cutting, is generally free form oxides of the parent material (Fig. 7.4).

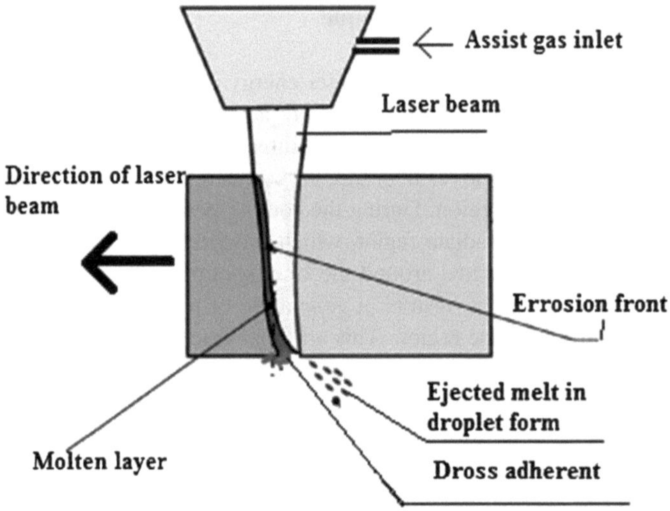

Fig. 7.4 Schematic diagram of the inert-gas laser beam fusion cutting process

7.3.1.4 Reactive Fusion Cutting

In some of the laser cutting processes, co axially flown high pressure reactive gas not only assists in removal of molten material from the machining zone, but also adds some energy on the machining process in terms of exothermic reaction to enhance the process. This aforesaid laser cutting process is a variation of laser fusion cutting and known as laser reactive cutting technique. During laser beam cutting, thermal energy source is added by the exothermic reaction at the erosion front when the impinged high pressure oxygen gas reacts with the molten material. Oxide of the base material is also formed on the top surface of the melt pool. Generation of enormous exothermic heat energy not only nullifies the conductive thermal loss to the parent work substrate, but also helps to enhance the cutting speed as a combine effect of higher fluidity and absorptivity of hot molten layer of oxide slag on the erosion front [12]. This increased cutting speed results in better cut quality due to less thermal diffusion to the adjacent base work substrate. The extra thermal energy added by the exothermic reaction at the irradiate region helps the rise of temperature of the melt pool up to evaporation temperature very rapidly results in ease of material removal by vaporization. Erosion front propagates through the melt pool-parent material interface in the direction of laser cutting by further melting into the unmachined region and material removal due to friction force works between impinged assist gas and melt pool to accomplish the cutting process. Heat balance at the machining zone is described by the heat gain at the melt pool by laser irradiation and exothermic reaction due to oxidization and heat losses due to conduction, vaporization and ejection of melt pool [13]. Generally, oxygen or oxygen enriched gas mixture is used as an assist gas for reactive fusion cutting. Thin layer of oxide above the melt pool have some good as well as adverse affect on quality of laser reactive fusion cutting. Sometimes oxide adherent on the kerf edges as dross which affects the cut quality adversely in terms of micro crack formation. Striation on the side wall of the kerf may be reduced sometimes due to blanket effect of the thin oxide film on melt pool. This process mainly used to cut thick section of hard to machine metals, i.e. stainless steel, aluminium alloys, titanium superalloy etc.

7.3.1.5 Laser Cold Cutting

Absence of thermal effects make this laser cutting technique known as laser cold cutting. In this technique, ultra-short pulsed laser, in the range of picosecond to femtosecond interact with the work substrate with a very high fluence, results material removal by molecular fragmentation due to direct bond breaking by multi photon absorption mechanism. Photon energy of 4.9 eV is very much sufficient to break the chemical bond directly [14]. To get debris free cutting of various material along with surface modification of biodegradable materials, this cutting process is used in industries i.e. medical industries, electronics manufacturing industries.

7.3.1.6 Laser Beam Microcutting

In the modern technologies, miniaturization is an important trend to fabricate microparts for biotechnological, microelectronics, telecommunication, MEMS, and medical applications. Laser beam micromachining is utilized for the generation of microproducts with better flexibility in the dimensional design. Photo-thermal or photochemical ablation is the mechanism behind the material removal in micron or in sub-micron range during laser beam micromachining operation. Laser beam with wavelength in the range in near infrared (1.06 μm) to deep Ultraviolet (150 nm) is commonly used in micromachining operation. Photon energy increases with shorter wavelength enhance the material removal, results in less thermal damages like melting, charring of adjacent unirradiated material. Thermal diffusivity, thermal relaxation time and absorption coefficient have a great effect on ablation process [15]. Efficiency of ablation increases by lower value of thermal diffusivity whereas the pulse duration shorter than thermal relaxation time facilitates restriction of temperature distribution on the surrounding material. Shorter pulse width (<100 ns to few femtosecond) is preferably utilized for micromachining application due to high peak intensity ($>10^{15}$ W/cm^2) to get précised dimensional accuracy with substantially low thermal damages [16].

7.3.1.7 Laser Cutting at Different Assisted Medium

Sometimes laser cutting process is performed at various atmospheric conditions to get desired cut quality. Laser cutting technique sometimes performed in submerged condition or in preheated condition or in vacuum or sometime mixture of various conditions.

7.3.2 Process Characteristics

Laser beam cutting is one of the most widely used laser material processing technique in industries. It is a complicated process dominant by various process parameters, i.e., laser beam properties, beam delivery system, assist gas properties, to get desired quality aspects in terms of narrow cut width, smoother inner kerf, striation and dross free cut along with less heat affected zone. The term laser beam properties indicate at laser fluence, wavelength, mode of operation (continuous or pulse), pulse frequency which have great impact on laser cutting as well as on physical, thermal and metallurgical properties of kerf generated.

7.3.2.1 Laser Fluence

Laser fluence or energy density determines the energy input at the irradiate region which is the most significant factor to effect the performance and quality of the machining. Laser fluence also the deciding factor to determine minimum cutting speed required for a through cut [17].

7.3.2.2 Mode of Operations

Laser beam cutting procedure generally performed either in continuous mode or in pulse mode. During continuous mode of operation laser, average power is considered, whereas laser peak power is taken into account for laser pulse mode operation. Mode of operation plays a vital role to get fine finishing of the kerf edges along with striation free kerf wall. Q-switching is one of the vastly used pulsing techniques. In industries, Nd:YAG laser generally used in pulsed mode, whereas CO_2 laser preferably used in continuous mode. During pulsing mode operations, pulsing time varies from 100 μm to few femto-seconds. This pulsing time effects the material removal mechanism from irradiate zone. Up to 1 ns pulse material removal takes place by photothermal ablation. Although, in pico or femto second range pulse operation, photochemical ablation takes place.

7.3.2.3 Wavelength

Wavelength has a linear relationship with absorption. Absorptivity increases with decrease in wavelength from infrared (IR) to near infrared (NIR). Wavelength is not only a predominant process parameter, but also effects the choosing of other process characteristics like mode of operation. The wavelength of Nd:YAG laser (1.06 μm) falls in NIR region and CO_2 (10.6 μm) falls in IR region. Wavelength in the ultraviolet region helps to machine transparent materials, i.e., polymer to glass. In very short wavelength (≤ 532 nm), the photon energy is 4.2 eV which is more than equal to bond energy of work substrate. Laser beam having wavelength in UV region is capable to break the chemical bond directly to machine transparent material with minimal or no thermal defects.

7.3.2.4 Beam Delivery System

Beam delivery or optical system determines the spot diameter, relative focal distance and beam energy distribution on the workpiece. Spot diameter is determined by measuring the area of single pulse laser irradiate zone. The laser power intensity on the irradiate zone is determined by laser spot diameter. Deep cut cannot be achieved by focusing the laser beam just on the top surface of the work substrate due to multiple reflections on the kerf wall. To get the optimum cut quality generally, laser beam is focused just beneath the top surface. Mainly, Gaussian energy distribution with the combination of very low spot diameter and very high energy density laser beam, is preferred in the industries. Similar interaction with work substrate in all direction due axial symmetry of the Gaussian beam, results in low roughness in and outside of the kerf.

Polarization is defined by the way of oscillation of electric and magnetic field vector respect to each other. In laser beam cutting operations, generally electric and magnetic vectors are oscillating in a same plane and perpendicular to each other. The angle between plane of incidence and electric field vector defined different types of polarization, such that S type or P type. Beam polarization is a control factor for absorption of incident laser beam energy by work substrate [18].

7.3.2.5 Assist Gas

Assist gas velocity not only act as an coolant but also helps to enhance the ejection of molten material from the erosion front to accelerate the cutting process. From the past research works, it has been found that velocity of assist gas enhance the cutting process up to a certain point, then the efficiency of the cutting process is degraded due to nonlinear effect of cooling by the assist gas velocity [19]. Generation of density gradient fields by non uniform assist gas velocity leads to change in refractive index which affects the performance characteristics [20]. The alignment of nozzle where gas jet is used co-axially has a great impact on surface roughness

[21]. Impact on cut quality by the composition of gasses is significant. It is found that with decrease in purity of oxygen utilized in reactive fusion cutting, the cutting performance reduces [22]. Generally, inert gases are used to perform cutting operations other than laser reactive fusion cutting technique to reduce the sidewall burning effect. In industries various kind of nozzles are used in which conical shape nozzles are commonly used. Selection of nozzle diameter is very important to get desired quality aspects as well as the thickness of the material to be cut.

7.3.3 *Quality Aspect*

Quality characteristics of laser beam cutting are a great concern with increased application in biomedical industries, aerospace industries and automobile industries. Geometric qualities like corner accuracy, kerf with good aspect ratio, smooth inner surface and surface roughness in the range of nanometer along with great metallurgical quality, i.e. less heat affected zone (HAZ) and thermal stress induced micro crack free cut is required to get during laser beam cutting.

7.3.3.1 Dross

Incomplete expulsion of molten material from the bottom side of the kerf during through cut is known as dross. It is required to understand mechanism of dross to get nonadherent dross by controlling the process variables. Arata et al. first explain the mechanism of dross during laser beam cutting of stainless steel. A small portion of molten material is ejected as droplet form whereas the other portion subsequently flows in reverse direction and solidified at the bottom corner edge of the kerf to form dross. This phenomenon is happened when the cutting operation takes place at lower cutting speed [23]. Dross free cut can be achieved when the maximum exit momentum thrust is equal to ambient gas pressure. This aforesaid condition controls the nature of the geometrical shape of ejected molten material from the bottom of the kerf [24]. Surface tension force dependent ejection speed of melt and insufficient melt expulsion due to blockage by excess molten material is the two mechanism of dross formation [25].

7.3.3.2 Striation

Formation of periodic patterns along the kerf inner edge. Frequency and amplitude of striation are the characteristics of striation on the inner edge which are mainly affects by cutting speed. Side wall burning and oscillation of melt pool phenomenon may be the most accepted mechanism of striation formation by the researchers [27]. Burning front is established by extra energy gain by the exothermic oxidation reaction apart from the laser irradiation. Burning fronts propagate faster than laser beam in radial direction from the centre of laser beam, end at the edge of the beam. Thus, single striation is created. Further movement of laser beam at the erosion front, reinitiate burning front to generate a striation pattern along the kerf edge. Temperature fluctuation on the melt pool due to sudden temperature drop by resistance to oxygen diffusion by oxide layer and subsequently temperature rise by sudden increase in reaction energy by extra oxide slag result in periodic striation [28]. Figure 7.5 illustrate different patterns of striation.

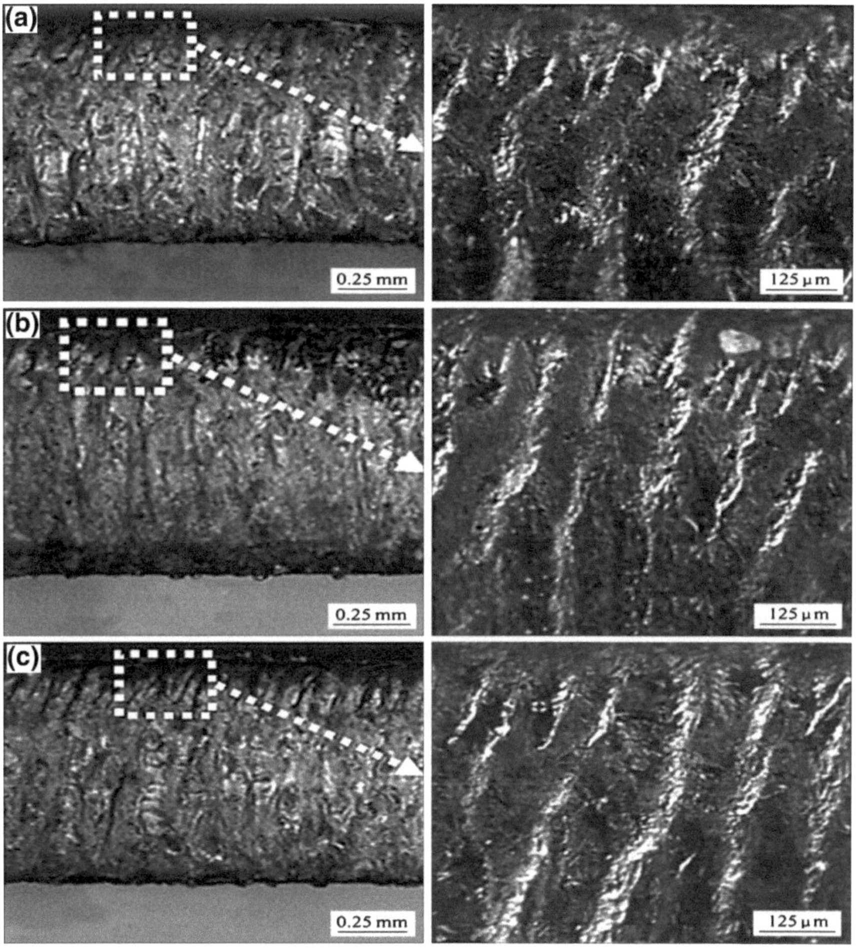

Fig. 7.5 Striation formation on cut section [26]

7.4 Application of Laser Beam Machining

Laser beam cutting process is on of the fastest, clean and green manufacturing technique applied in manufacturing industries with a precision and cost effective way. Laser cutting technique is first applied in industry to cut Die board [29]. This technique I used in job shop to produce complex profile like valve plates, gaskets, complex shape medical parts, i.e. coronary/pulmonary stent. In aerospace industry it is used to make titanium ducting, fuselage skin stiffeners an jet engine component with the dimensional profile. This process is utilized to produce micro electronics parts like circuit boards. For dismantling or repair nuclear power station, this technique is obvious choice. Fabrication of micro fluidic device is done by laser cutting process.

7.5 A Case Study on Laser Beam Micro Cutting of Inconel 625 Superalloy

7.5.1 Machining Conditions

A pulsed Nd:YAG laser-based CNC machining system, manufactured by M/s Sahajanand Laser Technology, India, is used for the experimental study. A specially designed workpiece holding unit is placed over the CNC controlled work table. Inconel 625 superalloy with 8 mm × 8 mm × 0.9 mm dimension has been considered as workpiece material for experimental study. Properties of Inconel 625 are listed in Table 7.1 [30]. Photographic view of machining set up is given in Fig. 7.6.

Lamp current, pulse frequency, pulse width, cutting speed and assist gas pressure are considered as the controllable process parameters to carry out the experiment. Kerf width is considered as process response. Here compressed air is selected as an assist gas. The ranges of these process parameters are selected on the basis of trial experiments conducted by using one factor at a time approach. Experiments have been carried out according to the central composite rotatable second-order design

Table 7.1 Properties of Inconel 625

Property	Unit	Value
Yield strength	MPa	414–645
Ultimate tensile strength	MPa	827–1034
Modulus of elasticity	GPa	207.5
Poisson's ratio	–	0.278
Elongation	%	60–30
Density	Gram/cc	8.44
Hardness	Brinell	145–220

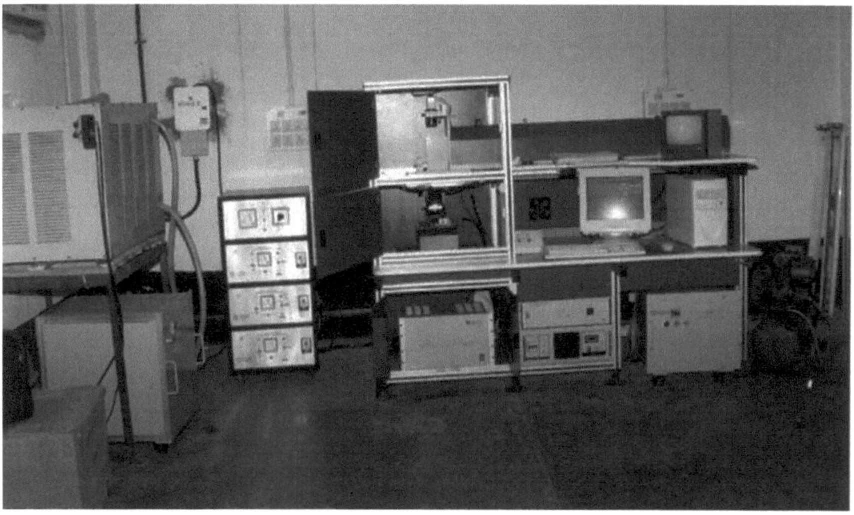

Fig. 7.6 Photographic view of the CNC-pulsed Nd:YAG Laser machining system [31]

Table 7.2 Process parameters levels

Process parameter	Unit	Symbol	Levels				
			−2	−1	0	1	2
Lamp current	A	x_1	20	22	24	26	28
Pulse frequency	kHz	x_2	2	4	6	8	10
Pulse width	%	x_3	2	4	6	8	10
Cutting speed	mm/sec	x_4	1.0	1.5	2.0	2.5	3.0
Assist gas pressure	kg/cm^2	x_5	0.1	1	1.5	2	2.5

based on response surface methodology. Range of all controllable input process variables are listed in Table 7.2.

Response surface modelling was used to establish the mathematical relationship between the response, y_u and the various machining parameters, with the eventual objective of determining the optimum operating conditions for the system. Usually, a second- order polynomial equation is used in RSM,

$$y_u = \beta_0 + \sum_{j=1}^{k} \beta_j x_j + \sum_{j=1}^{k} \beta_{jj} x_j^2 + \sum_{i<j} \sum_{j=2}^{k} \beta_{ij} x_i x_j \quad (7.1)$$

where y_u is the corresponding response, e.g., kerf width. x_i is the coded value of the i-th machining parameter, k is the number of machining parameters and β_i, β_{ii}, β_{ij} are the second order regression coefficients.

Olympus optical measuring microscope has been used to capture the image of cut region at 20× magnification from where kerf width is calculated by image

Table 7.3 Experimental results

Exp no.	Kerf width (mm)	Exp no.	Kerf width (mm)	Exp no.	Kerf width (mm)	Exp no.	Kerf width (mm)
1	0.0645	14	0.0645	27	0.0625	40	0.0753
2	0.0830	15	0.0830	28	0.0660	41	0.0828
3	0.0677	16	0.0677	29	0.0673	42	0.0793
4	0.0756	17	0.0756	30	0.0826	43	0.0878
5	0.0654	18	0.0654	31	0.0636	44	0.0877
6	0.0837	19	0.0837	32	0.0689	45	0.0882
7	0.0703	20	0.0703	33	0.0590	46	0.0878
8	0.0769	21	0.0769	34	0.0833	47	0.0881
9	0.0679	22	0.0679	35	0.0791	48	0.0906
10	0.0835	23	0.0835	36	0.0670	49	0.0909
11	0.0636	24	0.0636	37	0.0784	50	0.0906
12	0.0692	25	0.0692	38	0.0780	51	0.0905
13	0.0684	26	0.0684	39	0.0758	52	0.0901

analysis software. Measured value of kerf width is listed in Table 7.3. Minitab 17 Software is used for analysis of the measured responses and determining the mathematical models with best fits. The adequacy of the model is tested using the sequential f-test, lack-of-fit test and the analysis-of-variance (ANOVA) technique by the same software to obtain the best fit model.

7.5.2 Results and Discussion

The developed model second order polynomial equation is given below,

$$\begin{aligned} \text{Kerf width} = {} & 0.089295 + 0.005661x_1 - 0.003313x_2 + 0.000415x_3 - 0.000466x_4 \\ & + 0.000900x_5 - 0.004633x_1^2 - 0.004151x_2^2 - 0.002873x_3^2 - 0.003539x_4^2 \\ & - 0.002160x_5^2 - 0.002863x_1x_2 - 0.000424x_1x_3 - 0.000500x_1x_4 \\ & - 0.000078x_1x_5 + 0.000818x_2x_3 - 0.001048x_2x_4 - 0.001179x_2x_5 \\ & + 0.000047x_3x_4 - 0.000322x_3x_5 + 0.000089x_4x_5 \end{aligned} \tag{7.2}$$

ANOVA analysis of the quadratic model with other adequacy measures R2, adjusted R2 and predicted R2 are given in Table 7.4.

The associated *p*-value of less than 0.05 indicates that the model term can be considered as statistically significant at 95% confidence level. Value of lack of fit of the model is more than 0.05 which indicates at non significant as desired. From the table it is found that lamp current is the most dominating factor followed by pulse frequency, assist gas pressure, cutting speed and pulse width. It can be concluded

Table 7.4 ANOVA analysis results

Source	DF	Adj SS	Adj MS	F value	*P* value
Model	20	0.004573	0.000229	159.63	0.000
Linear	5	0.001769	0.000354	246.99	0.000
Lamp current	1	0.001282	0.001282	894.91	0.000
Pulse frequency	1	0.000439	0.000439	306.50	0.000
Pulse width	1	0.000007	0.000007	4.82	0.036
Cutting speed	1	0.000009	0.000009	6.07	0.019
Assist gas pressure	1	0.000032	0.000032	22.63	0.000
Square	5	0.002423	0.000485	338.33	0.000
2 way Interaction	10	0.000381	0.000038	26.60	0.000
Error	31	0.000044	0.000001		
Lack of fit	22	0.000027	0.000001	0.63	0.822
Pure error	9	0.000018	0.000002		
Total	51	0.004617			
Value of S 0.0011968		R sq 99.04%	R sq (adj) 98.42%	R sq (pred) 97.39%	

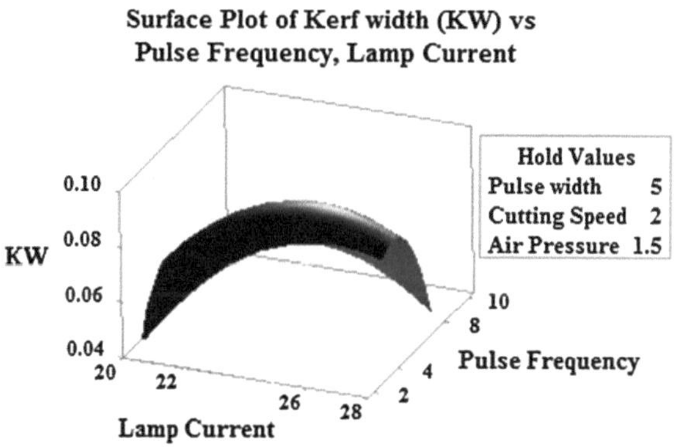

Fig. 7.7 Response surface plot of Kerf Width with respect to lamp current and pulse frequency

that the data for each response are well fitted in the developed model from the values of S, R-sq, R-sq(adj) and R-sq(pred) given in Table 7.4.

Parametric analysis:

From Fig. 7.7 it is observed that kerf width first increase with increase in lamp current but it marginally decreases with lamp current after 26 A. Firstly increase in lamp current generates more laser power for which amount molten material increases, results in increase in kerf width. But when lamp current exceeds 26 A value, excess amount of molten material is formed in machining zone which cannot properly removed by assist gas from the machining zone and results in decrease in kerf width. Kerf width marginally increases with increases in pulse frequency up to 4 kHz then gradually decreases. At low pulse frequency comparatively high energy beam is generated which removes more material from the top surface of the

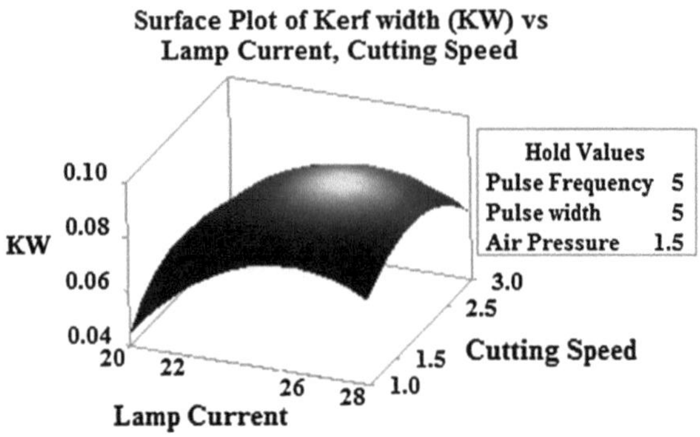

Fig. 7.8 Response surface plot of Kerf Width with respect to lamp current and cutting speed

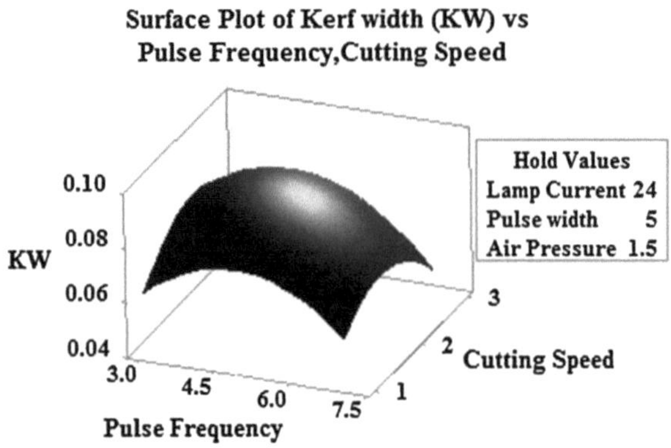

Fig. 7.9 Response surface plot of kerf width with respect pulse frequency and cutting speed

workpiece and vice versa. From Fig. 7.7 it can be concluded that low kerf width can be achieved at lowest lamp current with highest value of pulse frequency.

From Fig. 7.8 it is observed that kerf width first increases with increase in lamp current but it marginally decreases with increase in lamp current after 26 A whereas incase of cutting speed kerf width gradually increases up to mid value of cutting speed then decreases. In beginning due to increase in lamp current along with low cutting speed at that very point of operation, more laser power is transferred to machining zone results in increase in kerf width but further increase in cutting speed results in decrease in interaction time between laser beam and workpiece results in decrease in kerf width inspite of higher laser power at higher lamp current.

From Fig. 7.9 it is observed that lowest kerf width can be achieved at highest cutting speed along with highest value of pulse frequency. At highest value of pulse frequency comparatively low beam power generates which incorporates with less interact time at high cutting speed transferred lower amount of energy to the machining zone results lesser value of melt pool. For that lower kerf width may be achieved.

From Fig. 7.10 it is observed that, lowest kerf width is achieved at highest value of cutting speed and lowest value of assist air pressure. During aforesaid parametric condition interaction time between laser beam and work piece is very small but due

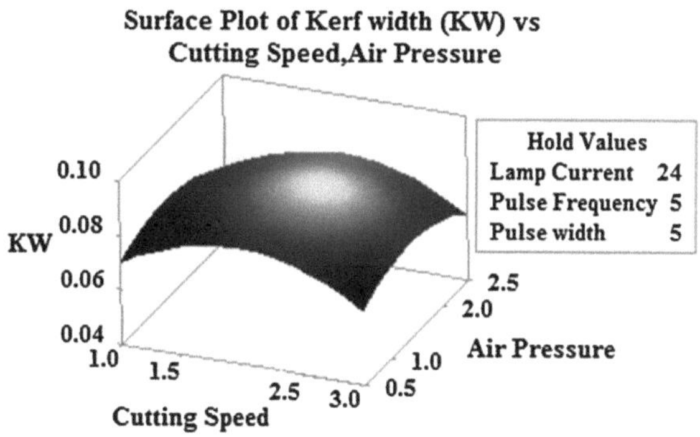

Fig. 7.10 Response surface plot of kerf width with respect to cutting speed and assist gas pressure

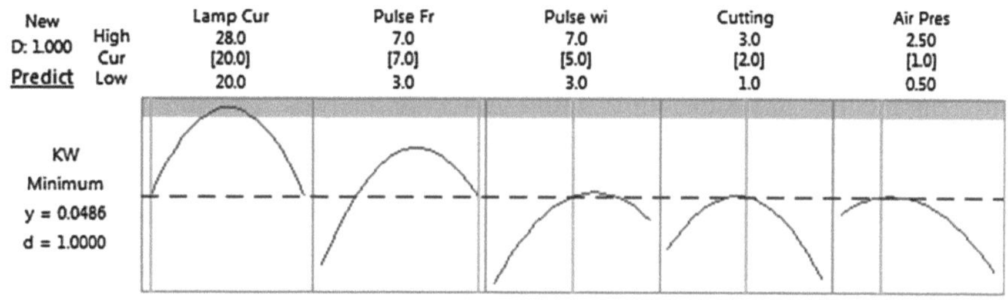

Fig. 7.11 Optimal process parameter settings

to moderate value (hold value) of lamp current some adequate amount of melt pool may be formed which may not be properly removed by assist gas at that lowest value. That result in resolidification of the molten material along kerf edges and inner wall of the cut profile, make the kerf width narrow.

7.5.3 Determination of Optimal Parameter Settings

Figure 7.11 shows the optimization results for the minimum kerf width based on the mathematical model developed using Eq. (7.2). To get ideal machining response, the value of the weight for linear desirability function (D) is considered as 1.

It is found from the figure that minimum kerf width of 0.0486 mm dimension can be achieved at process parameter settings of 20 A lamp current, 7 kHz pulse frequency, 5% pulse width, 1 mm/sec cutting speed and 1 kg/cm^2 of assist gas pressure (Fig. 7.12).

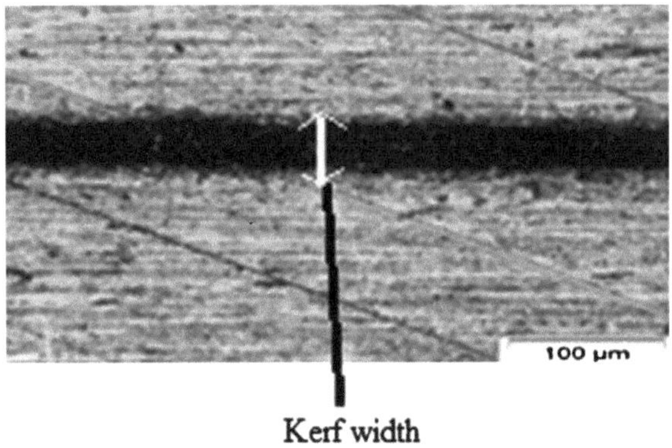

Fig. 7.12 Microscopic view of kerf width at optimum parameter settings

7.5.4 *Conclusion of Case Study*

Experimental study of pulsed Nd:YAG laser beam cutting is carried out successfully on Inconel 625 superalloy. From ANOVA analysis it is found that lamp current is the most dominating process parameter whereas pulse width is the least effecting process parameter. Optimal parameter setting is determined by RSM CCD technique at highest level of pulse frequency and pulse width with lowest level of lamp current and moderate level of cutting speed and assist gas pressure to get minimum kerf width.

7.6 Summary

Overview of laser beam cutting is illustrated in this chapter. Material removal by photo-thermal or photo-chemical is discussed here in details. Fundamentals of different kind of laser beam cutting technique are illustrated. Undesired laser cutting characteristics and controllable process variables are briefly discussed here. Defects like dross formation, striation may be removed and less heat affected zone can achieved by knowing the fundamentals of these defects and identifying the process variable which have a great impact. Also different assisted medium can be used to get desired result. An experimental study of nanosecond pulsed Nd:YAg laser on Inconel superalloy also carried out here. From the study, it can be concluded that laser beam cutting is fast, flexible manufacturing technique for processing of a wide range of engineering material from thickness of thin film to very thick material.

Acknowledgements The authors acknowledge the financial support from UGC, New Delhi under the scheme of Rajiv Gandhi national Fellowship scheme and technical equipment support from UGC, New Delhi under the CAS Ph-IV program of Production Engineering Dept. of Jadavpur University.

References

1. Küper, S., Stuke, M. (1992). Applied physics letters. **60**, 1633.
2. Roy, N., Kuar, A.S., Mitra, S, Acherjee, B. (2015). *Nd:YAG laser microdrilling of SiC-30BN nanocomposite: Experimental study and process optimization*. Lasers based manufacturing, Dixit, U.S. & Joshi, S.N. (Ed.) Topics in Mining, Metallurgy, Materials Engineering. Chapter 17, ISBN 978-81-322-2352-8, © Springer India 2015, PP 317–341.
3. Steen, W.M. (1991). *Laser Materials Processing*. Springer, London.
4. C.W. White, M.J. Aziz (1987), *Surface Alloying by Ion, Electron and Laser Beams*, eds. By L.E. Rehn, S.T. Picraux, H. Wiedersich, ASM, Metals Park, Ohio, p. 19
5. Perry M.D., Pennington D., Stuart B.C., Tietbohl G., Britten J.A., Brown C., Herman S., Golick B., Kartz M., Miller J., Powell H.T., Vergino M., Yanovsky V. (1999), Optics Letter. **24**, 160–162
6. Collins G.W., Celliers P.M., DaSilva L.B., Cauble R., Gold D.M., Foord M.E., Holmes N.C., Hammel B.A., Wallace R.J., Ng A., Physics Review Letters. **87**(16), art. no. 165504 (2001)
7. Bäuerle, D. (2000). *Laser Processing and Chemistry*. Springer, Heidelberg.
8. Biyikli, S., Modest, M.F. (1988). *Beam expansion and focusing effects on evaporative laser cutting*. ASME Journal of Heat Transfer 110:529–532.
9. Modest MF, Abakians H (1986) Evaporative cutting of a semi-infinite body with a moving CW laser. ASME Journal of Heat Transfer 108:602–607.
10. Tsai CH, Chen CJ (2004) Application of iterative path revision technique for laser cutting with controlled fracture. Optics and Lasers in Engineering 41:189–204.
11. Schulz W, Simon G, Urbassek HM, Decker I (1987) On laser fusion cutting of metals. Journal of Physics D: Applied Physics 20:481–488.
12. O'Neill W, Gabzdyl JT (2000) New developments in laser-assisted oxygen cutting. Optics and Lasers in Engineering 34:355–367.
13. Schuöcker D (1986) Theoretical model of reactive gas assisted laser cutting including dynamic effects. Proceedings of SPIE (International Society for Optical Engineering) 650:210–219.
14. Duley WW (1996) UV laser effects and applications in material science. Cambridge University Press, Cambridge.
15. Henyk M, Vogel N, Wolfframm D,Tempel A, Reif J. (1999), Femtosecond laser ablation from dielectric materials: Comparison to arc discharhe erosion, Applied physics A 69:1 pp: S355–S358.
16. Liu X, Du D, Mourou G; laser ablation and micromachining with ultrashort pulse, IEEE journal of electronics, vol 33:10, 1997.
17. Lamikiz A, Lacalle LN, Sanchez JA, Pozo DD, Etayo JM, Lopez JM (2005) CO_2 laser cutting of advanced high strength steels (AHSS). Applied Surface Science 242:362–368.
18. Olsen F (1981) Studies of sheet metal cutting with plane polarised CO2 laser. In: Proceedings of Laser'81 optoelectronics conference, Munich. Springer, Berlin, pp 227–231.
19. Fieret J, Terry MJ, Ward BA (1986) Aerodynamic interactions during laser cutting. Proc SPIE 668:53–62.
20. Man HC, Duan J, Yue TM (1999) Analysis of the dynamic characteristics of gas flow inside a laser cut kerf under high cut-assist gas pressure. Journal of Physics D: Applied Physics 32:1469–1477.
21. Gabzdyl JT, Steen WM, Cantello M (1987) Nozzle beam alignment for laser cutting. In: ICALEO'87 proceedings, San Diego. Springer, Berlin/IFS Kempston, pp 143–148.
22. Gabzdyl J (1996) Process assist gas for cutting of steels. Industrial Laser User Aug 23–24.
23. Arata Y, Maruo H, Miyamoto I, Takeuchi S (1981) Quality in laser-gas-cutting stainless steel and its improvement. Transactions of JWRI 10:129–139.
24. Duan J, Man HC, Yue TM (2001c) Modelling the laser fusion cutting process: III. Effects of various process parameters on cut kerf quality. Journal of Physics D: Applied Physics 34:2143–2150.

25. Tani G, Tomesani L, Campana G (2003) Prediction of melt geometry in laser cutting. Applied Surface Science 208–209:142–147.
26. Dong-Gyn AHN, Kyun-Won BYUN (2009), Influence on cutting parameters on surface characteristics of cut section in cutting of Inconel 718 sheet using CW Nd:YAG laser, Transactions of nonferrous metals society of China, vol.19, S32–S39.
27. Powell, J. (1993). *CO_2 laser cutting*. Springer, London.
28. Chen SL (1999) The effects of high-pressure assistant-gas flow on high-power CO2 laser cutting. Journal of Materials Processing Technology 88:57–66.
29. Dawson P (1996) The use of lasers in the die board industry. Industrial Laser User (3):21–22.
30. Dinda G.P., Dasgupta A.K., Mazumder J., Laser aided direct metal deposition of Inconel 625 superalloy: Microstructural evolution and thermal stability, Materials Science and Engineering A 509 (2009) 98–104.
31. Biswas R., Kuar A.S., Sarkar S., Mitra S. (2010),A parametric study of pulsed Nd:YAG laser micro-drilling of gamma-titanium aluminide, journal of Optics & Laser Technology, 42 23–31.

Chapter 8
Electrochemical Micromachining (EMM): Fundamentals and Applications

V. Rathod, B. Doloi and B. Bhattacharyya

Abstract Product miniaturization is the principal driving force for 21st century's industries because of the escalating demands for compact, intelligent, robust, multi-functional, and low cost products in all fields. As demand of miniaturized products is exponentially increasing, the need to manufacture such products from advanced engineering materials becomes more apparent. Micromachining plays significant role in miniaturization, and consist of machining different microfeatures on products. Design of microtools, tool wear, surface quality, burr and heat removal are the main challenges in various micromachining methods. Electrochemical micromachining is one of the important techniques because of its special material removal mechanism, better precision and control, environmentally acceptable, and mainly it permits machining of any metallic materials irrespective of its hardness. For better understanding of EMM process, the basic concepts such as electrochemistry, Faraday's laws of electrolysis, electrical double layer, equivalent electrical circuit, and material removal mechanism have been discussed. Significant process parameters which affect the process performance, need of EMM setup development, various subsystems, along with the challenges in setup developments, and important techniques for improving the machining accuracy have been highlighted. Machining, finishing, and surface engineering applications of EMM, as well as recent advancement in EMM for micro and nanofabrication have also been discussed.

V. Rathod (✉)
Department of Mechanical Engineering, Government Polytechnic Ratnagiri, Maharashtra, India
e-mail: vurathodju@gmail.com

B. Doloi · B. Bhattacharyya
Production Engineering Department, Jadavpur University, Kolkata 700032, India
e-mail: bdoloionline@rediffmail.com

B. Bhattacharyya
e-mail: bb13@rediffmail.com

G. Kibria et al. (eds.), *Non-traditional Micromachining Processes*,
Materials Forming, Machining and Tribology,
DOI 10.1007/978-3-319-52009-4_8

8.1 Introduction

Material removal as a means of making things dates back to prehistoric times, when ancient man started to carve wood and chip stones to make the hunting and farming tools. Human beings shaped copper by casting for metal and jewelry work before 5000 BC. Invention of copper, bronze and then iron age revolutionized the ancient civilizations. Development of modern machine tools is closely related to the Industrial revolution. Introduction of hard, temperature resistant materials in various fields of application led to the development of various non-conventional machining processes. The non-traditional processes have been developed since World War II, in response to new and unusual manufacturing requirements such as machining of complex part geometries with close tolerances, without any surface damage that could not be realized by conventional methods. There are various non-traditional machining processes, most of which are unique in their range of applications. Electrochemical machining (ECM) is one of the important non-conventional machining processes used for machining of extremely hard alloys, which cannot be machined satisfactorily by conventional methods. The metal removal phenomenon of controlled anodic dissolution was known early in 20th century, but it was until the 1960s that ECM came into use as a practical machining method. In last decades, ECM has been developed, and anodic dissolution can be effectively utilized for high-precision ultrafine machining to generate macro-as well as micro features on work pieces, due to its advantages like no tool wear, ability to machine any metallic surfaces irrespective of its hardness, stress/burr free surfaces, high MRR and accuracy, and capability to machine irregular 3D shapes.

Miniaturization is the principal driving force for 21st century's developments because of the escalating demands for compact, intelligent, robust, multi-functional, and low cost industrial products. As the demand of micro and nano products is exponentially increasing, the need to manufacture micro and nano products from advanced engineering materials becomes more apparent. Advanced micro machining may consist of machining different micro features on micro or macro work pieces. Micro features such as micro holes, micro slots, microgrooves, thin walls etc. are to be machined on different surfaces of a product to fulfill its functional requirements. Design of microtools and micro fixtures, tool wear, lack of tool rigidity, poor surface finish, scrap and heat removal etc. are the main challenges, when these micro features are machined with traditional machining techniques.

8.2 Electrochemical Machining (ECM): Basic Process

The principle of ECM was discovered long back in the nineteenth century by Michael Faraday (1791–1867). Faraday established the laws of electrolysis in 1833, which is the foundation of both the better-known electro deposition and electro dissolution techniques. In 1929, the Russian researcher W. Gussef first developed a process to machine metal anodically through electrolytic process. The significant developments in ECM occurred in the 1950s, when ECM was investigated as a method for shaping high-strength alloys. In 1959, Anocut Engineering, Chicago established the anodic metal machining techniques for commercial applications. In 1960s and 1970s the technique was applied mainly for machining of large components made of advanced and difficult-to-cut metals particularly in the gas turbine, aircraft and aerospace industries for shaping, finishing, deburring, and milling operations. Although ECM was initially developed to machine the hard-to-machine alloys, any metal can also be machined, only from the 1990s ECM was employed in applications such as automotive, biomedical, and aerospace firms, which are its major user until now.

Manufacturing processes such as electroplating, electro polishing, and other allied processes that are well established in the manufacturing industries were initiated from the basic concept of Faraday's laws. The basic mechanism of metal removal in ECM is based upon the electrolysis, in which metals are liberated from anode workpiece surface atom by atom. ECM is an anodic dissolution process in which work piece as anode and shaped tool as cathode are placed very close to each other and immersed in an electrolyte tank. When potential is applied across the electrodes, the work piece material dissolves locally so that the shape of the generated work piece is approximately negative mirror image of the tool, as shown in Fig. 8.1. The electrolyte, which is generally a concentrated salt solution, is pumped at high velocities through the machining gap in order to remove the reaction products and to dissipate the heat generated. The smaller the machining gap between preshaped tool and workpiece surface, the greater will be the current flow and rate of metal removal from the anode. It also promotes rapid generation of metal hydroxide and gas bubbles in the small gap between the electrodes and becomes a barrier to the current flow and further machining. Hence, for the continuation of the dissolution process, these products of the machining process are to be removed by circulating electrolyte at a high velocity through the gap between the electrodes. The initial gap increases in size as metal ions are removed from the anode, which increases the electrical resistance across the gap and in turn reduces the current flow. To maintain the initial current flow and rate of metal removal, the

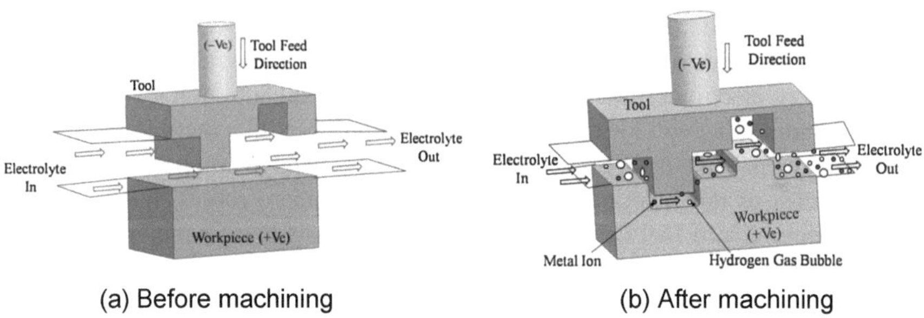

Fig. 8.1 Schematics of ECM process

gap between the electrodes should be maintained uniform by advancing the cathode toward the anode at the same rate at which the metal is being dissolved. As the cathode tool advances during machining operations the anode workpiece gradually attains a shape that is almost a replica of the cathode tool.

Controlled metallic dissolution with high accuracy is required to put the ECM technique in industry for practical applications. For this purpose the ECM setup needs electrical power source to supply the machining current to the electrodes, arrangement to circulate an electrolyte in the machining gap, and mechanical structure for regulating the tool movement towards the workpiece. In ECM, dissolution rate of anodic surface depends upon the machining current, machining time, atomic weight of the workpiece metal, and valency of the ion produced, whereas it does not depends on the hardness of material. Various process parameters such as applied voltage, machining current, electrolyte type, concentration, flow rate, and inter electrode gap (IEG) influences the major ECM machining criteria like metal removal rate, surface finish, and accuracy. The shape of the tool remains unchanged during machining, since only hydrogen gas is evolved at the cathode surface [1]. Significant technological development such as tool vibration, pulsed machining current, electrolyte microfiltration, CNC controller and application of computer aided design to predict cathode tool profiles, for improving machining accuracy and waste disposal make today's ECM technique a better choice than many other conventional as well as non-conventional machining techniques.

8.3 Electrochemical Micromachining (EMM): Focusing Area

Product miniaturization with integrated functions is the recent trend in manufacturing industries due to the multiple benefits of microproducts such as less space, light weight, less material and less energy consumption, and low cost of production etc. Microproducts has to perform multiple functions in adverse service conditions, which necessitates the micro components to be made from advanced engineering materials like super alloys, titanium, nickel, aluminum alloys, copper alloys, and stainless steel, which are hard and difficult to cut by conventional machining methods. With the development in MEMS, micro products are widely used in various fields namely electronics, aerospace, optical, biomedical, automotive, refrigeration and air conditioning etc. for fabrication of micro products like inertia sensor for air bag deployment systems in automobiles, read-write heads in computer storage systems, nozzles of ink jet printers, bio-medical filters, micro robots, micro reactors, micro pumps, micro fuel cells, and ultra precision machinery parts, etc. Different micromachining methods are available to machine the micro features on macro or micro sized components on advanced engineering materials, which can be selected based on the number of criteria's such as required machining accuracy, surface finish, type of workpiece material, production rate, and cost of production [2].

8.3.1 ECM and EMM

The term micromachining refers to the material removal of smaller dimensions ranging from 1 to 999 μm. When ECM is applied for machining of micro features in microscopic domain, it is called as Electrochemical Micromachining (EMM) [2]. EMM is an anodic dissolution process in which high frequency pulsed direct current with low voltage is applied between the metallic workpiece as anode and microtool as cathode, immersed in an electrolyte with few microns of inter-electrode gap. The anodic material dissolves into metallic ions by the electrochemical reactions generating hydrogen gas bubbles on the cathode surface. Micro features of the micro tool electrode such as shape, size and surface finish are directly transferred to the work surface, hence micro tool electrodes plays a vital role during machining of micro features by EMM. EMM is similar to the ECM in all respect, whereas these processes can be compared with each other on the basis of range of major machining characteristics as given in Table 8.1.

Table 8.1 Comparison between ECM and EMM [3]

Major machining characteristics	Electrochemical machining (ECM)	Electrochemical micromachining (EMM)
Voltage	10–30 V	1–12 V
Current density	20–200 A/cm^2	75–100 A/cm^2
Power supply	Continuous/pulsed	Pulsed
Electrolyte flow	10–60 m/s	<3 m/s
Electrolyte type	Natural salt solution	Natural or dilute acid/alkali
Electrolyte temperature	24–65 °C	37–50 °C
Electrolyte concentration	>20 g/l	<20 g/l
Size of the tool	Large to medium	Micro
Inter electrode gap	100–600 μm	5–50 μm
Operation type	Maskless	Mask/maskless
Machining rate	0.2–10 mm/min	5 μm/min
Side gap	>20 μm	<10 μm
Accuracy	±0.02–0.1 mm	±0.01 mm
Surface finish	Good, 0.1–1.5 μm	Excellent, 0.05–0.4 μm
Problems due to waste disposal/toxicity	Low	Low to moderate

8.3.2 Advantages and Limitations of EMM

The EMM process has numerous advantages and some limitations which can be categorized by product, material and machine advantages.

(a) Product Advantages
The EMM has the following advantages on the final product:

(i) Burrs free surfaces are machined by EMM.
(ii) In EMM, micro tool does not touch the workpiece surface, hence it does not cause thermal or physical strain in the product. Also, the upper layer of the workpiece surface is free from any deformations.
(iii) By designing a single tool cathode or adopting suitable microtool movement strategies, free formed 3-dimensional micro features can be processed in single step. Use of EMM techniques permits more freedom of design a micro product.
(iv) By adopting the advanced process monitoring capabilities, high dimensional accuracy and high surface quality ($R_a < 0.05$ μm) can be attained with EMM.

(b) Material Advantages

In EMM, the rate of anodic dissolution i.e. rate of machining depends only upon the atomic weight (A) and valency (z) of ions produced, the machining current passed through the electrodes, and the time (t) for which the current passes. The mechanical properties of metal such as hardness, and toughness, as well as thermal conductivity do not influence the material removal rate. The process has a complete freedom of machining the product before or after hardening step. Material removal rate is higher in comparison to the competing micromachining techniques and can be controlled by selecting appropriate process parameters.

(c) Machine Advantages
The EMM process has following advantages over other micro machining processes

(i) Initial investment on process design and micro tool fabrication is low. Also the, running cost and tooling costs are low.
(ii) The shape of the microtool remains unchanged during machining, since only the hydrogen gas is evolved at cathode microtool surface during electrolysis. Hence microtools of complicated shapes can be reused multiple times.
(iii) EMM is a process with high machining speed and relatively low cost.

(d) General Limitations

Although ECM process has following disadvantages:

(i) ECM was previously known as process that harms environment. After thorough developments in the treatment of electrolytes, the process has become less harmful to the environment. However, EMM also needs very less volume of electrolyte.
(ii) Each micro product and material requires new research. So, higher production numbers are essential for the cost effectiveness of the process.
(iii) Needs careful handling of micro tools with in situ fabrication.
(iv) EMM requires a relatively high knowledge base for operation.

8.3.3 Role of EMM in Micromachining

Micromachining is an application of specific techniques to produce micro features on macro or micro sized components with high precision and close geometrical tolerances. Micromachining technology will be an important technology in the future, since the miniaturization will continue as long as people require effective space utilization with more efficient and accurate products. Therefore attention needs to be given for further developments in traditional as well as non-traditional micro machining techniques to improve the machining precision. Machining of advanced engineering materials needs suitable micro machining techniques for the fabrication of precise and accurate micro features on products. Traditional micromachining techniques such as micro turning, micro grinding, and micro milling have dimensional limitations for the cutting tools and fixtures, high tool wear, and poor surface quality. Whereas, non-traditional micro machining methods like EDM and LBM are thermal based, and results tool wear and heat affected zone respectively. LIGA process is limited to the fabrication of 2D micro structures and chemical machining process cannot be controlled properly in micromachining domain [4]. Hence, EMM is the most suitable method for machining of micro features on difficult to cut metals because of its important benefits like better precision and control, good surface quality without affecting metallic properties of work piece irrespective of the hardness of the material, no residual stress, no heat affected zone, and no tool wear i.e. complex microtools can be used repeatedly for longer period of time [5].

8.4 Fundamentals of EMM

EMM is the application of basic ECM in microscopic domain for micromachining applications. Basically it is anodic dissolution process in which, workpiece as anode and microtool as cathode are submerged in an electrolyte with narrow inter-electrode gap, and low voltage pulsed direct current is applied in between the electrodes. The workpiece material dissolves into metallic ions by the electrochemical reactions according to Faraday's law of electrolysis. The dissolved material and other process by-products such as precipitates, sludge, hydrogen gas etc. are carried away from narrow machining zone by the electrolyte. The final shape, and size of the machined micro feature is approximately negative mirror image of the microtool electrode. Machining accuracy depends on different machining conditions in the narrow machining zone. Machining criteria like material removal rate, machined surface characteristics, and machining accuracy are influenced by the process parameters such as pulse amplitude i.e. applied voltage, pulse frequency, duty ratio, machining current, electrolyte type and its concentration, inter-electrode gap and tool feed rate etc. [6].

8.4.1 Electrochemistry of EMM

The chemical process which occurs when an electric current is passed between the two conductors immersed into electrolytic liquid solution is known as 'Electrolysis'. The metallic conductors immersed in liquid solution are termed as 'electrodes'. The electrode with positive polarity is called as 'anode' and the electrode with negative polarity is called as 'cathode'. Electrodes conduct electrical current by the movement of electrons. The liquid solution in which electrodes are immersed, and also conducts electricity through it is known as 'electrolyte'. The system of electrodes and electrolytes is called as 'electrochemical cell'. The reactions which occur at anode and cathode are called as 'anodic reaction' and 'cathodic reaction' respectively. Electrolytes carry electrical current by the movement of atoms or group of atoms, which have either lost or gained electrons, thus acquiring either positive or negative charge. Such atoms are called as 'ions'. Ions which carry positive charges and moves through an electrolyte towards cathode are known as 'cations'. Similarly, the negatively charged ions which move towards the anode are known as 'anions' [7].

When small electric current (DC) is applied across the workpiece as anode and the microtool as cathode, submerged in an electrolyte with very small inter-electrode gap, the transfer of electrons between ions and electrodes completes the electric circuit. Since the electrons cannot flow through the electrolyte, the electric current is maintained by the removal of electrons from the atomic structure of the workpiece, i.e. the current is carried by the ions. The metal material dissolves atom by atom from anodic surface and enters into the electrolyte as positive ions. The positively charged ions move through the electrolyte towards the cathode and the negatively charged ions travel towards the anode as shown in Fig. 8.2. The movement of ions is accompanied by the flow of electrons in the opposite sense outside cell and both the actions are a consequence of the applied potential difference. The least strongly bounded electrons are found at the surface of the

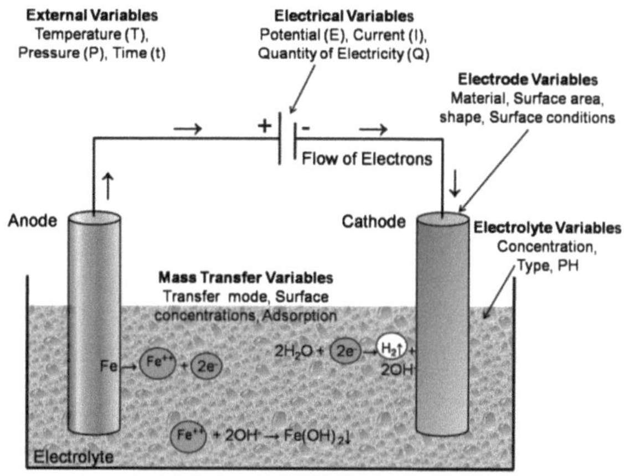

Fig. 8.2 Principle of anodic dissolution of metal

workpiece and flow into the electrolyte circuit. These electrons dissociate themselves from the workpiece and flow into the electric circuit.

In electrochemical machining, dissolved metal appears as precipitated solid of metal hydroxides. Chemical reactions occur at the cathode, and anode in the electrolyte. At cathode i.e. microtool, the reaction having the smallest oxidation potential takes place, and at anode i.e. the workpiece, the reaction having the largest oxidation potential occurs first. Hence, the factors such as (i) nature of metal being machined, (ii) type of electrolyte, (iii) current density and, (iv) temperature of electrolyte influences the oxidation potential and thus determine the kind of reaction that will occur. Different variables as shown in Fig. 8.2, also affects the rate of electrochemical reaction. The material removal in electrochemical reactions are governed by (i) electrical variables such as an applied potential, and quantity of electricity passed i.e. machining current passed for machining time, (ii) electrolyte variables like type of electrolyte, its concentration, pH, and temperature of electrolyte, (iii) mass transfer variables i.e. different modes of mass transfer such as diffusion, convection, surface concentrations and adsorption, (iv) electrode variables namely electrode material, surface area, surface conditions and geometry of electrode, (v) External variables such as room temp., pressure and machining time.

The electrochemical reactions taking place at cathode and anode during machining are as follows [8].

8.4.1.1 Cathode Reactions

Two possible reactions occurring at cathode are:

(i) Evolution of hydrogen gas,
(ii) Neutralization of positively charged metal ions.

The reactions causing the evolution of hydrogen gas at the cathode are:

$$2H^+ + 2e^- \rightarrow H_2 \uparrow \text{ (for acidic electrolyte)} \tag{8.1}$$

$$2H_2O + 2e^- \rightarrow H_2 \uparrow \text{ (for alkaline electrolyte)} \tag{8.2}$$

The neutralization of positively charged metal ion is caused by the reaction:

$$M^+ + e^- \rightarrow M \text{ (Metal)} \tag{8.3}$$

For example, when the workpiece is iron, the cathode reactions are:

$$Fe - 2e \rightarrow 2Fe^{2+} \tag{8.4}$$

$$2Fe^{2+} + 2OH^- \rightarrow Fe\,(OH)_2 \tag{8.5}$$

Metal ions form the metal hydroxides when neutral electrolytes are used. They appear as solid precipitates since these are insoluble in water. These precipitates do not affect the electrochemical reaction.

$$Fe + 2H_2O \rightarrow Fe(OH)_2 + H_2 \uparrow \tag{8.6}$$

Ferrous oxide may further react with water and oxygen to form ferric hydroxide such as:

$$4Fe(OH)_2 + 2H_2O + O_2 \rightarrow 4Fe(OH)_3 \downarrow \tag{8.7}$$

8.4.1.2 Anode Reactions

At anode also two possible reactions are occurring

(i) Evolution of oxygen or halogen gas,
(ii) Dissolution of metal ions.

The reactions leading to the evolution of oxygen gas are as follows:

$$2H_2O \rightarrow O_2 \uparrow + 2H^+ + 2e^- \text{ (for acidic electrolyte)} \tag{8.8}$$

$$4(OH)^- \rightarrow 2H_2O + O_2 \uparrow + 4e^- \text{ (for alkaline electrolyte)} \tag{8.9}$$

The reaction leading to the dissolution of the metal

$$M \rightarrow M^+ + e^- \tag{8.10}$$

$$M^{+} + (OH)^{-} \rightarrow M(OH) \quad (8.11)$$

The overall reactions occurring during ECM of iron can be given as:

$$\text{Anode: } Fe \rightarrow Fe^{++} + 2e^{-} \quad (8.12)$$

$$\text{Cathode: } 2H_2O + 2e^{-} \rightarrow 2(OH)^{-} + H_2 \uparrow \quad (8.13)$$

It has been observed that, the metal dissolution is the main or the only reaction that occurs at the anode, and the electrolyte acts as carrier of current only. The current efficiency, which is the ratio of the amount of dissolved metal to the amount that should be dissolved according to Faraday's law for the known current and time, is often lower than 100%. This is because apart from the dissolution of the metal, other anode reactions can occur, such as the oxidation of water, with the release of oxygen gas:

$$2H_2O \rightarrow 4H^{+} + O_2 \uparrow + 4e \quad (8.14)$$

The extent to which this reaction lowers the current efficiency depends greatly on the material of the workpiece, the electrolyte and the current density. Besides the oxidation of water further oxidation of metal ions can also occur at the anode.

8.4.2 Faraday's Law of Electrolysis

Michael Faraday introduced two fundamental laws of electrolysis in 1934, which governed the phenomenon of electrolysis. The most common statements of Faraday's law resemble as the following:

(i) The mass of any substance deposited or dissolved at an electrode during electrolysis is directly proportional to the amount of electricity passed through electrode. Quantity of electricity refers to the electrical charge, and typically measured in coulombs.
(ii) The amounts of different substances deposited or dissolved by the same quantity of electricity at an electrode is directly proportional to chemical equivalent weights.

Mathematically these two laws can be combined to give mass (m) removed from electrode or deposited upon the electrode, as:

$$m = \left(\frac{Q}{F}\right)\left(\frac{M}{z}\right) \quad (8.15)$$

where, 'Q' is the total electric charge passed through the metallic material, 'F' is Faraday's constant (96.485 C/mol), 'M' is the atomic weight of the substance, 'z' is the valancy number of the substance (electrons transferred per ion), and *(M/z)* is the chemical equivalent weight of the substance altered. As per Faraday's first and second law, *Q*, *M*, *F*, and *z* are the constants hence larger value of *Q*, as well as chemical equivalent weight of metal implies higher mass of material removal.

For constant current electrolysis, total electric charge passed through the metallic material can be given as:

$$Q = I.t \tag{8.16}$$

where, '*I*' is the current passed through the electrodes for time '*t*'Therefore, amount mass removed (m) can be given as

$$m = \left(\frac{I.t}{F}\right)\left(\frac{M}{z}\right) \tag{8.17}$$

When the product *(I.t)* is unity i.e. one coulomb charge is passed through the electrodes, the mass of material removed (*M/z.F*) is known as 'electrochemical equivalent of the metal'. From above equations, it can be concluded that rate of anodic dissolution depends upon atomic weight, valency of the ions produced, the current, and the time for which the current is passed. The metallic dissolution rate is not influenced by the hardness or any other mechanical properties of metal. During the electrochemical reaction only hydrogen gas is evolved at the cathode surface, hence the shape of cathode tool remains unchanged i.e. no tool wear takes place. This feature is very much useful during machining of complex micro features with micron sized tools in EMM applications.

8.4.3 Electrical Double Layer

When the metallic electrode is placed in electrolyte, an equilibrium potential difference is established between the metal electrode and electrolyte solution. This potential difference arises due to the transfer of metallic ions into the electrolyte from the metal electrode, and the simultaneous discharge of ions from the electrolyte at the electrode surface. Equilibrium is established when the electrons left in the metal contribute to the formation of a layer of ions whose charge is equal and opposite to that of the cations in solution at the interface. The negative charges of electrons just inside the metal electrode and positive charges of cations just outside the electrode surface in the electrolyte solution forms an array of positive and negative charges, as shown in Fig. 8.3, and is known as 'electrical double layer' [9]. The solution side of the double layer consists of several layers. The inner layer, which is closest to the electrode consist of solvent and other ions, known as specifically adsorbed ions. This inner layer is compact, called as Helmholtz layer, and locus of the electrical centers of this inner layer is called 'Inner Helmholtz plane (IHP)'. The locus of the centers of the nearest solvated ion is called as 'Outer Helmholtz plane (OHP)'. The interaction of solvated ion with metal electrode

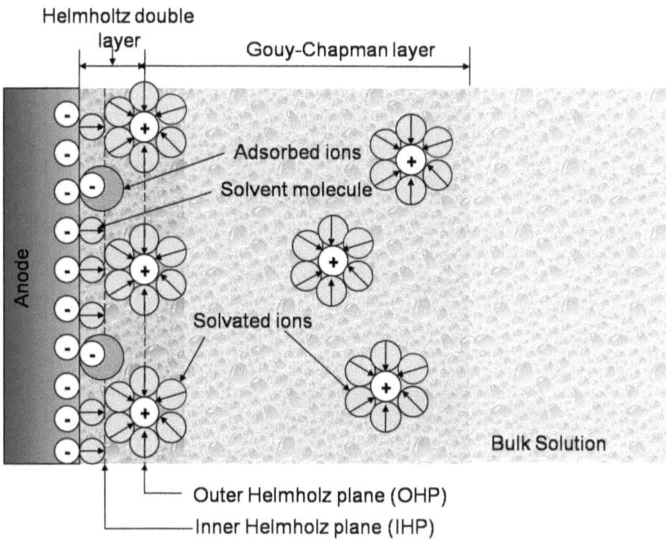

Fig. 8.3 Schematics of electrical double layer

involves only electro-static force and independent of the chemical properties of the ions. These ions are called as non-specifically adsorbed ions. These ions are distributed in three dimensional region called as diffusion layer or Gouy-Chapman layer and its thickness depends on ionic concentration in the electrolyte. The bulk electrolyte with its usual properties is located outside the Gouy-Chapman layer. The structure of the double layer affects the rate of electrode reactions. The transfer of ions will cease, when the energy required for an ion to dissolve is less than the work necessary to pass it across the double layer. The excess charge stored on both sides of the double layer depends on the electrode potential and acts like the charged plates of capacitor separated by very small distance. Therefore, double layers at the electrode-electrolyte interface can be represented by a capacitor in equivalent circuits and the interfacial capacity can be described as two capacitors in series. The resulting potential drop across the interface due to this capacitance is termed as 'double-layer potential'.

8.4.4 Equivalent Electrical Circuit

In EMM, workpiece as anode and microtool as cathode are immersed in electrolyte with very small inter-electrode gap and connected to the DC power source. The inter-electrode gap filled up with electrolyte, acts as resistance to the flow of

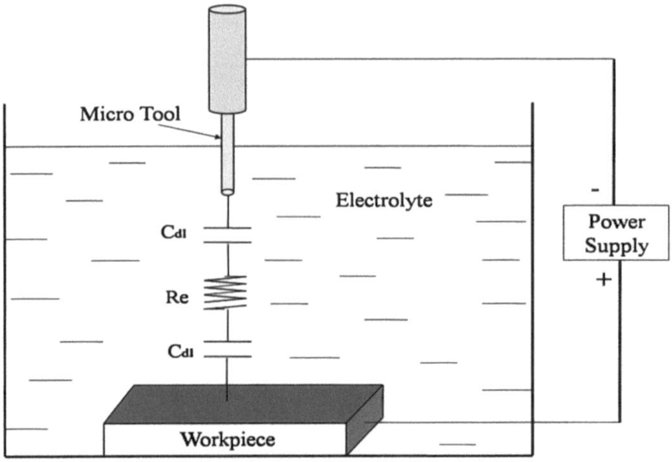

Fig. 8.4 Equivalent electrical circuit of EMM

current, and the double layers which form at electrode-electrolyte interfaces which act as capacitors.

Hence the electrodes i.e. anode and cathode, immersed in the electrolyte can be represented by the basic analogous equivalent R-C circuit as shown in Fig. 8.4, which consist of two capacitors (C_{dl}) representing double layers at both the electrode-electrolyte interfaces and resistance of the electrolyte (R_e) in between two electrodes.

At metal-electrolyte interface, the metal molecule dissolves into the elecrolyte. In the forward reaction, electrons enter the metal and metal ions diffuse into the electrolyte, called as 'charge transfer'. Whereas in backward reaction, metal ions are discharged to metal, and thus equilibrium condition is reached. For the dissolution of anodic metal, the reaction must be in the forward direction only i.e. irreversible. The resistance involved in irreversible reaction is called 'charge transfer resistance'. The electrolyte flow velocity is negligible or stagnant electrolyte is preferable in the case of EMM. So there is not sufficient transfer of mass from one electrode to the other electrode due to convection. This gives rise to another diffusion component called Warburg Impedance (R_w). Faradic reaction consists of an active charge transfer resistance (R_{ct}) and Warburg resistance. In EMM, the equivalent electrolyte resistance can be represented by considering the effect of flow of current from the lateral and longitudinal surfaces of the microtool to the workpiece. The inter-electrode gap in case of EMM is very small, hence current flow path is small across the front end of the microtool, and this electrolyte resistance is represented by R_{bottom}. Path of current flow from longitudinal surface of the microtool to the workpiece surface is much longer and this electrolyte resistance is represented by R_{side}. Therefore by considering the flow of a current along lateral and longitudinal surfaces of the microtool, their respective electrolyte

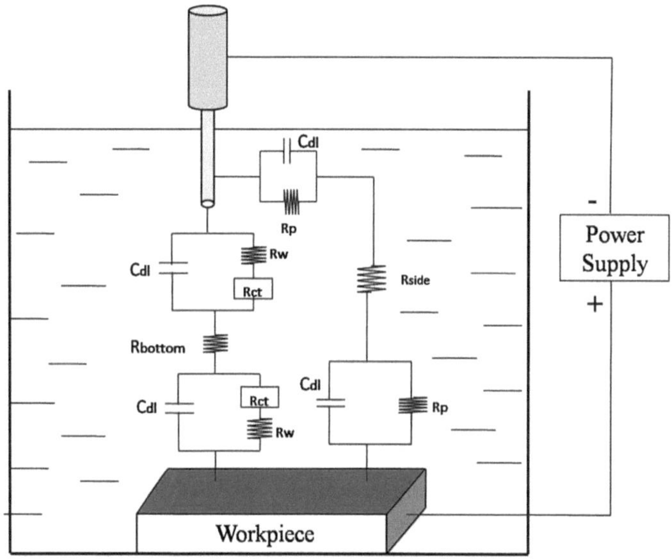

Fig. 8.5 Modified equivalent electrical circuit of EMM

resistances, and polarized resistance (R_p), the equivalent circuit of EMM can be modified as shown in Fig. 8.5 [10].

8.5 Material Removal Mechanism in EMM

In EMM, ultra short pulses of very low voltage amplitude are applied between the electrodes which are separated by very narrow machining gap. At narrow machining zone, in addition to the electrolyte resistance, other resistances which are not so prominent in conventional ECM have much more influence in EMM. Hence, all of these resistances at machining zone need to be considered in electrical circuit model of EMM. The inter-electrode gap electrolyte resistance in EMM is given as:

$$R = \frac{\rho_s . h}{A} \tag{8.18}$$

where, 'ρ_s' is specific resistance or resistivity of electrolyte, 'A' is active micro electrode surface area which is taking part in machining, and 'h' is inter-electrode gap. Faraday's two laws of electrolysis can be combined to predict the volume of material removed during electrochemical micromachining as:

$$V_m = \frac{\eta.M.I.t}{z.F.\rho_w} \tag{8.19}$$

where, '*M*' is the atomic weight of the material dissolved, '*I*' is the amount of current passed through the electrode, for machining time '*t*', 'ρ_w' is the density of workpiece material, and 'η' is the dissolution efficiency or current efficiency. The current efficiency is defined as the ratio of the actual amount of metal dissolved to the theoretical amount as predicted by Faraday's laws. The material removal rate or unit removal in electrochemical reaction basically depends on the following factors [11]:

(i) Anodic reaction and current efficiency,
(ii) Mass transport and controlled anodic dissolution, and
(iii) Current distribution and shape evolution.

(i) Anodic reaction and current efficiency

Based upon the various machining conditions and electrode-electrolyte combinations, different anodic reactions take place. Rate of these reactions depends upon the ability of the electrochemical cell to remove the reaction products as soon as they are formed, and supply of fresh electrolyte at machining zone. All of these factors influence the machining performance namely dissolution rate, shape control and surface finish of the workpiece. The current efficiency (η) of the metal dissolution is related to the weight loss (∇w), which is given as:

$$\eta = \frac{\nabla w.z.F}{I.t.M} \tag{8.20}$$

where, '*z*' is the valency of metal dissolved, '*F*' is Faraday's constant, '*I*' is machining current, '*t*' is machining time, and '*M*' is atomic weight of the metal.

Current efficiency for metal dissolution is the function of current density and local flow conditions, and varies as a function of distance from the tool. Though current efficiency is a commonly used factor in ECM, a more accurate evaluation of the process requires an estimate of electrical power efficiency. Power efficiency is the ratio between output power and input power of a device. Conductivity of electrolyte increases with the increase of temperature. Hence, for the same operating condition power requirements decreases and power efficiency increases.

(ii) Mass transport and controlled anodic dissolution

Electrolyte with negligible flow velocity or almost stagnant electrolyte is utilized in the case of EMM. Therefore, mass transport plays significant role in shaping and surface finishing of anode in dissolution process. Rate of electrochemical reactions at electrode surfaces are mainly influenced by the current passing through the electrodes, and also depends on mass transport, various surface effects, and kinetic variables. Current distribution and machining accuracy may get affected by mass transport conditions. The simplest electrode reactions are those in which the rates of all associated chemical reactions are faster as compared to those of the mass-transfer processes.

The mass transfer is the movement of material from one location to another, in electrolytic solution, and is achieved through the different modes as follows:

(a) Migration is the movement of a charged body under the influence of an electric field (electrical potential gradient).
(b) Diffusion is the movement of a species under the influence of a gradient of chemical potential (i.e. a concentration gradient).
(c) Convection is related to hydrodynamic transport. Generally fluid flow occurs because of natural convection i.e. convection caused by density gradients, and forced convection, and may be characterized by stagnant regions, laminar flow, and turbulent flow.

In EMM, smooth surface finish can be achieved only at limiting current density. An increase in current density leads to increase in the rate of metal dissolution at the anode. If the current density is too high, it may cause heating effect due to rise in temperature and it finally results improper surface finish and accuracy. The limiting current density is controlled by convective mass transfer, the anodic limiting current density 'J' is given by

$$J = \frac{v.F.D.C_{sat}}{\delta} \tag{8.21}$$

where, 'D' is the effective diffusion coefficient that takes into account the contribution from transfer by migration, 'C_{sat}' is surface concentration, and 'δ' is diffusion layer thickness.

(iii) Current distribution and shape evolution

The nature of current distribution pattern also influences the shape generation and degree of leveling in EMM. In through mask EMM, three different scales must be considered with respect to current distribution, i.e. workpiece scale or cell scale, pattern scale and feature scale. At the workpiece/cell scale, geometry of the workpiece and tool can be controlled by the current distribution. On the pattern scale, current distribution is achieved by carrying out dissolution under mass transport control. Current distribution also depends on the spacing of the features and on their geometry. On the feature scale, shape is evaluated through the current distribution. The current distribution at the anode depends on the geometry, anode reaction kinetics, electrolyte conductivity and hydrodynamic conditions.

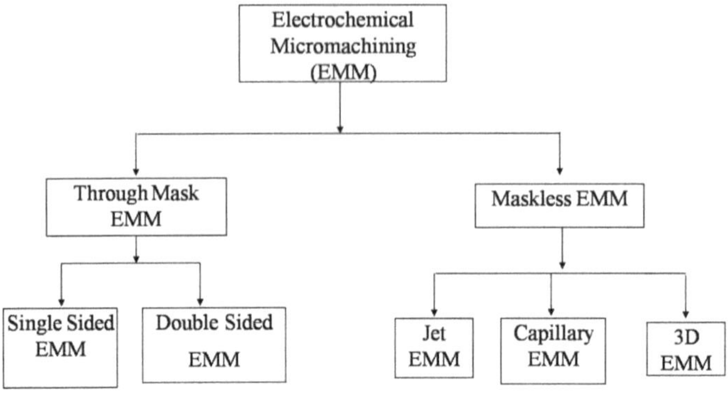

Fig. 8.6 Types of EMM

8.6 Different Types of EMM

Two-dimensional as well as three-dimensional micro features of different shapes, sizes, and surface qualities can be machined on micro or macro products by electrochemical micro-machining (EMM). In EMM controlled metallic dissolution can be achieved in microscopic domain utilizing high-frequency pulse power supply, and by the precise movement of microtool and workpiece. Flexibility in anodic material dissolution by applying the principle of electro-chemical technique is one of the main criteria to be considered during selection of the type of EMM. Depending upon the basis of the degree of localization effect of the material removal mechanism, EMM can be broadly classified into two groups as shown in Fig. 8.6.

8.6.1 Through-Mask EMM

Material removal is restricted by positioning a photoresist pattern i.e. masks on the metal surface thus, dissolution is allowed from the desired portions of the metal surface only. Through-mask EMM is more suitable for shaping and finishing of 2D micro features, as well as for fabrication of micro patterns. The photoresist is applied on metallic workpiece in required pattern, and made as anode in an electrochemical cell, so that metal surface is removed by anodic metal dissolution from the exposed surfaces only. In through-mask EMM, metal dissolution takes place at the workpiece surface that lies at the bottom of the cavity created by the photoresist mask [12]. Figure 8.7 shows the schematics for the shape evolution during through-mask EMM. Through-mask EMM process involves the careful implementation of various steps that include production of the master artwork, surface preparation, choice of suitable photoresist, and imaging. The metal removal is isotropic in nature, and hence leads to undercutting below the mask. During the designing of a photoresist mask, it is essential to have a knowledge of the metal removal rate and the undercutting of the photoresist in through-mask EMM at different electrolytes and metal combinations. The undercut and shape of the evolving surface are governed by aspect ratio, spacing to opening ratio, and film thickness ratio, etc. [13].

Through-mask EMM can be of two types (i) dissolution of the metal substrate from a single side, and (ii) simultaneous dissolution of material from both the sides. Micro nozzles can be fabricated on a thin metal foil by one-sided through-Mask EMM. Fabrication of nozzle plates by EMM involves cleaning of the metallic foil and application of photoresist on both sides of the foil. The photoresist on one side is exposed and developed to generate the initial pattern, consisting of an array of circular openings. A controlled one-sided through-Mask EMM process is employed to fabricate flat-bottomed V shaped nozzles on the sample as shown in Fig. 8.8. The taper angle of the micronozzle i.e. the ratio of the undercut (r) to thickness of metal film (b) can be controlled by regulating the EMM parameters [14]. By controlling EMM parameters, nozzles of desired shapes can be fabricated. The final nozzle shape is determined by several factors that include undercutting, etching factor,

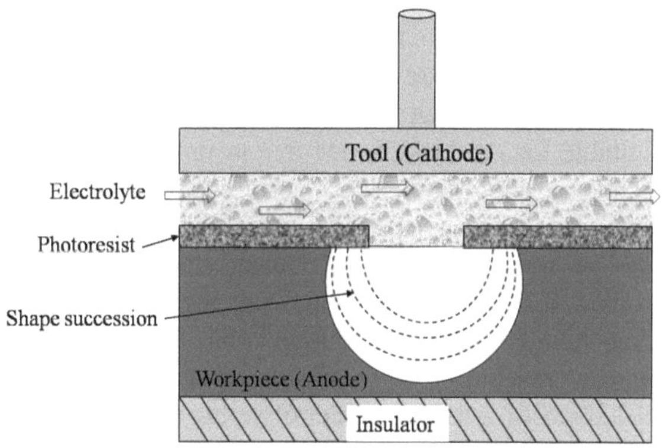

Fig. 8.7 Schematics for shape evolution during through-mask EMM

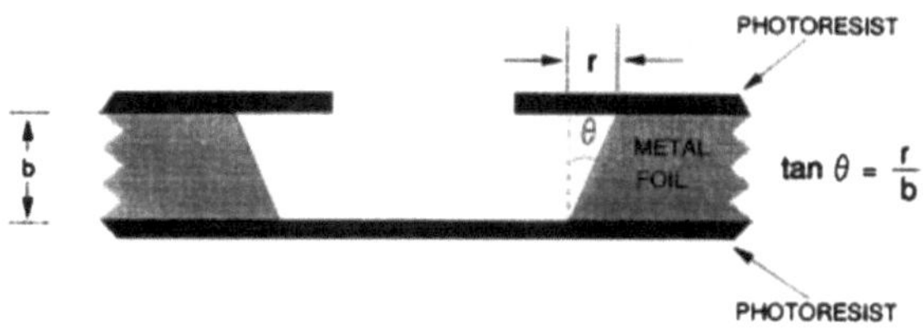

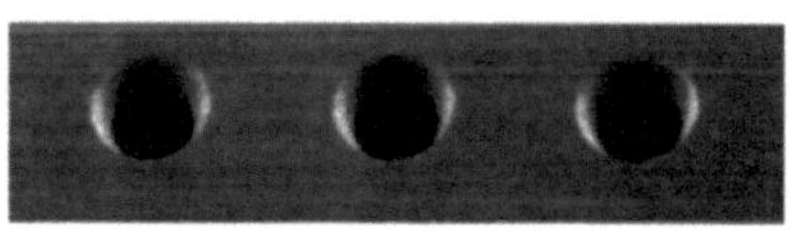

Fig. 8.8 Micronozzles fabricated by single sided through-mask EMM [11]

dissolution time, and dissolution conditions. Pulsed power supply is found to be effective in achieving dimensional uniformity of an array of nozzles during one-sided through-Mask EMM.

Through-mask EMM offers better control and flexibility for micro fabrication, as compared to chemical etching process. Higher machining rate and the use of less corrosive electrolyte are some of the advantages of through-mask EMM. Moreover, a wide range of materials, including high strength corrosion resistance alloys, can be machined by this technique. However, a limitation of through-mask EMM is the low aspect ratio of the produced micro features, due to the isotropic etching behavior.

8.6.2 Maskless EMM

Material removal from the workpiece surface is not limited by photoresist masking but is controlled by localized material dissolution mechanism. Highly localized selective metal dissolution from workpiece surface can generate the desired pattern or shape in 2D or 3D scale [15]. Anodic dissolution in maskless EMM is controlled by the current density, which depends on various predominant machining parameters. Inter electrode gap between the workpiece surface and tool is maintained at a very low value such that stray current effect is minimized. Passivating electrolyte is suitable for maskless EMM due to its ability to form transpassive oxide films and evolve oxygen in the stray current zone. To achieve highly localized anodic dissolutions, maskless EMM requires machining setup for precise movement of the tool as well as workpiece. During micromachining, to maintain and monitor the narrow IEG, the EMM setup needs highly sensitive sensing and controlling devices that should work on close loop control strategies. Circulation of fresh electrolyte at narrow machining zone is one of the important challenges in maskless EMM. Selection of a suitable electrolyte during micromachining of different materials is another essential task that influences the controlled metal removal process at different metal–electrolyte combinations. Another significant problem is the removal of machining by-products such as sludge, gas bubbles, and heat generated at fine IEG. For effective localization of metal removal, maskless EMM demands a higher frequency pulsed power supply.

Maskless EMM is competent to machine a high aspect ratio micro feature, which is one of the main limitations of through-mask EMM process. One of the main advantages of maskless EMM is the machining of 3D micro features. Different types of micro features can be fabricated by maskless EMM by exploiting its flexibility of material removal mechanism. It can be utilized to generate microstructures starting from a very low depth to very high aspect ratio, including surface structuring, micro patterns, deep micro holes, as well as 3D micro features. Depending on the depth of machining and geometrical complexity of the micro features, different techniques of maskless EMM may be selected. Maskless EMM can be classified into three categories, for example, Jet EMM, micro drilling, and

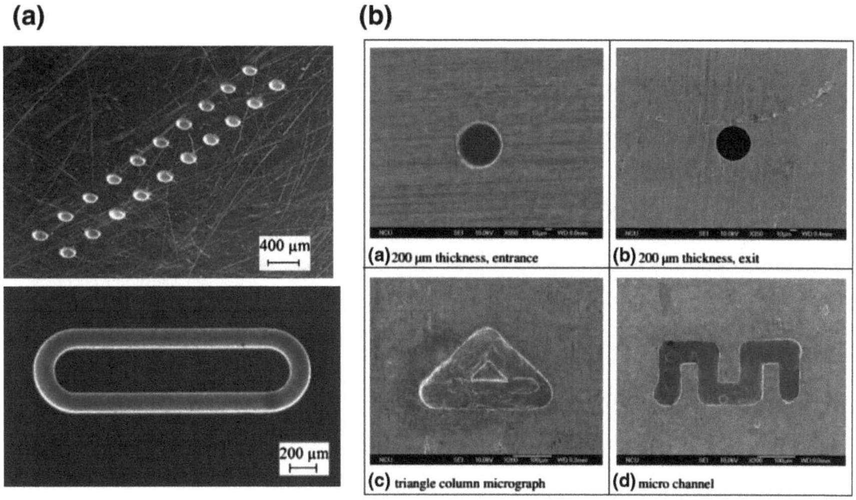

Fig. 8.9 Microholes and micro patterns machined by **a** jet EMM [15], **b** micro drilling, and 3D EMM [16]

3D EMM. Figure 8.9a shows micro hole array and micro pattern machined by jet EMM on stainless steel foil of 250 µm thickness with a nozzle diameter of 100 µm, voltage 56 V, 5 M $NaNO_3$ as an electrolyte, with at a feed speed of 300 µm/s. The cavity has a geometries consisting of lines having length of 1500 µm and semi-circles with a radius of 250 µm, depth of about 180 µm and width of approximately 190 µm. Figure 8.9b shows the micro hole and micro patterns machined on nickel at 1 MHz pulse frequency, 5% NaCl + 0.3 M HCl electrolyte concentration, 3 V applied voltage, 30 ns pulse duration, 50 µm tool diameter [16].

8.7 Important Process Parameters of EMM

Basic mechanism of material removal in EMM is based on anodic dissolution, in which metals are liberated from workpiece surface atom by atom. Various process parameters, namely applied voltage, machining current, electrolyte type, concentration, flow rate, inter-electrode gap (IEG), etc., influence major machining criteria of EMM such as metal removal rate, surface finish, profile accuracy, etc. [17]. In order to achieve the effective and high precision of machining in microscopic domain the process parameters needs to be controlled optimally [18, 19]. The machining accuracy and process performance can be improved by selecting and regulating the appropriate process parameters at narrow machining zone.

Figure 8.10 shows a fishbone diagram, also called as Ishikawa diagram or cause and effect diagram, indicating the influential EMM process parameters. Some of the

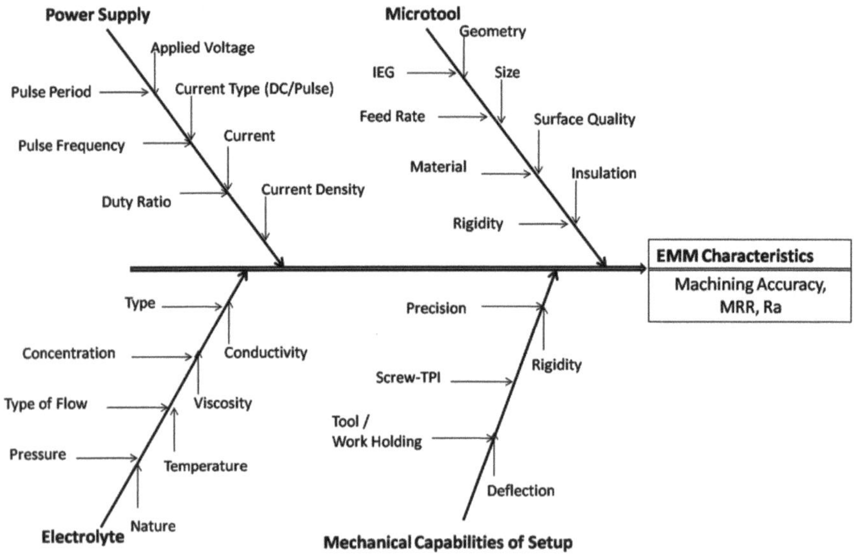

Fig. 8.10 Cause and effect diagram for EMM

predominant process parameters, which have major influences on EMM criteria are identified as follows:

(i) Nature of power supply i.e. DC or pulsed DC, its amplitude, and pulse frequency etc.
(ii) Microtool parameters like shape, size, surface quality, IEG, tool feed rate.
(iii) Electrolyte parameters such as type of electrolyte, its concentration, temperature, flow rate, and density.
(iv) Mechanical capabilities of machining setup such as resolution, rigidity, damp proof structure etc.

8.7.1 *Nature of Power Supply*

Electrical power is the driving force for the flow of charged particles inside the electrolyte, which is essential for continuation of electrochemical reaction. The applied power supply may be direct current or pulse direct current. With direct current supply workpiece material dissolves continuously. It may give very high concentration of reaction products, which can only be partly removed by the electrolyte, especially if the inter-electrode gap is very narrow. The increasing contamination can cause a deposit to form on the microtool surface, so that the workpiece material no longer dissolves uniformly. Furthermore, changes in the electrolyte composition rise in electrolyte temperature, and increase in electrical resistivity can also affect the machining accuracy. These problems can be largely minimized by applying the pulsed direct current instead of continuous one [20]. Pulse duration and the intervals between the pulses are properly matched to the current density, so that the machining gap can be completely swept clean during machining resulting in a continuous EMM process. The pulse off-time should be long enough to ensure a complete flushing of the electrolyte in narrow machining gap. The current efficiency is much more dependent on the current density when pulsed current is used than the use of continuous current. With continuous supply, the efficiency decreases gradually when the current density is reduced, where as with pulsed supply the decrease is much more rapid. A steep fall in efficiency with decreasing current density improves the machining accuracy. This improvement depends upon the pulse duration and to somewhat lesser extent on the interval. By using pulsating current one can apply extremely high instantaneous current densities to the workpiece without the need for an elaborate electrolyte pumping system and rigid machine frame. This is possible since each current pulse is followed by a relaxation time of zero current, which allows removal of reaction by-products and heat from the inter electrode gap. Compared to direct current dissolution where only average current can be chosen, pulsating current has three parameters, i.e. pulse on-time, pulse off-time, and peak current density which can be varied independently in order to achieve a desired machining rate. By proper selection of these parameters, it is possible to minimize variations of electrolyte conductivity in the machining region and to achieve high instantaneous mass transport even at low electrolyte flow rates. On the other hand, the average current density in pulse EMM is lower than the direct ECM. Pulsating current is therefore, particularly suitable for high precision micromachining of delicate micro features where high electrolyte flow velocity cannot be tolerated. Shorter pulse period is preferable for achieving higher accuracy. Anodic dissolution becomes more localized, the throwing power is more restricted and the machining accuracy is improved with the application of shorter pulses.

8.7.2 Microtools for EMM

In electrochemical micromachining, the machined micro features are the mirror image of the microtool, and somewhat greater than the microtool size called an oversize or overcut. Hence, in machining of micro features by EMM, shape and size of the microtool plays an important role, as the geometry of machined feature is dependent on the geometry of microtool. The design of microtool depends upon microtool geometry, as well as suitable microtool material. The development of microtools for micro manufacturing is still a challenging task, and mainly deals with (i) selection of suitable microtool material, (ii) determination of the accurate microtool geometry, and (iii) fabrication of the microtools for actual machining. Microtool handling is also an important issue to be considered, especially when the microtool are fabricated by one machining method and to be used by another machining method for actual micro machining applications. Some of the important issues related to the micro tool are as discussed below:

(a) Microtool material

EMM demands micron sized tools to be used in working conditions like extremely corrosive environment, at elevated temperature due to Joules effect. Hence, microtool materials should have properties like high electrical and thermal conductivity, good wear and corrosion resistance, mechanical strength i.e. stiffness to withstand the pressure of electrolyte, and be easily machinable. Considering these requirements of the microtool materials, the microtools are generally made of chemically inert materials. The best choice of the materials limits to tungsten, platinum, titanium, and some super alloys. Other metals, such as gold, nickel, copper, silver, molybdenum, and steel are also used as electrode materials in connection with specific applications. Among a wide choice of metals available, tungsten and its alloy, platinum and its alloy, and titanium are most widely used as metallic electrodes. Such electrodes offer a very favorable electron-transfer kinetics, large anodic potential range and low hydrogen overvoltage. Platinum wire is a soft metal but alloying with iridium and rhodium increase its hardness. Tungsten is widely used as tool material because of its important properties like high rigidity, toughness and resistance to chemicals [21].

Tungsten carbide (WC) with cobalt binder is also used as tool material. Titanium and its alloys have also been used in electrochemical micromachining due to the high strength, high melting point and has proven suitable for harsh environment where the corrosion resistance plays a major role. Tungsten wires are commercially available in a long pieces and cutting-off the tungsten wire with wire cutter or micro grinder may leave burrs or micro-cracks. Therefore, the cut ends should be grind and clean in order to get a good microtool tip shape. No single material has been proven to be superior over all others, no single material can satisfy all the microtool design purposes, since each material has its benefits and limitations.

(b) Microtool shape, size, and surface finish

In EMM, geometry i.e. shape of the microtool plays an important role in machining accuracy, as well as machining performance. Microtool size is also an important parameter to be considered in EMM, which helps to improve the machining accuracy, as well as machining performance [22]. Microtools of higher diameter increases the effective surface area of microtool, increasing machining current and tool polarization area, and finally generates micro features of higher overcuts. Use of micron sized tool, not only generates accurate micro features but also takes less time to machine the micro features [23]. Microtools with different end shapes have already proven their effectiveness in micromachining applications. Cylindrical microtools with flat end, conical end, reverse conical end, spherical end, and disc shaped microtools, can be developed by different methods and utilized in EMM [24]. Use of disc shaped microtool generates straight walled micro features of higher depths, with minimum stray current effects. Surface quality of the microtool also plays an important role during machining of micro features by EMM.

The minor defects on the surface of microtool, directly affects the surface quality of machined micro features [25]. This can be observed on features machined with a microtool having a surface not finished to a high degree of surface smoothness. The presence of nicks, notches, scratches, lines, burrs, or other similar type of surface defects will be reproduced as the mirror image on the work surface. Therefore, special care must be taken to maintain the surface of the microtool very smooth.

(c) Microtool insulation

While machining micro features of high aspect ratio by EMM, workpiece material dissolves from the front end, as well as the lateral surface of the microtool. This results a micro features whose entry side is wider than exit side i.e. taper along the depth and also poor surface quality due to stray current effects [26]. Therefore sidewalls of the microtools are insulated to minimize the stray current effects and taper formation along the walls of the micro features.

(d) Microtool feed rate

In EMM, material dissolution depends upon machining time also. Tool feed rate regulates the time for which microtool is available at particular position during machining of micro features, hence microtool feed rate is one of the important process parameter in EMM. The micro tool feed rate should be always less than material removal rate to avoid short circuit during machining, since short circuit can seriously damage both the microtool and delicate surface of workpiece. The maximum value for the microtool feed rate for a particular machining condition can be determined depending on the occurrence of sparks or short-circuits between microtool and the workpiece during machining operation.

8.7.3 Electrolytes for EMM

Electrically conductive solution which carries electricity through ions is termed as electrolyte. The electrolyte not only completes the electric circuit between the tool and workpiece, but also allows the desired machining reactions to occur. The conductivity of an electrolyte solution depends on the concentration of the ions and behaves differently for concentrated and dilute electrolytes. The electrochemical processes take place at the interface of electrode-electrolyte solution, usually bulk solution. In electrochemical cell, the electrode potential is used to dissolve workpiece materials. The mass transfer inside the electrochemical cell i.e. the dissolution of material depends on the hydrodynamic conditions for a given metal-electrolyte combination, and the dissolved mass transfer occurs in three different modes namely diffusion, convection, and migration. In electrochemical micromachining processes, the migration mode dominates the process mostly when very low concentration electrolytes are used, and the use of electrolyte circulation system is very limited. Electrolyte selection is an important task in EMM [27], and following points needs to be considered before electrolyte selection for specific application.

(a) Types of electrolytes

An electrolyte contains free ionic species that make it electrically conductive. Depending upon pH value electrolytes can be classified into three categories namely acidic, neutral, and alkaline. pH is the negative logarithm of hydrogen ions concentration in the solution. If the electrolyte solution has pH value equal to 7 then it is neutral solution. If the electrolyte solution has pH value less than 7 then the solution is acidic, and if the pH value exceeds 7, then electrolyte solution is alkaline. Neutral electrolytes such as NaCl, $NaNO_3$ are commonly used in EMM. However, for microhole drilling acidic electrolytes are preferred. Electrolytes can also be classified based on the degree of dissociation and nature of passivation factor. The term degree of dissociation of an electrolyte is the fraction of solute which is dissociated into ions that are free to carry current at a given concentration. Electrolytes are also classified as strong electrolyte and weak electrolyte based on degree of dissociation of solute. Strong electrolytes dissociates greatly for concentration ranging from very low to high values, whereas, for weak electrolytes dissociation of an electrolyte tends to unity at very low, limiting concentration, and reduces to about zero at high concentration. Strong electrolytes such as NaCl are generally preferred for anodic dissolution. Whereas the electrolytes based on passivation are classified into two categories: passivating electrolytes containing oxidizing anions, i.e., sodium nitrate, sodium chlorate, and nonpassivating electrolytes containing relatively aggressive anions such as sodium chloride. Passivating electrolytes are known to give better machining precision. The pH value of electrolyte solution is chosen to ensure the good dissolution of the workpiece material during EMM process without affecting the microtool. In some applications acidic electrolytes are preferable for EMM because it produces soluble reaction products which can be removed easily from narrow inter-electrode gap without affecting the micro features of microtool. Although the precipitate has no direct effect on the process, it definitely increases the possibility of damage of microtool from short circuit. Hence it is advisable to use fresh and clean electrolyte for micromachining instead of re-circulation. For electrochemical micromachining of various metallic materials different electrolytes or mixed electrolytes are suitable. For machining of aluminum and aluminum alloys, NaCl or $NaNO_3$ is suitable; for copper, copper alloys, and tungsten, NaOH can be used; for titanium and titanium alloys, electrolytes such as NaBr, NaCl or mixture of NaCl and $NaNO_3$ is suitable, for Stainless steel, electrolyte such as H_2So_4, NaCl and $NaNO_3$ can be used; and for machining of tungsten carbide, mixture of NaCl, NaNOH and triethanolamine can be used [10]. Machining accuracy can be improved by using additives such as $NaHSO_4$ in the electrolyte. Also to prevent to prevent formation of metal hydroxide precipitates, complexing agents such as citric acid are added in the electrolyte.

(b) Electrolyte properties

The electrolyte at narrow inter-electrode gap facilitates the electrochemical reaction. Therefore the electrolyte solution should:

(i) Ensure a uniform and high speed anodic dissolution,
(ii) Avoid the formation of a passive film on the anodic surface,
(iii) Not deposit on the cathode surface,
(iv) Have a high electrical conductivity and low viscosity to reduce the power used,
(v) To have good flow conditions in the narrow inter-electrode gap,
(vi) Be safe, nontoxic, and less erosive to the machine body,
(vii) Maintain its stable ingredients and PH value during machining,
(viii) Have minimum variation in its conductivity and viscosity due to temperature rise during machining,
(ix) Be less expensive and easily available,
(x) Possess less throwing power apart from basic properties like good chemical stability, high electrical conductivity, low viscosity, non-corrosive and inexpensive to increase the machining accuracy.

Apart from facilitating for electrochemical reactions, electrolytes have to perform various functions such as:

(i) Create an environment for anodic dissolution of workpiece material,
(ii) Conduct the machining current,
(iii) Remove the process by-products formed during machining,
(iv) Carry away the heat generated at narrow inter-electrode gap, during machining to maintain the constant temperature.

(c) Working life of electrolytes

The composition of the electrolyte begins to change with the progress of time during electrochemical machining. The major changes that may occur and their effects are enumerated as follows:

(i) Loss of hydrogen, which may cause a reduction in electrical conductivity of the electrolyte and increase its pH value.
(ii) Loss of water, either by evaporation or carried off by evolved hydrogen gas, which may increase the concentration of the solution and thus may affect its electrical conductivity and its viscosity.
(iii) Formation of precipitate, which will reduce the concentration of the electrolyte and may affect its electrical conductivity,
(iv) Metal ions from the anode may pass into the solution and may be deposited on the cathode.

The aforesaid changes mean that the electrolyte has got finite life and in practice the life may be limited because of the following reasons:

(i) The need to maintain a reasonably constant electrical conductivity so as to facilitate the control over the process and to ensure the machining accuracy.
(ii) The need to prevent plating out of dissolved material on the microtool surface to ensure the machining accuracy.
(iii) The need to avoid formation and accumulation of excessive quantities of precipitate and sludge at narrow machining zone.

The first of the aforesaid consideration applies to all type of electrolytes, the second consideration applies mainly to acidic electrolytes and the third one applies mainly to neutral type of electrolytes.

(d) Electrolyte concentration, temperature and flow

(i) Electrolyte concentration

The electrolytes in electrochemical cell carry electrical current by the movement of ions and the number of ions available for electrochemical reaction increases with increases in concentration. Therefore increased concentration of an electrolyte offers low resistance to flow of current resulting increased electrical conductivity. The magnitude of conductivity is determined by the type and number of ions present in the electrolyte. Hence, to compare the conductivities of the different electrolytes, the term electrolyte concentration is generally used, which can be applied to all electrolytes. Electrolyte concentrations are measured as weight of the solute per unit (w/w) or weight per unit volume (w/v) of solution, or the volume of the solute per unit volume (v/v) of the solution. Molecular weight in grams i.e. gram-molecules, or moles of solute per liter of solution (M), is generally used to measure the concentration of the electrolytes. Electrical current is carried out by the movement of ions in electrochemical cell. The rate of ionic movement is termed as the ionic mobility. Increase in electrolyte concentration increases machining current due to reduced electrolyte resistance, however further increase in electrolyte concentration reduces the electrical conductivity due to reduced ion mobility. Also increased density of electrolyte at very high electrolyte concentration makes it difficult to remove the process by-products from narrow machining zone. Therefore dilute electrolytes are preferred in electrochemical micromachining of micro features [28].

(ii) Electrolyte temperature

During electrochemical micromachining of micro features, electrolyte concentration and temperatures are expected to be constant throughout the process for machining the micro structures with uniform features. Temperature of electrolyte at narrow inter-electrode gap increases due to Joule's heating effect. Electrical conductivity of an electrolyte increases with increase in temperature because of increase in mobility of ions at increased temperature. Finally, increase in machining current because of reduced electrolyte resistance may result micro features with varied characteristics like increased surface roughness, reduced machining accuracy, and higher MRR.

Therefore, maintaining constant temperature of electrolyte during machining of micro features has its own importance.

(iii) Electrolyte flow

Machining of precise micro features by EMM demands micron sized tools with few micron of inter-electrode gap during machining. As per the requirements of the EMM process, fresh electrolyte needs to be supplied continuously at narrow inter-electrode gap by flushing out used electrolyte, for effective machining of micro features. Size of the microtools and IEG of few microns limits the movement i.e. flow rate of an electrolyte during machining. Higher flow rate of an electrolyte may vibrate the microtool in turn results in reduction of machining accuracy or result short circuit due physical contact with workpiece. Therefore, generally steady electrolytes are used in machining of micro features by EMM, since very small amount of material is removed precisely during micro machining [29].

8.7.4 Mechanical Capabilities of Setup

Mechanical capabilities of machining setup also play an important role in machining of accurate micro features. Resolution of linear stages representing X, Y and Z axes directly affects the machining accuracy of machined features. Arrangement for rotating the microtool, as well as an attachment of microtool vibration system enhances the supply of fresh electrolyte at machining zone, as well as helps to remove the machining by-products from narrow machining gap, and improves the machining accuracy of EMM [30]. Hall current sensor for detecting the short circuit is very much useful in detecting the physical contact of the microtool with workpiece surface, which prevents the damage of the micron sized tool or precious workpiece surface [31]. CNC controller for synchronizing the movements of linear stages representing X, Y and Z axes facilitates the machining of complex or free-formed surfaces or micro structures. Damp proof structure helps to improve the machining accuracy by isolating the machining setup from the mechanical vibrations generated in surrounding.

8.8 EMM Setup Development

8.8.1 Need of Setup Development

The escalating demand of microproducts in various fields have forced manufacturing industries to fabricate microproducts of best surface qualities from advance engineering materials, in very short period of time. In recent years, applications of microproducts have been considerably increased in electronics, optical, medical, automotive, aerospace, and telecommunication fields. The demand for microproducts will increase exponentially in the coming years also due to the vast applications of microproducts and global competition among industries. Electrochemical micromachining offers unique advantages over competing micromachining technologies which has a great potential to machine microfeatures like surface structuring, patterning, microholes, microgrooves, microchannels, microcavities, and 3D microstructures of complex shapes with a high aspect ratio for various industrial and household microproducts. To fulfill the increasing market demand of micro products, an industrial EMM machine needs to be developed, which will be robust, will have a low price, maintenance free, and will have a higher production rate.

8.8.2 Challenges in Setup Development

Anodic dissolution of workpiece material by electrolysis demands several technical specifications of the setup. Precise microtool movement, narrow IEG and regulating it uniform throughout the machining, arrangement and rigidness of mechanical structure, power supply requirements, in-process monitoring, microtool and workpiece holding arrangements etc. are important criteria to be considered for the effective utilization of EMM. There has been few micromachining systems developed in research institutes, academic universities, as well as by commercial companies. Some of them are multipurpose systems or dedicated systems for EMM as well as for micro electro-discharge machining. Many research papers describes the machining capabilities of EMM for microtool fabrication, machining of micro holes and microstructures, still as of today there is no commercial, ready to use EMM machining set up for micromachining applications. This is because of the electrochemical micromachining parameters which do not remain same for the different electrode-electrolyte combinations, and required shape and size of micro features and machining accuracy etc. Therefore for machining specific micro feature of desired machining accuracy, specific electrode-electrolyte combination is required. Also various microtool movement strategies namely layer by layer machining, milling, sinking and milling, wire cut EMM demands different arrangements of machining setup [32]. Therefore, EMM setup development is facing the problem of combining the various requirements in single machine setup, and is under the developmental stage. Smaller inter-electrode gap in the range of few microns is preferred in EMM, since machining localization improves with reduction reduced IEG. Whereas, accumulation of sludge and generation of bubbles at narrow IEG hinders the EMM process that may reduce IEG gradually causing short circuit microtool damage. It is very much difficult to maintain the narrow IEG throughout the machining. Therefore, regulation of minimum IEG during machining demands online monitoring of the process to avoid short circuit, physical contact of the microtool with workpiece [33].

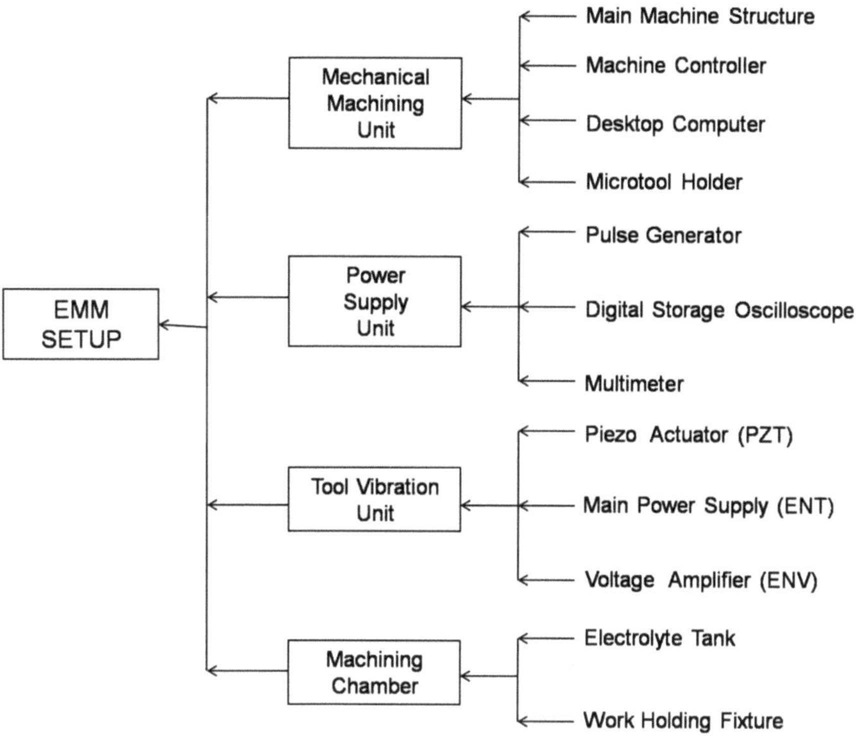

Fig. 8.11 Various subsystems of the developed EMM setup

8.9 EMM Subsystems

Controlled machining of micro features by EMM necessitates different tasks to be done simultaneously in-tune with time or in coordination with each other. Hence, to develop an electrochemical micromachining system setup, various subsystems along with their sub-components such as mechanical machining unit, power supply unit, microtool vibration unit and machining chamber along with work holding fixture have been interconnected as shown in Fig. 8.11. Individual subsystems are described in detail in succeeding sections.

8.9.1 Mechanical Machining Unit

The basic function of mechanical machine unit is to provide precise movements to the tool or workpiece and to provide mechanical support to tool holder, machining chamber etc. The mechanical machining unit of the developed EMM system setup

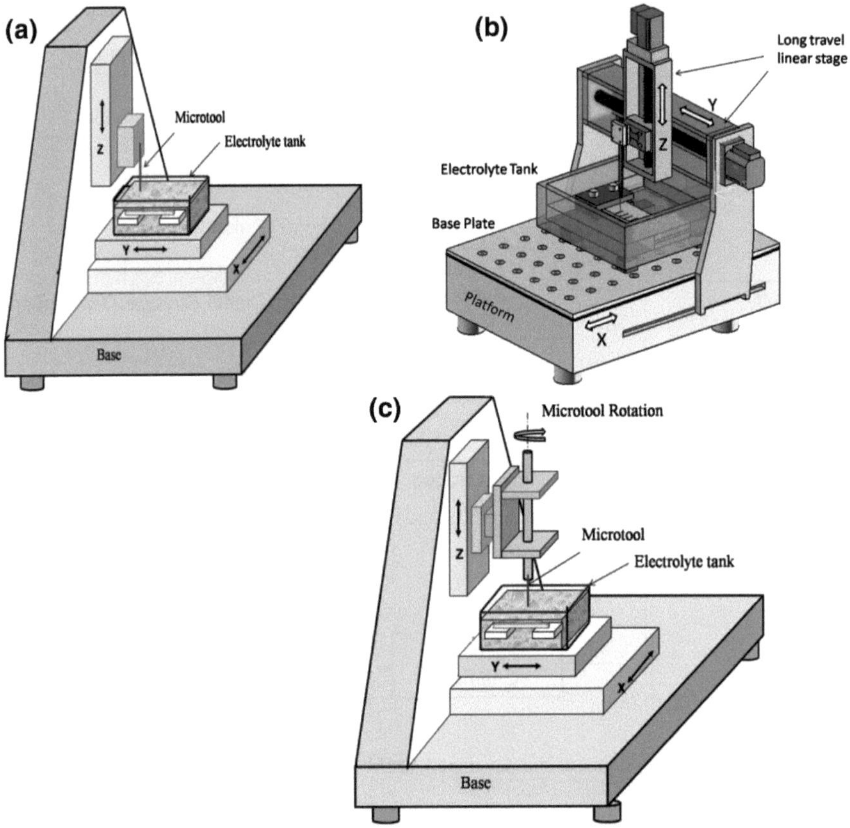

Fig. 8.12 Different types of Mechanical Machine units **a** column and Knee type, **b** gantry type, and **c** structure with rotational axis

comprises of the various machine elements like main machine structure, machine controller unit, and desktop computer with graphics user interface software.

(a) Main machine structure

Main machine structure provides rigid support to various components during machining, and space to mount the different components such as base plate, linear stages, tool holder at required positions. The main structure comprises of platform with support, vertical or horizontal supports, Stepper motor or servomotor operated long travel linear stages with precise movements to represent X, Y, Z axis. Figure 8.12 shows the different combinations of linear stages to fabricate mechanical machine unit.

(b) Machine controller unit

Machine controller unit controls the long travel linear stages through stepper motor or servomotor movements. It also synchronizes the movements all stages according to instructions of the program for a specified path. It is interfaced in between desktop computer and mechanical machine setup. Various feeds can be given to all or any of the three motors at a time through a motor controller unit using position controller software or labview programs installed on a desktop computer. The software enables the user to program movement of each stages in the form of commands executed sequentially. For machining of complex microprofiles and 3D microstructures, Computer Numerical Controlled (CNC) controllers are preferred.

8.9.2 Power Supply Unit

Power supply unit is very important unit in EMM system setup, since the nature of pulse and pulse parameters directly affect the machining accuracy. Machining takes place during pulse 'on' time only, and sludge removal from narrow machining zone mainly takes place during pulse 'off' time, therefore pulsed DC supply with short pulse period is preferred to machine the micro features by EMM. Better control over process needs pulse monitoring and control of the pulse parameters during machining operation. Main components of the power supply unit can be elaborated as:

(a) Pulse generator

Pulse generator generates DC pulses of required parameters and is the heart of power supply system. Continuous supply of the stable pulse patterns during machining is the prime function of a pulse generator. Various pulse parameters such as pulse period, pulse amplitude, pulse 'on-time' and 'off-time' i.e. duty ratio, pulse rise/fall times, positive or negative bias, as shown in Fig. 8.13 can be adjustable from a minimum to maximum value, resulting in different outputs for the same frequency.

(b) Digital storage oscilloscope

Digital storage oscilloscope provides the online image of the supplied pulse along with the detail information during machining. Digital storage oscilloscope can also be used during initial IEG setting and to monitor the machining conditions at narrow machining zone to detect the occurrence of short circuit if any between electrodes. As soon as the short circuit is detected, microtool feed and power supply is turned off, and microtool is retracted back by few microns to clear the IEG, and machining is continued again with required parameter settings. Digital storage oscilloscope can also be interfaced to the desktop computer for online monitoring, and pulse parameters can be directly stored using external storage devices for further analysis.

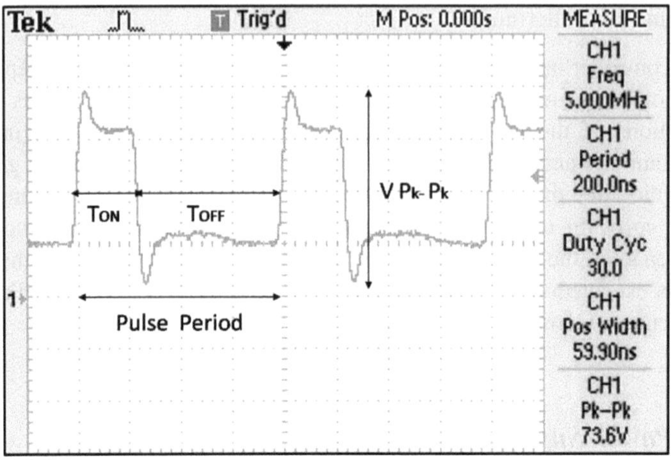

Fig. 8.13 Pulse parameters

8.9.3 Microtool Vibration Unit

Main objective of vibration system is to provide vibrations to the microtool to improve the machining performance by enhancing the availability of fresh electrolyte at narrow machining zone by displacing machining by-products, especially during machining of deep micro features. The microtool vibration produces pressure waves which help in flushing away the reaction product from the machining zone leads to generation of better quality micro features [34]. Microtool vibration system may consist of Piezoelectric transducer (PZT) actuated compact single axis translational stage and voltage amplifier system.

8.9.4 Machining Chamber

It is the space where electrochemical reaction takes place in between anode and cathode that are submerged in an electrolyte with very small IEG. It consists of an electrolyte tank and workpiece holding fixture. Material used to fabricate the electrolyte tank must be non-conductive, non-corrosive, light in weight, transparent, machinable into the required size. Perspex is one of the most suitable materials to satisfy all of these requirements [35].

8.9.5 *Process Monitoring and Control*

While fabricating micro features of desired dimensions by EMM, its process parameters, path of machining, microtool movement and control strategy has to be predetermined before machining. During machining, a micron-sized tool electrode moves with a constant feed rate toward the workpiece to maintain the IEG. Based on the machining strategy, the microtool tracks the scheduled path to fabricate the desired microfeatures such as microhole, microgroove, etc. Machining strategy followed during machining directly influences the machining accuracy, as well as process performance. IEG controls the dimensional accuracy of the machined feature in micro-machining. Minimum IEG needs to be maintained throughout the machining, for continuation of the process effectively. Accumulation of sludge and generation of bubbles at narrow IEG hinders the EMM process, and may reduce IEG gradually causing short circuit. Smaller IEG in the range of few microns results higher machining accuracy, whereas it is very difficult to maintain the narrow IEG throughout the machining. Physical contact of microtool with workpiece surface during machining may lead short circuit and may damage microtool or workpiece surface also. Sparking phenomena may damage both the microtool and the workpiece. When the tool feed rate is smaller than the material removal rate, the IEG increases gradually, and generates a higher machining gap that may affect the machining accuracy. Therefore, to maintain a constant IEG, the microtool feed rate should synchronize with the material removal rate, known as equilibrium speed. While machining microfeatures by EMM, the material removal rate varies according to various process parameters as well as the equilibrium speed of the microtool. Hence, to maintain a constant IEG during micromachining, different strategies have been developed and adopted by researchers. Various IEG control strategies for EMM adopted by researchers are reported from Figs. 8.14, 8.15, 8.16 and 8.17 as follows.

8.10 Accuracy Improvement Techniques in EMM

To exploit full potential of EMM, research is still needed to improve the machining accuracy by controlling different factors such as the effect of overcut and taper formation during the machining of various micro features. Geometrical shape, size and surface quality of the microholes, microslots and grooves affect the performance and service life of the various micro components. Hence, fabrication of micro features consisting of various types of microholes; microgrooves and slots with good shape accuracy and surface quality are the focused areas of research in EMM. To improve the machining accuracy of EMM, the control of stray current phenomenon is one of the major challenges. The proper removal of sludge, precipitates and gas bubbles from the very narrow machining gap in EMM is another major challenge to the researchers. Because of the improper flushing of machining

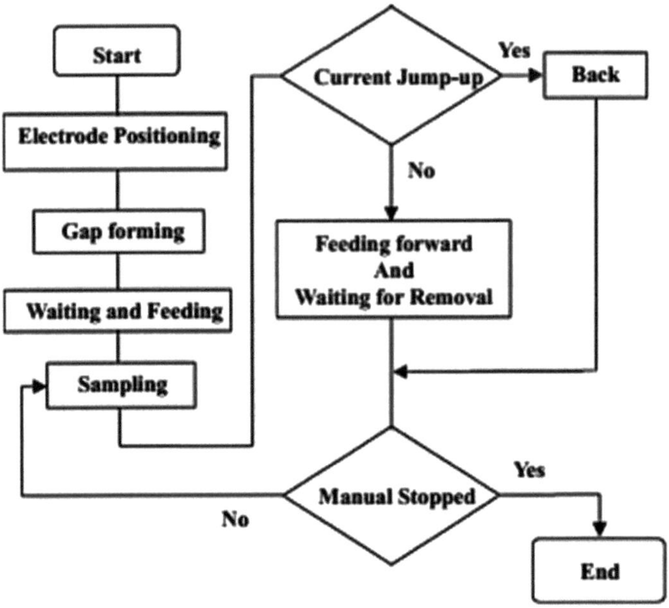

Fig. 8.14 Flow chart for IEG control [36]

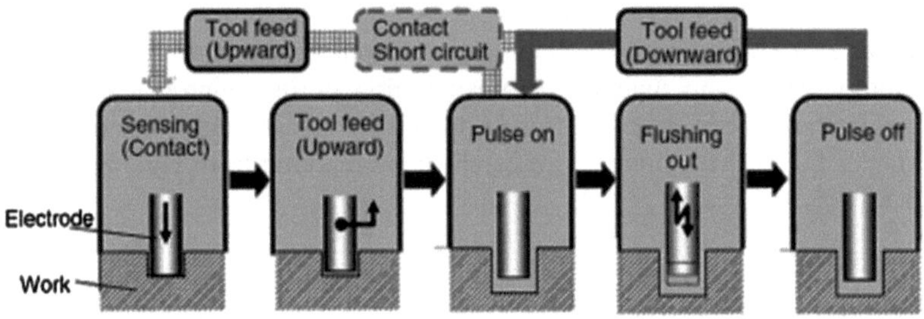

Voltage	Sensing Voltage	×	Machining voltage	×	(Sensing voltage)
Electrode movement	Downward	Upward	×	Piezo quick up and down	×

Fig. 8.15 IEG control sequence [37]

zone as well as inadequate control of microtool feeding movement, micro sparks may generate between microtool and workpiece. Machining accuracy can also be improved by developing newer machining strategies and increasing the competence of anodic dissolution by incorporating other special methods and effects. The machining accuracy of EMM can also be improved by controlling and optimizing the combination of different influencing EMM parameters. Some of the important aspects by which accuracy of EMM can be improved have been discussed as:

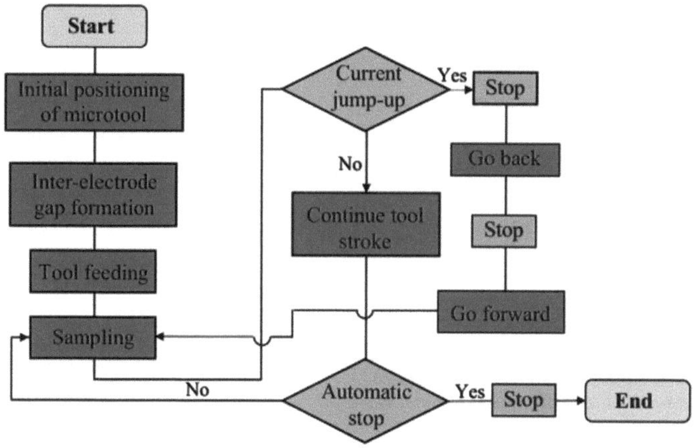

Fig. 8.16 Gap control strategy [38]

8.10.1 Geometry of Microtools

Geometry of microtools i.e. shape and size plays an important role during controlled anodic dissolutions to achieve the desired shape, size, and surface finish of the micro-features. In many application microfeatures like straight cylindrical microhole, taper free microgrooves and vertical walled 3D microstructures are essential, whereas during machining of such microfeatures of higher depths using straight cylindrical microtools, taper is formed on the sidewalls of the structure because of the machining time difference between top face and bottom face of the microfeature. Microtools of different shapes or end shapes such as reverse conical, spherical based, disc shape microtools as shown in Fig. 8.18 can be successfully used to improve the machining accuracy.

Cylindrical microtools with a flat end are suitable for machining of microholes, microgrooves, and microcavities by controlling the movement of the tool. For machining straight cylindrical microholes, cylindrical microtools with a flat end and cut edge electrode can be effectively used with a rotary arrangement. Cut edge microtools enhance the supply of fresh electrolyte flow and improve the machining accuracy as compared to microtools with flat ends. Conical microtools are suitable for machining of conical microholes to fabricate micronozzles required for the applications in inkjet printers, etc. Reverse microtool electrodes are suitable for taper reduction of microfeatures, Disk shape, and spherical end-shaped microtools are suitable for reducing the taper formation on vertical walls of the microfeatures with straight edges as shown in Fig. 8.19. Use of disc shape microtool restricts the dissolution of the workpiece along the disc height only which minimizes the taper along the vertical wall of the microfeature.

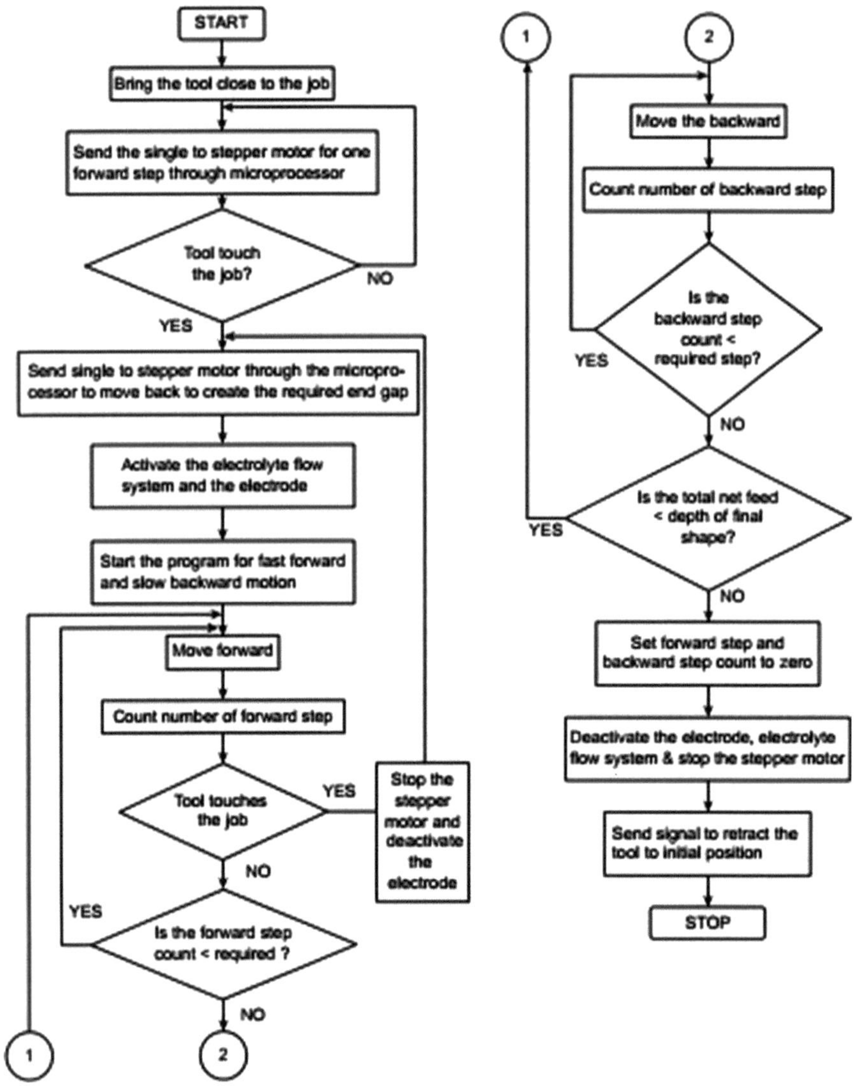

Fig. 8.17 Flow chart for IEG control [39]

8.10.2 *Microtool Insulation*

High aspect ratio micro features are commonly used in many applications. For the fabrication of these micro features, cylindrical microtools of a few microns in diameter are used in EMM, and scanning type machining strategies are followed for the microtool movement. The microstructures fabricated by un-insulated microtool are not capable of producing higher aspect ratio micro features, because of the

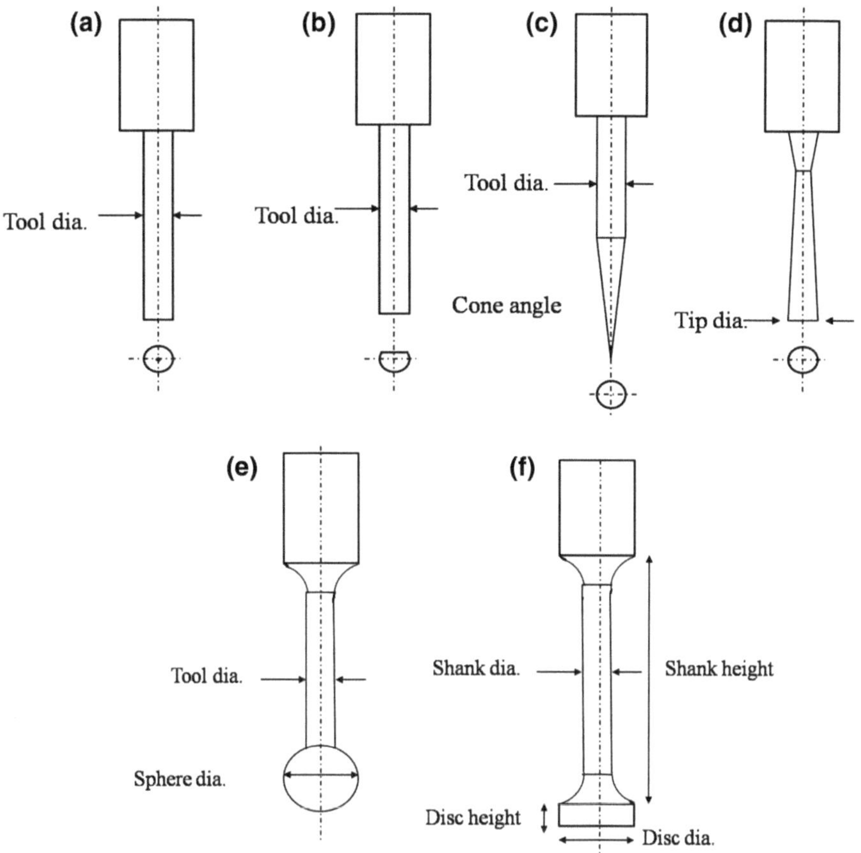

Fig. 8.18 Microtools of different end shapes **a** cylindrical **b** cut edge **c** conical **d** reverse conical **e** spherical and **f** disc shape

material dissolution from front as well as sidewalls of the microtool. Sidewall insulation of microtool restricts the material dissolution along the lateral surface and improve the machining accuracy in terms of reduced taper angle, stray current effects, with better surface quality [40]. Figure 8.20 illustrates the effect of sidewall insulation on vertical wall profile of microhole machined with cylindrical microtool.

There are various methods of insulating the microtools such as physical vapour deposition, chemical vapour deposition, spin coating, and dip coating etc. with uniform film thickness of few microns [41]. Uniform coating thickness in the range of few microns, difficulty in handling of the microtools, suitability of the insulating film in different electrolytes, method of opening the front end of the microtool for EMM, etc. are the major challenges in sidewall insulation of the microtools. Figure 8.21 illustrates the micro holes machined with and without sidewall insulations of the microtool at applied voltage 3 V, pulse frequency 8 MHz, 35% duty ratio, tool feed rate 0.1 μm/s, using 0.2 M H_2SO_4. Similarly microgrooves of 2000 μm length

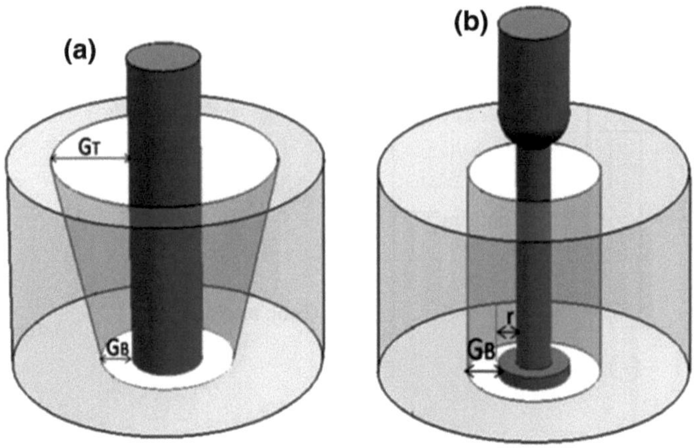

Fig. 8.19 Effect of microtool shape on wall profile of microhole machined by **a** cylindrical **b** disc microtool

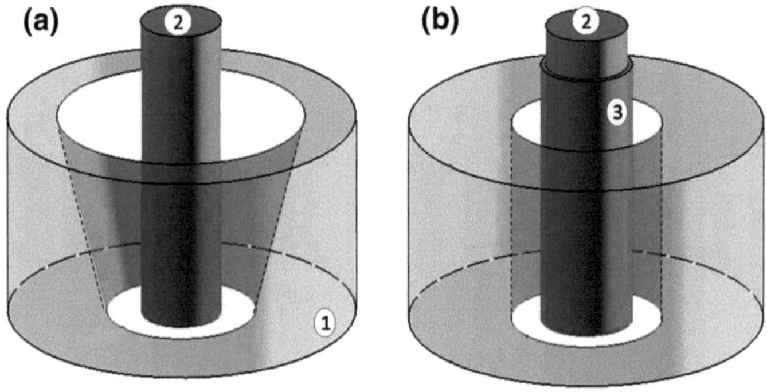

1. Workpiece 2. Microtool 3. Insulation

Fig. 8.20 Effect on wall profile of microhole machined **a** without **b** with, sidewall insulation

machined at 3 V, 8 MHz, 35% duty ratio and 0.2 M H_2SO_4. with scanning speed 93.75 μm/s and microtool feed of 0.3125 μm at the end of each scan, with and without sidewall insulations of the microtool are shown in Fig. 8.22 [41].

8.10.3 Electrolyte Circulation

In EMM, presence of fresh electrolyte at narrow machining zone is essential for the continuation of the process. Practically, IEG between tool and workpiece is very small i.e. of few microns and the circulation of electrolyte through machining gap is difficult task. Fresh electrolytes cannot be circulated with high velocity through

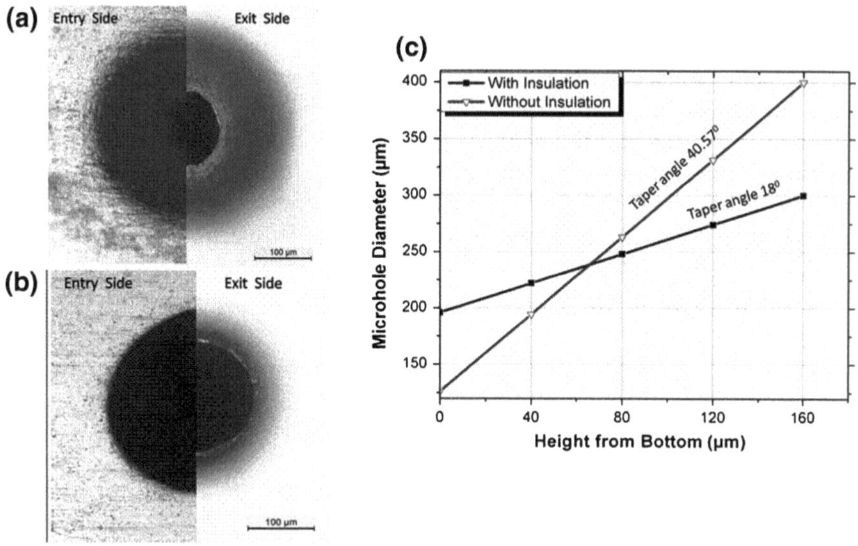

Fig. 8.21 Micro holes machined **a** without, **b** with sidewall insulations [41] and **c** variation in microhole diameter

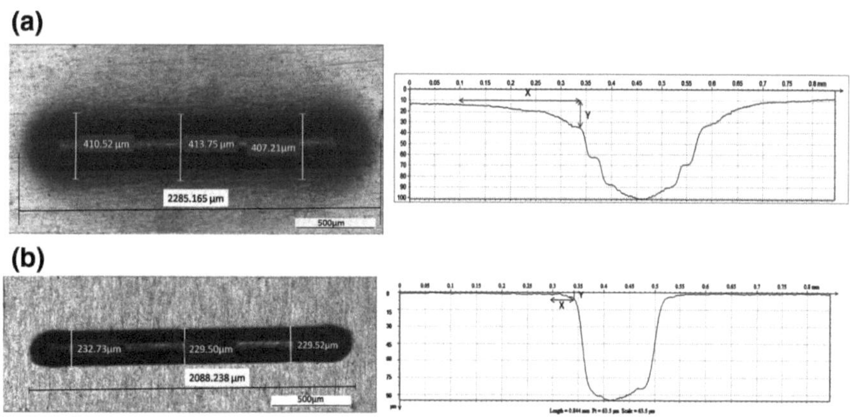

Fig. 8.22 Microgrooves and its depth profiles machined **a** without and **b** with sidewall insulations [41]

machining zone, since it may vibrate the microtool and delicate workpiece which will affect the stability of dissolution process and may deter the machining accuracy. However, it is essential to remove process by-products, heat, precipitates as well as gas bubbles from the narrow machining gap to achieve higher machining accuracy. There are various techniques by which electrolyte circulation in the machining zone can be improved. Electrolyte circulation can be improved by hybridized EMM with low-frequency tool vibration. The microtool vibrates longitudinally with the definite combination of amplitude of vibration and frequency of vibration. Vibrations within the machining zone, in the stagnant electrolyte, have considerable influence on the diffusion and convection of dissolved metal ions. As the microtool vibrates, the generated bubble collapse, which will be maximum very near to the microtool [42]. The impact of the micro jets in the machining zone, results in the enhanced convective mass transport of dissolved ions, the disruption of diffusion layer and supply of fresh electrolyte. To improve the machining accuracy in EMM, circulation of fresh electrolyte by completes flushing of sludge and precipitates from the very narrow gap between tool and workpiece is essential.

8.10.4 Microtool Movement Strategy

The achievable machining accuracy of EMM changes depending on the tool movement strategy. The effect of uncontrolled machining due to the improper flushing of the machining zone as well as the poor localization effect of current can be greatly minimized by regulating the movement of the microtool. During machining, microtool may move horizontally along X–Y plane and vertically along Z axis as well as it rotates. During EMM by scanning, microtool moves mainly along the combination of XY plane and Z axis following the specific tool movement strategy for three-dimensional shape generation. Microtool may move with faster or slower feed rate which directly controls the time of interaction between microtool and workpiece; thus controlling the anodic dissolution in the machining zone results in the degree of machining accuracy. The tool movement also controls the flow of electrolyte in the narrow machining zone thus facilitating the removal of machined products. Rotary movement to the microtool while it is fed along the desired path enhances the effective circulation of electrolyte in the machining zone. The rotary tool generates centrifugal force in stagnant electrolyte along its peripheral surface which may create micro-pumping action in the confined machining zone. This helps to stir up the micro machined products and gas bubbles to eject from the machining area to the periphery and create cavitations effect in the machining zone. Due to the availability of fresher electrolyte in the machining zone, the efficiency of the anodic dissolution improves which in turn increasing the machining accuracy.

8.10.5 Micro-sparks Phenomena in EMM

Micro sparks in the narrow machining zone cannot be eliminated completely, which deteriorates the machining accuracy of EMM. The micro sparks are occurring in narrow machining zone due to the variation in machining parameters apart from tool feed rate, heat generation across IEG accumulation of sludge and gas bubbles in the very small IEG. The increase in gap resistance due to various reasons such as generation of gas bubbles, sludge formation etc. leads to the occurrence of micro sparks, causing higher overcut as well as micro spark affected zone that results in the poor quality of final products.

8.11 Applications of EMM

EMM has improved in reliability through continuous developments, permitting its industrial implementation for automated large-scale micro manufacturing. The various applications of electrochemical micro fabrication technologies covering micromachining, finishing, as well as surface engineering applications have been illustrated through selected industrial examples. It has also some technical limitations that may constrain its full-fledged applicability in the micro fabrication field. Typical applications of EMM technologies for micro fabrication of components can be categorized in three domains such as machining, finishing, as well as surface structuring applications. Some of the machining applications are discussed as:

8.11.1 Machining Applications

Machining of micro features on micro or macro devices has become an important issue in recent technologies. EMM is one of the best techniques which can be successfully utilized to machine different micro features with high accuracy and surface quality. Very few techniques are available to machine microstructures with three-dimensional (3-D) features on advanced engineering materials like copper, aluminum, nickel, titanium, steel, and their alloys. Some of the important machining applications of EMM are as:

8.11.1.1 Micronozzles

Micronozzles with high surface quality can be fabricated on advanced high strength temperature resistance (HSTR) materials such as titanium, nickel alloys, and stainless steel utilizing the EMM technique. These micronozzles can be successfully utilized in various applications such as injectors for automobile and aerospace applications, microfluidic applications for heat transfer devices, as well as for various biomedical applications. Figure 8.23 shows a magnified view of a micronozzle fabricated by EMM utilizing a conical microtool [43].

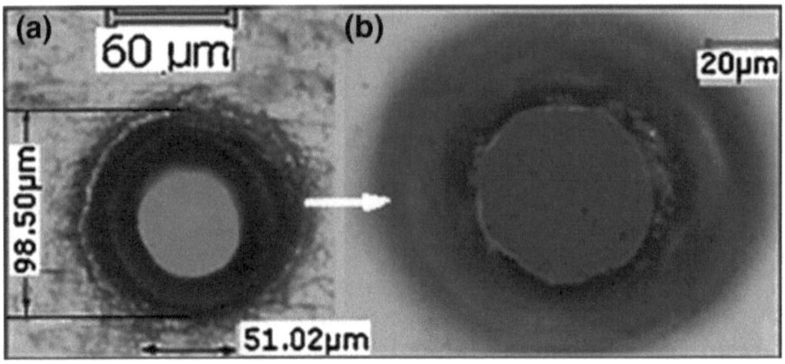

Fig. 8.23 **a** Micronozzle fabricated by EMM. **b** Surface quality of internal wall [43]

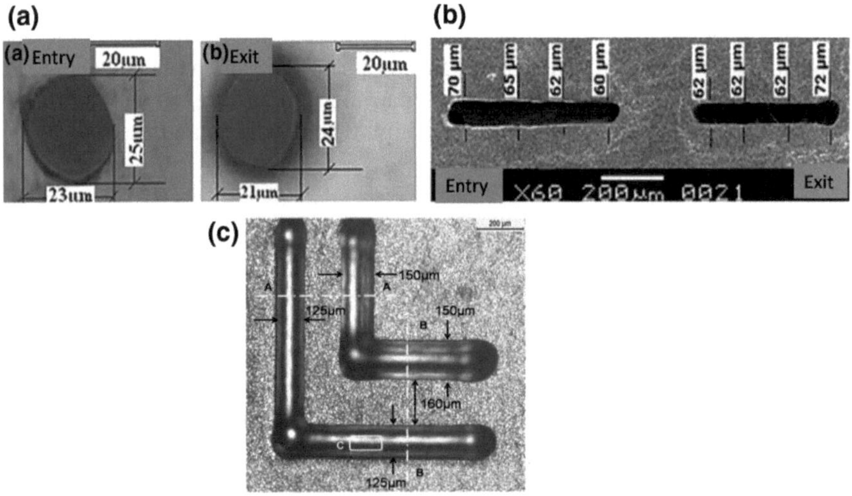

Fig. 8.24 Microscopic image of **a** microhole [50], **b** micro slots [45] and **c** microgrooves by EMM

8.11.1.2 Microholes, Slots, and Channels

Microholes, slots, and channels can be fabricated successfully by EMM with high precision and quality. Figure 8.24 shows the microhole of average entry diameter 24 µm and exit diameter 22.5 µm drilled on an SS-304 plate. Conical microtool can be used for the generation of taperless micro channel. Initial drilling followed by milling, i.e., sinking and milling method, is used for the generation of micro channel. Microchannels separated by very thin fins can also be fabricated by EMM [44].

8.11.1.3 Three-Dimensional Micro Features

Three-dimensional microstructures can be machined successfully by EMM utilizing scanning movement of the microtool at the desired path. Figure 8.25 shows the microscopic images of the various three dimensional micro features machined by EMM. Micro features have been machined utilizing cylindrical microtool adopting layer by layer machining.

Figure 8.26a shows the SEM micrograph of micro hemisphere machined on SS. Micro hemisphere has been machined on SS with 6 V, 60 ns pulse on-time, 1 μs pulse period. This structure has been machined in three steps. As the rough cut, the cylinder has been machined and the hemisphere with 100 μm diameter was machined on the cylinder. As the finish cut, the hemisphere with 60 μm diameter was machined [46]. Figure 8.26b shows the single crystallites of the steel that is cut without changing the texture. Micro cube has been machined into SS foil with flat

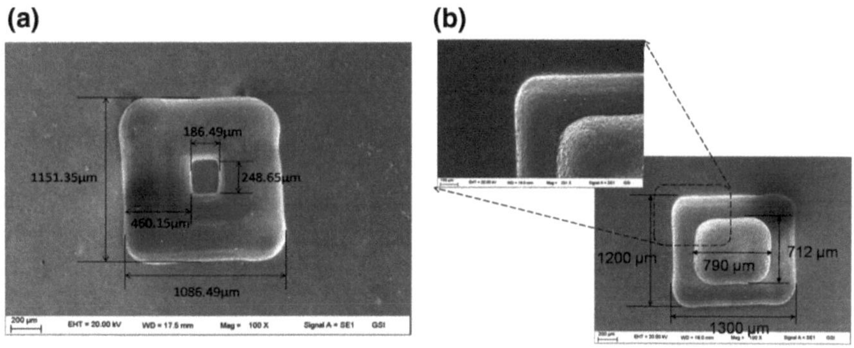

Fig. 8.25 3D microstructures with plane surfaces

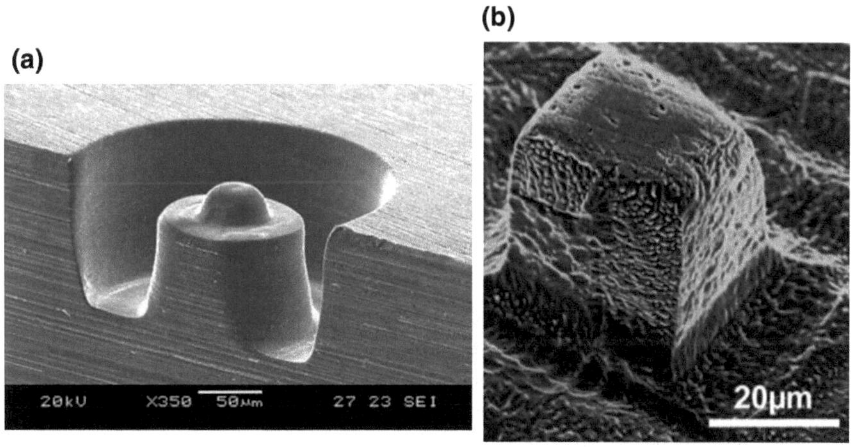

Fig. 8.26 Complex 3D microstructures **a** micro hemisphere [46] **b** micro cube [58]

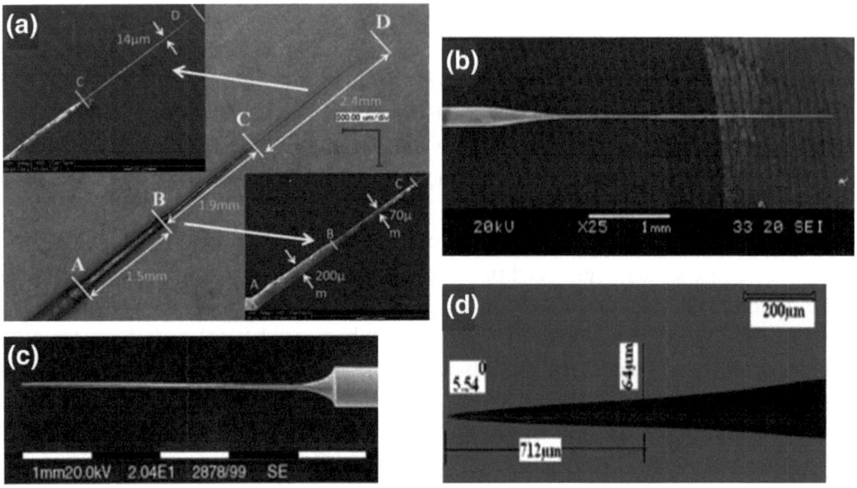

Fig. 8.27 Microrods fabricated by EMM **a** microelectrode of 5.8 μm length with diff. diameters [47] **b** WC micro-shaft of ϕ 5 μm diameter and 3 mm length [48] **c** micro electrode of ϕ 50 μm [49] **d** sharp conical microtool [50]

end cylindrical wire having 50 μm diameter. Upon applying pulse frequency 1 MHz, 100 ns pulse period, voltage 2 V, the wire was fed 30 μm into the work-piece vertically, followed by lateral movement along a rectangular path. Sharp grain boundary to a crystal with a different orientation can be seen in the lower left part on the front face of the cube.

8.11.1.4 Micropins or Microtools

Micropins or microrods of different shape and size are used for different applications, such as ultra high aspect ratio penetrating metal microelectrodes can be used in biomedical applications for painless surgery [47]. Microrods can also be used as microtools in EMM for machining of various micro features. These types of micropins or microrods of different features can be easily fabricated by reverse EMM by regulating various process parameters [48, 49]. Figure 8.27 shows the various microtools fabricated by EMM.

8.11.1.5 Disc Shape Microtools

Disc shape microtools have already proven their effectiveness in EMM for machining of taper free micro features. Disc shape microtools of different features namely disc height, disc diameter, shank diameter, and shank height can also be fabricated by EMM. Figure 8.28 shows the microscopic image of precise disc shape microtool fabricated by EMM from microrod of 300 μm diameter, under the machining conditions of applied voltage 1 V, 1 MHz, 80% DR, and 1 M NaOH

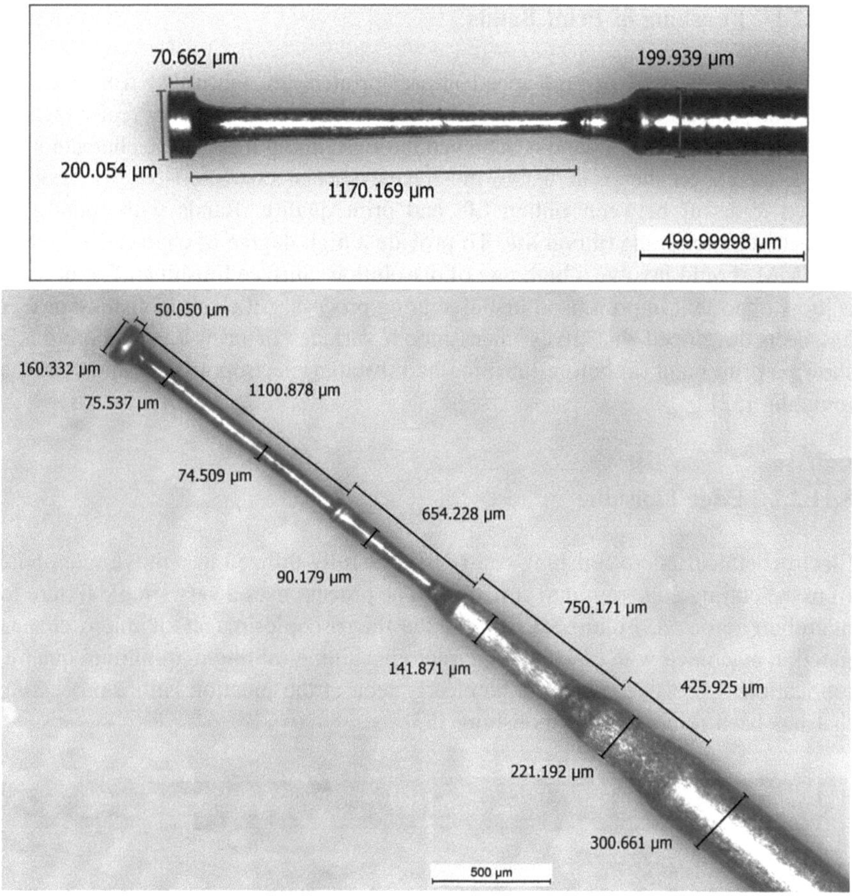

Fig. 8.28 Disc shape microtools fabricated by EMM

electrolyte [51]. For achieving strength and rigidity to the microtool, disc microtool has been fabricated in multiple steps of diameter reduction.

8.11.2 Finishing Applications

Micro components fabricated by different methods may consist of uneven and sharp edges, microburrs, and rough surfaces, which needs further finishing operations to make them suitable for various applications. Finishing of such micro components is a challenging task due to the size limitations. EMM can be effectively utilized for finishing of different micro components. Some of the finishing applications using EMM are specified as:

8.11.2.1 Finishing of Print Bands

The print bands used in high-speed impact printers are fabricated from stainless steel. The print band system consists of group of formed characters. Precise location of all the characters on a band is achieved through timing marks. The characters and timing marks on the print bands must have special characteristics to meet the desired trade-off between ribbon life and print quality. Bands with round-edge characters increase the ribbon life. To provide a high degree of character rounding, the EMM should involve a high rate of dissolution. Surface finishing of print bands is most important in print band manufacturing process. An electropolishing process has been developed that gives microsmooth surfaces of print bands. Figure 8.29 shows a print band (a) before finishing and (b) after electropolishing and character rounding [52].

8.11.2.2 Edge Finishing

Electrochemical micro finishing can be successfully utilized to remove microburrs from the cutting edge of micro structures. The process uses a very simple fixture for microburr removal. Figure 8.30 shows the microscopic images of micro channel junction machined with conventional micromachining milling in minimum quantity lubrication. Excessive burrs can be clearly seen at the junction before processing, that has been removed after polishing [53].

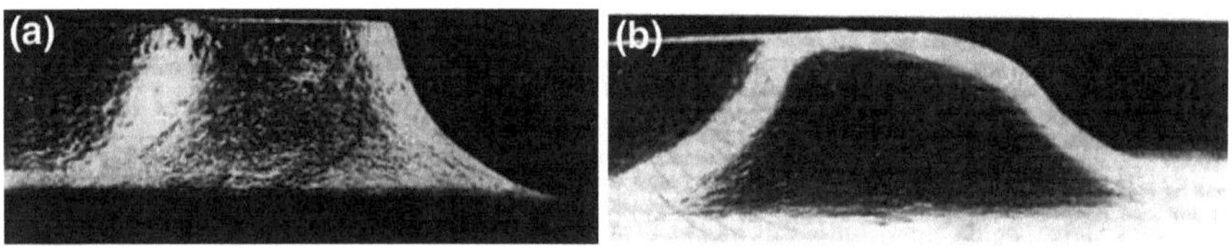

Fig. 8.29 SEM micrograph of a print band character

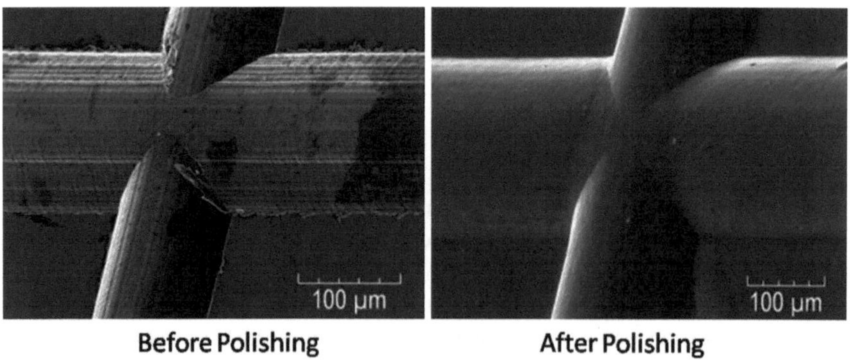

Fig. 8.30 Micro channel junction before and after polishing

8.11.3 Surface Engineering Applications

Surface quality is an important property of the microproducts, as it affects the performance and service life of products. Different types of surface topography can be engineered on industrial products by EMM. Microscopic features on the surface of the product may provide different advantages during various engineering applications such as biomedical, tribological, and aesthetic. EMM offers viable means to produce surface engineering applications, since it involves the material dissolution at atomic level. EMM does not bring in mechanical and thermal residue stresses that are generally accompanied by laser micromachining or dry etching processes. Some of the surface engineering applications achieved by EMM can be described as:

8.11.3.1 Generation of Micro Pattern on Stainless Steel

Surface phenomena, particularly at the microscale, have played an elementary role in the development of many advanced fields such as energy, machining, optics, tribology, and biomedicine. Surface engineering features such as shape of the micro fabricated pattern, size, depth, and its functionality can be varied easily by changing the process parameters of EMM. Surface texturing with micro-dimple is a well known method for friction reduction under lubricated sliding contact. Figure 8.31 shows the micrographs of micro-dimple patterns machined on carbon steel machined with applied voltage 9.5, 2 μs pulse on-time and 10 s of nominal machining time, in 2 M sodium nitrate solution [54].

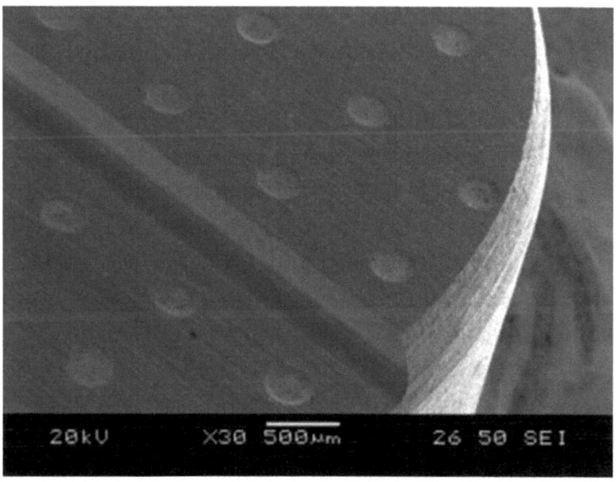

Fig. 8.31 Micro-dimple pattern on the step-shaped specimen [54]

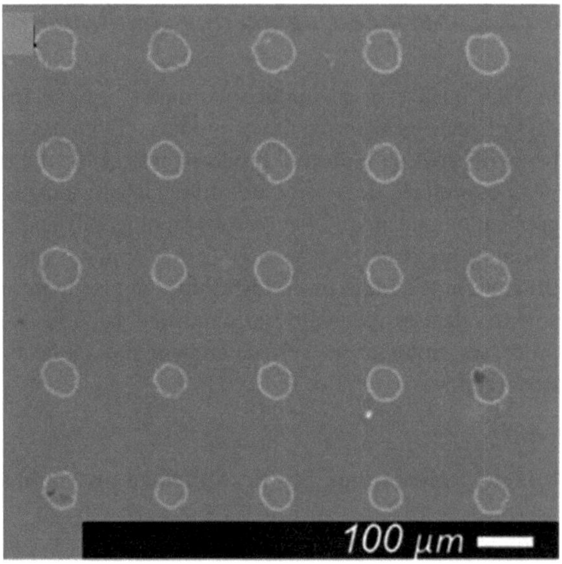

Fig. 8.32 SEM micrograph of a regular pattern machined on a titanium surface [55]

8.11.3.2 Surface Structuring of Titanium by EMM

Titanium and its alloy have very good mechanical and chemical properties, because of which they are in high demand for various engineering applications. Surface structuring of titanium is a main requirement in micro engineering and biomedical applications. EMM is one of the important process for surface structuring of titanium, whereas micro structuring of titanium surface is a challenging task due to the rapid formation of passive oxide layer, which obstructs the smooth anodic dissolution of titanium. Figure 8.32 shows SEM micrograph of a regular pattern machined on a titanium surface by 20 V, 200 ns pulse length and 20% duty cycle, using ethylene glycol as an electrolyte.

8.11.3.3 Surface Structuring for Biomedical Implants

Microscopic features on surfaces are critical factors that affect successful application of titanium as a load-bearing implant in orthopedic surgery. EMM offers a means to produce patterns on titanium surfaces. Figure 8.33 shows the microscopic image of microhole patterns machined on Ti6Al4V for biomedical applications. Jet electrochemical micromachining (jet-EMM) is used to machine microholes with a high aspect ratio on titanium surfaces. The etching process was con-ducted with a voltage of 200 V and average current of 45 mA.

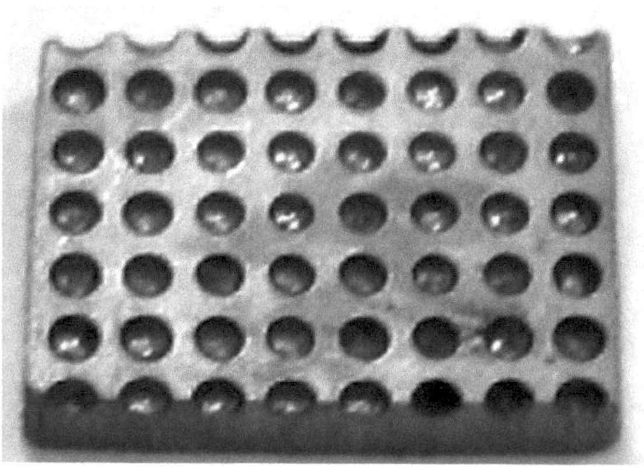

Fig. 8.33 SEM micrographs of microholes produced on the Ti6Al4V [56]

8.12 Recent Advances in EMM

Anodic dissolution in EMM can be utilized for metal removal in micro as well as nanoscopic level, because of which EMM has emerged as one of the best alternatives for fabrication of micro components with high quality and precision economically. In recent years EMM technology has been introduced to meet the increasing demands of advanced micro as well as nanoscale applications in various fields of applications. Over the years through research and development in this area, EMM has diversified its operational capabilities by the introduction of different features to exploit its potential in the area of submicron and nanorange fabrication. Different variants of EMM such as wire EMM, solid-state EMM, surface structuring by EMM, micropatterning and stamping by EMM, and nanofabrication by applying EMM have been established. Recent advancements in the area of EMM demonstrating the potential of EMM as a solution to the various challenging problems especially those that arise in the micro and nanofabrication areas have been presented as:

8.12.1 Fabrication of Micro Features for MEMS

Basic properties and characteristics of the materials are to be retained after fabrication of microdevices for MEMS. Advanced engineering such as copper, aluminum, steel, nickel, titanium, and their alloys, are some of the most suitable materials for the fabrications of mechanical microdevices. Machining of these materials is a difficult task, when features as well as components are in the microscale size. EMM can be successfully utilized for micromachining of various microfeatures made of different metals without altering the basic material characteristics and properties because of the unique mechanism of material removal by anodic dissolution at atomic level. Therefore the process can be precisely controlled

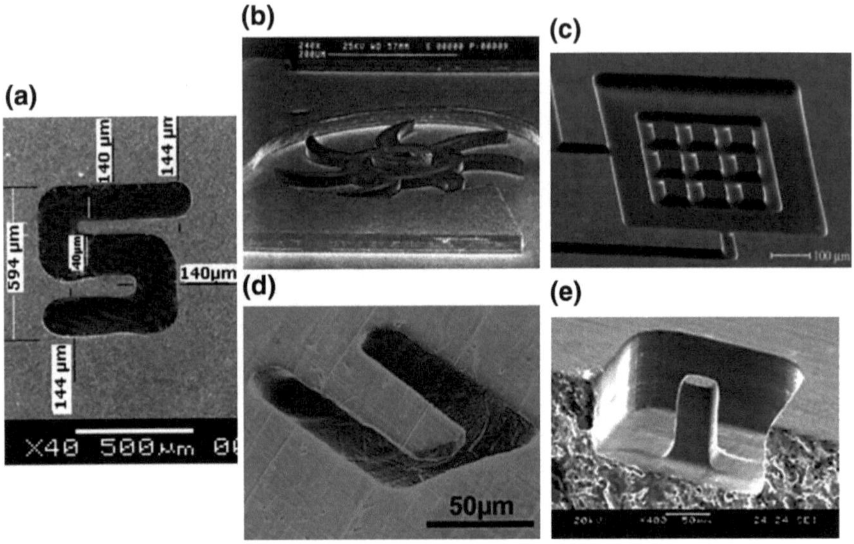

Fig. 8.34 **a** Channel-net of symmetrical micro cantilevers [45], **b** magnetically driven micro turbine of nickel [11], **c** two-level structure machined in titanium [57], **d** freestanding micro cantilever [58], **e** high-aspect-ratio micro column [46]

with submicrometer resolution to achieve highly accurate microfeatures that are most acceptable for MEMS. Electrochemical micromachining of various microfeatures proposed for microdevices that are well suited for functional applications of MEMS have been attempted by various researchers detailed as:

Figure 8.34a shows microscopic image of taperless and through, microchannel band generated by the straight cylindrical micro tool of 115 μm diameter in SS sheet of 35 μm thickness. Machining parameters were 5 MHz pulsed frequency, 0.2 M H_2SO_4, 40% duty ratio, sinking depth 75 μm, and feed rate along the path was 0.781 μm/s [45]. Figure 8.34b shows the magnetically driven micro turbine fabricated by electro-deposition of magnetic alloys through ultraviolet-patterned photo-resist on nickel. The casing and the axis in the middle have been fabricated by electro deposition of 50 μm thick Ni, whereas the wheels have been fabricated in a separate substrate by electro deposition of 40 μm thick NiCo alloy having diameter of 400 μm. Figure 8.34c shows SEM image of multilevel microstructure machined on titanium using oxide film laser lithography which can be utilized for MEMS applications. Figure 8.34d shows the freestanding micro cantilever of 80 μm long, 32 μm wide, and 11 μm thick fabricated in stainless steel by EMM using ultrashort voltage pulse and microtool of 10 μm Pt wire as microtool. Use of EMM milling strategy with a disk-type electrode, ultrashort pulses tens of nanosecond duration, and 0.1 M sulfuric acid, enables the machining of 3-D microstructures of high aspect ratio on SS as shown in Fig. 8.34e.

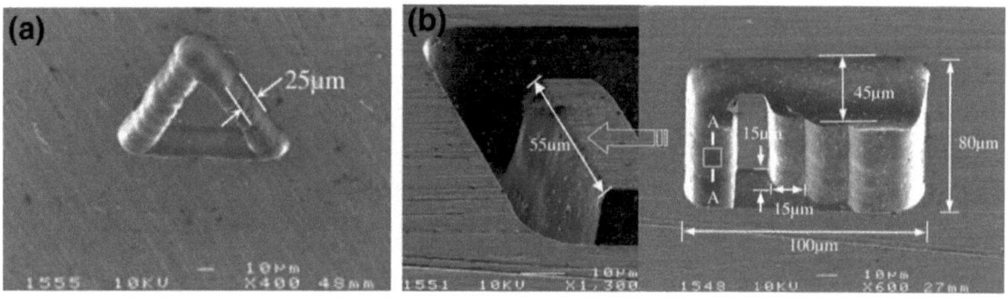

Fig. 8.35 Complex microstructures by micro electro-chemical milling **a** 2D micro complex shapes, **b** 3D micro structure with three step [59]

Micro-scale metal complex structures have a wide range of application in many fields, including biomedicine and aviation. Micro electrochemical milling can be applied to the fabrication of micro metal parts also. Figure 8.35 shows complex 2D and 3D micro metal structures machined on the hard-to-cut material super alloy plate, utilizing microtool of 10 μm diameters. 2D micro shape having width of about 25 μm, and 3D structure with three steps has been machined with machining feed rate of 1 μm/s, nanosecond pulses with pulse amplitude of 4.5 V, pulse on time of 95 ns, and pulse period of 1 μs, with milling layer thickness of 5 μm. The total depth of the cavity is about 45 μm, and each single-step size was about 15 × 55 × 15 μm [59].

Micro features like microgrooves and microholes are to be machined on majority of the microdevices in various applications such as microactuators, micropumps, and microdies etc. Presently the demands of microgrooves and microholes with internal features have increased, since specific cross-sectional shape i.e., geometry and internal surface quality, improves product functionality as well as product performance. If the microgrooves and microholes contain different internal features, the fabrication of microgroove becomes difficult. These types of microfeatures can be easily machined by EMM by regulating the process parameters during machining. Figure 8.36 shows microgrooves and microholes with different internal features machined by EMM using sidewall insulated microtool. EMM has successfully proved its capability as an alternative technique for machining different micro features, high-aspect-ratio 3-D microstructures with high resolutions for MEMS and various micro engineering applications.

8.12.2 Solid-State EMM

Solid state EMM is an application of anodic electrochemical reaction at micro contact between the ion conducting microelectrode and metal substrate as a result of continuous application of direct current. The metal substrate is locally incorporated into the ionic conductor in the form of metal ions through the micro contact [61]. This method can easily control the machining size and depth by adjusting the

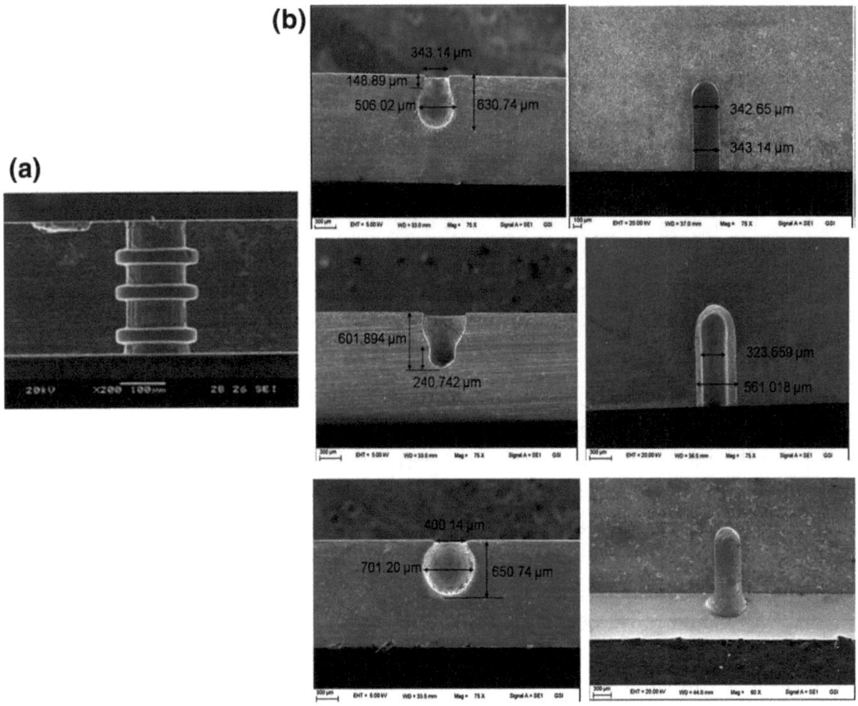

Fig. 8.36 **a** Micro hole with array of internal microgrooves [60], **b** microgrooves with internal features, machined by EMM

contact areas, i.e., shape of the apex or other electrochemical machining parameters. Figure 8.37a shows the basic arrangement of solid-state EMM cell silver with pyramid like microelectrode as cathode and target metal substrate M: Ag or Zn as anode in the system. SEM image of the Ag surface after micromachining under a constant current of 100 μA for 7 min at 873 K, is as shown in Fig. 8.37b.

8.12.3 Wire-EMM

Wire electrochemical micromachining is a novel method for machining of high aspect ratio metallic microstructures by electrochemical micro machining using very thin micro wire as a tool. Figure 8.38 illustrates the principle of Wire EMM, utilizing tungsten micro wire as tool instead of complex shape electrodes. During machining, ultra short voltage pulses are applied between workpiece as anode and wire electrode as cathode that are separated by narrow IEG ranging from 1 to 10 μm. Electrolyte flows around the micro wire electrode accompanying bubbles up and process by-products. Workpiece material dissolves electrochemically and narrow groove is produced as the cathode wire moves towards the anode

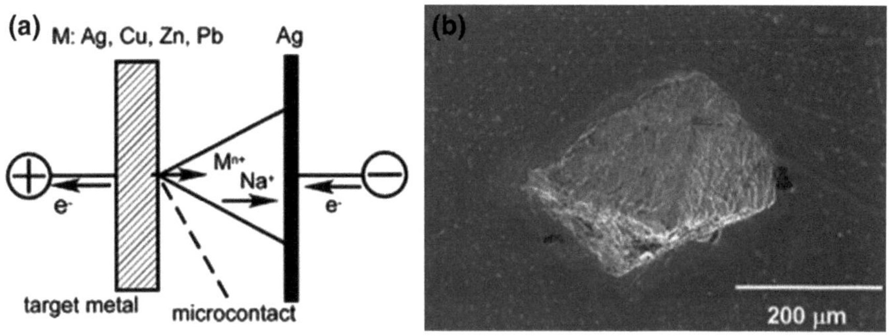

Fig. 8.37 **a** Basic arrangements of solid-state electrochemical micromachining **b** pyramid-shaped micro-features fabricated by SSEM [62]

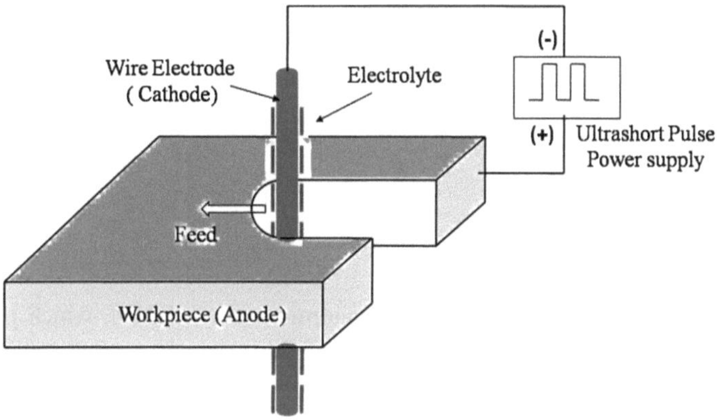

Fig. 8.38 Schematics of wire EMM

workpiece. The motion and path of the micro wire electrode is controlled by computerized numerical control system, and complex shape parts of high aspect ratio can be fabricated. Different process parameters of WECM such as feed rate, pulse voltage, and pulse on time influence the size of the machining gap that affects the machining accuracy. Very thin wires of diameter 5–10 μm are used in this process [63]. Platinum, tungsten, and copper are some of the common types of wires used in wire-EMM.

In WEMM microstructures with high aspect ratio up to 30 could be fabricated by electrolyte flushing and wire traveling. However, with further increase in aspect ratio, machining gap becomes extremely narrow and long, during which electrolyte flushing is less effective. Mass transport inside the machining gap plays an important role in WEMM. The machining stability, the material removal rate and surface quality can be significantly improved by introducing microwire vibration along with electrolyte flushing and wire traveling during machine [64].

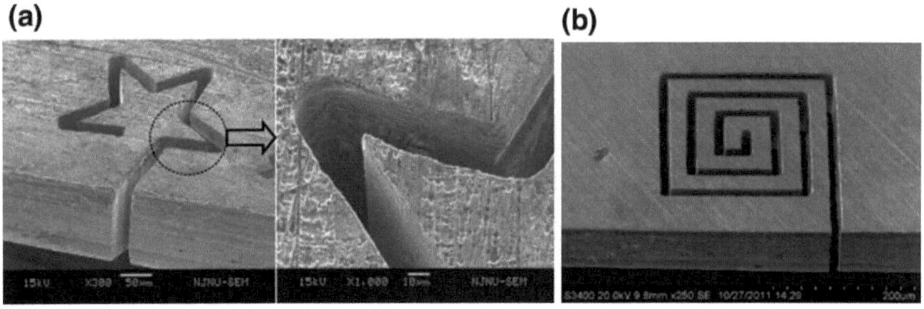

Fig. 8.39 Micro structures by WEMM **a** complex micro structure [63], **b** micro square helix [64]

Figure 8.39a shows the SEM image of with the slit width of 20 μm machined using micro tungsten wire with the diameter of 5 μm, wire feed rate 0.125 μm/s., slight vibration of workpiece with the frequency of 5 Hz, nanosecond pulses with pulse amplitude of 4.2 V, on-time of 50 ns and off-time of 1 μs [63]. Figure 8.39b shows the SEM image of micro square helix fabricated with 10 μm tungsten wire and 2 Hz wire vibration [64].

8.12.4 Nanofabrication by EMM

Anodic dissolutions can be controlled to atomic scale in EMM, which permits the use of EMM for nanofabrication in many applications. Use of ultrashort voltage pulses of nanosecond duration, material dissolution can be localized by spatial confinement of electro-chemical reactions which leads to high precision machining of metallic materials with nanometer accuracy. Use of few pulses of ultrashort duration, complicated shapes in the nanoscale can be fabricated on the workpiece by utilizing an appropriate shaped microtool containing nanofeatures, which allows parallel nano-fabrication that is difficult to attain by traditional techniques.

Figure 8.40a demonstrates the improvement in machining resolution improves with reduction in pulse duration, while machining troughs by EMM on nickel sheet with different pulse durations, machining voltage 2.2 V, 10% duty ratio, 0.2 M HCl and feed rate of 2 μm/min with tungsten tool tip of 2 μm diameter [65]. Figure 8.40b shows a spiral trench machined into Ni sheet with 3 ns pulses and 2 V. The depth of the spiral trench is about 5 μm and surface roughness is less than 100 μm. It proves that application of ultrashort voltage pulse in EMM may improve the precision up to the nanometer range. Figure 8.41 shows SEM images of high aspect ratio nanometer features fabricated by EMM in nickel sheet with 2 ns pulse duration, machining voltage of 2.2 V, 10% duty ratio, and 0.05 M HCl electrolyte [66]. The width of parallel lines in the center of the fabricated structure is 90 nm. The nanometer features are machined in very short period of time, hence the same microtool or array of microtools can be repeatedly utilized for mass production.

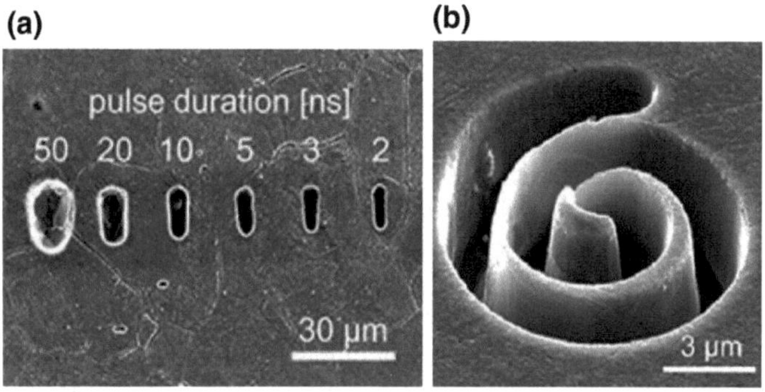

Fig. 8.40 **a** Influence of pulse duration **b** spiral trench on Ni sheet by EMM [65]

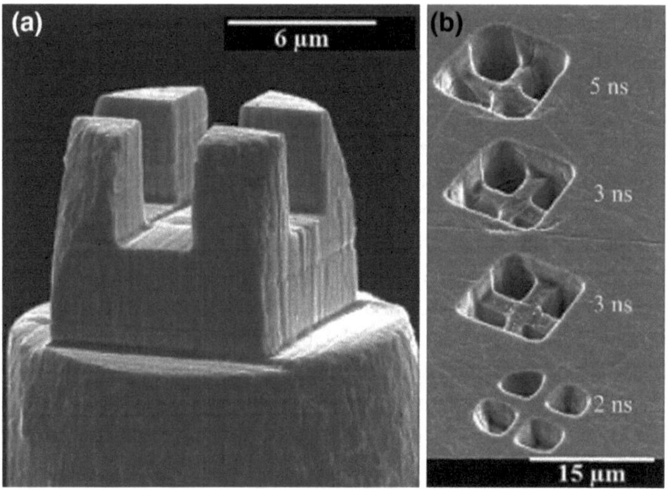

Fig. 8.41 **a** Tungsten microtool fabricated by FIB, **b** nano-features fabricated by EMM [66]

Application of ultrashort voltage pulses with very small IEG in EMM improves the machining precision to nanoscale and provides an alternative to the existing nanoscale fabrication techniques which are mostly limited to two-dimensional structures. EMM can also be successfully utilized for nano-fabrication of three-dimensional structures economically, and in shorter time which is still a challenge to the researchers. However, the area of EMM requires in-depth research to make it commercially successful in various nanotechnology applications.

8.13 General Conclusions

Electrochemical micromachining has become one of the most important micromachining process because of its numerous advantages. This chapter on electrochemical micromachining highlights the fundamental concepts in EMM process, role of EMM in micromachining, material removal mechanism, types of EMM, important process parameters, challenges in EMM setup development, and applications of EMM in various fields along with the recent advancements of it. The EMM technique can be efficiently applied for precise machining applications such as fabrication of micronozzles, microholes, slots, three dimensional micro structures, and micropins or microtools of high aspect ratio. EMM can also be applied effectively for surface finishing applications like edge finishing of print bands, and surgical blades, and also for the generation of micro pattern on stainless steel, and for biomedical implants, etc. Extensive research efforts and continuous developments are required for effective utilization of EMM in various fields of applications. This may include the improvements in machining setup, microtool design and development, monitoring and control of IEG, control of material removal and accuracy, power supply, elimination of micro sparks generation in IEG and selection of electrolyte to enhance the applications of EMM technology in modern industries. For better control over the material removal i.e. improving the machining accuracy in EMM, geometry of microtool, sidewall insulation of the microtool, supply of fresh electrolyte at narrow machining zone, and eliminating micro sparks at IEG during machining are the important factors to be focused. EMM offers many opportunities that have been unexplored till now. Further research in EMM will open up many challenging possibilities for effective utilization of ECM in the microscopic domain. Electrochemical Micromachining technique will be more successful and will be able to play a key role in micro and nano fabrication considering its advantages i.e., quality, flexibility, productivity and ultimately cost effectiveness.

References

1. J.A. McGeough, Principles of Electrochemical Machining, Chapman and Hall, 1974.
2. K.P. Rajurkar, G. Levy, A. Malshe, M.M. Sundaram, J. McGough, X. Hu, R. Resnick, and A. De Silva, Micro and nano machining by electro-physical and chemical processes, Annals of the CIRP, 55 (2) (2006) 643–666.
3. B. Bhattacharyya, J. Munda, M. Malapati, Advancement in electrochemical micro-machining, Int. J. Mach. Tool. & Manuf. 44 (2004) 1577–1589.
4. V. K. Jain, Micro manufacturing processes, CRC Press, USA, 2013.
5. T. Masuzawa, State of the Art of Micromachining, Annals of the CIRP 49 (2), (2000) 473–488.
6. K.P. Rajurkar, M. M. Sundaram, A. P. Malshe, Review of Electrochemical and Electro discharge Machining, The Seventeenth CIRP Conference on Electro Physical and Chemical Machining (ISEM), Procedia CIRP 6 (2013) 13–26.

7. V. S. Bagotsky, Fundamentals of Electrochemistry, John Wiley & Sons, Inc., Hoboken, New Jersey, USA, 2006.
8. John F. Wilson, Practice and Theory of Electrochemical Machining, John Wiley and Sons Inc., USA, 1971.
9. Allen J. Bard, Larry R. Faulkner, Electrochemical Methods – Fundamentals and Applications, second ed., John Wiley & Sons, Inc, 2001.
10. Bijoy Bhattacharyya, Electrochemical Micromachining for Nanofabrication, MEMS and Nanotechnology, Elsevier Publications, Waltham, MA 02451, USA, 2015
11. M. Datta, D. Landolt, Fundamental aspects and applications electrochemical micro fabrication, Electrochimica Acta 45 (2000) 2535–2558.
12. G.J. Kwon, H.Y. Sun, H.J. Sohn, Wall profile developments in through mask electrochemical micro machining of invar alloy films, J. Electrochem. Soc. 142 (9) (1995) 3016–3020.
13. R.V. Shenoy, M. Datta, Effect of mask wall angle on shape evolution during through mask electrochemical micromachining, Journal of Electrochemical Society, 143 (2) (1996) 544–549.
14. M. Datta, Fabrication of an array of precision nozzles by through-mask electrochemical micro machining, Journal of Electrochemical Society, 142 (11) (1995) 3801–3805.
15. Matthias Hackert-Oschätzchena, Gunnar Meichsner, Mike Zinecker, André Martina, Andreas Schubert Micro machining with continuous electrolytic free jet, Precision Engineering 36 (2012) 612–619.
16. Zhi-Wen Fan, Lih-Wu Hourng, Ming-Yuan Lin, Experimental investigation on the influence of electrochemical micro-drilling by short pulsed voltage, Int. J. Adv. Manuf. Technol., 61 (9) (2012), 957–966
17. B. Bhattacharyya, S.K. Sorkhel, Investigation for controlled electrochemical machining through response surface methodology-based approach, Int. J. Mater. Process. Technol. 86 (1999) 200–207.
18. J. Kozak, K.P. Rajurkar, Y. Makkar, Selected problems of micro-electrochemical machining, Int. J. Mater. Process. Technol. 149 (2004) 426–431.
19. J. Munda, B. Bhattacharyya, Investigation into electrochemical micromachining (EMM) through response surface methodology based approach, Int. J. Adv. Manuf. Technol. 35 (2008) 821–832.
20. J. Kozak, K.P. Rajurkar, B. Wei, Modeling and analysis of pulse electrochemical machining, Trans. ASME 116 (1994) 316–323.
21. P.I. Ortiz, M.L. Teijelo, M.C. Giordano, Electrochemical behavior of tungsten in alkaline media, J. Elec-troanal. Chem. 243 (1988) 379–391.
22. B.J. Park, B.H. Kim, C.N. Chu, The effects of tool electrode size on characteristics of micro electrochemical machining, CIRP Ann. 55 (1) (2006) 197–200.
23. M. A. H. Mithu, G. Fantoni, J. Ciampi, How microtool dimension influences electrochemical micromachining, Int J Adv Manuf Technol (2014) 70:1303–1312.
24. Yong Liu, Di Zhu, Yongbin Zeng, Hongbing Yu, Development of microelectrodes for electrochemical micromachining, Int. J. Adv. Manuf. Technol. 55 (2011) 195–203.
25. V. Rathod, B. Doloi, B. Bhattacharyya, Influence of electrochemical micromachining parameters during generation of microgrooves, Int. J. Adv. Manuf. Technol. 76 (2015) 51–60.
26. Malapati, M.; Bhattacharyya, B. Investigation into electrochemical micromachining process during micro-channel generation. Materials and Manufacturing Processes 26 (8), (2011) 1019–1027
27. S. Anasene, B. Bhattacharyya, Experimental investigation on suitability of electrolytes for electrochemical micromachining of titanium, Int. J. Adv. Manuf. Technol. 86 (5), (2016) 2147–2160.
28. B. Bhattacharyya, Electrochemical micromachining, in: V.K. Jain (Ed.), Introduction to Micromachining, Narosa Publishing House, New Delhi, 2014.
29. B. Bhattacharyya, S. Mitra, A.K. Boro, Electrochemical machining: new possibilities for micromachining, Robotics and Computer Integrated Manufacturing 18 (2002) 283–289.
30. Alexandre Spieser, Atanas Ivanov, Recent developments and research challenges in electrochemical micromachining (μECM), Int. J. Adv. Manuf. Technol., 69 (2013) 563–581.

31. Zhaoyang Zhangz, Yaomin Wang, Fei Chen, and Weiping Mao, A micro machining system based on electrochemical dissolution of material, Russian Journal of Electrochemistry, 47 (7) (2011) 819–824.
32. B. Ghoshal, B. Bhattacharyya, Micro electrochemical sinking and milling method for generation of micro features, IMechE, Part B: J. Eng. Manuf. 227 (11) (2013) 1651–1663.
33. Alexandre Spieser, Atanas Ivanov, Design of an electrochemical micromachining machine, Int. J. Adv. Manuf. Technol., 78 (5) (2015) 737–752.
34. B. Bhattacharyya, M. Malapati, J. Munda, A. Sarkar, Influence of tool vibration on machining performance in electrochemical micro-machining of copper, International Journal of Machine Tools & Manufacture, 47 (2007) 335–342.
35. B. Bhattacharyya, B. Doloi, and P.S. Sridhar, Electrochemical micromachining: New possibilities for micro manufacturing, Journal of Materials Processing Technology, 113 (2001) 301–305.
36. L. Yong, Z. Yunfei, Y. Guang, P. Liangqiang, Localized electrochemical micro machining with gap control, Sens. Actuators-A Phys. A108 (2003) 144–148.
37. T. Kuritaa, K. Chikamorib, S. Kubotac, M. Hattoria, A study of three-dimensional shape machining with an ECmM system, Int. J. Mach. Tools Manuf. 46 (2006) 1311–1318.
38. M.A.H. Mithu, G. Fantoni, J. Ciampi, M. Santochi, On how tool geometry, applied frequency and machining parameters influence electrochemical microdrilling, CIRP Journal of Manufacturing Science and Technology 5 (2012) 202–213.
39. B. Bhattacharyya, J. Munda, Experimental investigation on the influence of electrochemical machining parameters on machining rate and accuracy in micromachining domain, International Journal of Machine Tools & Manufacture, 43 (2003) 1301–1310.
40. Hu, M.; Li, Y.; Yue, Z.; Jian, W.; Xiaogu, Z. Experimental study of micro electro-chemical milling with side insulated electrode. Applied Mechanics and Materials, 159, (2012) 127–131
41. V. Rathod, B. Doloi, B. Bhattacharyya, Sidewall insulation of microtool for electrochemical micromachining to enhance the machining accuracy, Int. Journal of Materials and Manufacturing Processes, 29 (3) (2014) 305–313.
42. Natsu Wataru, Nakayama Hisashi, and Yu Zuyuan, Improvement of ECM characteristics by applying ultrasonic vibration, International Journal of Precision Engineering and Manufacturing, 13 (7) (2012) 1131–1136.
43. B. Ghoshal, B. Bhattacharyya, Shape control in micro borehole generation by EMM with the assistance of vibration of tool, Precis. Eng. 38 (2014) 127–137.
44. V. Rathod, B. Doloi, B. Bhattacharyya, Experimental investigations into machining accuracy and surface roughness of microgrooves fabricated by electrochemical micromachining, Proc. Inst. Mech. Eng. Part B: J. Eng. Manuf. (2014) 1–22.
45. B. Ghoshal, B. Bhattacharyya, Generation of microfeatures on stainless steel by electrochemical micromachining, Int J Adv Manuf. Technol. (2015) 76:39–50.
46. B.H. Kim, et al., Micro electrochemical machining of 3D micro structure using dilute sulphuric acid, Ann. CIRP 54 (1) (2005) 191–194.
47. A.B. Kamaraj, M.M. Sundaram, and R. Mathew, Ultra high aspect ratio penetrating metal microelectrodes for biomedical applications, Microsystem Technology, 19 (2013) 179–186.
48. S. H. Choi. S. H. Ryu. D. K. Choi. C. N. Chu, Fabrication of WC micro-shaft by using electrochemical etching, Int J Adv Manuf Technol, 31, (2007) 682–687.
49. Young-Mo Lim, Hyung-Jun Lim, Jang Ryol Liu, Soo Hyun Kim, Fabrication of cylindrical micropins with various diameters using DC current density control, Journal of Materials Processing Technology 141 (2003) 251–255.
50. B. Ghoshal, B. Bhattacharyya, Influence of vibration on micro-tool fabrication by electrochemical machining, International Journal of Machine Tools & Manufacture, 64 (2013) 49–59.
51. V. Rathod, B. Doloi, and B. Bhattacharyya, "Parametric investigation into the fabrication of disk microelectrodes by electrochemical micromachining", Journal of Micro and Nano-Manufacturing-ASME Transactions, 1(041005) (2013) 1–11.

52. M. Datta, J.C. Andreshak, L.T. Romankiw, L.F. Vega, Surface finishing of high speed print bands: I. A prototype tool for electrochemical microfinishing and character rounding of print bands, J. Electrochem. Soc. 145 (9) (1998) 3047–3051.
53. D. Berestovskyi, M.P. Soriaga, P. Lomeli, J. James, B. Sessions, H. Xiao, W.N.P. Hung, Electrochemical Polishing of Microcomponents, Proceedings of the 8th International Conference on Micro Manufacturing University of Victoria, Victoria, BC, Canada, March 25–28, 2013.
54. Jung Won Byun, Hong Shik Shin, Min Ho Kwon, Bo Hyun Kim, and Chong Nam Chu, Surface Texturing by Micro ECM for Friction Reduction, International Journal of Precision Engineering and Manufacturing, 11 (5), 747–753.
55. Terje Sjöström, BoSu, Micropatterning of titanium surfaces using electrochemical micromachining with an ethylene glycol electrolyte, Materials Letters 65 (2011) 3489–3492.
56. X. Lu, Y. Leng, Electrochemical micromachining of titanium surfaces for biomedical applications, J. Mater. Process. Technol. 169 (2005) 173–178.
57. D. Landolt, P.-F. Chauvy, O. Zinger, Electrochemical micromachining, polishing and surface structuring of metals: fundamental aspects and new developments Electrochimica Acta 48 (2003) 3185/3201.
58. V. Kirchner, L. Cagnon, R. Schuster, G. Ertl, Electrochemical machining of stainless steel microelements with ultrashort voltage pulses, Appl. Phys. Lett. 79 (11) (2001) 1721–1723.
59. Yong Liu, Di Zhu, Linsen Zhu, Micro electrochemical milling of complex structures by using in situ fabricated cylindrical electrode, Int J Adv Manuf Technol, 60 (9), (2012) 977-984.
60. C.H. Jo, B.H. Kim, C.N. Chu, Micro electrochemical machining for complex internal micro features, CIRP Ann. – Manuf. Technol. 58 (2009) 181–184.
61. K. Kamada, K. Izawa, Y. Tsutsumi, S. Yamashita, N. Enomoto, J. Hojo, Y. Matsumoto, Solid-state electrochemical micromachining, Chem. Mater. 17 (2005) 1930–1932.
62. K. Kamada, M. Tokutomi, N. Enomoto, J. Hojo, Electrochemical micromachining using a solid electro-chemical reaction at the metal/β-Al2O3 microcontact, Electrochim. Acta 52 (2007) 3739–3745.
63. D. Zhu, K. Wang, N. S. Qu, Micro Wire Electrochemical Cutting by Using In Situ Fabricated Wire Electrode, CIRP Annals - Manufacturing Technology, 56 (1) (2007), 241–244.
64. Yong-Bin Zeng, Qia Yu, Shao-Hua Wang, Di Zhu, Enhancement of mass transport in micro wire electrochemical machining, CIRP Annals - Manufacturing Technology, 61 (2012) 195–198.
65. M. Kock, V. Kirchner, R. Schuster, Electrochemical micromachining with ultrashort voltage pulses-a versatile method with lithographical precision, Electrochim. Acta 48 (2003) 3213–3219
66. A.L. Trimmer, J.L. Hudson, M. Kock, R. Schuster, Single-step electrochemical machining of complex nanostructures with ultrashort voltage pulses, Appl. Phys. Lett. 82 (19) (2003) 3327–3329.

Chapter 9
Electrochemical Micromachining of Titanium and Its Alloys

Sandip S. Anasane and B. Bhattacharyya

Abstract Miniaturization has covered every area of modern world. Micromachining is one of the key technologies that can enable the realization of almost all requirements of the microproducts and related domain. However, materials which can micromachined easily and used in present day Microsystems, MEMS and microengineering applications have some limitations such as low strength to weight ratio, corrosion resistance and biocompatibility. Titanium is one of the potential material and arising as alternative to the conventional MEMS materials particularly silicon or silicon based materials or glass. Titanium is known as super metal due to its high strength to weight ratio, excellent corrosion resistance and superior biocompatibility. This chapter highlights the challenges in micromachining of titanium and its alloys as well as potential use of electrochemical micromachining (EMM) technique for micromachining of titanium. Utilization of masked i.e. TMEMM as well as maskless electrochemical micromachining techniques for generation of various microfeatures on titanium has been presented in this chapter. Effect of various EMM process parameters on machining accuracy of microfeatures of titanium as well as most suitable EMM process parameters for fabrication of various complex microfeatures on titanium has also been discussed. This chapter provides comprehensive information on electrochemical micromachining of titanium and its prospective applications in the area of MEMS, Bio-MEMS and microengineering fields.

S.S. Anasane (✉)
Department of Production Engineering and Industrial Management,
College of Engineering, Pune, Pune 411005, India
e-mail: anasane@gmail.com

B. Bhattacharyya
Production Engineering Department, Jadavpur University, Kolkata 700032, India
e-mail: bb13@rediffmail.com

G. Kibria et al. (eds.), *Non-traditional Micromachining Processes*,
Materials Forming, Machining and Tribology,
DOI 10.1007/978-3-319-52009-4_9

9.1 Introduction

Miniaturization of the product and systems is in increasing demand in today's world. The miniaturization has covered every area of modern world. To name a few are micro-electromechanical systems (MEMS), micro-sensors, micro-actuators and also to machine micro size features. For example in electronic devices, e.g. computers, cell phones, CD players, etc. Relays and switches are required to be assembled to produce functional micro sized mechanical parts. In aerospace industries lightness of aircraft or satellite is one of the most vital requirements, which demands to design and manufacture components in smallest possible size on advanced materials with high accuracy. Another example is the fuel injection nozzles for automobiles; several factors of environmental safety have forced manufacturers to improve the nozzle. In addition to that a new potential area requiring microproducts is the field of biotechnology. Since the objects in this field include biological cells and genes, the tool that handles them must have micro effectors. Similarly in medical applications miniaturization of medical tools is one of the effective approaches to arrive at this target of inspection and surgery without pain as well as micro devices for drug delivery system. Micromachining is one of the key technologies that can enable the realization of all of the above requirements for microproducts and fields with such requirements are rapidly expanding. The potential advantages of microproducts are less material requirements, lesser space requirements, compactness, minimum energy consumption etc. Hence, due to this growing demand of society for micro products, scientists and engineers forced to develop stable micro-manufacturing process which can efficiently produce micro components, mechanisms and micro features by machining engineering material precisely in micron scale by utilizing various micromachining methods. However, materials which can micromachined easily and used in present day Microsystems, MEMS and microengineering applications have some limitations such as low strength to weight ratio, corrosion resistance and biocompatibility therefore these materials may not cope up with the demands of next generation microproducts. Hence, there is an urgent need to identify such materials which can have edge over conventional MEMS materials and importantly these materials should successfully micromachined. Titanium is the potential material as alternative to the conventional MEMS materials particularly silicon or silicon based materials or glass. Titanium is known as super metal due to its high strength to weight ratio, excellent corrosion resistance and superior biocompatibility. Therefore, titanium is ideal material for fabrication of microfeatures or microcomponents of new age microstructures. Introduction of advanced metals such as titanium, forces researchers to develop newer efficient and stable micromachining processes which can machine this advanced material efficiently and cost effectively.

9.2 Electrochemical Micromachining (EMM)

The non-conventional machining processes discussed above are thermal oriented machining processes except Chemical machining process e.g. Electro discharge machining (EDM), Laser beam machining (LBM), Electron beam machining (EBM), etc., which may cause thermal distortion of the machined surface [1]. Chemical machining and electrochemical machining are thermal free processes, but chemical machining cannot be controlled properly in this micromachining domain. However, accuracy level of ECM can be highly improved if the inter-electrode gap i.e. gap between microtool and workpiece is reduced in the order of less than fifty microns. Electrochemical machining (ECM) process can be applied in the microscopic domain for manufacturing of ultra precision micro features, is called electrochemical micromachining process (EMM).

Electrochemical micromachining process is based on the principles of electrolysis and governed by Faraday's laws of electrolysis. In this process high frequency DC voltage of pulsed current is applied between the cathode of micron scale size, acts as a microtool and metallic work piece which acts as an anode. Both the electrodes i.e. cathode and anode are immersed in an electrolyte usually stagnant in nature and are separated by small gap of few microns generally termed as inter electrode gap (IEG) as shown in Fig. 9.1. As soon as the pulsed current passed through these electrodes anodic material dissolves into metallic ions by the electrochemical reactions, and hydrogen gas bubbles are generated on the cathode surface. In order to achieve anodic dissolution in desired direction and shape, the microtool moves with constant feed rate towards the workpiece to maintain the predetermined inter-electrode gap (IEG). Ideally, anodic dissolution rate or material removal rate (MRR) must be synchronized with microtool feed rate to obtain controlled dissolution with shape accuracy.

EMM appears to be a very potential technology for micro machining due to its advantages that include high machining rate, better precision and control, rapid machining time, reliable, flexible, environmentally acceptable and it also permits machining of chemically resistant materials like titanium, copper alloys, super

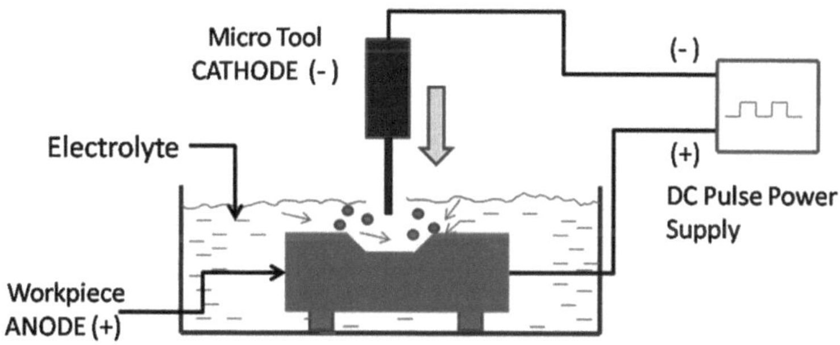

Fig. 9.1 Schematic diagram of electcrochemical micromachining process

alloys and stainless steel, which are widely used in biomedical, electronic and MEMS applications. EMM can be advantageously employed in most applications related to micromachining of metallic parts due to its cost effectiveness and achievable high precision, which are previously fabricated by chemical micromachining.

Neutral solution like aqueous solution of $NaNO_3$, NaCl, NaBr etc. are mostly used as electrolyte in electrochemical micromachining process and hence it does not cause any harmful effect to the operators. Electrochemical micromachining process generates sludge during electrochemical dissolution. Most of the cases the sludge is formed during the process are neutral salts, which are not harmful to the human beings as well as to the environment. But some time these sludge materials may contains small amount of metal ions, acids, nitrates, oils and even traces of heavy metal ions which may be harmful to the environment and finally to the human beings. Electrolyte is filtered during the EMM operation, which reduces the amount and frequency of electrolyte disposal. During electrochemical micromachining material removed only from the specified area, which also reduces the material removal rate as well as the generation of sludge. Furthermore, the harmful sludge is treated before disposal to the atmosphere to make the process environmental friendly. Since miniaturization will continue as long as people require effective space utilization with more efficient and better accuracy products without introducing any harmful material to the environment, electrochemical micromachining process will be more important in the future.

Furthermore, the use of EMM will widen the range of materials application for electronic industries, MEMS etc. The role of convective mass transport and current distribution on the surface finish and shape evolution is very important. Effective EMM process can be achieved by optimal combination of the process parametric conditions. In order to achieve the effective and highly precise material machining in the order of microns, the predominant process variables of the EMM system will have to be optimally controlled.

9.3 Titanium and Its Alloys: Types and Usage

Titanium was first discovered in 1791 by William Gregor, who was the mineralogist and chemist. William Gregor observed the magnetic sand from the local river, Helford, in the Menachan Valley in Cornwall, England, and isolated “black sand”, now known as “ilmenite”. By removing the iron with a magnet and treating the sand with hydrochloric acid he produced the impure oxide of a new element, called as “mechanite”. Subsequently, Martin Heinrich Klaproth, who is the Berlin chemist, isolated titanium oxide from a mineral, identified as “rutile”. Greek mythology provided him a new name the Titanium from the children of Uranos and Gaia, the titans. Matthew Albert Hunter from Rensselaer Polytechnic Institute in Troy, N.Y., was able to isolate the metal in 1910 by heating titanium tetrachloride ($TiCl_4$) with sodium in a steel bomb. Finally, Wilhelm Justin Kroll from Luxembourg is recognized as father of the titanium industry. In 1932 he produced significant quantities of titanium by combining $TiCl_4$ with calcium and then changed reducing agent from calcium to magnesium. Today this is still the most widely used method and is known as the “Kroll process”. In 1948 the DuPont Company was the first to produce titanium commercially [2]. Titanium is the ninth element available in the earth’s crust, and the fourth metallic element. Titanium is naturally available in the forms of rutile (titanium dioxide, TiO_2) and ilmenite (titanium iron oxide, $FeTiO_3$). These two mineral forms are the most common and commercially exploitable. Titanium is a material known for its superior physical and mechanical properties such as high strength to weight ratio, high compressive and tensile strength, low density, high fatigue resistance in air and seawater, and exceptional corrosion resistance. Therefore, titanium is a perfect material for micro engineering and allied applications. Fabrication of micro features on titanium has great potential in the area of micro electromechanical systems (MEMS) because titanium provides superior properties to traditional semiconductor materials since it has excellent fraction toughness and corrosion resistance. Titanium also possesses exceptional biocompatibility. Today titanium and its alloys are widely used in aerospace chemical processing, medicine, power generation, marine and offshore, sports and leisure, and transportation industries.

Titanium naturally resists corrosion from acids, alkalis, natural salt and polluted waters. This tendency of titanium metal is achieved by formation of hard, protective oxide film when it is exposed to oxygen present in the air or water. This thin tenacious film makes titanium resistant to erosion. Titanium is the only metal, which is completely immune to microbiological induced corrosion (MIC) in seawater. Titanium has a low modulus of elasticity about half that of steel. This gives it excellent flexibility. It is the most biocompatible of all metals. It is non-toxic, it resists attack from bodily fluids, it’s strong and light, and its flexibility is close to bone. Commercially pure titanium finds variety of applications from aerospace industry to medical industry. Typical applications of titanium in various fields are discussed in brief:

(a) **Aerospace Applications**

The exceptional properties of titanium alloys include high specific strength and excellent corrosion resistance. Therefore, titanium alloys are found in aerospace applications where the combination of weight, strength, corrosion resistance, and/or high temperature stability of aluminum alloys, high strength steels, or nickel based super alloys are insufficient. The main drivers for use of titanium in aerospace applications are:

i. Weight reduction (substitute for steels and Ni-based super alloys)
ii. Application temperature (substitute for Al alloys, Ni-based super alloys, and steels)
iii. Corrosion resistance (substitute for Al alloys and low-alloyed steels)
iv. Galvanic compatibility with polymer matrix composites (substitute for Al alloys)

v. Space limitation (substitute for Al alloys and steels).

The most common and prime use of titanium is in aircraft fuselage applications, for hydraulic tubing of modern aircraft. Aircraft floors surrounding on-board kitchens and toilets where the corrosive environment is present. Aircraft landing gear are manufactured from forged titanium, it has also find its application in the engine bay of fighter aircraft, where temperatures can quickly exceed. The main area of application for aerospace titanium and its alloys are in the gas turbine engine. Approximately one third the structural weight of modern turbine engines is made up of titanium. Compressor blades were the first engine components to be made from titanium. For helicopters, titanium alloys are used in the most highly stressed component i.e. the rotor head. In the space vehicles titanium and its alloys are also used for fuel and satellite tanks and high-pressure piping in the hydrogen pumping systems of the Space Shuttle.

(b) **Process Industry Applications**

Other than aerospace industry titanium found suitable in the areas where harsh environment is very common such as chemical, process and power generation industry. Preferred applications are in heat exchangers in which the cooling medium is seawater, brackish water, and also polluted water. Titanium is also employed in millions of meters of welded and seamless titanium tubing in steam turbine power plants, refineries, chemical plants, air conditioning systems, multi-stage flash distillation, desalination and vapor compression plants, offshore platforms, surface ships and submarines, as well as for swimming pool heat pumps. Commercially pure titanium grades and alloys are applied in production facilities for acetaldehyde and acetone, acrylic fibers, and urea. Another interesting application of titanium is in architecture, titanium has been increasingly used as exterior and interior cladding material for roofing, curtain walls, column covers, soffits, fascias, canopies, protective cladding for piers, artwork, sculptures, plaques, and monuments.

(c) **Sports and Leisure Applications**

Tennis Racquets, Baseball Bats and Pool Cues, head of the golf club, titanium accessories for racing bikes, Scuba Diving Equipment, for trekking and hiking applications such as High strength climbing gear like snap shackles, hooks, rings and eyes, latches, locking carabineers, pins, clips, eye bolts, cliffhangers, and straps are usually manufactured from forged titanium and its alloys.

(d) **Medical Applications**

Another popular application of titanium is in medical industry due to the excellent compatibility with the human body titanium is generally used as the biocompatible metallic material. Additionally, titanium is extremely resistant to corrosion from body fluids, and is compatible with bone and living tissue, and is elastically deformable as thin foil material. Thus, pure titanium combines many of the attributes desirable for heart pacemaker cases and as the carrier structure for replacement heart valves. Titanium possesses excellent fatigue behavior which is decisive for the choice of material for orthopedic devices such as various implants for hip, knee, as well as it is used to make substitute parts of the shoulder, the spine, the elbow, and the hand. Titanium has a relatively low modulus of elasticity, which reduces the differences in stiffness between the human bone and the implant.

The introduction of a titanium implant into the jawbone favors osseointegration before the superstructure is built onto the implant. Today, titanium is successfully used in prosthetic dentistry for implant screws, crowns, bridges, dental posts, inlays, and removable partial dentures.

9.3.1 Challenges in Titanium Machining

Titanium possesses two electrons in the third shell and two electrons in the fourth shell. Titanium has arrangement of electrons, in which outer shells are filled before the inner shells are completely occupied, which makes titanium transition metal. This arrangement of electrons is responsible for the unique physical properties of titanium. Due to the versatile physical and mechanical properties such as high strength to weight ratio, high compressive and tensile strength, low density, high fatigue resistance in air and seawater, and exceptional corrosion resistance. Titanium, machining is always a concern to the researchers. Owing to poor thermal properties of the titanium as well as low modulus of elasticity with its ability to maintain high strength at elevated temperature made machining of titanium an intricate task.

9.3.2 Machining of Titanium by Conventional and Non conventional Processes

Titanium proved its wider range of applications in different domains due to its excellent properties. However, machining of titanium is difficult either by conventional or non-conventional machining methods. In conventional machining process, where cutting tool removed material by shearing due to deformation through physical contact, in this mode, Titanium and its alloys are considered as difficult to cut materials due to the high cutting temperature and the high stresses at and/or close to the cutting edge during machining. The poor thermal properties of the materials forced the larger portion of heat generated in machining process which will be absorbed by the tool thus causes the rapid wear of tools. Both the high temperature and the high stresses developed at the cutting edge of the tool may create plastic deformation and/or accelerate the wear of the tools [3].

Low modulus of elasticity of titanium can cause chatter, deflection, and rubbing problems [4]. During conventional machining forces perpendicular to the workpiece may increase three to four times as a result of a build up of titanium on the wear land of the tool [5]. Because of this high thrust force and the low elastic modulus of titanium, the deflection of the workpiece can be a serious problem. Cutting tool materials undergo severe thermal and mechanical loads when machining titanium alloys due to the high cutting stresses and temperatures near the cutting edge, which greatly influence the wear rate and hence the tool life. Flank wear, crater wear, notch wear, chipping and catastrophic failure are the prominent failure modes when machining titanium and its alloys. Flank and crater wear may be attributed to dissolution-diffusion, abrasion and plastic deformation, depending on the cutting conditions and the tool material, whilst notch wear is caused mainly by a fracture process and/or chemical reaction.

Non-conventional machining techniques such as Electric Discharge Machining (EDM), Laser Beam Machining (LBM) and Electron Beam Machining (EBM) are thermal oriented machining process which put constraints to machine titanium and its alloys beyond i.e. formation of pear shaped and tapering effects on holes as well as possibilities of hot spot due to lower heat dissipation rate due to low thermal conductivity of titanium [6]. In EDM, formation of thermal stresses in a small heat-affected zone is a serious issue which can lead to micro-cracks, decrease in strength and fatigue life and possibly catastrophic failure of the component [7]. In case of USM, serious tool wear as well as lower material removal rate has been encountered during machining of titanium and its alloys [8]. Electrochemical machining (ECM) technique based on the principle of electrolysis proved to be partial successful in machining of titanium and its alloys. But passive oxide layer present on titanium surface is a concern in ECM and limits the process up to macroscopic level. However, electrochemical micromachining (EMM) has proved its capability to produce metal microstructures based on anodic dissolution principle. In EMM material removal takes place at atomic levels irrespective of their hardness and toughness. EMM of titanium by electrochemical process could be one of the promising alternative technique for machining of titanium.

Titanium machining is always a complex task before the researchers and manufacturing engineers because the pace of material research could not be maintained with machining or machinability research of titanium and its alloys.

9.4 Machining of Titanium by Anodic Dissolution

Material removal based on the principle of anodic dissolution was started in early twentieth century. The Russian engineer V.N. Gusev was the first to develop Electrochemical machining set up and patented all basic principles of ECM. The ECM process is conducted in the working chamber (electrochemical cell) of the machine, where a workpiece (Anode) and a tool electrode (Cathode) are positioned. The anode is connected to the positive polarity of a power supply, and the tool electrode to the negative polarity of the power supply. The electrolyte (usually an aqueous solution of an inorganic salt is pumped through the inter-electrode gap (IEG) in order to remove the electrode reaction products (gases, hydroxides) and the heat generated by the current. Electric current is passed through the electrodes by the movement of electrons. Flow of electric current in the electrolyte is takes place by the movement of ions. As the electrons transferred between the electrodes and the electrolyte, physical changes occur at the electrodes and anodic dissolution begins. To continue the process of material removal by anodic dissolution tool electrode is moved in the direction of the intended anodic dissolution (metal loss) in order to maintain a prescribed machining gap. As the anode dissolves, the cathode shape is reproduced on the anode i.e. work piece.

Machining of titanium by anodic dissolution is different than anodic dissolution of other metals such as Stainless steel, copper and mild steel etc. According to the position of titanium in periodic table and its electrochemical behaviour the titanium is metal, which always contain a natural oxide film, when it is exposed to air, water or media containing oxygen. Thickness of this natural oxide film of titanium ranges from 5 to 70 Å [9]. Titanium gains its excellent corrosion resistance due to the existence of this persistent and passivating surface film of oxide. This corrosion resistance property of stable oxide layer obstructs the electrochemical dissolution of titanium. Hence, to achieve controlled anodic dissolution of titanium it is necessary to dissolve protective oxide film of titanium.

9.4.1 Difficulties Encountered with Anodic Dissolution of Titanium in Microscopic Domain

Successful anodic dissolution process occurred when ionic conductivity between the electrode and electrolyte is sufficient. This ionic conductivity is depends on the metal and electrolyte combination. Dissolution of the metal electrode can be influenced by the formation of the oxide film on the anode surface. If the oxide film is passive in nature then its ionic conductivity is weaker. Hence, this passive oxide film acts as a barrier between the surface of the electrode and electrolyte. Development of this passive film depends on the relationship between current density and applied potential, i.e., polarization curve of the used electrolyte. This Passivating nature of oxide film present on the surface of titanium may terminate anodic dissolution process. The strength of this passive oxide film and its transpassive dissolution behaviour is based on how this passive film is formed on the titanium. In the case of macroscopic dissolution i.e. with ECM, the electrolyte is circulated with external pressure which creates high turbulence and thus reduces the effect of concentration polarization, which limits the passivation. In addition to that higher anode potential also causes anodic dissolution at higher valency by breaking down the passive oxide film. However, in microscopic domain i.e. with micro-ECM, the electrolyte flow is much lower or almost stagnant electrolyte is preferred as well as potential difference is also lower compared to ECM, this creates favourable situations for passivation. Selection of appropriate electrolyte combination as well as power parameters may minimize the chances of Passivating effect of oxide film. As discussed, titanium always possesses highly passive oxide film.

The nature of this thin tenacious film of oxide is highly passive in nature, which makes the anodic dissolution of titanium difficult compared to most of the other metals.

9.4.2 *EMM as a Potential Process for Titanium Micro Machining*

Electrochemical micromachining is appears as effective micromachining process due to its several advantages discussed in sect. 9.2. EMM is also capable of micromachining of chemically resistance material due to its ability to generate localized current density. Machining of titanium through EMM is relatively different from EMM of common metals. Titanium gains resistance to corrosion because of its protective surface oxide film. When titanium undergoes electrochemical actions, the role of transfer of Ti^{2+} and O^{2-} is to contribute for the development of anodic film. Formation of passive oxide layer on titanium surface during electrochemical process with aqueous environment was initiated by reaction of Ti^{2+} with hydroxide ions (OH^-) ionized from aqueous solution. Following electrochemical reactions represents stable titanium oxide (TiO_2) [10]. At the interface of anode workpiece and electrolyte, the reactions taking place are:

$$Ti \rightarrow Ti^{2+} + 2e^- \tag{9.1}$$

$$Ti^{2+} + H_2O \rightarrow TiO_2{}^+ + H_2 \uparrow \tag{9.2}$$

The oxocation, $TiO_2{}^+$ is acidic in nature and subsequently reacts with OH^- to form stable TiO_2 [11].

Following chemical reaction represents the formation of stable TiO_2:

$$TiO_2{}^+ + 2OH^- \rightarrow TiO_2 + H_2O \tag{9.3}$$

Throughout the anodic dissolution process, development of oxide layer with the help of titanium and hydroxide ions has been accelerated by the application of electric field. This thin oxide film is highly passive in nature, causes anodic dissolution of titanium difficult. The controlled anodic dissolution titanium contains this oxide film is difficult by EMM process parameters generally utilized for micromachining of other metals especially in terms of machining voltage and type of electrolyte. Hence, in order to attain uniform transpassive dissolution of titanium, the oxide film that obstructs the controlled dissolution of pure titanium in the passive potential region must lose its passivation phenomenon. Removal of oxide film is possible when the applied potential is adequately high [12]. The passive oxide film develops linearly with potential until a significant value is attained and the breakdown of the film takes place from random pitting at higher current densities and then shape controlled dissolution begins [13]. An additional factor, which plays vital role in rupturing this passive oxide film, is the electrolyte type. EMM offers both the advantages i.e. to achieve higher localized current density as well as possibility to utilize various non toxic eco-friendly aqueous as well as non aqueous electrolytes. Hence, EMM can be a potential micromachining method to achieve controlled anodic dissolution of pure titanium as well as titanium alloys.

9.5 Effect of Various EMM Process Parameters on Maskless EMM of Titanium

Material removal in electrochemical micromachining is based on anodic dissolution where metals are liberated from anode surface in atomic level. In order to achieve efficient and precise machining in the sub-microns level, the various process parameters of the EMM system play crucial role. The accuracy of the micro machined product in EMM is highly influenced by process parameters e.g. applied machining voltage, duty ratio, pulse frequency, concentration and type of electrolyte and micro tool vibration etc. During electrochemical micromachining of titanium these process parameters should be optimally selected and controlled due to Passivating phenomenon of titanium. Hence, appropriate selection along with controlling of all these EMM parameters play a vital role in attaining the preferred results during titanium micro machining utilizing EMM. Some of the predominant factors, which have major influences on EMM of titanium, are discussed to obtain fundamental information about how to control the EMM process most optimally while micromachining of titanium.

9.5.1 Role of Electrolyte

When potential is applied between the electrodes, electrolyte plays crucial role in flowing the electric current from electrode to electrolyte solution to achieve anodic dissolution through electrochemical reactions by completing the electric circuit. Selection of electrolyte is governed by material properties of workpiece and micromachining criteria such as machining rate, accuracy, surface texture, and surface integrity. Type and concentration of electrolyte proves crucial in anodic dissolution of materials which posses passive oxide film such as titanium.

Electrolytes of different combination and concentrations have been employed by various researchers during masked as well as maskless electrochemical micromachining of titanium. Regular pattern of up the depth of 30 μm on pure titanium by electrochemical dissolution through the patterned photoresist using sodium bromide and methanol solution containing sulfuric acid and the etching performance of these two electrolytes has been tested [14]. With the aqueous solution of bromide irregularly shaped cavities with a rough surface have been appeared. However,

Regular shape and smooth surface textures were obtained when titanium was dissolved in the methanol electrolyte. Solution of methanol sulfuric acid has also effective in electro polishing of titanium. It has been reported that the electro polishing of titanium in methanol sulfuric acid electrolytes is mass transport controlled. The limiting current density vary with electrolyte composition. This concluded that the dissolution of titanium tetravalent species from anode surface to the bulk solution is rate limiting due to the presence of compact salt film at the anode surface under limiting current conditions [15]. Crucial role of mass transport phenomenon in anodic dissolution process and presence of thin salt film at the limiting current has confirmed the linear growth of passive film with potential until a critical value reached where the film breakdown occurs and initiates the transpassive dissolution [13]. The porous anodization of titanium to create nano-scale features on titanium surfaces has been possible by utilizing sulfuric acid and H_3PO_4 electrolyte with potential sweep [12]. Electrochemical anodization of titanium through patterned photoresist mask using 0.5 M sulfuric acid and Phosphoric acid (H_3PO_4) were also performed [16]. Electrolyte combination containing Sodium chlorate ($NaClO_3$), Sodium nitrite ($NaNO_2$) and sodium fluoride (NaF) can be suitable for the generation of etchants of Hydrofluoric acid (HF) and Nitric acid (HNO_3) for machining of Ti alloy (Ti6Al4V) with the help of confined etchant layer technique (CELT). Its micromachining resolution depends mainly on the rate of the scavenging reaction. Sodium hydroxide (NaOH) is an effective scavenger to obtain sub-micrometer resolution [17]. Use of high concentration of aqueous sodium bromide with working voltage of as high as 200 V through jet electrochemical micromachining (Jet-EMM) technique were also successful to create deep holes in titanium alloy [18]. Anodic dissolution of titanium has been performed in NaCl containing ethylene glycol solution. Anodic polarization of titanium electrode in NaCl-containing ethylene glycol solutions involves dissolution of titanium as tetravalent species with the gas being evolved initially [19]. The electrochemical micromachining of pure Titanium with Ethylene glycol and Sodium Bromide of higher molar concentration up to 5 M can deliver satisfactory results however, machining time is very high due to the slower etch rate [20]. Aqueous solution of sodium bromide has been suitable for carrying out anodic dissolution of titanium with the help of conventional ECM in macroscopic domain [21].

Majority of electrolytes which can dissolve titanium in microscopic domain are Sulphuric acid, Hydrofluoric acid (HF), Nitric acid (HNO_3) and combination of Methanol and Sulphuric acid. All these electrolytes are either hazardous as well as toxic and may create problem to environmental issues. Aqueous solution of Sodium Bromide (NaBr) and sodium perchlorate can be suitable for anodic dissolution of titanium. These two electrolytes can be effectively utilize for controlled electrochemical micro machining of titanium by employing lower range of machining voltage and duty ratio. However, formation of black oxide film is predominant in all these aqueous base electrolytes. In order to overcome the effect of regeneration of oxide film due to aqueous base electrolytes non-aqueous base electrolytes can be effectively employed. Non-aqueous base electrolytes, which are non-toxic and environment friendly are combination of ethylene glycol and sodium bromide

(EG + NaBr) as well as combination of ethylene glycol, sodium bromide and sodium chloride (EG + NaBr + NaCl) were demonstrated excellent controlled anodic dissolution of titanium based on the criteria of lesser radial overcut and lower taper angle of the micro holes machined with these electrolytes [10]. Electrolyte solution includes bromide ions can promote anodic dissolution of titanium with valance of 4 i.e. Ti $\rightarrow$ Ti_4 + Electrolytes containing bromide ions have been effective in breakdown of oxide film [22].

9.5.2 Effect of Machining Voltage

In order to dissolve the titanium anodically, passive oxide film must lose its passivity. To overcome this difficulty suitable composition and type of electrolyte plays vital role to attack on the film and to break it through pitting. However, mere pitting or breaking the passive film at random points is not adequate. Phenomenon of pitting should be controlled to achieve uniform dissolution of base metal i.e. titanium. Another factor which is crucial to attain controlled anodic dissolution of titanium is applied voltage i.e. voltage between cathode and anode. To achieve uniform transpassive dissolution of titanium, the anodic oxide film that usually protects the titanium in the passive potential region must lose its protective properties. Rapid removal of oxide film is obtained when the applied potential is sufficiently high. The passive oxide film develops linearly with potential until a critical value is attained and after that the breakdown of the film takes place from local pitting at higher current densities and then shape controlled dissolution begins. Hence, to achieve proper current density with the application of appropriate potential range is crucial to overcome passivity of the oxide film. In order to attain shape controlled dissolution the desired current density in the appropriate potential region should maintain until the controlled dissolution gains stability. To make the process stable so as to attain desired accuracy of the machined product, influence of applied machining voltage, play vital role.

Study based on the breakdown voltage in terms of a change in efficiency of film growth the change in slope of the voltage/time relation at constant current density, are deceptive in the case of Ti in that such changes are not always accompanied by a change in columbic efficiency. This indicates that the higher voltage is necessary to breakdown the oxide film because the film consists of amorphous oxide which gives way to a new anodic behaviour low efficiency oxide growth and gas evolution associated with crystalline oxide in the film [23]. Based on the changes in the slopes of the voltage/time relations at constant current density occurred at ~100% efficiency for growth indicated that there is variable oxide field strength. During through-mask electrochemical machining, regular shape and smooth surface textures were obtained when titanium was dissolved at 8 V with mass transport limiting current density [14]. Micro-dimples with a diameter of 110 μm and depth of 20 μm could be generated with applied voltage of 24 V, pulse duty cycle of 10% and frequency of 100 Hz [24]. In masked EMM of titanium feature aspect ratio is limited as well as toxic and reactive electrolytes can be used hence, applied potential can vary from 5–30 V. However, in maskless EMM machining voltage can be limited to the range of 8–20 V, in few exceptional cases such as jet-electrolyte titanium machining voltage can be elevated to 200 V [18]. Maskless EMM of titanium, machining voltage is the major parameter in breaking down the passive oxide layer. Removal of passive oxide film under the influence of increased machining voltage is governed by the induced current density, as the current density increases passive layer breaks randomly at some weaker points and exposes the base metal. Further, increase in machining current rupture the passive oxide film and turns into transpassive leads to initiation of controlled anodic dissolution of titanium. Increase in machining voltage tends to rise in machining current. According to Faraday's law material removal rate increases with machining current and hence, higher material removed at higher machining voltage. With the increase in machining current in the narrow machining zone Joule heating effect generated which leads to elevation in the temperature in the narrow machining zone. This causes variation in electrolyte conductivity results in non-uniform current distribution in the inter electrode gap. Hence, reduces localization of current flux flow leads to random material removal with higher stray machining. Therefore, higher stray current flows in the micromachining zone causes more material removal from the larger area of workpiece, results in an increase in overcut. Figure 9.2 shows typical behaviour of radial overcut and material removal rate of micro holes machined on pure titanium sheet of 100 μm thickness, machining voltage is varied in the range of 8–14 V, utilizing maskless EMM process.

The micro holes machined with machining voltage of Fig. 9.3 shows the SEM micrographs of micro holes machined at 8 and 14 V machining voltage. From the

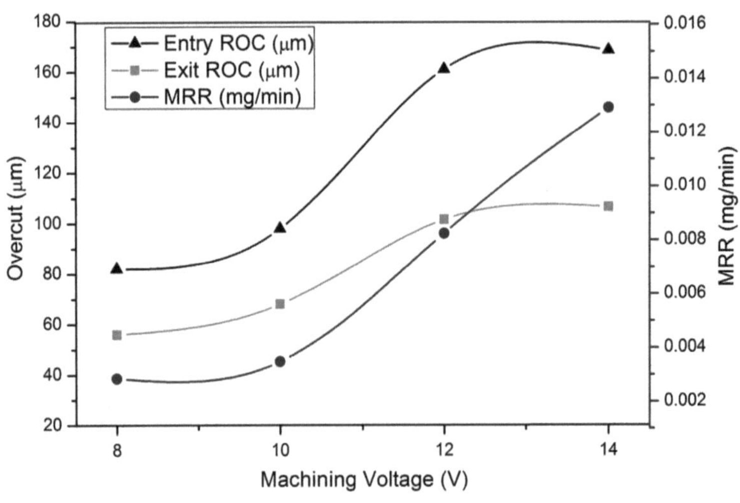

Fig. 9.2 Effect of machining voltage on radial overcut and MRR of micro hole

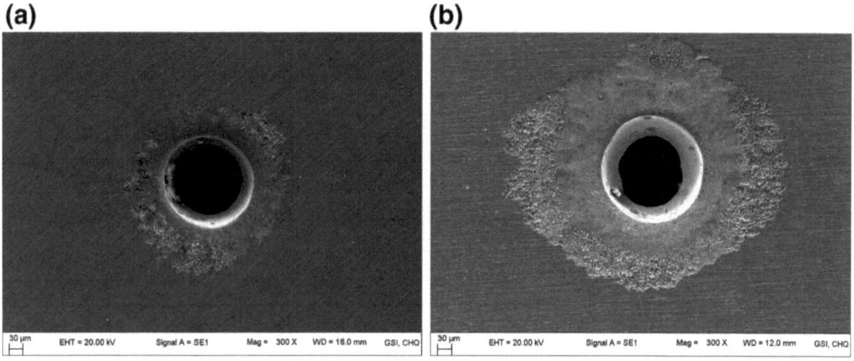

Fig. 9.3 Micro holes machined at **a** 8 V and **b** 14 V machining voltage [32]

SEM micrograph it is also confirmed that the effect of stray machining is predominant in micro hole machined at 14 V.

Stray machining induced by higher machining voltage can also increase conicity or tapering effect during microfeature fabrication on titanium. The effect of stray machining is higher at higher machining voltage, majority of stray machining takes place at the entry of the micro hole. When micro tool further advances into the workpiece machining depth increases, circulation of fresh electrolyte becomes crucial and accumulation of sludge in the narrow machining gap increases which in turn lower down material removal rate result in generation of smaller diameter at exit of micro hole compared to entry. Hence, tapering effect takes place and micro hole becomes taper in shape. However, this tapering effect can be utilized for fabrication of micro nozzles. To reduce the effect of stray machining during micro hole generation lower machining voltage is preferred or suitable microtool insulation has been applied.

9.5.3 *Effect of Pulse Duty Ratio*

Duty ratio represents the percentage of time for which pulse remains on i.e. percentage of time available for both faradic and non-faradic current. Increase in duty ratio in turn increases time available for faradic current. Hence, the amount of faradic effect and current density increases results in more material removal and causes increase in overcuts and MRR. Micro holes machined with varied duty ratio keeping other process parameters fixed at machining voltage 8 V, pulse frequency 200 kHz, tool feed rate 0.2 µm/s and keeping electrolyte concentration as 2 M/L. the influence of duty ratio on overcut and MRR during micro holes generation on titanium by EMM has been shown. From Fig. 9.4, it can be observed that radial overcut of micro hole at entry and exit is increases as duty ratio increases. Figure shows the influence of duty ratio on overcut and MRR during micro holes

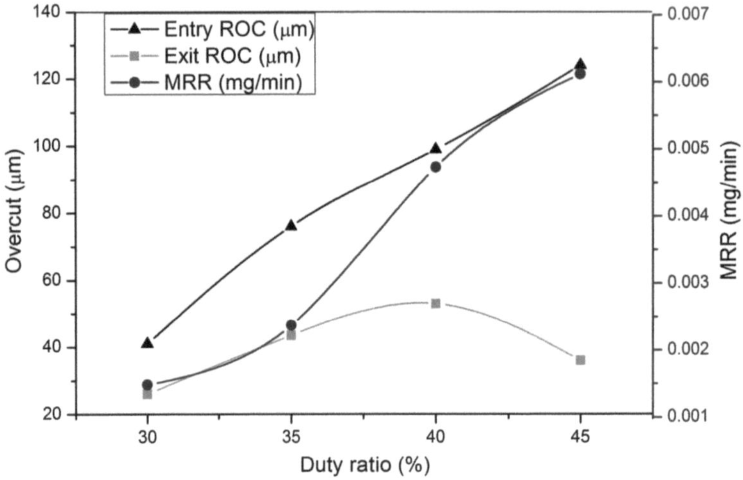

Fig. 9.4 Influence of duty ratio on radial overcut and MRR of micro hole

generation by EMM. From Fig. 9.4 it can be observed that radial overcut of micro hole at entry and exit is increases as duty ratio increases.

Increase in radial overcut at entry follows linear trend. The lowest radial overcut at 30% duty ratio at entry and exit is found to be 41 and 26 μm respectively. The highest radial overcut is found to be 124 and 36 μm at entry and exit at 45% duty ratio. As the duty ratio increases pulse ON time increases and machining current flows for higher time, which causes higher material removal as well as stray machining effect is also increases at the entry. Hence, radial overcut at entry is much larger than exit, which in turn increases taper of micro hole. The micro hole with highest taper angle of 41° has been obtained at 45% duty ratio. The SEM micrograph represented in Fig. 9.5 exhibits dominant effect of larger metal removal right from the entry of micro hole, machined at higher duty ratio of 45%.

9.5.4 Effect of Pulse Frequency

In EMM, pulse frequency directly governs pulse period. Total pulse period decreases as the pulse frequency increases, results in proportionate reduction in pulse ON time. Pulse ON time plays crucial role in charging and discharging of double layer capacitance. In the EMM, during pulse ON time of pulse period, the total current available consists of two components i.e. non faradic current and faradic current. Charging and discharging of double layer capacitance has been performed by Non faradic current and faradic current governs material dissolution rate. Thus, material dissolution performed only during faradic time of every cycle. Increase in pulse frequency results in lower faradic time tends to lesser material

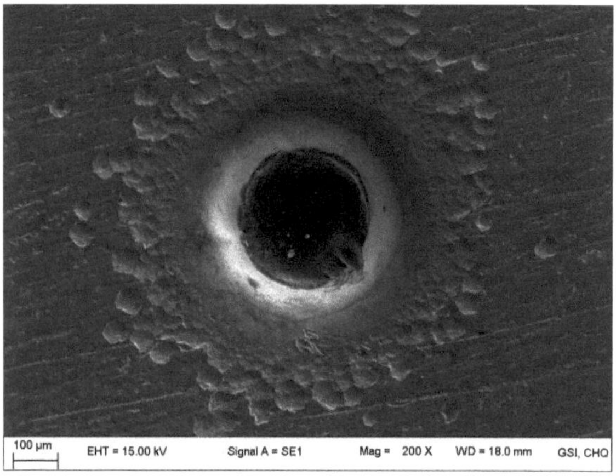

Fig. 9.5 Micoholes machined with 45% duty ratio

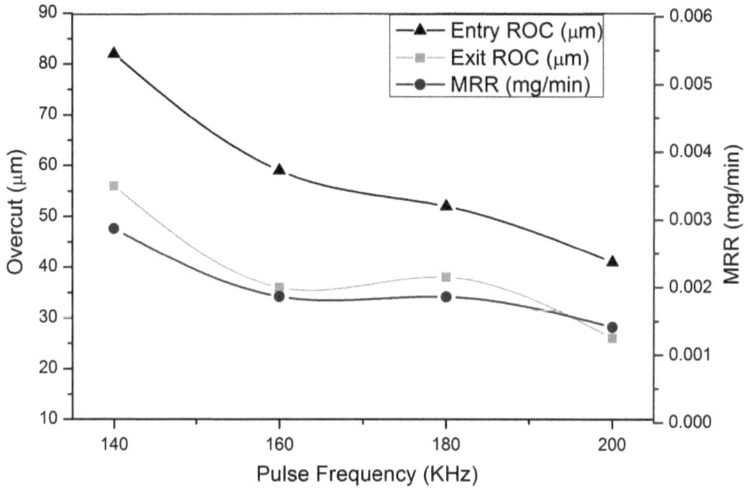

Fig. 9.6 Influence of pulse frequency on radial overcut and MRR of micro hole

removal per cycle as a result amount of sludge and gas bubbles are also lesser which will washed out completely from the narrow machining zone during pulse OFF time. Hence, it facilitates completely clean machining zone to accomplish more controlled dissolution. Therefore, controlled and localized material removal takes place which in turn improves the geometric accuracy of the microfeature. This is clearly observed from the experimental results of MRR as plotted against pulse frequency as shown in Fig. 9.6.

Micro hole machined with lower range of pulse frequency i.e. at 140 kHz has 282 µm entry diameter and 226 µm exit diameter. Whereas, micro hole machined with higher pulse frequency of 200 kHz has entry as well as exit diameter is reduced to 196 and 167 µm respectively. Hence, remarkable improvement will

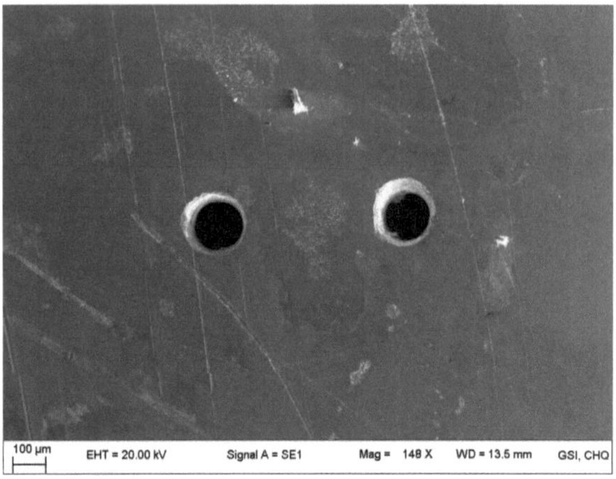

Fig. 9.7 SEM micrograph of micro holes machined at 200 kHz pulse frequency

takes place in machining accuracy of microfeatures machined at higher frequency compare to lower frequency. The micro holes machined at higher frequency is shown in Fig. 9.7. From the SEM micrograph, it is clear that the micro hole exhibits controlled geometrical shape and minimum overcut.

9.5.5 Effect of Micro Tool Vibration

During micromachining material dissolution takes place in very narrow gap between the cathode and anode electrodes. To maintain smooth anodic dissolution in this constricted machining zone flow of fresh electrolyte is important to flow off sludge and gas bubbles. If, external pressure applied to electrolyte to create the flow in this highly narrow machining zone may displace the microtool. However, vibrations of very small amplitude with fairly high frequency have been applied to microtool, made substantial effect on enhancing the dissolution process by circulating electrolyte in the form of micro-jets produced due to microtool vibration. Microtool vibration creates hydrodynamic effects on the bubble behavior, which has been utilized for the effective removal of sludge, hydrogen bubbles and replenishment of fresh electrolyte [25]. The application of microtool vibrations made significant influence on the diffusion and convection of dissolved metal ions because of hydrodynamic effects on the bubble performance. This phenomenon of enhancing anodic dissolution in EMM has also effectively utilized in maskless electrochemical micromachining of titanium. Machining voltage has been lowered down as well as controlled dissolution of titanium could be achieved by introduction of microtool vibrations. Machining current density induced by applied voltage is important in dissolution of material during electrochemical reaction. Machining current density has been considerably elevated by the effect of microtool vibrations. Microtool vibrations within the narrow machining zone will help to

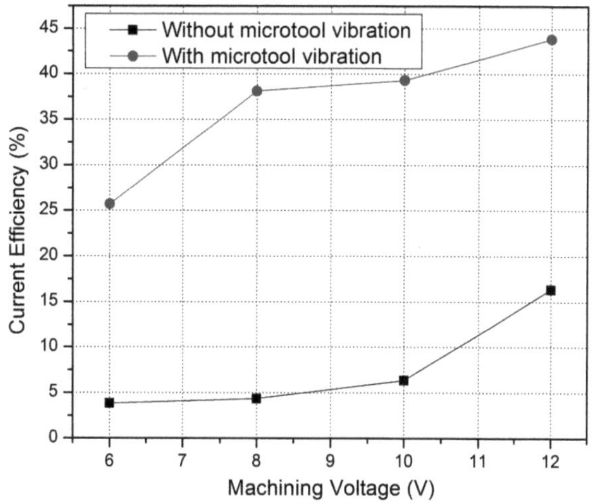

Fig. 9.8 Effect of microtool vibration on change in current efficiency

circulate the fresh electrolyte and hence increase the convective mass transport which in-turn increase rate of diffusion as well as conductivity results in increase in current density.

The current efficiency, which is the ratio of observed amount of metal dissolved to the theoretical amount predicated from Faraday's law has been influenced by dissolution of titanium with the application of microtool vibration. Figure 9.8 shows the change in current efficiency with respect to voltage during micro hole machining on pure titanium.

The major change in current efficiency proves that the application of microtool vibration facilitates circulation of electrolyte, which causes increase in the utilization of considerable portion of machining current for the metal dissolution and leads to increase in current efficiency which further enhances mass of metal removal results in increase in machining depth. Figure 9.9 shows the comparison of machining depth achieved by two conditions of with microtool vibration and without microtool vibration. It is clear that the microtool vibration significantly improves the controlled anodic dissolution of titanium even at lower machining voltage.

SEM micrograph shown in Fig. 9.10 exhibits the micro holes machined on titanium with and without microtool vibrations with machining voltage of 8 V.

9.6 Through Mask EMM of Titanium

Through-mask EMM (TMEMM) is a typical masked electrochemical micromachining process. This process is commonly employed for fabrication of micropatterns, as well as for shaping and finishing of 2D microfeatures. In TMEMM anode workpiece is covered with photoresist patterned in the form required shape pattern.

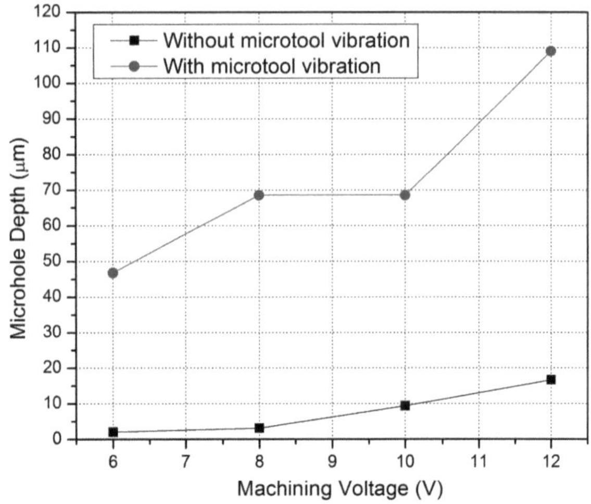

Fig. 9.9 Effect of microtool vibration on change in machining depth

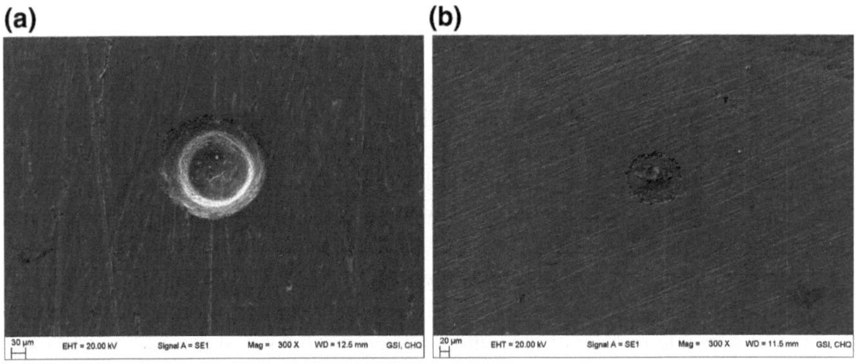

Fig. 9.10 SEM image of microhole machined at **a** with microtool vibration **b** without microtool vibrations [32]

Uncovered or unmasked metal surface is removed by high rate anodic metal dissolution. In through-mask EMM, metal dissolution takes place at the workpiece surface that lies at the bottom of the cavity created by the photoresist mask [26]. The metal removal in TMEMM can be possible by two ways i.e. on one side of masked metal or from two sides of masked anode workpiece which is as shown in Fig. 9.11.

Through mask electrochemical micromachining of titanium has been carried out in solutions containing hydrofluoric acid. Figure 9.3 shows a SEM micrograph of an individual cavity etched for 2 min in a solution containing aqueous hydrofluoric acid. An irregular shape and a rough surface texture result from crystallographic attack by the acid solution. Anodic dissolution of titanium through the patterned photoresist was also accomplished with the NaBr electrolyte at applied potentials

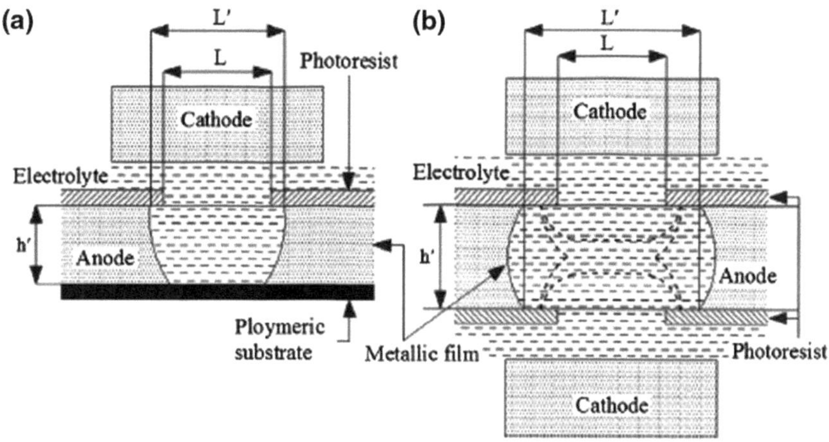

Fig. 9.11 Through-mask EMM: **a** one sided and **b** two sided [29]

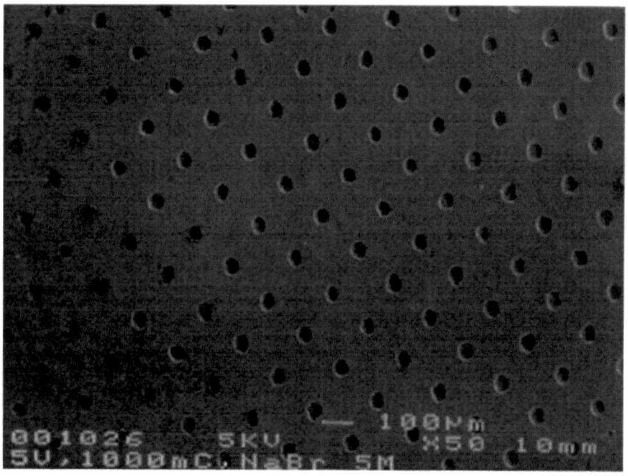

Fig. 9.12 SEM micrograph of a regular pattern etched on a titanium surface at 5 V in the NaBr electrolyte by passing a charge of 4 C. The pattern consists of 50 m diameter cavities, separated by 130 m [14]

ranging between 2 and 8 V. At 2 V only a few cavities dissolved randomly across the pattern while the remaining features of the pattern were not etched; moreover, the etched cavities were large and irregular. The patterns achieved are of pitting nature of titanium dissolution in NaBr electrolytes at the lower potential. However, at 8 V all the cavities dissolved across the pattern. When dissolution was performed at a potential of 5 V all the cavities were etched uniformly across the surface. The etched pattern were enhanced geometric feature with better surface quality when etched with 5 V. Figure 9.12 shows a regular pattern etched on a titanium surface at 5 V in the NaBr electrolyte by passing a charge of 4 C. Dissolution time corresponded to 2 s. The pattern consists of 50 μm diameter cavities, separated by 130 μm.

Electrochemical dissolution of titanium through the patterned photoresists was performed with the 3 M sulphuric acid containing electrolyte at 0 °C. Figure 9.13

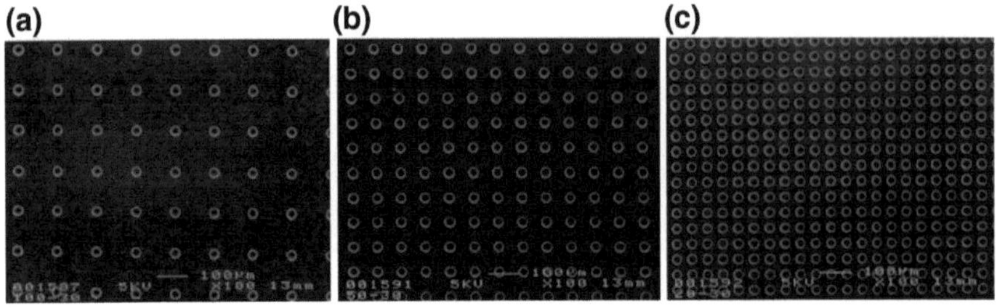

Fig. 9.13 SEM micrograph of a pattern etched on a titanium surface at 8 V in 3 M H2SO4 in methanol. The pattern consists of 30 μm diameter cavities, separated by 130 μm (**a**), 80 μm (**b**), and 50 μm (**c**) [14]

shows regularly shaped and smooth patterns fabricated by electrochemical etching at 8 V. The pattern consists of 30 μm diameter cavities, separated by 130, 80 and 50 μm.

The extended application of through mask electrochemical micromachining (TMEMM) is by combining maskless UV and electron beam lithography in combination with robust SU-8 photoresist technology for increasing the flexibility in pattern shape and potential scale down of feature size on planar substrates [16]. In this technique, features with a gradient in etch depth have been possible in one single micromachining process. Electrolyte of 3 M H_2SO_4 in methanol at −10 °C using a 2-electrode setup in a jacketed and tempered glass cell was utilized. Maskless Ultraviolet and e-beam patterned SU-8 is highly suitable for planar electrochemical surface micromachining from submicron to several hundred microns feature scale, excellent chemical stability of SU-8 together with the flexibility of maskless UV lithography providing great freedom in the pattern design. This technique is suitable for patterning of highly curved surfaces with substrate material choice, good mask patterning speed and freedom in feature shape. Figure 9.14 exhibits Ti surface with well-defined hemispherical cavities by TMEMM via maskless UV patterned SU-8.

Through mask electrochemical micromachining is an effective way to produce regular patterned microfeatures on titanium. However, in TMEMM processes relatively low achievable aspect ratio is the main limitation.

9.7 Micro Features Generation on Titanium

Masked as well as maskless EMM has proved its compatibility in fabrication of various microfeatures on titanium. Masked EMM i.e. TMEMM or oxide film laser lithography (OFLL) are the methods popular for machining regular patterned microfeatures such as micro dimples or microholes on titanium. As discussed, TMEMM can be effective in generation of micro patterns by applying photoresist

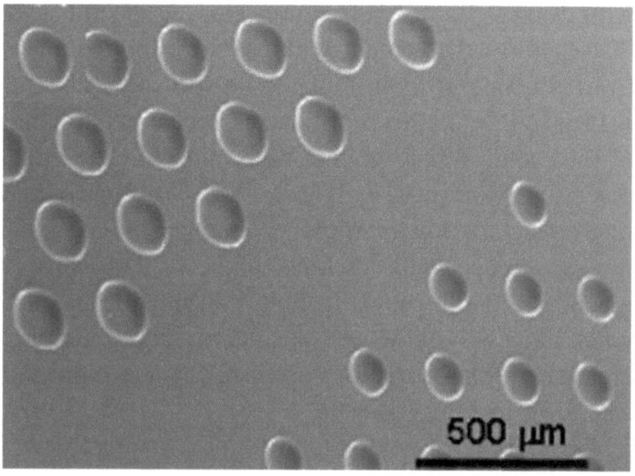

Fig. 9.14 Hemispherical cavities by TMEMM via maskless UV patterned SU-8 [16]

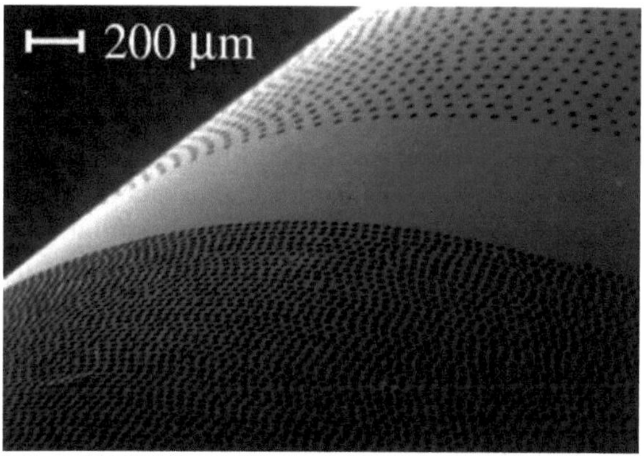

Fig. 9.15 Titanium cylinder surface electrochemically microstructured using OFLL [13]

mask on the anode workpiece. However, in oxide film laser lithography oxide layer on the surface of titanium is developed through anodic oxidation of the titanium in an anodising electrolyte such as sulphuric acid, thereafter patterning of the oxide layer has been carried out by excimer laser irradiation in air to form desired pattern in such a way that the irradiated area of oxide layer is exposed for anodic dissolution. Electrochemical dissolution of the exposed titanium metal from the irradiated areas has been performed with an electropolishing electrolyte such as H_2SO_4 in methanol. After dissolution, protruded oxide film due to under-etching is removed by ultrasonic cleaning [27]. The OFLL technique is better adapted for fabricating multilevel structures since it does not require application of a photoresist. Figure 9.15 shows microstructure fabrication by EMM utilizing OFLL. Various microfeatures which will find potential applications in MEMS can be effectively produced by this method.

Electrochemical bulk micromachining method named the confined etchant layer technique (CELT) for micromachining of titanium and its alloys. In this process the etchant is generated electrochemically on the surface of a machining tool or a mold with desired 3D microstructures [17]. A specific scavenger added to the electrolyte that captures the etchant within a very short duration so as to prevent the etchant from diffusing away from the mold surface. Thus, the etchant layer around the mold is kept so thin that its profile takes approximately the contour of the microstructures of the mold. When moving the mold until the etchant layer contacting the workpiece, the workpiece will be etched. An approximate mirror-image replica of the microstructures of the mold is obtained by continuously approaching the mold to the workpiece with etching process. This process can control precisely the machining depth by controlling the moving distance of mold. It can be used in a batch process with fewer steps than in photolithography. Authors demonstrated the creation of microstructures consists of trapezoidal slots fabricated on Ti6Al4V alloy as shown in Fig. 9.16.

Generation of microfeatures into titanium substrates of various thicknesses, ranging from 0.5-mm sheet to 10 μm free-standing titanium foils has been fabricated by Metal Anisotropic Reactive Ion etching with Oxidation (MARIO) process [28]. This process has capability to fabricate arbitrarily high-aspect-ratio structures with straight sidewalls micromachined structures free of residual stresses. The MARIO process permits the creation of bulk titanium MEMS, which offers potential for the use of a set of material properties beyond those provided by traditional semiconductor-based MEMS.

Maskless electrochemical micromachining can also be effectively utilized for fabrication of microfeatures on titanium. Various microfeatures such as microholes, micrnozzles and microslots can be successfully machined by maskless electrochemical micromachining. Microholes and micronozzles can be fabricated by

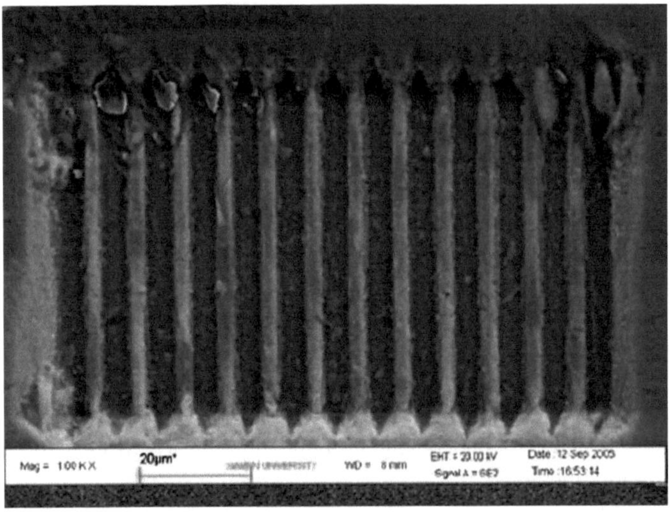

Fig. 9.16 Trapezoidal slot microstructure fabricated on Ti alloy surface in the solution composed of 0.2 M NaF + 0.4 M $NaClO_3$ + 0.6 M $NaClO_4$ + 0.3 M $NaNO_2$ + 0.1 M NaOH [17]

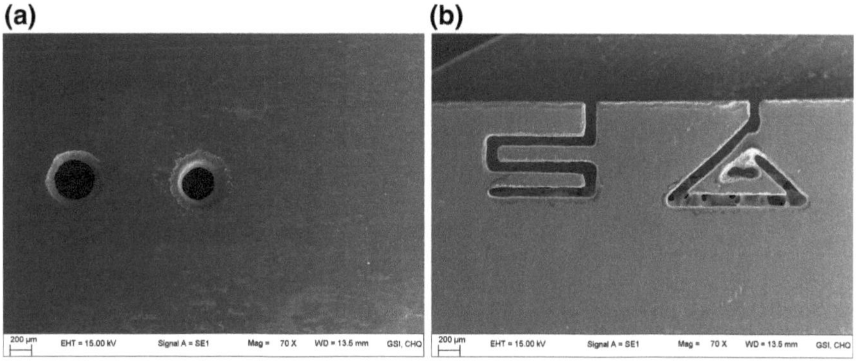

Fig. 9.17 Microfeatures fabricated by maskless EMM **a** microholes **b** microcantilevers

utilizing stagnant electrolyte of mixture of ethylene glycol and sodium bromide by advancing microtool into the titanium workpiece made as anode. Microfeatures other than microholes can fabricate by employing micromilling strategy i.e. cylindrical microtool which acts as cathode removes material in layer by layer travelling mode or by sinking and milling method in the sinking and milling method initially microdrilling is performed up to the required depth of the microfeature, followed by milling along the path of the microprofile [29]. Effect of stray machining and pitting of oxide layer is still an Issue in maskless EMM. However, utilizing various shaped microtools such as disc, cone, conical disc shape and insulated microtools or optimizing various process parameters can minimize these issues to some extent. Figure 9.17 shows microholes and microcantilevers fabricated on titanium by maskless EMM.

Microfeature such as micro slits were also fabricated on titanium alloy using Wire electro chemical machining (WECM) [30]. In this process tungsten wire of diameter in the range of 10–50 μm were utilized. The electrolyte has been introduced between the machining zone by axial electrolyte flushing system to facilitate removal of electrolysis products and renewing electrolyte. Various process parameters such as wire feed rate, machining voltage, electrolyte concentration for cutting microslots or micro slits with various structures were optimized and various micro slits as well as slots can be machined with this process. Instead of using single wire, multi-wire can be used to improve the machining productivity of WECM. Figure 9.18 shows typical microfeature fabricated with WECM.

9.7.1 Suitable Range of EMM Process Parameters for Fabrication of Micro Features on Titanium

Fabrication of microfeatures on titanium by EMM is relatively different than microfeatures fabricated on other metals such as stainless steel, copper etc.

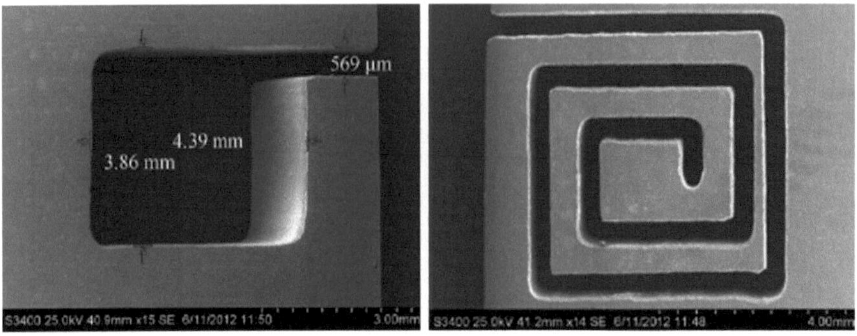

Fig. 9.18 Microfeatures fabricated by WECM on titanium alloy (Ti6Al4V) [30]

especially in terms of machining voltage and electrolyte. However, identification and selection of suitable range of EMM process parameters either in masked or in maskless EMM process parameters play crucial role in achieving machining accuracy of microfeatures generated on titanium. In masked EMM, dissolution of metal is confined to the exposed metal surface among the patterned photoresist therefore, anodic dissolution must occurs within the patterned area. Hence, machining time and machining current induced due to applied potential plays vital role. In addition to that electrolyte combination is also an important factor, mostly toxic acidic base electrolytes are preferred in masked EMM or TMEMM of titanium. Generally, in reported work of the masked EMM of titanium, machining voltage ranges from 8 to 150 V with machining temperature as low as −10 °C with electrolyte combination of sulphuric acid in methanol of higher concentration up to 3 M or hydrofluoric acid were also employed [14, 16, 29]. However, in maskless EMM major challenge during micromachining of titanium lies in controlling the dissolution process by localizing the machining area. Localization of dissolution in maskless EMM is inversely proportional to stray machining effect. Therefore, highly localized controlled dissolution is important to achieve higher machining accuracy of microfeature generated by maskless EMM. Selection of suitable process parameters can directly influence accuracy of microfeature. In order to generate precise microfeature in pure titanium machining voltage in the range of 8–10 V with pulse frequency 160–200 kHz keeping duty ratio 30–35% and optimum microtool feed rate of 0.6–0.8 µm/s has been found suitable process parameters of maskless electrochemical micromachining [31]. Machining accuracy of microfeatures generated by maskless EMM can be further improved by utilizing electrolytes which can break passive layer of titanium in lower current density which will further reduce machining voltage in turn achieving more localized dissolution of titanium.

9.7.2 Potential Applications of Titanium Micro Features

Titanium and titanium alloys gained popularity in aerospace, chemical process and biomedical industry due to their biocompatibility, good mechanical properties and excellent corrosion resistance. Titanium has also possesses plasticity-based failure when subjected to external loading which makes titanium superior device safety capability with increase safety and reliability. These properties qualifies the titanium as potential material to MEMS and BioMEMS applications. Titanium has also a established superior biocompatible metal with a proven applications in dental and orthopaedic implants, which confirms its physiological compatibility for novel biomedical microdevices. To state few examples of such devices are microneedles for transdermal drug delivery [22], thin-foil devices for biomolecule separation and characterization [23], large-area thermal ground planes for electronics cooling [24, 25], and rationally nanopatterned substrates for improved cellular response [26]. Titanium has also proved its compatibility for sophisticated area such as aerospace. In many aerospace applications, array of micro holes, micro-nozzles in titanium alloy plays vital role in critical aerospace applications. EMM can provide better and reliable micromachining solution for titanium micromachining so as to meet these emerging titanium microfeatures requirements.

9.8 Future Scope and Challenges in Titanium Micromachining

Titanium, due to its versatile physical and mechanical properties is a highly demanding material in MEMS and biomedical engineering applications therefore titanium will replace traditional MEMS materials such as Silicon and/or semiconductor material. These new opportunities of this material will pose a new challenge of effective micromachining of titanium and its alloys. EMM has proved its compatibility to micromachine various microfeatures on titanium and its alloys. However, few issues are still to be addressed to make EMM more stable and more precise technique for titanium micromachining. Masked EMM or TMEMM have to be improved in terms of effective photoresist masking technique in order to achieve more confined precise material dissolution so as to minimize undercut material dissolution below the edges of mask. Masking of cathode needs to be done instead of anode workpiece which will enhance the productivity in terms of minimizing the complexity of workpiece masking. Use of non-toxic and environment friendly electrolytes should be promoted during EMM of titanium. During maskless EMM total control of anodic dissolution of titanium can be achieved by lowering down the machining voltage in the range of 2–5 V. This will allow application of higher range of pulse frequency and thus more localized machining. Micromachining of high to very high aspect ratio microfeatures is still a challenge which needs to be addressed.

9.9 Conclusions

Electrochemical micromachining is the only process stress free and clean machining in the microscopic domain irrespective of material hardness. Titanium and its alloys known for its excellent material properties and difficulty in machining either in macro or in microscopic domain can be smoothly micromachined by EMM. To generate precise regular micro patterns on titanium and its alloys by the application of Masked EMM or TMEMM technique is highly suitable, which is proved by generating various micro patterns on titanium surface. Maskless EMM for titanium microfeature generation through stagnant non-aqueous base electrolyte by employing various micromilling strategies is found effective. These techniques established suitable process parameters for EMM of titanium either by masked or maskless techniques. Considering the emerging potential of titanium microfeatures in various MEMS and BioMEMS applications EMM will appear as a most reliable and stable micromachining process for titanium micromachining.

References

1. Snoeys, R., Staelens, F., Dekeyser, W., 1986, Current trends in non-conventional material removal processes, Annals of the CIRP, 35, 2, 467–480
2. M. Peters, J. Hemptenmacher, J. Kumpfert and C. Leyens, "Structure and Properties of Titanium and Titanium Alloys," In: Titanium and Titanium Alloys: Fundamentals and Applications, C. Leyens and M. Peters (eds.), 2003 WILEY-VCH Verlag GmbH & Co. KGaA, Weinheim. (2003)
3. W. Konig, "Applied Research on the Machinability of Titanium and Its Alloys," Proceedings of the Forty-seventh Meeting of AGARD, AGARD, CP256, 1.11.1 O(1979), AGARD Structural and Material Panel
4. A. R. Machado and J. Wallbank, "Machining of Titanium and Its Alloys-A Review," Proceedings of Institution of Mech Engrs, Part B: Journal of Engineering Manufacture,204 (1990) 53–60
5. N. Zlatin and M. Field, "Procedures and Precautions in Machining Titanium Alloys," Titanium Science and Technology, 1(1973)489–504
6. Kumar Jatinder, Khamba J.S., Mohapatra S.K., An investigation into the machining characteristics of titanium using ultrasonic machining, International Journal of Machining and Machinability of Materials, Vol. 3, Nos. 1/2, 2008
7. Vinod Yadav, Vijay K. Jain, Prakash M. Dixit, Thermal stresses due to electrical discharge machining, International Journal of Machine Tools & Manufacture 42 (2002) 877–888
8. Yan Cherng Lin, Biing Hwa Yan, Yong Song Chang, Machining characteristics of titanium alloy (Ti6Al4 V) using a combination process of EDM with USM, Journal of Materials Processing Technology 104 (2000) 171–177
9. Aladjem A., Anodic Oxidation of titanium and its alloys, J.O. Material Science 8 (1973) 688–704
10. Anasane S.S., Bhattacharyya B., Experimental investigation on suitability of electrolytes for electrochemical micromachining of titanium, International Journal of Advanced Manufacturing Technology, DOI 10.1007/s00170-015-8309-2
11. Cotton FA, Wilkinson G, Murillo CA, Bochmann M (1999) Advanced inorganic chemistry, 6th edn. Wiley, New York

12. Zinger O.,Chauvy P.-F, Landolt D., Scale-Resolved Electrochemical Surface Structuring of Titanium for Biological Applications, Journal of The Electrochemical Society, 150 ~ 11 B495-B503 ~ 2003
13. Landolt D., Chauvy P-F.Zinger O., Electrochemical micromachining, polishing and surface structuring of metals: fundamental aspects and new developments, Electrochemica Acta 48 (2003) 3185–3201
14. Madore C., Landolt D., Electrochemical micromachining of controlled topographies on titanium for biological applications, J. Micromech. Microeng,**7** (1997) 270–275
15. Piotrowski O., Madore C. and Landolt D., The mechanism of electropolishing of titanium in Methanol-Sulfuric acid electrolytes, J. Electrochem. Soc., Vol. 145, No.7, July 1998 The Electrochemical Society, Inc.
16. P. Kern, J. Vehand J. Michler, New developments in through-mask electrochemical micromachining of titanium, J. Micromechanics and Microeng. 17 (2007) 1168–1177
17. L. M. Jiang, W. Li, A. Attia, Z. Y. Cheng, J. Tang, Z. Q. Tian, Z. W. Tian, “A potential method for electrochemical micromachining of titanium alloy Ti6Al4 V”, J Appl Electrochemistry (2008) 38:785–791
18. Xiong Lu, Yang Leng “Electrochemical micromachining of titanium surfaces for biomedical applications”, J of Material processing technology 169 (2005) 173–178
19. Fushimi Koji, Habazaki Hiroki, Anodic dissolution of titanium in NaCl containing ethylene glycol, Elechtrochimica Acta 53 (2008) 3371–3376
20. Terje Sjöström, Bo Su, Micropatterning of titanium surfaces using electrochemical micromachining with an ethylene glycol electrolyte, Materials Letters 65 (2011) 3489–3492
21. Dhobe Shirish D., Doloi B., Bhattacharyya B., Surface characteristics of ECMed titanium work samples for biomedical applications, Int J Adv. Manuf. Technol (2011) 55:177–188
22. Bannard J., On the electrochemical machining of some titanium alloys in bromide electrolytes, J. Applied electrochemistry 6 (1976) 477–483
23. C. K. Dyer and J. S. L. Leach, Breakdown and Efficiency of Anodic Oxide Growth on Titanium, J. Electrochem. Soe.:ELECTROCHEMICAL SCIENCE AND TECHNOLOGY July 1978 pp-1032–1038
24. Xiaolei Chen & Ningsong Qu, &Zhibao Hou, International Journal of Advanced Manufacturing Technology, DOI 10.1007/s00170-016-8807-x April 2016
25. Ghoshal B., Bhattacharyya B., Influence of vibration on micro-tool fabrication by electrochemical machining, Int. J. Mach. Tool. Manuf. 64 (2013) 49–59
26. G.J. Kwon, H.Y. Sun, H.J. Sohn, Wall profile developments in through mask electrochemical micro machining of invar alloy films, J. Electrochem. Soc. 142 (9) (1995) 3016–3020
27. P.F. Chauvy, P. Hoffmann, D. Landolt, Applications of laser lithography on oxide film to titanium micromachining, Appl. Surf. Sci. 208–209 (2003) 165–170
28. M.F. Aimi, M.P. Rao, N.C. Macdonald, A.S. Zuruzi, D.P. Bothman, High- aspect-ratio bulk micromachining of titanium, Nat. Mater. 3(2004) 103–105
29. Bhattacharyya B., “Electrochemical Micromachining for Nanofabrication, MEMS and Nanotechnology”, William Andrew publications 2015
30. Qu Ningsong, Fang Xiaolong, Li Wei, ZengYongbin, Zhu Di, Wire electrochemical machining with axial electrolyte flushing for titanium alloy, Chinese Journal of Aeronautics, 2013,26(1): 224–229
31. Anasane S.S., Bhattacharyya B, Investigation on micromilling of through microslots on titanium by electrochemical micromachining, Int. J. Precision Technology, Vol. 6, Nos. 3/4, 2016
32. Anasane S.S., Bhattacharyya B, Experimental investigation into fabrication of microfeatures on titanium by electrochemical micromachining, Advances in Manufacturing (2016) 4:167–177

Chapter 10
Electrochemical Discharge Micro-machining of Engineering Materials

B.R. Sarkar, B. Doloi and B. Bhattacharyya

Abstract Electrochemical discharge micro-machining (micro-ECDM) process appears to be very promising as a future micro-machining technique, since in many areas of applications it offers several advantages, which include machining of variety of electrically non-conducting hard, brittle materials including glass, ceramics and composites etc. It is an advanced hybrid micro-machining process that combines the techniques of electrochemical machining (ECM) and electrodischarge machining (EDM). This book chapter focuses on the current researches and developments in micro-ECDM process. The chapter discusses in details about the micro-ECDM system, which includes the mechanical hardware unit, electrolyte supply unit and electrical power supply unit etc. The effects of various factors on different machining performance characteristics such as material removal rate, accuracy, heat affected zone, gas film quality, machining depth, surface topography and tool wear etc. during its application mainly for micro-drilling and micro-cutting operations on engineering materials are represented in this chapter. The chapter also gives a glimpse on the fundamentals, problematic areas and applications of micro-ECDM process and highlights the challenges and future possibilities of research in this area. The recent advancements for improvement of performance of μ-ECDM process by using the rotating and travelling micro-tool, controlling the gap between the tool and the workpiece, changing the shape of micro-tool and also controlling the surface texture and material of tool for micro-spark discharge for required micro-machining operations are also depicted in this chapter. The chapter is expected to open up new insights into the process characteristics for successful application of electrochemical discharge micro-machining (micro-ECDM) process and provides valuable guidance to the applied researchers and manufacturing scientists for setting up unique platform for micro-machining electrically non-conducting engineering materials.

B.R. Sarkar · B. Doloi (✉) · B. Bhattacharyya
Production Engineering Department, Jadavpur University, Kolkata 700032, India
e-mail: bdoloionline@rediffmail.com

B.R. Sarkar
e-mail: sarkarbiplab_s@rediffmail.com

B. Bhattacharyya
e-mail: bb13@rediffmail.com

G. Kibria et al. (eds.), *Non-traditional Micromachining Processes*,
Materials Forming, Machining and Tribology,
DOI 10.1007/978-3-319-52009-4_10

10.1 Introduction

The electrochemical discharge in electrolyte cell was observed in 1968 by Kurafuji and utilized for machining soda lime glass. The electrochemical discharge machinings (ECDM) [1–3] technology has been developed by combining two machinings i.e. electrochemical machining (ECM) and electrodischarge machining (EDM). This technology has been utilized for machining electrically non-conducting engineering materials in the product range of meso as well as micro scale. *Electrochemical Discharge Micro-machining* (μ-ECDM) is an advanced hybrid micro-machining process comprising the techniques of electrochemical reactions and electrodischarge phenomena. The process is also referred as "*Electrochemical Spark Machining (ECSM)*" [4, 5] and "*Spark Assisted Chemical Engraving (SACE)*" [6]. The process is important since it can machine a variety of electrically non-conducting hard, brittle materials including glass, ceramics and composites etc. in the micron level. Non-conventional micro-machining processes such as μECM, μEDM, etc., are not able to machine electrically non-conducting engineering materials. Micro-ECDM technology is used for micro-machining operation using micro dimension tool (less than 500 μm) as cathode electrode. Depending upon the shape and size of micro-tool, the micro-features are generated on workpiece. Micro hole/cut (less than 500 μm) can be generated on micro-components. The travelling wire can be used as cathode electrode for profile cutting and micro-grooving (width of cut or groove less than 500 μm) using ECDM technology and this process is called as *Travelling Wire Electrochemical Discharge Machining (TW-ECDM)*. Research works on μ-ECDM and TW-μECDM processes are going on to solve the problems during its application in the area of micro-cutting, micro-grooving and micro-drilling etc. on various engineering materials like glass, quartz, ceramics (alumina, zirconia, silicon nitride etc.) and composites etc. There are recent advancements for improvement of performance of μ-ECDM process by rotating the micro-tool, controlling the gap between the tool and the workpiece, changing the shape of micro-tool, controlling the surface texture and material of tool and also controlling the bubble layer thickness for micro-spark discharge for required micro-machining operations.

10.1.1 Fundamentals of Electrochemical Discharge Micro-machining (Micro-ECDM) Process

Electrochemical Discharge Micro-Machining (μ-ECDM) is a hybrid and non-traditional micro-machining process consisting of electro-chemical machining and electro-discharge machining. The material removal in μ-ECDM process occurs in the same way as that takes place in ECDM process. Although the basic mechanism of the process is not yet completely understood and is still a matter of research investigations yet it can be revealed that material is removed due to the combined effects of electrical spark discharge (ESD) action and electro-chemical reactions.

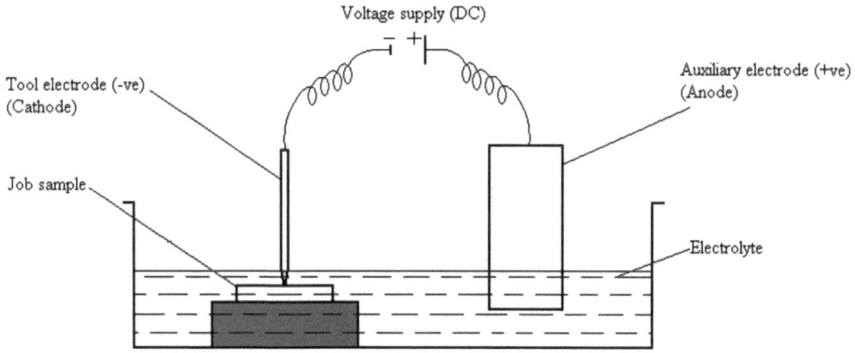

Fig. 10.1 Electrolytic cell of micro-ECDM system

In μ-ECDM the electrolyte cell as shown in Fig. 10.1 is similar to that used in ECM. The workpiece is dipped in an appropriate electrolyte solution (typically sodium hydroxide or potassium hydroxide) and the level of electrolyte is just 1 mm above the tool tip and it is controlled by electrolyte supply unit. A constant D.C. voltage is applied between the machining tool or tool-electrode and the auxiliary electrode with proper polarity, generally positive terminal as *anode* (*auxiliary electrode*) and negative terminal as *cathode* (tool). The auxiliary electrode is in general a flat metal plate. The tool-electrode surface is always significantly smaller than the auxiliary electrode surface (by a factor of 100). When tool touches the job, there are micro-gaps present both in the surface of tool and job due to surface irregularities. The electrolyte present in micro-gaps is responsible for the generation of gas and vapour bubbles in the micro-gaps and surrounding tool surface. The generation of the combined gas and vapour bubbles takes place due to electro-chemical reactions in ECM and joule heating of electrolyte respectively. From the analysis of the ECM, it has been found that there are two types of reactions usually occur in the system: electrochemical reactions at the electrodes and chemical reactions in the bulk of the electrolyte [3]. When the applied voltage is increased hydrogen gas bubbles grow in size. Their nucleation site density increases, current path gets restricted between cathode and electrolyte interface causing discharge to occur at this interface instantly. Thus these bubbles ultimately form an insulating gas film layer around the tool-electrode as shown in Fig. 10.2. When voltage gradient is sufficient to breakdown this gas bubbles layer between the tool and workpiece i.e. beyond a threshold value, a conductive path is developed for the spark discharge owing to the ionisation of the gas bubbles, which thereby causes the flow of a large amount of current. The electro-discharge action takes place between the tool (cathode) and the electrolyte across the gas bubble layer. Each electric discharge causes a focused stream of electrons to move with a very high velocity and acceleration from the tool (or cathode) towards workpiece. Ultimately it creates the compressive shock waves on the workpiece surface as depicted in Fig. 10.3. This discharge phenomenon occurs within few micro-seconds and as a result temperature of the spot hit by

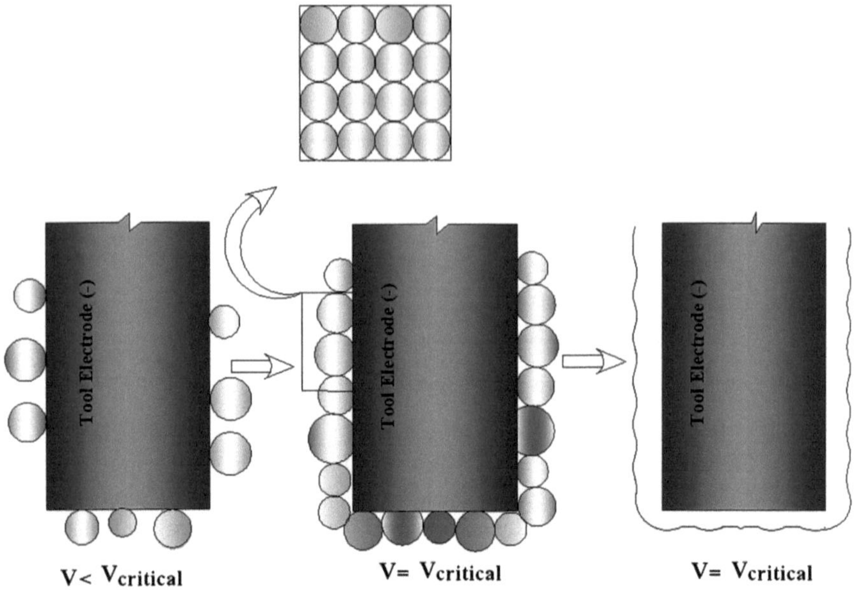

Fig. 10.2 Gas bubble formation as in the micro-ECDM process

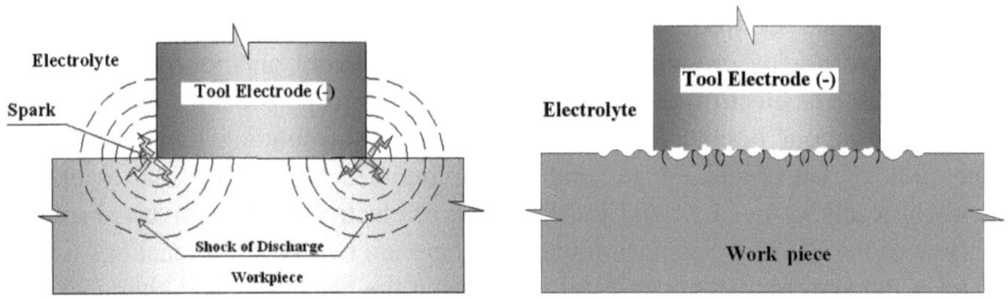

Fig. 10.3 Mechanism of material (tool and workpiece) removal in micro-ECDM

the focused stream of electrons may rise to a very high value. When this temperature is above the melting temperature of the workpiece, material is melted and finally vapourised.

10.1.2 Need of Electrochemical Discharge Micro-machining of Engineering Materials

In modern manufacturing industries the use of advanced engineering materials such as ceramics is increasing due to some favorable characteristics and at the same time some advanced machining methods are required to manufacture the products with these materials. In Abrasive Water Jet Machining (AWJM) the cutting ability is hazardous due to the traverse cutting speed and requires high investment as well as maintenance cost. Also the quality of the product is not so good due to the machined surface roughness. Ultrasonic Machining (USM) is generally used for machining electrically non-conductive material but this process has some inherent limitations such as tool wear, high capital cost and there is a chance of tool bending. In case of Laser Beam Machining (LBM) a high intensity of monochromatic light is used during machining operation. But this process requires very high investment and the formation of very large undesirable heat affected zone degrades the quality of the product. ECM requires high capital investment, skilled labour to operate and large space for installation. Also the disposal of used electrolyte and effect of stray current are the major drawbacks of ECM process. Electro-discharge machining (EDM) has some drawbacks like difficult to fabricate various shapes, long machining time required to produce micro-products and high cost of equipments. Also, the process is useful mainly for electrically conductive materials.

Hence, there is a need of special machining process, which will be helpful for fabricating products of electrically non-conductive materials and can cope up the adverse effects of above machining processes. Electrochemical Discharge micro-Machining (micro-ECDM) process has a great ability to machine electrically non-conductive materials. It utilizes the electro-discharge phenomena to machine advanced engineering materials such as ceramics, glass etc. In comparison to above machining processes, thermal effects i.e. formation of heat affected zone is small. The process is independent of physical and chemical properties of the material. The investment cost is very low.

10.1.3 Problems in Electrochemical Discharge Micro-machining of Engineering Materials

From the basic study of micro-ECDM process for machining of engineering materials like ceramics and glass etc., it is found that the micro-ECDM process performances depend on the fundamental process parameters such as applied voltage, electrical conductivity of tool and auxiliary electrode, type and concentration of electrolyte, which ultimately control the electrochemical reactions and electro-spark discharge actions. It is also revealed that the material removal rate can be increased by the proper control of level of the process parameters in micro-ECDM process. Figure 10.4 shows the cause and effects exhibiting the mix of various levels of factors for achieving desired major micro-machining criteria in electrochemical discharge micro-machining process.

But the μ-ECDM process has several problems. The following chart (Fig. 10.5) shows the problematic areas of μ-ECDM process. In the μ-ECDM process there are two electrodes named cathode and anode as they are connected to the negetive and positive terminals of the power supply respectively. The shape and size of those

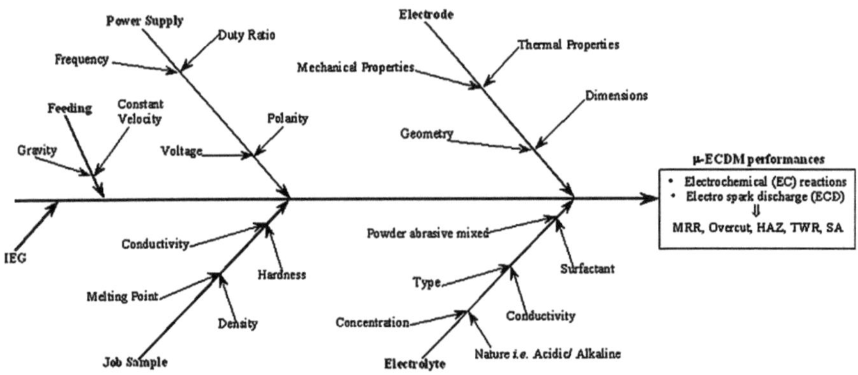

Fig. 10.4 Cause and effect (fishbone diagram) of μ-ECDM process

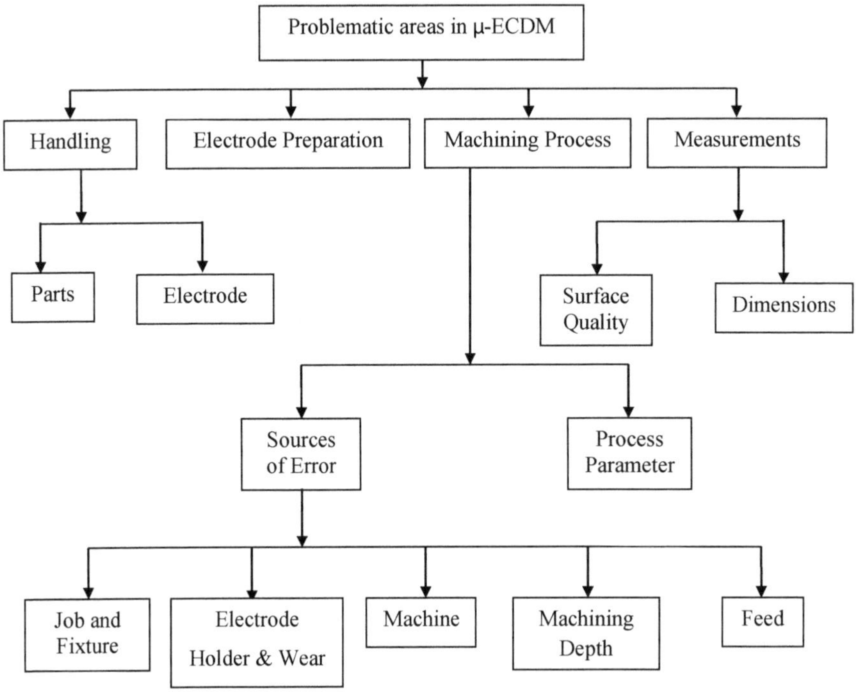

Fig. 10.5 Problematic areas of μ-ECDM process

electrodes depend upon the different types of operation like micro-drilling, micro-profile cutting etc. The preparation of auxiliary electrode is one of the most important tasks in micro-ECDM and controlling the inter-electrode gap (IEG) by special devices is also the challenging task to the researchers. Also the devices, which are used for performing the above operations, are difficult to handle.

The accuracy in positioning of the machining zone with respect to the measuring point should be high. Inaccurate positioning of the workpiece may damage the tool electrode. Hence, workpiece and its holding device preparation have some effects on machining accuracy. The electrode holder is manipulated by turning, conventional drilling and milling etc. In micro-ECDM the material is removed from both the tool and workpiece due to the thermal effect. This causes the errors in the machining process. Also, plating and etching of workpiece take place due to chemical reactions. When machining is done by micro-ECDM, side sparking lies among the problematic areas observed during machining the workpiece materials. Controlling this side sparking is not easy task. By using the heat absorbed material or insulator the side sparking may be controlled to some extent. Again, the controlling of feed rate in micro machining by electro-chemical discharge phenomenon is a difficult task. Feed can be given by two ways: one is where workpiece is fed and tool is kept stationary, another is where the tool is fed keeping the workpiece stationary. To modify the feeding arrangement some researchers used various types of special devices such as gravity and spring feeding mechanisms, encoder, dial indicator and load cell.

One of the major problems in micro-ECDM process is to control the DC power supply, which provides the breakdown voltage and essential current flow that causes sparking. The ranges of voltage, frequency and duty ratio are provided from the power supply unit. Selection of critical process parameters for micro-machining is not also an easy task. Carefully selected machining parameters and lots of pilot experiments with different electrolyte concentrations can give better result for finding the range. From various research works, different process parameters and their ranges have been investigated and with the help of experimental design the combination of various process parameters are chosen to carry out various machining operations. For example, the suitable concentration is not same for NaOH and KOH electrolytes to perform the machining operation by using micro-electrochemical discharge machining process. Also, mixing of these two electrolytes may enhance the performance of the machining process though the necessity of suitable concentration range is hard to find.

10.1.4 Possibilities and Applications of Electrochemical Discharge Micro-machining

Eletrochemical discharge micromachining (micro-ECDM) process has a great ability to machine electrically non-conducting materials. In comparison to other nonconventional micro-machining processes thermal effect i.e. formation of heat affected zone is small. There is no requirement of skilled labour to operate the machine. The investment cost is very low, as simple equipments are required to run the machining operation. The machining can be performed irrespective of the electrical and mechanical properties of workpiece materials and any shape of the tool-electrode can be produced on the workpiece surface. The performance of the machining process totally depends on the influences of some significant factors especially the thickness of insulating bubbles' layer formed around the tool-electrode. This film thickness is influenced by various parameters such as applied voltage, electrolyte's type and concentration etc. Again, the area of machining zone can be varied from several millimetres to microns either by changing the dimension of tool-electrode or by varying other predominant process parameters. If the diameter of tool is small then tiny shape or components can be fabricated since the area of machining zone can be restricted to micro-machining range. Also, no large force as found in all common conventional machining processes is experienced by tool-electrode during micro-machining operation. The machining is done only by the use of physico-chemical energy. Therefore, electrochemical discharge machining process can be applied as a micro-machining process like other micro-machining processes by controlling various process parameters.

The hybrid technology of micro-ECDM process consisting of EC reactions and ESD helps in micro-drilling on electrically non-conductive advanced ceramics like Silicon Nitride (Si_3N_4) [7], Alumina (Al_2O_3) [8], Zirconia (ZrO_2) [8] and Silicon Carbide (SiC) [9] etc. This method can also be used for slicing Quartz [10], micro-grooving on e-glass epoxy [11] etc. Its unique feature of using thermal energy to machine engineering materials such as electrically non-conductive parts regardless of their hardness has been its distinctive advantage in the manufacture of mould, die, automotive, aerospace, and surgical components. Figure 10.6 represents various micro operations performed by micro-ECDM process.

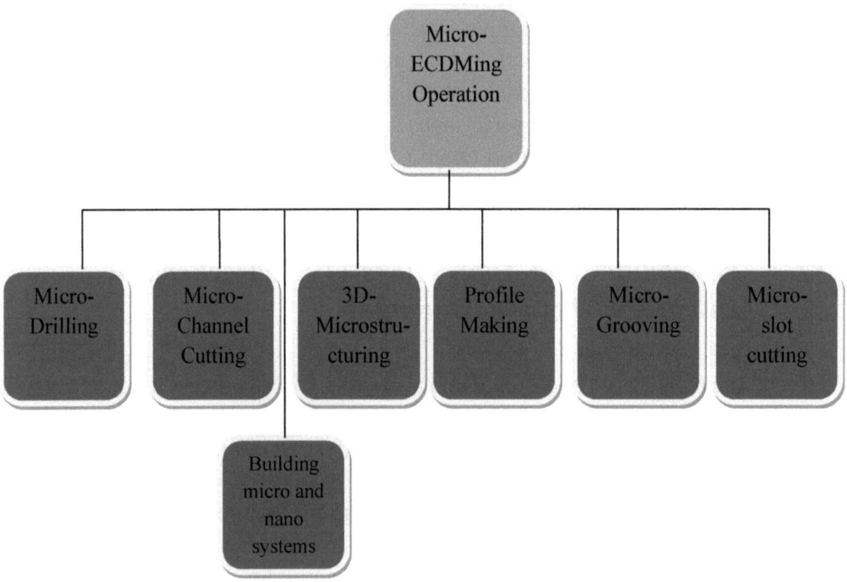

Fig. 10.6 Various micro-machining operations of ECDM

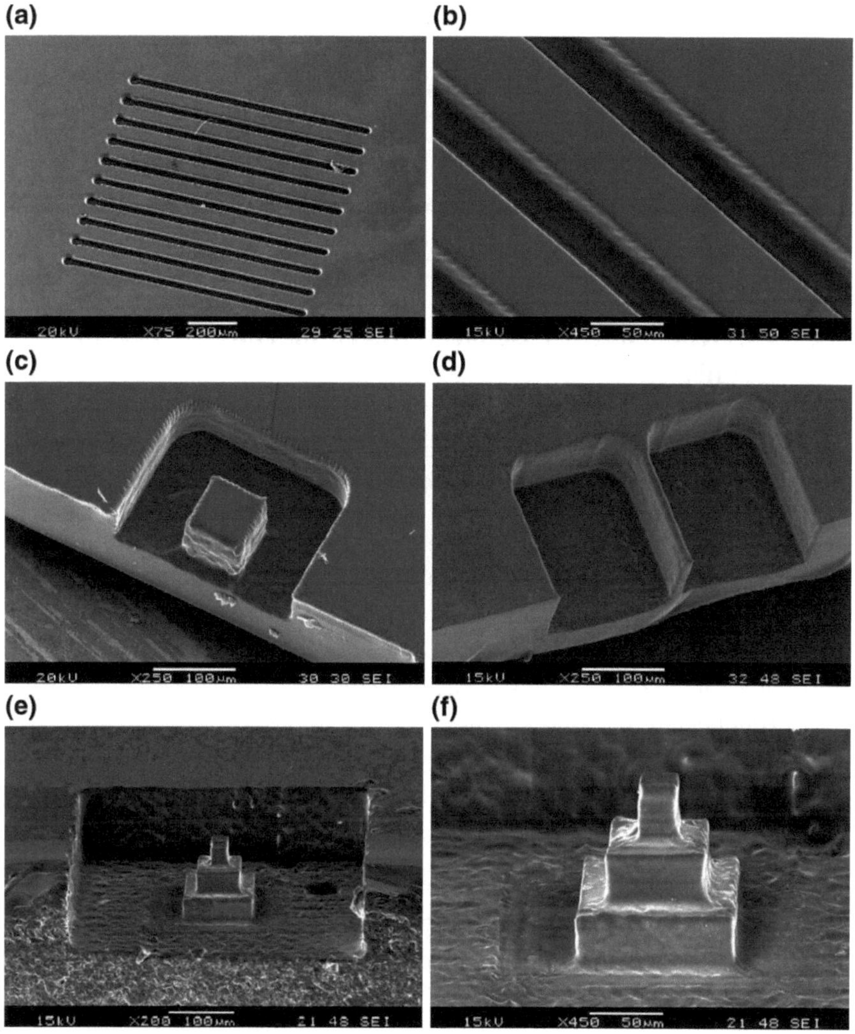

Fig. 10.7 **a**, **b** Micro-grooves, **c** micro-pillar, **d** micro-wall and **e**, **f** micro-pyramid machined on glass by μ-ECDM [12]

The micro-ECDM can be effectively used for high precision machining operations such as making grooves, micro-channels and 3D micro-machining etc. [12, 13]. Figures 10.7 and 10.8 exhibit some examples of micro-features generated by micro-ECDM process on glass. Micro channel formation and surface treatment can be done by micro-ECDM [14]. Glass-based micro-fluidic systems can be produced by μ-ECDM [15].

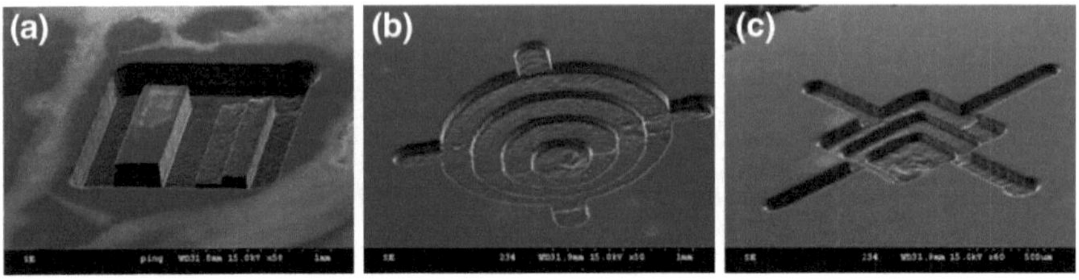

Fig. 10.8 3D microstructures in Pyrex glass machined using μ-ECDM [13]

10.2 Electrochemical Discharge Micro-machining System Details

In order to carry out the experimentation on electrochemical discharge μ-machining of engineering materials and also to control different process parameters such as applied voltage, machining current, inter-electrode gap, feed movement of workpiece as well as configuration of electrical power circuit etc., the μ-ECDM experimental set-up includes following sub-systems:

(i) Machining chamber details
(ii) Job holding unit
(iii) Micro-tool development for micro-ECDM
(iv) Tool holding and guiding unit
(v) Auxiliary electrode unit
(vi) Inter-electrode gap control unit
(vii) Feeding unit
(viii) Electrolyte supply unit
(ix) Electrical power supply unit.

The schematic diagram of the μ-ECDM experimental system is shown in Fig. 10.9. Each subsystem is equally important and has been designed separately to fulfil the requirement of the research work and then integrated to develop the total experimental set-up. A detailed discussion of each subsystem has been made in the following sections.

10.2.1 Machining Chamber Details

The machining chamber is the main component of μ-ECDM system where micro-machining of electrically non-conducting workpiece takes place. It is a rectangular box having dimension 250 × 250 × 150 mm and its wall thickness is 2 mm. It has been fabricated with stainless steel. Stainless steel is selected for construction of machining chamber due to its high strength, toughness and corrosion resistance property. The chamber is fixed on a vice for providing linear motion

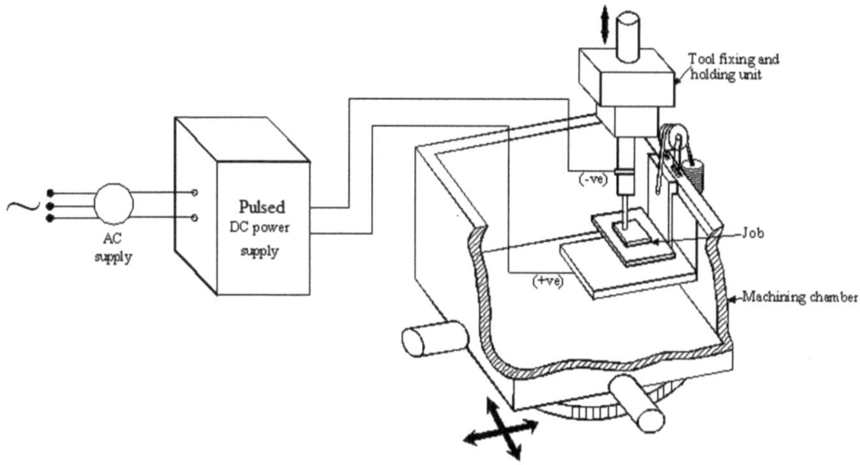

Fig. 10.9 Schematic diagram of micro-ECDM system

in one direction and the vice is placed on a rotating table for proper positioning of the job with tool. Within the machining chamber, job-holding unit, tool holding unit, inter-electrode gap controlled unit, auxiliary electrode device and feeding unit have been attached. At the bottom of the chamber there is an exit passage for the electrolyte to flow out.

10.2.2 Job Holding Unit

The job, which will be machined, should be hold rigidly for getting better micro-machining performance and the developed job holding unit fulfils this purpose. In the system the job is placed on a supporting plate, which itself rests on a support structure connected to the machining chamber of μ-ECDM system. The supporting plate is also made of stainless steel. A rectangular shaped cover plate, made of Perspex, is placed on the job in order to restrict the movement of work-piece during machining operation. The cover plate is tightened at four corners with the supporting plate by nuts. There is a slot on the cover plate for the easy movement of tool towards the job during micro-machining. The displacement of the job from the original job holding position may lead to the variation in the dimensional accuracy of the job.

10.2.3 Tool Holding and Guiding Unit

Since the rigidity of micro-tool as well as its holding unit is one of the main factors in micro-machining processes. Special attention has been given during the development and fabrication of such tool-holding unit. In μ-ECDM system, tool holder not only holds micro-tool during machining operation but also provides enough rigidity to micro-tool as well its holding unit. The micro-tool holding unit is made up of mild steel and consists of a small drill chuck having internal thread at one end. Lower portion of the support column is made of insulator to insulate the remaining part of the machine from tool holding arrangement. To generate micro-channel on the workpiece a template with a curved profile on its outer edge is made and fitted with the tool holding unit so that the tool can move along the same profile guided by the template. The tool holder along with micro-tool can be rotated by a handle. The tool holding unit is fitted with the cover plate by means of screw-nut mechanism so that the position of the tool can be adjusted according to the requirement. There is also a provision for connecting the micro-tool holding unit to the negative terminal of power supply unit. The diameter of micro-tools used for experiments is ranging from 300 to 400 μm and the material is stainless steel or other. The schematic diagram of the micro-ECDM set-up for micro-structuring is shown in the Fig. 10.10.

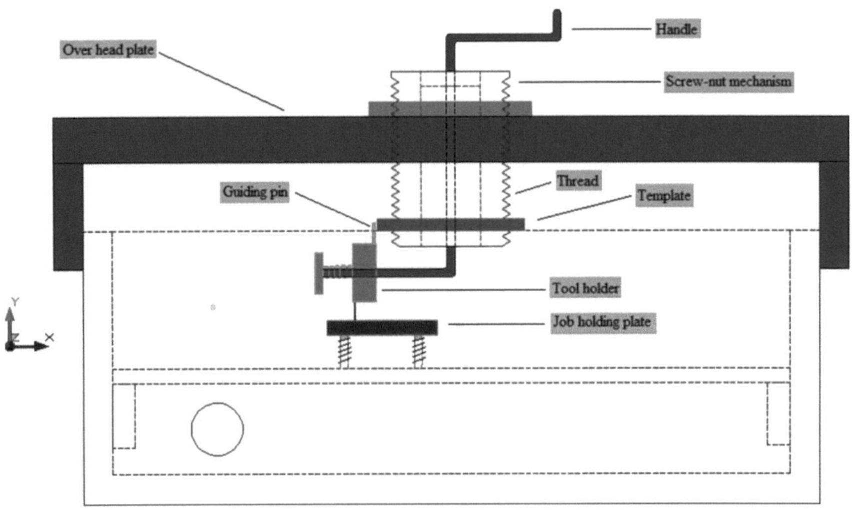

Fig. 10.10 Schematic diagram of μ-ECDM machining chamber

10.2.4 Micro-tool Development for Micro-ECDM

The tool material is worn out due to thermal effect of the process and it influences the process performances specially the accuracy and repeatability. So, the development of micro-tool for micro-ECDM process is a hill task and depends on the physical and chemical properties of micro-too material. There are several types of micro-tool such as cylindrical shaped with flat end, conical end and spherical end etc. are generally used in micro-ECDM process. The fabrication techniques of the micro-tools and their materials in micro-ECDM are quite similar with the fabrication techniques and materials of micro-tool used in micro-EDM process. The micro-tool is developed by wire electro-discharge grinding (WEDG) technique, block electro-discharge grinding (BDG) technique, electrical discharge grinding using two block electrodes (EDG-TBE) and also by rotary disc with wire electrode-discharge grinding technique etc.

10.2.5 Auxiliary Electrode Unit

Two electrodes are immersed into the electrolyte to complete the electrolytic cell. Micro-tool is considered as one of the electrodes in micro-ECDM process. μ-ECDM process machines electrically non-conducting materials. Workpiece cannot be used as another electrode because it is electrically non-conducting. Therefore, a special electrode known as *auxiliary* or *counter electrode* is necessary to complete the electrolytic cell and highly electrically conductive material is used to act as auxiliary electrode. In μ-ECDM system auxiliary electrode is a flat rectangular metal plate and placed parallel to the job holding plate. It is made of stainless steel or graphite. Also, in order to fulfil the criteria of electrochemical discharge (ECD) phenomenon the size of auxiliary electrode is generally hundred times greater than the area of smaller electrode i.e. micro-tool. It is connected to positive or negative terminal of the power supply unit.

10.2.6 Inter-electrode Gap Control Unit

As discussed in the previous section, the inter-electrode gap plays a vital role in μ-ECDM process. Therefore, special attention is given during its development and fabrication in such a way that inter-electrode gap can be varied during micro-machining operation. The distance between the tip of tool (*cathode*) and top surface of auxiliary electrode (*anode*) is known as *inter-electrode gap* (IEG). In order to study the effect of the gap on different machining performance characteristics, the inter-electrode gap should be adjusted with the help of inter-electrode gap control mechanism. Two long vertical screws are joined to the auxiliary electrode holder block and screwed to the plates projected from the top of main machining chamber. The movement of the screws in upward or downward direction can be adjusted by rotating the nuts. Due to this, the auxiliary electrode unit moves in the upward or downward direction and the desired inter-electrode gap can be set during the μ-ECDM process.

10.2.7 Feeding Unit

Generally tool feeding is not done in micro-ECDM process so as to protect the disruption of gas bubbles layer surrounding the micro-tool. Micro-tool is held at a fixed height and the feeding is given only to the workpiece. So, a gravity-controlled job feeding arrangement is designed and developed to have the vertical upward feed during the machining process. A string is connected to the guide block of the job holding unit. The string passes over a pulley and a pan is placed at the other end of the string. Two rectangular bars guide the upward movement of the guide block and a support plate limits the upward motion of the job holding unit. A suitable weight is placed on the pan so that job just touches the micro-tool tip. The feeding to job sample can also be given by the use of spring, which is placed under the job holding unit. In this unit the job is placed on a plate of size $60 \times 60 \times 5$ mm, which itself rests on four springs guided by four stainless steel rods of diameter 5 mm each located at four different corners of the plate. The stainless steel rods guide the spring in such a way that the job holding plate can move up and down along the axis of the rods i.e. only vertical motion is possible for the job holding unit as well as the workpiece. Micro-tool feeding can also be done keeping the workpiece stationary. Here, tool-feeding can be observed and measured by the use of encoder, dial indicator and load cell.

10.2.8 Electrolyte Supply Unit

In electrochemical discharge micro-machining process electrolyte plays significant role. The electrochemical reactions and formation of insulating layer of the gas bubbles and also the initiation of sparking during machining operation depend on the types and concentration of electrolyte. Therefore, to get better machining performance in terms of material removal rate (MRR) and radial overcut, the generation of gas bubbles due to electrochemical reactions is very important and it should be accumulated at the bottom edge of the micro-tool. For accumulation of gas bubbles under the tool a stagnant electrolyte is preferred and the upper level of electrolyte is always kept at a constant height generally 1 mm above the surface of workpiece. As heat is generated during this process, the electrolyte is heated up and evaporates; as a result the level of electrolyte is reduced. So, to maintain the desired level of electrolyte in the machining chamber the slight movement but steady supply of electrolyte in the machining chamber is needed. The outlet pipe is fitted with control valve at the bottom level of the machining chamber for removing the used electrolyte containing sludge and precipitates. This can also be used for cleaning of the machining chamber.

10.2.9 Electrical Power Supply Unit

The polarity of DC power supply during micro-ECDM process plays a significant role and straight polarity is always maintained during micro-machining operation. The voltage required for machining operation ranges from 10 to 70 V. The machining current is very small in the range of 0–5 A. Conversion from A.C. supply to D.C. is got from a step down transformer and silicon-diode controlled rectifier unit. The voltmeter and ammeter are connected with the circuitry to measure the applied voltage and machining current. The voltage is varied and set to the desired value by controlling variac connected with power supply unit. Though the power supply is D.C. but it is not smooth D.C. rather pulsed D.C., which helps the machining operations. The electrical circuitry of pulsed D.C. power supply unit

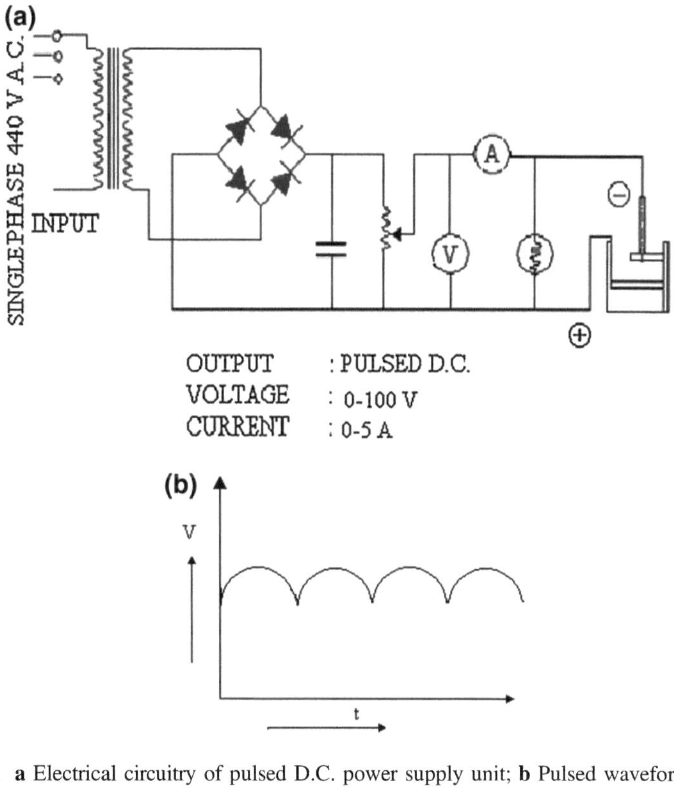

Fig. 10.11 **a** Electrical circuitry of pulsed D.C. power supply unit; **b** Pulsed waveform

is shown in Fig. 10.11. The arrangements for adequate protective circuit for the transformer rectifier and machine have been provided to avoid the short-circuit and over-load conditions. But, the common μ-ECDM system is not fully compatible to control the sparking phenomenon when it is used for micro-machining operation, although it has some provisions such as worktable movement, tool movement, etc. Therefore, the modification in power circuit configuration of the μ-ECDM system is essential in order to machine non-conductive engineering materials in the micro-machining domain with high machining rate.

An inductor does the filtering of supply voltage/current and generates high current spikes when it is connected to the positive terminal of the power supply. Therefore, inductance (L) is used in the power circuit of micro-ECDM set-up to achieve controlled discharge during micro machining operation. There is a provision to vary the values of inductor during experimentation. A resistance (R) is added to the negative terminal of power supply in the ECDM system to decrease the load at the inter-electrode gap between tool electrode and auxiliary electrode. As in case of micro-machining a controlled sparking as well the less energy per sparking is desirable. There is a provision in the power circuit configuration where the value of the resistance can be varied. A filter for light loads is observed in the capacitor, which is connected directly across the load. The capacitor charges up to the peak value of the input voltage and tries to maintain this value so that D.C. power supply behaves as a smooth D.C. power supply. The capacitance (C) is connected across the two terminals of the power supply of μ-ECDM system during micro-machining operations. The values of the capacitor can also be varied in the power circuit configuration of the developed micro-ECDM system. The schematic diagrams of these various power circuit connections have been exhibited in Fig. 10.12.

10.2.10 Specification Details of the Micro-ECDM System

The detailed specifications for the various subsystems of the μ-ECDM system are as follows:

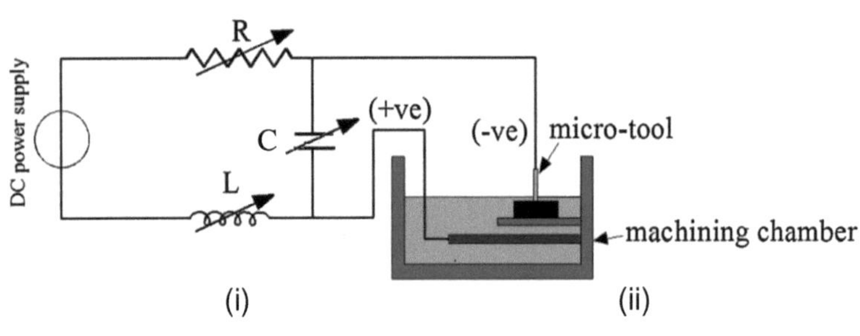

Fig. 10.12 Schematic diagram of modified power circuit

I. ***Specification details of main machine set-up***

The specifications of the important mechanical hardware modules are:

(i) Size of the machining chamber: 250 mm × 250 mm × 150 mm.
(ii) Maximum inter-electrode gap: 50 mm.
(iii) Auxiliary Electrode Material: Stainless steel, Graphite
Size: 80 mm × 50 mm × 1 mm.
(iv) Maximum work-platform movement in horizontal direction: 50 mm.
(v) Feeding mechanism: Gravity feed with counter weight, Spring loaded job holding table.
(vi) Tool material: Stainless steel and other.
(vii) Tool diameter: 300–400 μm.
(viii) Maximum dimension of the job that can be machined:
For square job: 20 mm × 20 mm × 5mm.
For circular job: Ø60 mm × 5 mm thick.
(ix) Capacity of the electrolyte reservoir tank: 5 lit.

II. ***Specification details of electrical power supply unit***

Main input power supply: 3 phases 440 V A.C.
Output Power supply: Pulsed D.C.

(i) Voltage range: 0–100 V.
(ii) Current range: 0–5 A.
(iii) Frequency: 200 Hz to 1 kHz.
(iv) Duty factor: 0.3–0.8.
(v) External Inductance: 5–25 mH.
(vi) External Resistance: 40–120 Ω
(vii) External Capacitance: 50–200 μF.

10.3 Parametric Studies on Electrochemical Discharge Micro-machining of Engineering Materials

Basic principle of electrochemical discharge micro-machining process can be used to perform various operations like micro-drilling, micro-cutting and micro-structuring. These operations are very effective to fabricate different micro-profiles on hard and brittle engineering materials, irrespective of their electrical conductivity. Therefore, next discussions focused on the parametric studies on electrochemical discharge micro-machining of engineering materials for the above mentioned operations.

10.3.1 Micro-drilling

Electrochemical discharge micro-drilling is the primary application of micro-ECDM process and being driven by the demand of high aspect ratio precise micro-holes on thick or thin substrate [16] of engineering materials. To fulfill the demands of micro-fabrication, many researchers has drilled through and blind micro-holes by using micro-ECDM process. This process has been used to drill silicon nitride ceramics [7], Alumina and Zirconia [8], Silicon Carbide [9], borosilicate glass [17, 18], Pyrex wafer [19], soda lime glass [20], other glass wafers [21] and e-glass-fiber-epoxy composite [11] etc. The Electrochemical discharge micro-drilling operation is performed either by die-sinking tool or through the controlled and progressive movements of the tool electrode along the z-axis. There are various process parameters related to the electrolyte, tool electrode and power supply that effectively control the discharge energy. Applied voltage and electrolyte concentration have strong effect on material removal rate (MRR) of micro-ECDM process. MRR increases with applied voltage while electrolyte concentration is fixed [7]. Initially MRR is observed to be decreased with applied voltage at higher electrolyte concentration due to the possibility of material re-deposition at the machining zone. As shown in Fig. 10.13 MRR increases with electrolyte concentration due to the fact that the conductivity of the electrolyte medium increases with electrolyte concentration and as a result, greater current flows through the path between the tool and the auxiliary electrode. An increase in the inter-electrode gap (IEG) lowers the flow of electric current hence MRR decreases [7, 8, 22]. Here, the resistance in the current flow path increases as a result of increase in IEG. Also material removal rate

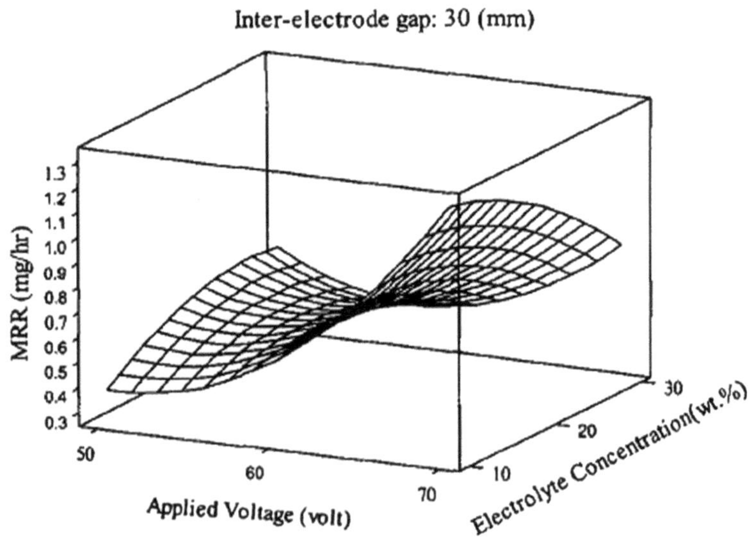

Fig. 10.13 Effects of applied voltage and electrolyte concentration on MRR [7]

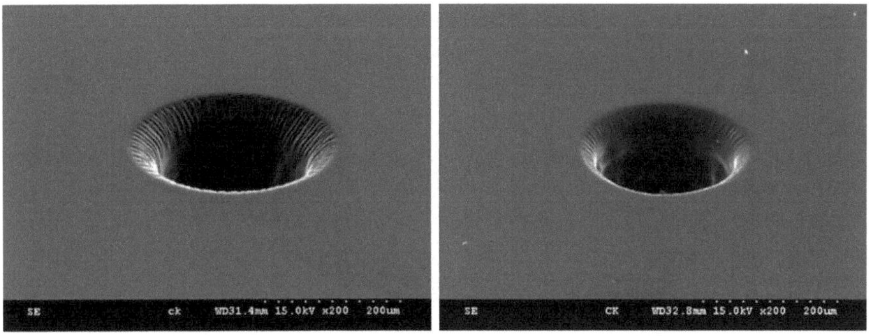

Fig. 10.14 SEM images of micro-holes machined by different tool electrodes [26]

varies with the pulse frequency because the duration of the discharge increases as the frequency decreases, even though the total time when voltage is applied remains the same [23]. Also, MRR is increased significantly with the help of the vibration of gravity feed tool [24] and due to increase in drilling speed [25]. But, it decreases as the duty ratio decreases [22].

Overcut (OC) during micro-ECDM operations increases with an increase in applied voltage because of the fact that at high voltage a large number of gas bubbles are generated at the tool's sidewall. Overcut is observed small under high voltage and low inter-electrode gap compared to that under high applied voltage and high inter-electrode gap [7, 8]. Resistance-capacitor circuit is found to be suitable in micro-ECDM process for low overcut [22]. A tool electrode with spherical tip exhibits higher machining rate and accuracy as compared to the conventional cylindrical tool electrode [26] as shown in Fig. 10.14. The combination of pulse voltage and side insulation demonstrates a significant effect on the reduction of overcut in the ECDM micro-drilling operation [27]. By utilizing a side-insulated electrode the overcut and roundness error of the holes are significantly reduced [28]. Tool wear behavior for different tool materials at various voltage levels plays an important role to preserve the machining accuracy in electrochemical discharge micro-drilling. It is observed that the steel and tungsten carbide (WC) tools result in better performance over brass tool [27].

Large amount of heat is generated while machining the ceramic materials by micro-ECDM process. Heat-affected zone (HAZ) is formed around the machined zone due to the heat energy conducted to the workpiece during sparking and micro-cracks are produced within this zone. These cracks degrade the quality of products. HAZ increases with an increase in applied voltage due to a higher intensity of sparking [7, 8]. HAZ also varies with electrolyte concentration as shown in Fig. 10.15. HAZ decreases initially with increase in concentration, gets maximized and then starts to increase with increase in concentration. HAZ (Fig. 10.16) is reduced for tungsten carbide tool than tungsten and stainless steel tool [26].

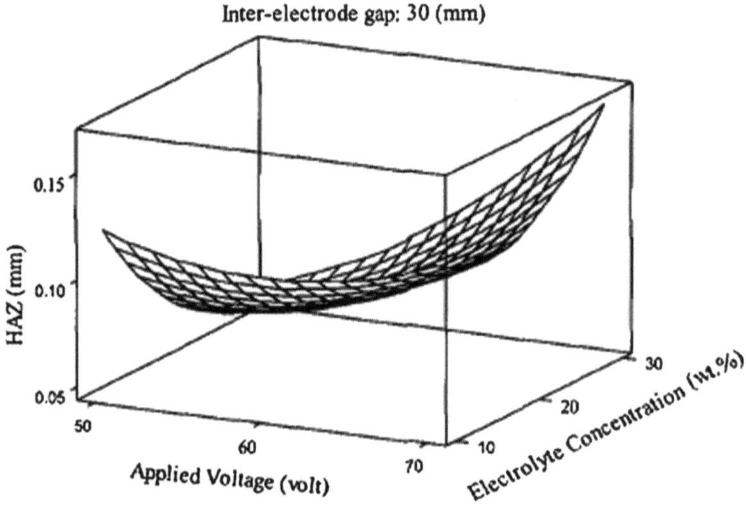

Fig. 10.15 Effects of applied voltage and electrolyte concentration on HAZ [7]

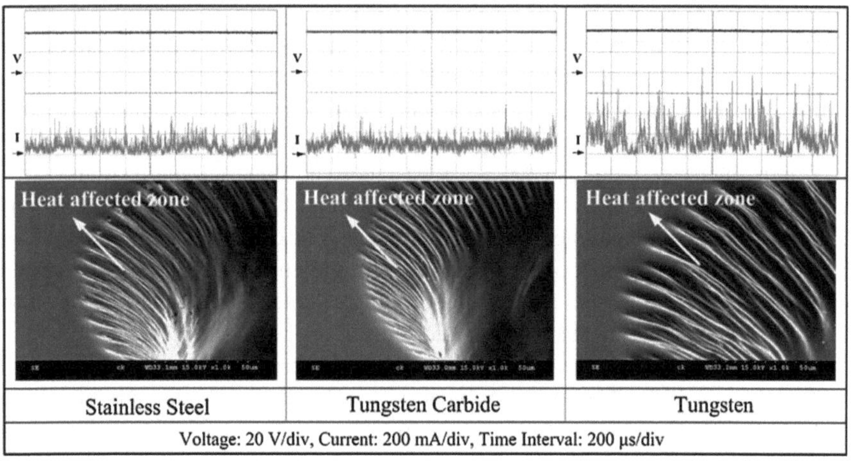

Fig. 10.16 Current responses at optimal voltage and SEM images of HAZ [26]

A powder-mixed electro-chemical discharge machining method has been used to fabricate glass micro surface [18, 29]. As a result the presence of conductive particles within the hydrogen film reduces the critical breakdown strength resulting in the decreased spark energy per single discharge pulse. The hydrogen film at the moment of discharge is modeled as breakdown of the insulating gas between two parallel plates for which distance of the gap becomes the theoretical thickness of the film. The presence of conductive particles reduces the dielectric strength of the film. The thinner gas film thickness will result in lower fluctuations of discharge

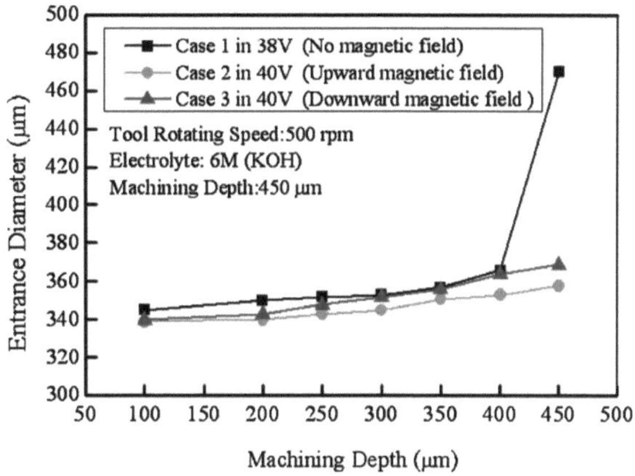

Fig. 10.17 Relationship between the entrance diameter and the machining depth under different magnetic field configurations [30]

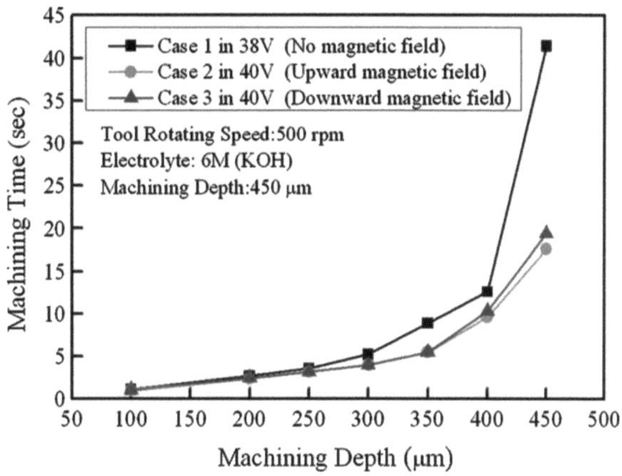

Fig. 10.18 Relationship between the machining time and the machining depth under different magnetic field configurations [30]

activities [17]. Figures 10.17 and 10.18 show the relationships between the machining time and the entrance diameter, respectively, with machining depth under different magnetic field configurations [30]. Further a tool-surface bonding reduces the machining rate during micro-drilling in gravity-feed or constant-velocity-feed mechanism [31].

10.3.2 Micro-cutting

Many researchers have combined wire electric discharge machining (WEDM) concept with electrochemical discharge machining process to cut/slice hard and brittle engineering materials [32, 33]. In WECDM, electrolyte dipped traveling wire serves as a cutting tool (cathode) as shown in Fig. 10.19. But keeping the work-piece in contact with wire is very difficult. Two different workpiece feed

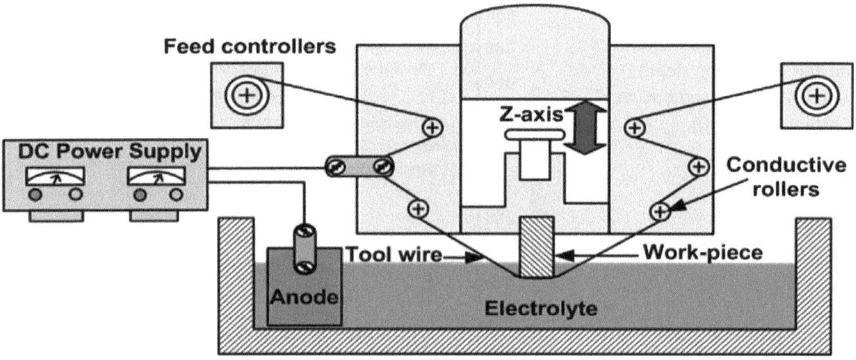

Fig. 10.19 Schematic view of TW-ECDM set-up [36]

Fig. 10.20 Influence of feed rate on slit mean width and depth [33]

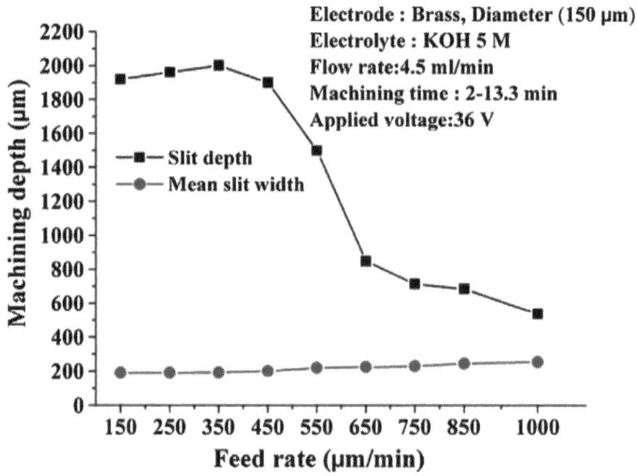

mechanisms, namely, weight-loading mechanism and reciprocating mechanism are used to ensure the physical contact between workpiece and tool wire [17]. Between these two mechanisms, reciprocating mechanism exhibits better surface quality and accuracy of microgrooves. In reciprocating (to-and-fro) mechanism of the work-piece across the tool wire facilitates debris removal whereas in weight-loading mechanism, the tool wire and workpiece are kept close to each other by using mechanical force, which makes flushing of debris from the machined surfaces very difficult. This un-flushed debris promotes secondary discharges and results in poor surface quality within accurate microgroove profiles. In order to enhance the machining rates as well as geometrical accuracy of the microgrooves, SiC abrasive particles are added into the electrolytic bath [17]. Feed rate is the most important process parameter during the fabrication of micro-grooves. It significantly affects the slit depth as shown in Fig. 10.20. In titrated flow method electrolyte flows in the form of droplets. Thus, the quantity of electrolyte available at machining zone is reduced. As a result lesser environmental pollution is found [33]. The optimum wire breakage ranges are recorded at 10% electrolyte concentrations and less than 60 V.

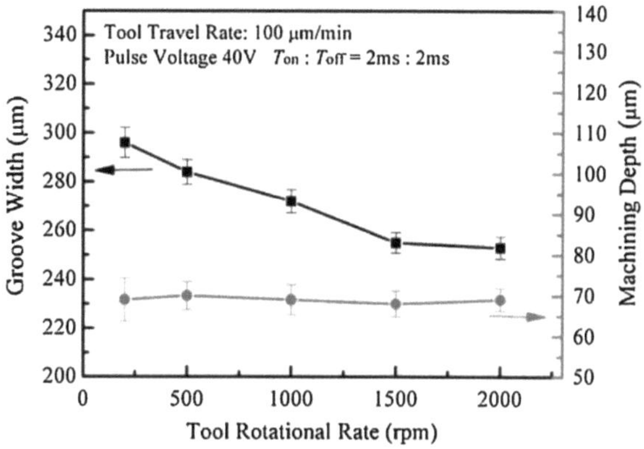

Fig. 10.21 Influence of tool rotational rate on groove width and depth [13]

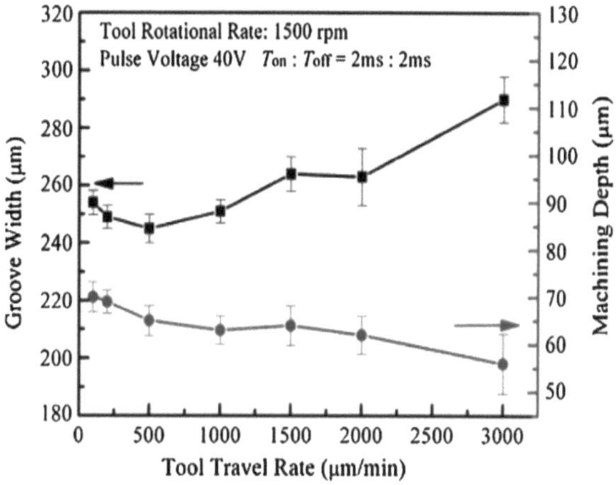

Fig. 10.22 Influence of tool travel rate on groove width and depth [13]

Electrochemical discharge micro-machining process is demonstrated as a potential method for the fabrication of complex 3D micro-structures on glass and quartz materials [12, 13]. Many investigations have been reported on surface texturing of the micro channels [34] and fabrication of micro channels [35] and micro grooves [16] etc. Figure 10.21 shows the individual effect of tool rotational rate on groove width and depth [13]. A higher tool travel rate can produce shallower microgrooves with larger width as shown in Fig. 10.22. In order to develop the deep micro grooves, layer-by-layer material removal approach with small depth of cut has been introduced in electrochemical discharge micro-structuring. This layer-by-layer approach facilitates electrolyte flushing at deeper and narrower gaps. Additionally, deeper micro grooves with smooth surface finish can be machined. For the fabrication of precise shaped microgrooves on glass substrates, the parameter settings: i.e. 40V DC supply, pulse on/off time, 2 ms/2 ms, tool rotation speed, 1500 rpm and tool travel rate, 1000 mm per minute are found suitable as machining condition [13].

10.4 Challenging Areas of Electrochemical Discharge Micro-machining

Miniaturisation of products continues to be the trend of the present technological advancement and micro-ECDM will take a vital role in the area of micro manufacturing. However, authors believe that few areas still have to be considered for improvement for wide acceptance of the electrochemical discharge micro-machining (ECDM) process for micro-machining applications in modern manufacturing industries. Therefore, some of the areas of micro-ECDM have been found out as challenging areas of the research for successful utilisation and these include:

(i) To increase machining depth of micro-hole in micro-ECDM process for fabricating high aspect ratio of holes.
(ii) To control the tool-feeding and gap sensing facility for micro-machining operation in order to facilitate the availability of electrolyte in the machining zone.
(iii) To develop CNC based micro-ECDM system with all control facilities and in situ micro-tool development arrangements for accurate and precise micro-machining operations.

10.5 Summary

Electrochemical discharge micro-machining (micro-ECDM) process is a prominent non-conventional hybrid machining process for machining of electrically non-conducting advanced engineering materials. The chapter will provide valuable guidance to the applied researchers and manufacturing scientists for machining electrically non-conducting materials such as ceramics and glass, etc. with desire machining rate, accuracy and other surface characteristics, which are unexplored till now. Extensive research efforts and continuous advancements in the area of micro-ECDM for effective utilization in micro-fabrication require monitoring and control of the gas film, material removal and accuracy. Improvement in HAZ and surface roughness are expected to enhance the application of micro-ECDM technology in modern industries. The increasing demands for precision manufacturing of micro-parts for biomedical and automotive components will lead modern manufacturing engineers to utilize micro-ECDM process more successfully considering its advantages, i.e. quality, productivity and ultimately cost efficiency, which are still vital for success in a competitive environment.

Acknowledgements The authors acknowledge the financial support provided by UGC under CAS-IV programme of Production Engineering Department, Jadavpur University, Kolkata-32, India.

References

1. I. Basak, and A. Ghosh, Mechanism of spark generation during electrochemical-discharge machining-A theoretical model and experimental verification, Journal of Materials Processing Technology, 62 (1996) 46–53.
2. A. Kulkarni, R. Sharan, and G. K Lal, An experimental study of discharge mechanism in electro-chemical discharge machining. International Journal of Machine Tools & Manufacture. 42 (2002) 1121–1127.
3. B. Bhattacharyya, B. N. Doloi, and S. K. Sorkhel., Experimental investigations into electro-chemical discharge machining (ECDM) of non-conductive ceramic materials, Journal of Materials Processing Technology, 95 (1999) 145–154.
4. V.K. Jain, P.M. Dixit, and P.M. Pandey, On the analysis of the electrochemical spark machining process, International Journal of Machine Tools and Manufacture, 39 (1999) 165–186.
5. V.K. Jain, P.S. Rao, S.K. Choudhury, K.P. Rajurkar, Experimental investigations into travelling wire electrochemical spark machining (TW-ECSM) of composites, Transactions of ASME, Journal of Engineering for Industry, 113 (1991) 75–84.
6. R. Wüthrich, and V. Fascio, Machining of non-conducting materials using electro-chemical discharge phenomenon—an overview, International Journal of Machine Tools & Manufacture, 45 (2005) 1095–1108.
7. B. R. Sarkar, B. Doloi, and B. Bhattacharyya, Parametric analysis on electrochemical discharge machining of silicon nitride ceramics, International Journal of Advanced Manufacturing Technology, 28 (2006) 827–881.
8. B. R. Sarkar, B. Doloi, and B. Bhattacharyya, Experimental investigation into electrochemical discharge micro-drilling on advanced ceramics, International Journal Manufacturing Technology and Management, 13 (2/3/4) (2008) 214–225.
9. B. R. Sarkar, B. Doloi, and B. Bhattacharyya, Analysis on the machining characteristics of electrochemical discharge micro-drilling process, 1st and 22nd AIMTDR, IIT Roorkee, (2006), pp 893–898.
10. W.Y. Peng and Y.S. Liao, Study of ECDM for slicing nonconductive brittle materials, Journal of Materials Processing Technology, 149 (2004), 363–369.
11. A. Manna, and V. Narang, A study on micromachining of e-glass fibre–epoxy composite by ECSM process, International Journal of Advanced Manufacturing Technology, 61 (2012) 1191–1197.
12. X. D. Cao, B. H. Kim, and C. N. Chu, Micro-structuring of glass with features less than 100 μm by electrochemical discharge machining, Precision Engineering, 33 (2009) 459–465.
13. Z-P. Zheng, W-H. Cheng, F-Y. Huang, and B-H. Yan, 3D micro structuring of Pyrex glass using the electrochemical discharge machining process, Journal of Micromechanics & Microengineering, 17 (2007) 960–966.
14. E.S. Lee, D. Howard, E. Liang S. D. Collins and R. L. Smith, Removal of tubing interconnects for glass-based micro-fluidic systems made using ECDM, Journal of Micromechanics & Microengineering, 14 (2004), 535–541.
15. J. West, and A. Jadhav, ECDM methods for fluidic interfacing through thin glass substrates and the formation of spherical microcavities, Journal of Micromechanics & Microengineering, 17 (2007) 403–409.
16. R Wüthrich, K Fujisaki, Ph Couthy, L A Hof and H Bleuler, Spark assisted chemical engraving (SACE) in micro factory Journal of Micromechanics & Microengineering, 15 (2005) S276–S280.
17. R. Wüthrich, and L. A. Hof, The gas film in spark assisted chemical engraving (SACE)—A key element for micro-machining applications, International Journal of Machine Tools & Manufacture, 46 (2006) 828–835.
18. M.S. Han, B.K. Min, and S.J. Lee, Improvement of surface integrity of electro-chemical discharge machining process using powder-mixed electrolyte, Journal of Materials Processing Technology,191 (2007) 224–227.

19. C.P. Cheng, K.L. Wu, C.C. Mai, Y.S. Hsu, and B.H. Yan, Study of gas film quality in electro-chemical discharge machining, International Journal of Machine Tools & Manufacture, 50 (2010) 689–697.
20. M.S. Han, B.K. Min, and S.J. Lee, Geometric improvement of electrochemical discharge micro-drilling using an ultrasonic vibrated electrolyte, Journal of Micromechanics & Microengineering, 19 (2009) 8.
21. R. Wüthrich, L. A. Hof, H. Lal, K. Kujisaki, H. Bleuler, P. Mandin, and G. Picard, Physical principle and minituarisation of spark assisted chemical engraving (SACE), Journal of Micromechanics & Microengineering, 15 (2007) S286–S275.
22. B. R. Sarkar, B. Doloi, and B. Bhattacharyya, Investigation into the influences of the power circuit on the micro-electrochemical discharge machining process, Proceedings of the Institution of Mechanical Engineers, Part B: Journal of Engineering Manufacture, 223 (2) (2009), 133–144.
23. D. J. Kim, Y. Ahn, S. H. Lee, and Y. K. Kim, Voltage pulse frequency and duty ratio effects in an electro-chemical discharge micro-drilling process of Pyrex glass, International Journal of Machine Tools & Manufacture, 46 (2006) 1064–1067.
24. R. Wuthrich, B. Despont, P. Maillard, and H. Bleuler, Improving the material removal rate in spark-assisted chemical engraving (SACE) gravity-feed microhole drilling by tool vibration, Journal of Micromechanics & Microengineering, 16 (2006) N28–N31.
25. M. Jalali, P. Maillard, and R. Wuthrich, Toward abetter understanding of glass gravity-feed micro-hole drilling with electrochemical discharges, Journal of Micromechanics & Microengineering, 19 (2009) 045001 (7 pp).
26. C. K. Yang, K. L. Wu, J. C. Hung, S. M. Lee, J. C. Lin, and B. H. Yan, Enhancement of ECDM efficiency and accuracy by spherical tool electrode, International Journal of Machine Tools & Manufacture, 51 (2011) 528–535.
27. C. K. Yang, C. P. Cheng, C. C. Mai, A. C. Wang, J. C. Hung, and B. H. Yan, Effect of surface roughness of tool electrode materials in ECDM performance, International Journal of Machine Tools & Manufacture, 50 (2010) 1088–1096.
28. P. Maillard, B. Despont, H. Bleuler, and R. Wüthrich, Geometrical characterization of micro-holes drilled in glass by gravity-feed with spark assisted chemical engraving (SACE), Journal of Micromechanics & Microengineering, 17 (2007) 1343–9.
29. C. T. Yang, S. L. Song, B. H. Yan, and F. Y. Huang, Improving machining performance of wire electro-chemical discharge machining by adding SiC abrasive to electrolyte, International Journal of Machine Tools & Manufacture, 46 (2006) 2044–2050.
30. C.P. Cheng, K.L. Wu, C.C. Mai, Y.S. Hsu, and B.H. Yan, Magnetic field-assisted electrochemical discharge machining, Journal of Micromechanics & Microengineering, 20 (2010) 7.
31. T.F. Didar, A. Dolatabadi, and R. Wüthrich, Characterization and modelling of 2D- glass micro-machining by spark-assisted chemical engraving (SACE) with constant velocity, Journal of Micromechanics & Microengineering, 18 (2008) 9.
32. B.K. Bhuiyan, and V. Yadava, Experimental study of traveling wire electrochemical spark machining of borosilicate glass, Materials and Manufacturing Processes, 29 (2014) 298–304.
33. K.Y. Kuo, K.L. Wu, C.K. Yang, and B.H. Yan, Wire electrochemical discharge machining (WECDM) of quartz glass with titrated electrolyte flow, International Journal of Machine Tools & Manufacture, 72 (2013) 50–57.
34. D. Jana, A. Ziki, T.F. Didar, and R. Wüthrich, Micro-texturing channel surfaces on glass with spark assisted chemical engraving, International Journal of Machine Tools & Manufacture, 57 (2012) 66–72.
35. L. Paul, and S.S. Hiremath, Characterisation of micro-channels in electrochemical discharge machining process, Applied Mechanics and Materials, 490–491 (2014) 238–242.
36. T. Singh, and A. Dvivedi, Developments in electrochemical discharge machining: A review on electrochemical discharge machining, process variants and their hybrid methods, International Journal of Machine Tools & Manufacture, 105 (2016) 1–13.

Chapter 11
Travelling Wire Electrochemical Spark Machining: An Overview

Anup Malik and Alakesh Manna

Abstract Travelling wire electro chemical spark machining (TWECSM) is a hybrid machining process; TWECSM is combination of electro chemical machining (ECM) and wire electric discharge machining (WEDM). This machining process can be effectively used for machining of difficult to machine non-conductive materials that are not possible to machine by non-conventional machining methods i.e. EDM, ECM and WEDM etc. Now a days the utilization of hybrid machining techniques are very much important to process advanced materials as the different types of advanced materials are developing by the modern research community. The non-conductive materials can be machined with the help of conventional machining processes but compromise with surface texture, accuracy, micron level slicing etc. Keeping in view, this chapter presents the detail overview about the hybrid TWECSM process. The design and development of TWECSM process, potential and effectiveness of the process, parametric optimization for effective application of this process etc. are also explained in the chapter.

11.1 Introduction

Developments in today's materials demand suitable processes for the machining of such advanced materials. Advanced machining processes are now being effectively used in industries for manufacturing of components made up of electrically conducting materials which are difficult to machine. With the speedy industrial progress in the field of engineering ceramic and composite materials, the machining of these materials is mandatory for applied researchers and manufacturing engineers. The non-conductive ceramic and composite materials can't machine by any well-known non-conventional machining methods such as EDM, WEDM etc. used

A. Malik (✉) · A. Manna
Department of Mechanical Engineering, PEC University of Technology, Chandigarh, India
e-mail: anupmalik321@gmail.com

A. Manna
e-mail: kgpmanna@rediffmail.com

G. Kibria et al. (eds.), *Non-traditional Micromachining Processes*,
Materials Forming, Machining and Tribology,
DOI 10.1007/978-3-319-52009-4_11

in industries. Again, these materials become more difficult to machine once they are highly sintered. Therefore, it is crucial to develop an accurate and efficient machining method for machining of advanced ceramic and composite materials. Non-conductive materials e.g. glass fibres, ceramic, composites etc. can be machined by utilizing the hybrid machining technique i.e. Travelling Wire Electro Chemical Spark Machining (TWECSM) process. Advanced Ceramics and glass fibre composites is having good strength, low weight, flexibility in shape and sizes, good corrosion, high thermal shock resistance and good creep resistance than that of monolithic metals. Advanced glass fibre reinforced composite materials can be used in place of aluminium, steel, their alloys and other monolithic materials. Advanced ceramics and glass fibre composites materials have extensive range of applications in industries due to their high performance properties particularly in the water and wastewater processing or storage, corrosive areas such as chemical processing and storage, aeronautical and defence, automotive industry, building construction, electrical utility, marine industry etc. Keeping in view, the hybrid machining methods can be successfully utilized for machining of non-conductive materials i.e. ceramics and composites. A TWECSM was designed, developed and utilized for effective machining of e-glass fibre epoxy composite materials [1]. Authors explained the effect of various parameters of the developed TWECSM on the machining performance characteristics. In this hybrid machining process, removal of material takes place because of combined actions of electro chemical reaction and electrical spark discharge action. TWECSM process has two electrodes immersed in an electrolytic solution which may be acidic or alkaline in nature. The workpiece material required to be machined is immersed in the electrolyte and placed adjacent to the cathode (tool). A constant direct current voltage supply is applied between the auxiliary electrode (anode) and the electrode (cathode). The auxiliary electrode used usually is a flat plate has larger surface as compared to wire surface i.e. about 100 times larger. When voltage applied is below 25 V (i.e. critical voltage) then electrolysis occurs. H_2 gas bubbles are formed at cathode (tool electrode) and O_2 gas bubbles at anode (counter electrode). It has been witnessed that if the two electrodes (cathode and anode) both are of different sizes then after certain applied voltage, electric sparks appear on the smaller electrode at the electrode-electrolyte interface and the cell current drops. As voltage increases, current density also rapidly increases. The mean radius and number of the bubbles increases, and finally bubbles are merging in form of a gas film around the tool-electrode.

However, the main principle of material removal in TWECSM process is electro chemical dissolution of the workpiece and it is due to the collective effects of electro chemical reaction and electrical spark discharge action. This collective action takes place when applying enough direct current (DC) supply voltage to the electrolytic cell in appropriate polarity; i.e. auxiliary electrode (anode) with positive terminal and travelling wire (cathode) with negative terminal. When voltage applied to tool-electrode interface reaches beyond a certain value then dissociation of electrolyte take places and formation of H_2 gas bubbles at wire tool surface (i.e. cathode) and generates electrolytic gas film around electrode surface. There is also

O_2 gas evolution and formation of oxide films on auxiliary electrode (i.e. anode) surface during machining of non-conducting materials. When number of H_2 gas bubbles formed at the cathode becomes larger thereby resistance at tool-electrolyte interface becomes very high, it is due to the restriction caused by the gas bubbles insulating effect. Because of the above actions there is an increase in the Joule heating around the wire electrode which turned electrolyte in form of vapours. So, there is a formation of a spark in the wire-electrolyte circuit because of the inductance similar to an electrical circuit.

11.2 Literature Review

Some of the researchers already worked on electro chemical spark machining and travelling wire electro chemical spark machining. They published their research outcome as well. Basak and Ghosh [2] investigated the electro chemical discharge machining and concluded that material removal rate can be substantial increased by introducing an additional inductance. Gautam and Jain [3] investigated on electrochemical spark discharge machining using various tool kinematics for enhancement of process capabilities. Jain et al. [4] concluded that electro chemical spark machining along with abrasive cutting tools improved the machining performance of electrically non-conducting materials, borosilicate glass and alumina. Kulkarni et al. [5] studied the discharge mechanism in electro chemical discharge machining and identified the basic mechanism of rise in temperature and material removal with the help of experimental observations for time-varying current in the circuit. Mediliyegedara et al. [6] studied the electro chemical discharge machining process and concluded that the ECDM process has different advantages over ECM and EDM because of higher machining performance. Manna and Bhattacharyya [7] studied a dual response approach for parametric optimization of CNC wire cut EDM during machining of particulate reinforced aluminium silicon carbide metal matrix composite. They used Taguchi method of experimental design and the significant factors were identified for machining performance characteristics during WEDM of Al/SiC-MMC. Mohen and Shan [8] reviewed the electro chemical micro-to-micro-hole drilling process and concluded that advanced hole-drilling process like jet-electrochemical drilling can be accepted in producing large number of quality holes in difficult to machined materials. Different authors such as Kulkarni and Jain [9], Hwa and Ling [10], Singh et al. [11], Manna and Khas [12], Manna and Narang [13] carried out experimental investigation on micro EDM and electrochemical spark machining processes; they explained the effects of various parameters on response characteristics. Theory proposed by Jain et al. [4] applied to ECM process that H_2 gas liberated and accumulated at the cathode (i.e. travelling wire). The density of bubbles at tool increases with increase in supply voltage, and when it becomes sufficiently high, substantial constriction of the current path takes place at the interface of the tool and electrolyte. The constriction cause an increase in the resistance of the machining region and ohmic heating of electrolyte solution becomes significant. This leads to the onset of vapour bubble nucleation on the electrode surface in addition to the presence of hydrogen bubbles and it increases very rapidly with the applied voltage. It is hypothesized that as the density of nucleation reaches a critical value, vapour blanketing of the electrode occurs. The point of contact between the electrolyte and tool is known as 'Bubble Bridge', which blow-off instantly due to intense heating. Consequently, the current through the circuit, within a very short time span, drops to zero.

Jain et al. [14] experimentally investigated the electro chemical spark machining (ECSM) process and described that SEM photograph of the machined piece showed micro-cracks when machining operation were performed at high applied voltages i.e. 70 V and above. It was due to the high thermal energy input per spark which consequently leads to thermal cracking in the glass-epoxy composite workpiece. Authors also claimed the visibility of globules on the protruding fibres of glass fibre epoxy composite (Fig. 11.1) when machined at 85 V supply voltage.

The evolution of heat because of sparking melts and evaporates workpiece material. That's seen by the existence of globules (i.e., resolidified material) on protruding fibres (Fig. 11.1) as explained by Jain et al. [14]. Bhuyan and Yadava [15] experimentally investigated the performance of the TWECSM through examined the surface finished and machined kerfs width of machined samples by SEM (Fig. 11.2). They claimed that the shape is irregular and also large heat affected zone (HAZ) produced on the bottom side of sliced workpiece. They also claimed that it may be due to the high applied voltage and low wire tension, which are responsible for formation of this kind of unusual shape. Authors also claimed

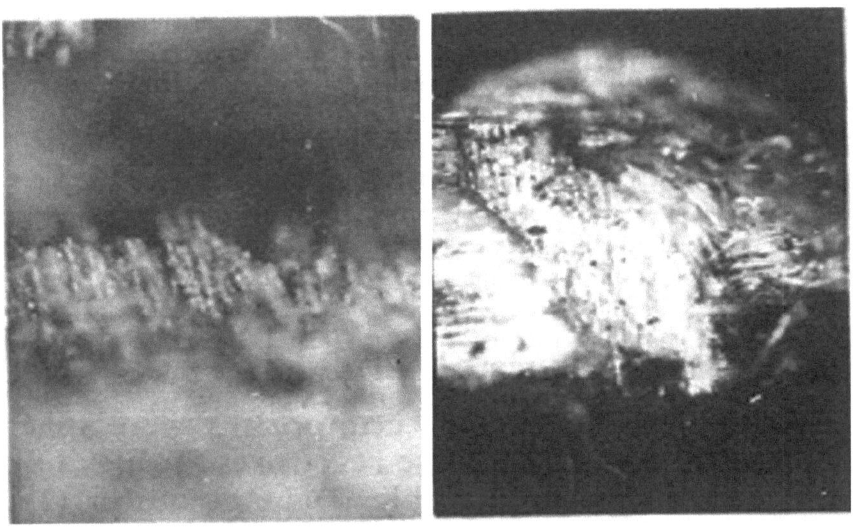

Fig. 11.1 Globules visible on the protruding fibres of glass-epoxy composites at 85 V, 25% concentration, magnification = 100× [14]

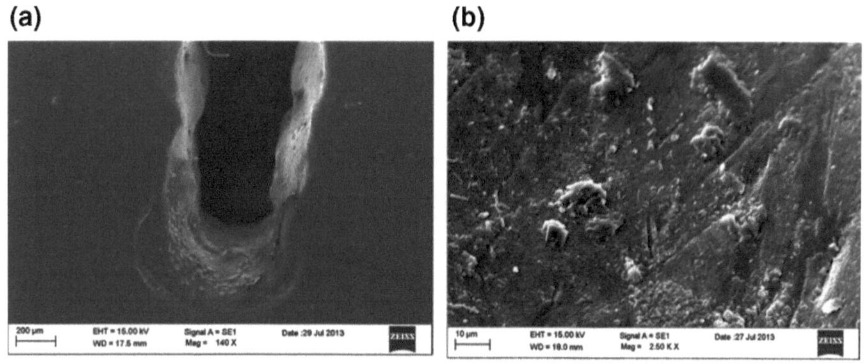

Fig. 11.2 SEM micrographs of **a** kerf width and **b** surface finish at 55 V applied voltage, 300 μs pulse-on time, 250 μs pulse-off time, 250 g/l electrolyte concentration and 1.8 m/min wire feed velocity [15]

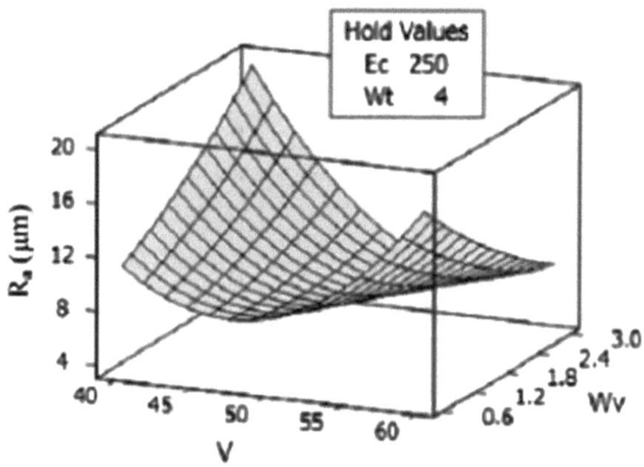

Fig. 11.3 Response surface plot of R_a with voltage and wire feed velocity [16]

the formation of small craters and shallow cracks on the machined surface because of higher concentration of electrolyte.

Bhuyan and Yadav [16] optimized the travelling wire electrochemical process parameters based on response surface methodology and claimed that the effect of wire feed velocity and voltage on R_a keeping constant value of electrolyte concentration (i.e. 250 g/l) and workpiece thickness (i.e. 4 mm) respectively, authors noticed that at the beginning decreases R_a and then gradually increase in curvilinear pattern with change in wire feed velocity as well as change in supply voltage (Fig. 11.3). This is because of high wire feed velocity, which is not helping for formation of bubbles at sidewall of the tool, hence formation of low crater taking place on the machined workpiece surface. They also explained the effect of workpiece thickness and voltage on MRR at constant value of electrolyte concentration (i.e. 250 g/l) and wire feed velocity (i.e. 1.8 m/min) respectively. The variation of MRR with voltage and workpiece thickness is quite similar in nature

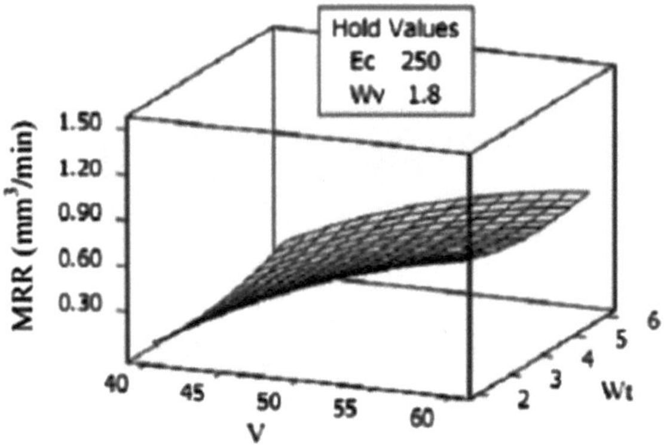

Fig. 11.4 Response surface plot of MRR with voltage and workpiece thickness [16]

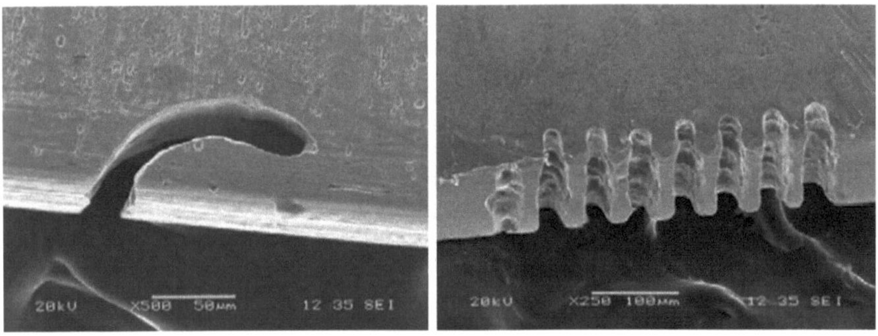

Fig. 11.5 Micro slits by wire electrochemical cutting [17]

(Fig. 11.4). The MRR increases with increase in supply voltage between electrodes. It is due to the fact that at high voltage, the generation and accumulation of gas bubbles at the tool sidewall are very large, which leads to generate large numbers of sparks in the sparking zone and enhances MRR (Fig. 11.4). As workpiece thickness increases MRR also increases the reason behind this is the shifting of discharging zone in the vicinity of the wire to the top of the workpiece and more bubble concentration take place on the wire passing through groove resulting in higher discharge energy per spark and thus enhances MRR as claimed by Bhuyan and Yadav [16].

Zhu et al. [17] used micro tungsten wire of 5 μm diameter as a tool (i.e. cathode) for the wire electro chemical micro machining. Micro slits were machined of 15 μm width (Fig. 11.5). They also generated several complex structures with the slit width (20 μm) by wire electrochemical micro machining process (Fig. 11.6) at 0.125 μm/s feed rate. A small workpiece vibration of 5 Hz frequency was used for successfully circulating electrolyte in the machining zone. The nanosecond pulses (ns) with 4.2 V pulse amplitude, 50 ns on-time and 1 μs off-time was used for machining of complex micro structures.

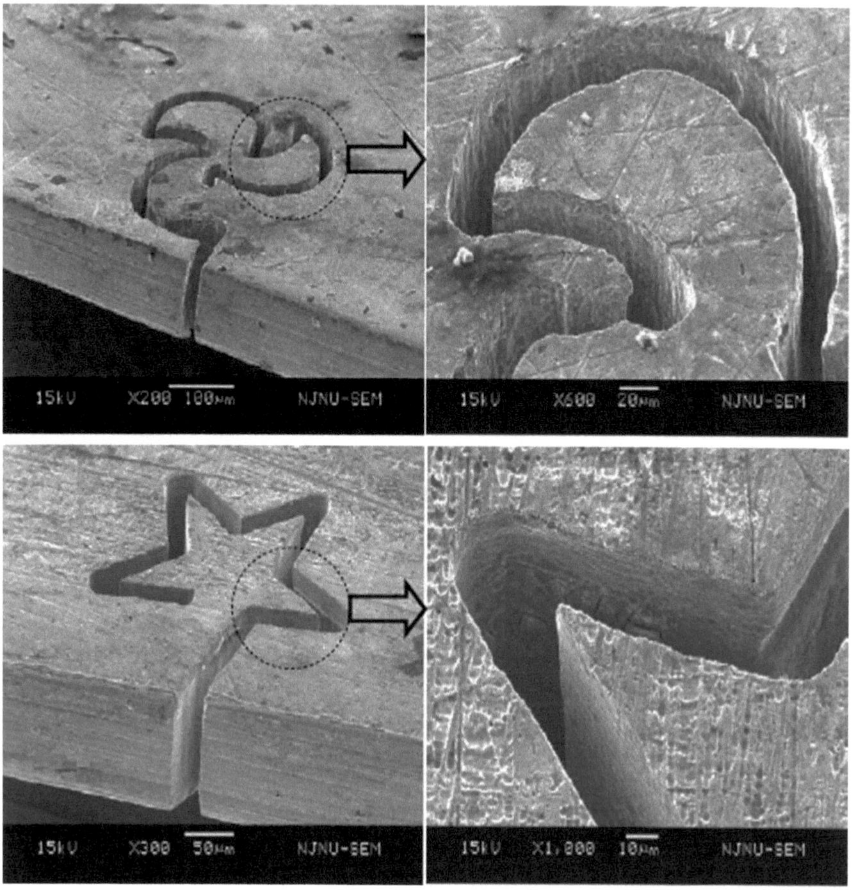

Fig. 11.6 Micro structure cutting by micro wire electrochemical cutting [17]

Manna and Kundal [18] experimentally investigated on slicing of Al_2O_3 ceramic materials by TWECSM process and claimed that at the beginning micro slice width is slightly more than the diameter of wire but after some minutes of continuous cutting the micro slice width increases along the depth of cut. They also claimed that there is an irregular micro slice surface and generated circular shape of large gap at the end of cutting when machining operation carried out with 200 μm wire diameter and at 60 V D.C. supply voltage, 240 mm gap between auxiliary anode and wire cathode, 90 g/l concentration of electrolyte and 0.15 m/min cutting wire speed for 70 min of continuous machining (Fig. 11.7a). Author also explained the condition of sliced workpiece by different SEM photographs. One of such SEM (Fig. 11.7b) machined with wire electrode (diameter 250 μm) at 0.20 m/min cutting wire speed. At the beginning micro slice width is more than the diameter of wire but after some minutes of cutting the width decreases along the depth of cut. They also concluded that micro slicing surface is irregular with poor surface finish.

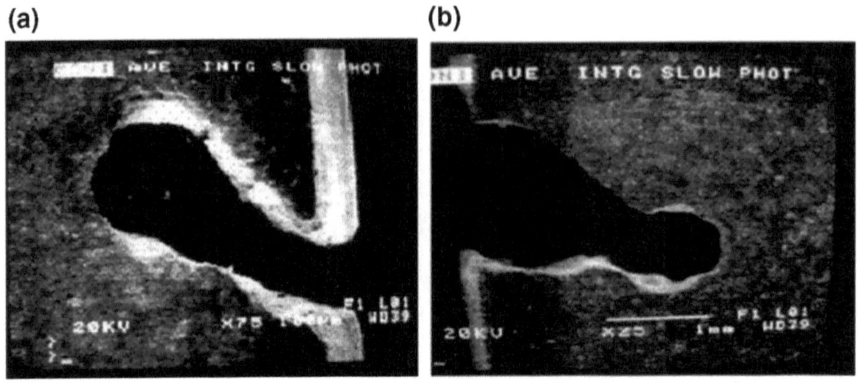

Fig. 11.7 SEM of micro slice work-piece [18]

Jain et al. [14] concluded that travelling wire electrochemical spark machining (TW-ECSM) of composites using NaOH as electrolyte has been found to be a feasible and effective process. Higher machining accuracy is obtained at lower values of voltage and electrolyte concentration. Thermo mechanical phenomenon has been identified as the main mechanism responsible for material removal in TW-ECSM. There is an increase in MRR at higher voltage along with the presence of thermal cracks, large heat affected zone and irregular machined surfaces. For higher machining efficiency, the distance between the anode and the cathode, and the distance between the tool and the work needs to be optimized. Also, a minimum distance between the point of power supply and the point of contact of wire with the workpiece is expected to improve the process performance.

Bhuyan and Yadav [16] concluded that the parameters, electrolyte concentration, workpiece thickness, and interaction effect of parameter workpiece thickness and wire feed velocity have significant factors for R_a. Material removal rate also affected by wire feed velocity, voltage and interaction effect of electrolyte concentration and workpiece thickness and square effect of workpiece thickness.

Manna and Kundal [18] concluded that the machining characteristics gradually declines with increase in tool depth and formed micro cracks on the sliced surface because of thermal shocks at high voltage. Three most influential parameters are identified i.e. DC supply voltage, electrolyte concentration and gap between cathode and anode. According to the authors, experimentally acquired results revealed that moderate DC supply voltage has reduced the tendency of cracking, and at low gap between cathode and anode generated debris during micro slicing of Al_2O_3 ceramics. They claimed that the sparking starts when the travel wire touches the work-piece immersed under the electrolyte in the machining chamber. Fins and burrs along the sliced surfaces were witnessed by the authors this is due to the adhering of particles dispersed from the workpiece surface during machining, which was not flash out by the flow of di-electric fluid. They explained that it's may be due to low flashing pressure of electrolyte; again at high flashing pressure of electrolyte may vibrate the wire during traveling and generates poor surface. Scattered and fins along the sliced surfaces were formed because of the adhering of small particles dispersed from the work surface during slicing operation, this is due to improper flashing of di-electric fluid in machining zone.

11.3 Experimental Investigation

A travel wire electro chemical spark discharge machine (TWECSM) has been designed, fabricated and used for experimental investigation.

11.3.1 Designed and Fabrication of TWECSM

The developed TWECSM has main four units such as (i) power supply unit, (ii) wire feed unit and (iii) electrolyte flow control unit and (iv) workpiece motion control unit. A voltage rectifier (i.e. converter cum regulator) is used to convert single phase 220 V A.C. supply into D.C. supply voltage range of 20–200 V. Auxiliary copper plate is connected with the positive terminal of the converter cum regulator, which is functioning as an auxiliary anode (i.e. auxiliary electrode). Tool (i.e. wire electrode) is connected with the negative terminal of the converter cum regulator, which is functioning as a cathode. A step down transformer with regulator is used to step down the voltage; this step down voltage is used to operate a stepper motor which runs the wire take up unit. A Direct Current supply power of 12 V is used to operate another stepper motor which rolls the used wire around the pulley mount on the main frame. Figure 11.8 shows the schematic diagram of indigenously developed TWECSM.

11.3.2 Experimental Planning

Utilized fabricated TWECSM machine different micro slicing experiments have been carried out for cutting of electrically non-conductive e-glass fibre composite; it is a light weight and highly corrosive resistant material. Table 11.1 represents the detail of fabricated wire electro chemical spark machine (TWECSM), tool electrode (travelling wire), workpiece and type of electrolyte used during experiments. Material removal (MR) is calculated with the difference of workpiece weights before and after each slicing operation. An electronic Balance having resolution of 0.01 mg is used to calculate the weight of work-pieces before and after each slicing operation. Scanning electron microscopic images are used to analyze the surface texture of the sliced surface. The machining time is recorded with the help of digital stop watch of accuracy 0.1 s.

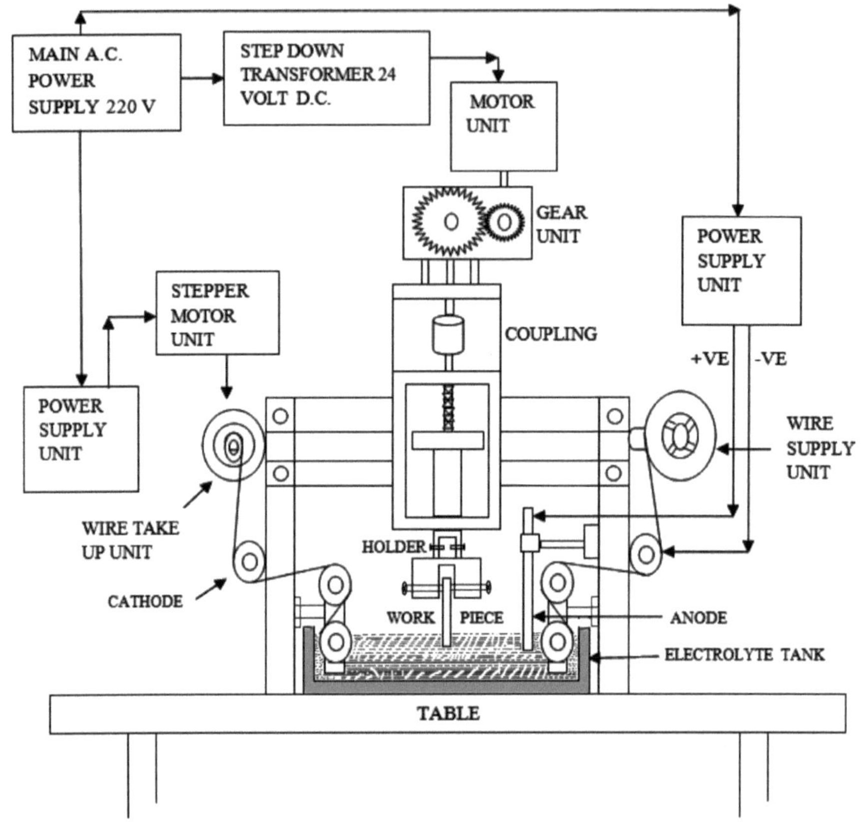

Fig. 11.8 Schematic diagram of TWECSM setup

Table 11.1 Details of experimental conditions

Machine used	Indigenously developed travel wire electrochemical spark machine (TWECSM)
Electrolyte used	Sodium hydroxide (NaOH)
NaOH and natural water concentration	(i) 40 g/l (ii) 80 g/l (iii) 120 g/l (iv) 160 g/l
Work-piece material	e-glass fibre composite
Work-piece thickness	2 mm
Tool used	Brass wire of 200 μm diameter

Table 11.2 represents the different input parameters of the developed TWECSM and their parameter levels used during experiments. Taguchi method based design of experiment $L_{16}(4^5)$ orthogonal array is employed to carry out the experiment for cutting of e-glass fibre epoxy composite on developed machine.

Table 11.2 Developed TWECSM parameters and their levels

Parameters (symbols)		Units	Parametric levels			
			1	2	3	4
A	DC supply voltage (X_1)	Volt	40	50	60	70
B	Electrolyte concentration (X_2)	gm/l	40	80	120	160
C	Wire speed (X_3)	m/min	0.15	0.20	0.25	0.30
D	Gap between tool and anode (X_4)	mm	40	80	120	160
E	Supply current (X_5)	amp	1	1.5	2	2.5

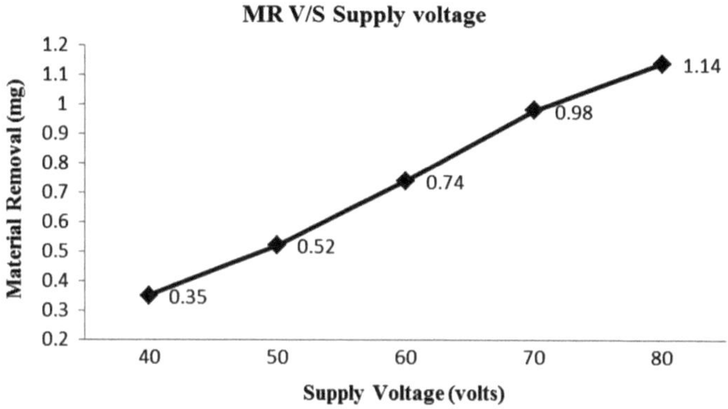

Fig. 11.9 Effect of supply voltage on material removal (MR, mg)

11.3.3 Results and Discussions

A series of experiments has been carried out with variation of different parametric setting value and the results are utilized for further analysis. Different graphs have been drawn to examine the various parametric effects on the different machining characteristics of fabricated TWECSM. Different SEM photographs demonstrate the machining characteristics (i.e. surface texture and spark gap width) during TWECSM operation. Figure 11.9 represents a curve showing the investigated results obtained during slicing of e-glass fibre epoxy composite, which is electrically non-conductive high corrosive resistant, less weight, high strength and high temperature resistant material on developed TWECSM. Figure 11.9 shows that material removal (MR, mg) increases with increase of supply DC voltage. It may be due to the increase of potential difference between the tool (i.e. cathode) and anode (i.e. auxiliary anode) which may discharge or liberates more numbers of electrons per unit area of cathode and simultaneously act for making crater with dissolution of non-conductive anode i.e. work-piece. Maximum material removal is observed at 80 V DC supply voltage. Figure 11.9 is drawn on the basis of results obtained by slicing of workpiece (i.e. e-glass fibre epoxy composite) for continuously 60 min of machining at constant electrolyte concentration (80 gm/l), wire speed (0.15 m/min),

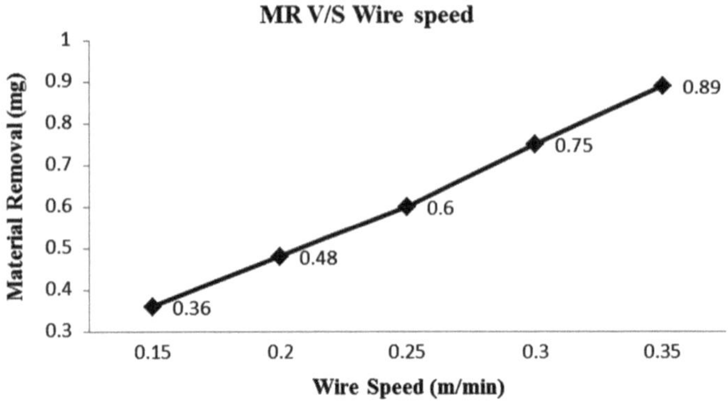

Fig. 11.10 Effect of wire speed on material removal

gap between tool (i.e. cathode) and auxiliary anode (120 mm), supply current (1.5 amp) with supply voltage variation from 40 to 80 V.

Figure 11.10 represents a curve showing the investigated results obtained during slicing of e-glass fibre epoxy composite on developed TWECSM. Figure 11.10 shows that there is increase in material removal (MR, mg) with increase in wire speed. Effect of wire speed is smaller on material removal. Material removal is increases with increase in wire speed because as wire speed increases fresh wire comes quickly in the machining zone and this may cause higher number of spark discharge which helps to form more crater per unit time. Maximum material removal is found at 0.35 m/min wire speed. Figure 11.10 shows effect of wire speed based on acquired results from continuously 60 min of machining at constant supply voltage (40 V), electrolyte concentration (80 gm/l), gap between tool (i.e. cathode) and auxiliary anode (120 mm), supply current (1.5 amp) with variation of wire speed from 0.15 to 0.35 m/min.

Figure 11.11 shows the effect of electrolyte concentration on spark gap width. This is drawn on the basis of investigated results acquired during slicing of e-glass fibre epoxy composite on developed TWECSM. Figure 11.11 shows that there is increase in spark gap width (Wg, μm) with increase in electrolyte concentration. Effect of electrolyte concentration on spark gap width is significantly high. It may be due to the increase of charge carriers in the electrolytic solution due to increase in electrolyte concentration which may discharge or liberates more numbers of ions per unit area of cathode and simultaneously act for making crater with dissolution of non-conductive anode i.e. work-piece. Maximum material removal is observed at electrolyte concentration (200 gm/l). Figure 11.11 is drawn based on the results acquired during slicing of workpiece for continuously 60 min of machining at constant supply voltage (40 V), wire speed (0.15 m/min), gap between tool (i.e. cathode) and auxiliary anode (120 mm), supply current (2 amp) with electrolyte concentration variation from 40 to 200 gm/l.

Taguchi method based $L_{16}(4^5)$ orthogonal array is employed and detail experiments are carried out. The acquired results are utilized for optimization of

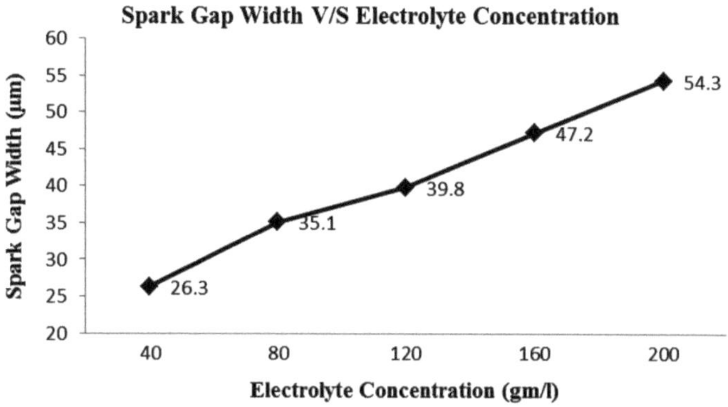

Fig. 11.11 Effect of Electrolyte conc. on spark gap width

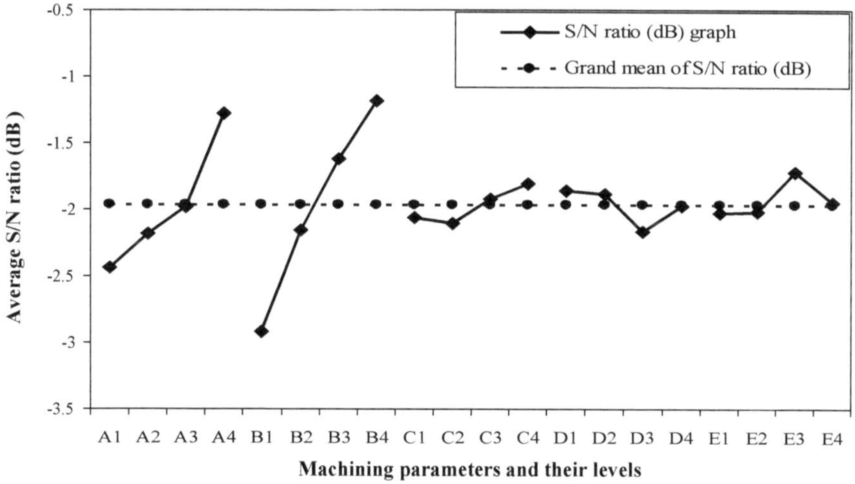

Fig. 11.12 S/N ratio by their factor level for material removal (MR, mg)

developed TWECSM process parameters. Figure 11.12 demonstrates the average signal to noise (S/N) ratio and grand mean of S/N ratio (dB) graphs for MR. Figure 11.12 clearly shows that the optimal combination of parameters for maximum material removal (MR, mg) is $A_4B_4C_4D_1E_3$ (Table 11.2).

11.3.4 Scanning Electron Micrographs

Figure 11.13 shows Scanning Electron Micrograph (SEM) of a micro sliced section of e-glass fibre epoxy composite workpiece using travelling wire electro chemical

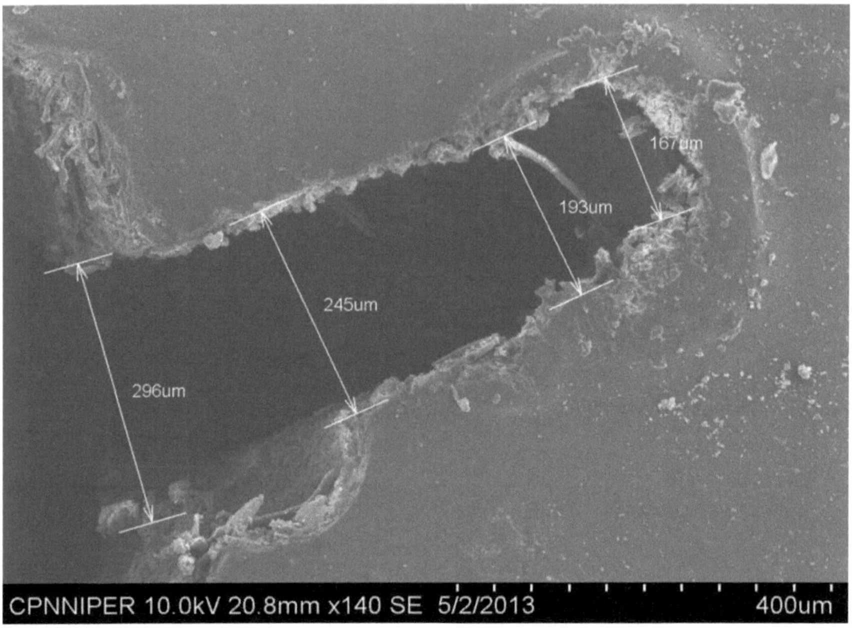

Fig. 11.13 SEM of micro sliced workpiece utilized developed TWECSM

spark machine (TWECSM). The SEM graph demonstrates the real condition of the micro sliced surface machined at supply voltage (40 V), electrolyte concentration (40 gm/l), wire speed (0.15 m/min), gap between tool and anode (40 mm) and supply current (1.5 amp). Figure 11.13 shows SEM photographs sliced workpiece for continuous 60 min machining with brass 200 μm diameter wire. From SEM graph, it is observed that micro slicing mainly proceeds in the conical manner. From Fig. 11.13, it is clear that there are continuous burrs and fins on the surface of the micro sliced workpiece. Over all slicing surface finish is not good when seen at micron level. At entry level the slicing width is 296 μm but after continuous cutting for 60 min it is found to be only 167 μm, measured just at the end of slicing. Hence it is proved that slicing is about conical in nature. Some fins and burs are also found during cutting in both sides of micro sliced surface. That's due to the irregular and insufficient electrolyte flow. During micro slicing it is also witnessed that there is formation of foam and gas bubbles which causes a barrier for electron's flow and decreases electrolyte dissolution strength.

Figure 11.14 shows SEM image of a micro sliced section of e-glass fibre epoxy composite machined on developed TWECSM. The SEM image demonstrates the real condition of the micro cutting surface. This machining operation was carried out at supply voltage (40 V), electrolyte concentration (120 gm/l), wire speed (0.25 m/min), gap between tool and anode (120 mm) and supply current (2 amp). Figure 11.14 shows SEM photographs sliced workpiece for continuous 40 min machining with brass wire of 200 μm diameter. It is also found that there is some

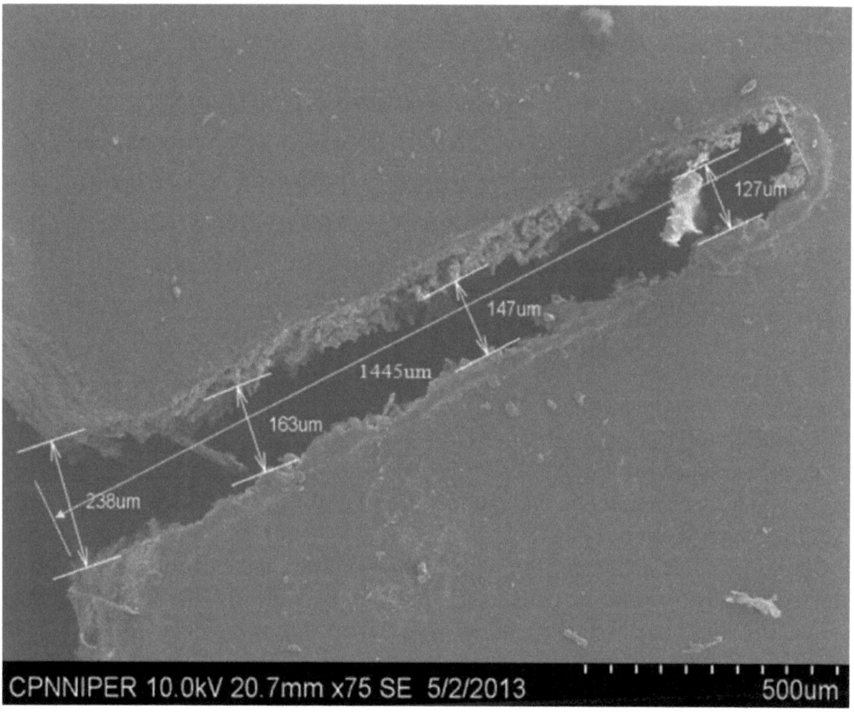

Fig. 11.14 SEM of another micro sliced

presence of continuous burrs and fins on the surface of the micro slicing. Over all slicing surface finish is not good when seen at micron level. At entry, the slicing width is 238 μm but after continuous cutting for 40 min it is found that this size is being 127 μm only, measured just at the end of slicing. It's proved that slicing is about conical. Some fins and burs are also found during cutting in both sides of micro sliced surface. That's due to the irregular and insufficient electrolyte flow. During micro slicing it is also witnessed that there is formation of foam and gas bubbles which causes a barrier for electron's flow and thereby decreases the electrolyte dissolution strength.

11.3.5 Mathematical Modelling

By taking the parameters such as supply voltage (X_1), electrolyte concentration (X_2), wire speed (X_3), gap between tool and auxiliary anode (X_4), and supply current (X_5) developed mathematical models for different response characteristics e.g. material removal (MR, mg) and spark gap width (Wg, μm). The developed mathematical models can be used for selection of parameters setting in advance for effective machining on developed TWECSM.

$$\begin{aligned} Y_{MR} = & -0.0002648 + 0.0204605\,X_1 + 0.0042091\,X_2 + 0.000105\,X_3 \\ & + 0.0041159\,X_4 + 0.0000121\,X_5 - 0.00011627\,X_1X_2 - 0.0941467\,X_1X_3 \\ & - 0.0001587\,X_1X_4 - 0.022536\,X_1X_5 + 0.029705\,X_2X_3 - 0.0000461\,X_2X_4 \\ & + 0.0017518\,X_2X_5 + 0.0000101\,X_3X_4 + 0.0000112\,X_3X_5 + 0.009647\,X_4X_5 \\ & + 0.000756\,X_1^2 - 0.0000138\,X_2^2 + 0.0000104\,X_3^2 - 0.0000451\,X_4^2 \\ & - 0.076381\,X_5^2 \end{aligned} \tag{11.1}$$

$R^2 = 0.9915$
where X_1 = Supply voltage (Volt)
X_2 = Electrolyte conc. (gm/l)
X_3 = Wire speed (m/min)
X_4 = Gap between anode and cathode (mm)
X_5 = Supply current (amp).

Figure 11.15 demonstrates the supply voltage (Volt) and electrolyte conc. (gm/l) effect on material removal (MR, mg). Figure 11.15 clearly shows the effect of parameters when machining operation was carried out at high electrolyte concentration (i.e. 200 gm/l) and high supply voltage (i.e. 70 V) which reflected high material removal. It is also observed that MR is high when the electrolyte concentration is 50 g/l and voltage is 40 V. There is increase in material removal with increase in electrolyte concentration and supply voltage. This is due to generation of higher potential difference between tool and anode at higher supply voltage, which may discharges more number of electrons per unit area and at the same time increases electrolyte concentration thereby increases rate of dissolution of ions from the anode. This is because of higher current density and thereby strong dissolution effect.

The developed mathematical model for spark gap width is

$$\begin{aligned} Y_{Wg} = & -0.330621 - 1.653169\,X_1 + 0.039677\,X_2 + 0.01279\,X_3 + 1.682246\,X_4 \\ & + 0.16910\,X_5 - 0.006801\,X_1X_2 + 0.398961\,X_1X_3 + 0.0061927\,X_1X_4 \\ & + 0.916407\,X_1X_5 - 0.118467\,X_2X_3 + 0.0000355\,X_2X_4 + 0.124377\,X_2X_5 \\ & + 0.001284\,X_3X_4 + 0.01496\,X_3X_5 - 0.9502047\,X_4X_5 - 0.0041215\,X_1^2 \\ & + 0.001516\,X_2^2 + 0.14036\,X_3^2 - 0.002237\,X_4^2 + 7.046385\,X_5^2 \end{aligned} \tag{11.2}$$

$R^2 = 0.9984$.

Figure 11.16 shows the electrolyte conc. (gm/l) and wire speed (m/min) effect on spark gap width (Wg, µm). Figure 11.16 clearly shows that at higher electrolyte concentration (i.e. 200 gm/l) and at higher wire speed (i.e. 0.30 m/min) the spark gap width is high. This is also noticed that spark gap width is low when the

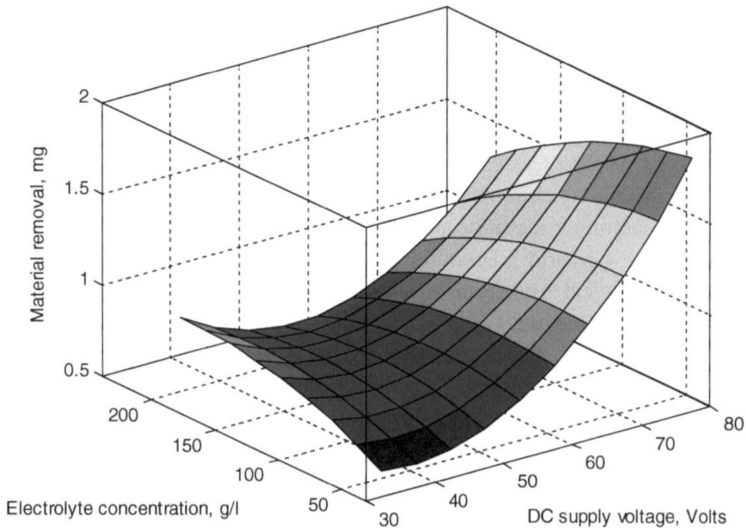

Fig. 11.15 Effect of supply voltage and electrolyte conc. on (MR, mg)

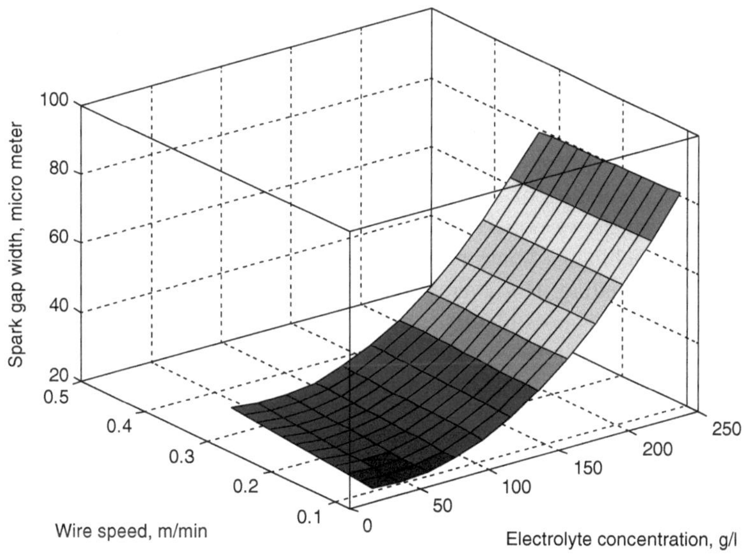

Fig. 11.16 Effect of electrolyte conc. and wire speed on spark gap width

electrolyte concentration (i.e. 50 gm/l) and wire speed (i.e. 0.15 m/min). There is increase in spark gap width with increase in electrolyte concentration and wire speed.

11.4 Conclusions and Future Scope

Glass fibre epoxy composites, ceramics etc. are non-conductive materials which are difficult to machine by any known unconventional machining methods like ECM, EDM and WEDM etc. These non-conductive materials can be machined with the help of conventional machining processes but compromise with surface texture. But conventional machining methods are not suitable even not possible to micro slicing as well. But a new hybrid machining process which is combination of both ECM (i.e. electro chemical machining) and WEDM (i.e. wire electric discharge machining) can be effectively used to machine such materials. This chapter presents some investigated published results during machining of composites and ceramics which proved that this hybrid machining technique can effectively use for machining of non-conductive materials. This chapter also presents the brief about the design and development of the particular hybrid travelling wire electro chemical spark machine. This developed machine is utilized for slicing of non-conductive workpiece (i.e. e-glass fibre epoxy composite). By studying different experimental results during machining (i.e. slicing) of workpiece (i.e. e-glass fibre epoxy composite) on developed TWECSM, it is believe that the developed travel wire electro chemical spark machine can be effectively use to machine non-conductive workpiece (i.e. e-glass fibre epoxy composite). However, it is identified that during starting of micro cutting the micro slice width is slightly more than wire diameter (i.e. 200 μm) but after some minutes of continuous slicing the micro slice width decreases along the depth of cut. The micro sliced surface is irregular and poor if experiments are carried out without proper plan and optimization of parameters. Fins and scattered along the cutting edges of the sliced workpiece are observed because of the adhering of small particles dispersed from the workpiece surface during slicing of the workpiece. This is due to insufficient or improper electrolyte flow during machining. Some burrs are noticed along the edge of micro sliced workpiece after cutting. However, experimental results also revealed that the very good micro slicing is possible. By optimizing machining parameters generation of high surface texture can be possible. This hybrid machining technique can be used for machining of PSZ, zirconia, $YBa_2Cu_3O_7{-}x$, Si_3N_4, PZT etc. difficult to machine nonconductive materials.

References

1. Malik Anup, and Manna Alakesh (2016), An experimental investigation on developed WECSM during micro slicing of e-glass fibre epoxy composite, Int. J. Adv. Manuf. Technol; 85: 2097–2106.
2. Basak I. and Ghosh A. (1996); Mechanism of spark generation during electrochemical discharge machining: A theoretical model and experimental verification; J. Mater Process Technol, Vol. 62, pp. 46–53.

3. Gautam N. and Jain V. K. (1998); Experimental investigations into ECSD process using various tool kinematics; Int. J. of Mach Tool Manuf, Vol. 38, pp. 15–27.
4. Jain V. K.; Choudhury S. K.; Ramesh K. M. (2002); On the machining of alumina and glass; Int. J. of Mach Tool Manuf, Vol. 42, pp. 1269–1276.
5. Kulkarni A.; Sharan R.; Lal G. K. (2002); An experimental study of discharge mechanism in electrochemical discharge machining; Int. J. of Mach Tool Manuf, Vol. 42, pp. 1121–1127.
6. Mediliyegedara T.K.K.R.; De Silva A.K.M; Harrison, D.K.; McGeough, J.A. (2004); An intelligent pulse classification system for electro-chemical discharge machining (ECDM)-a preliminary study; J. Mater Process Technol, Vol. 149, pp. 499–503.
7. Manna A. and Bhattacharyya B. (2006); Taguchi and Gauss elimination method: A dual response approach for parametric optimization of CNC wire cut EDM of PRAl/SiC-MMC; Int. J. Adv. Manuf. Technol, Vol. 28, pp. 67–75.
8. Mohen S. and Shan H. S. (2005); A review of electrochemical micro-to micro-hole drilling processes; Int. J. of Mach Tool Manuf, Vol. 45, pp. 137–152.
9. Kulkarni A. and Jain V. K. (2007) "Electrochemical Discharge Machining Process" Defence Science Journal, Vol. 57, Page 765.
10. Hwa Y. B. and Ling W. K. (2007); A study on the mirror surface machining by using a micro-energy EDM and the electrophoretic deposition polishing; Int. J. Adv. Manuf. Technol, Vol. 34, pp. 96–103.
11. Singh Y. P; Jain V K; Prashant Kumar and Agrawal, D. C.; (1996), Machining piezoelectric (PZT) ceramics using an electrochemical spark machining (ECSM) process, J. Mater Process Technol, 58: 24–31.
12. Manna Alakesh. and Khas Kanwaljit; (2009), Micro machining of electrically non-conductive Al_2O_3 ceramic, Journal of Machining and Forming Technology, 1(1–2): 101–112.
13. Manna Alakesh; and Narang, Vivek; (2012), A study on micro machining of glass-fibre-epoxy composite by ECSM process, Int. J. Adv. Manuf. Technol, 61: 1191–1197.
14. Jain V. K., Rao P. Sreenivasa, Choudhary S. K., Rajurkar K. P. (1991) "Experimental investigations into travelling wire electrochemical spark machining (TW-ECSM) of composites", Journal of Engineering for Industry, Vol. 113/75.
15. Bhuyan B. K. and Yadava V. (2014) "Experimental study of traveling wire electrochemical spark machining of borosilicate glass", Journal of Materials and Manufacturing, Vol. 29, pp. 298–304.
16. Bhuyan B. K. and Yadava V. (2013) "Experimental modeling and multi-objective optimization of travelling wire electrochemical spark machining (TW-ECSM) process", Journal of Mechanical Science and Technology, Vol. 27(8), pp. 2467–2476.
17. Zhu D., Wang K., Qu N. S. (2007) "Micro wire electrochemical cutting by using in situ fabricated wire electrode", Annals of the CIRP Vol. 56, pp. 241–244.
18. Manna Alakesh and Kundal Amandeep; (2015), An experimental investigation on fabricated TW-ECSM setup during micro slicing of nonconductive ceramic, Int. J. Adv. Manuf. Technol, 76: 29–37.

Index

G. Kibria et al. (eds.), *Non-traditional Micromachining Processes*,
Materials Forming, Machining and Tribology,
DOI 10.1007/978-3-319-52009-4

MIX
Papier aus verantwortungsvollen Quellen
Paper from responsible sources
FSC® C105338

If you have any concerns about our products,
you can contact us on
ProductSafety@springernature.com

In case Publisher is established outside the EU,
the EU authorized representative is:
Springer Nature Customer Service Center GmbH
Europaplatz 3, 69115 Heidelberg, Germany

Printed by Libri Plureos GmbH
in Hamburg, Germany